T0142965

Lecture Notes in Business Information Processing 513

LNBIP reports state-of-the-art results in areas related to business information systems and industrial application software development – timely, at a high level, and in both printed and electronic form.

The type of material published includes

- Proceedings (published in time for the respective event)
- Postproceedings (consisting of thoroughly revised and/or extended final papers)
- Other edited monographs (such as, for example, project reports or invited volumes)
- Tutorials (coherently integrated collections of lectures given at advanced courses, seminars, schools, etc.)
- Award-winning or exceptional theses

LNBIP is abstracted/indexed in DBLP, EI and Scopus. LNBIP volumes are also submitted for the inclusion in ISI Proceedings.

João Araújo · Jose Luis de la Vara ·
Maribel Yasmina Santos · Saïd Assar
Editors

Research Challenges in Information Science

18th International Conference, RCIS 2024
Guimarães, Portugal, May 14–17, 2024
Proceedings, Part I

Editors
João Araújo
NOVA University Lisbon
Caparica, Portugal

Maribel Yasmina Santos
University of Minho
Guimarães, Portugal

Jose Luis de la Vara
University of Castilla La Mancha
Albacete, Albacete, Spain

Saïd Assar
Institut Mines-Télécom Business School
Evry, France

ISSN 1865-1348 ISSN 1865-1356 (electronic)
Lecture Notes in Business Information Processing
ISBN 978-3-031-59464-9 ISBN 978-3-031-59465-6 (eBook)
https://doi.org/10.1007/978-3-031-59465-6

This Springer imprint is published by the registered company Springer Nature Switzerland AG
The registered company address is: Gewerbestrasse 11, 6330 Cham, Switzerland

Paper in this product is recyclable.

Preface

This volume of the Lecture Notes in Business Information Processing series contains the proceedings of the 18th International Conference on Research Challenges in Information Science, RCIS 2024, held in Guimarães, Portugal, during May 14–17, 2024. Guimarães is a scenic historical city located in the north of the country and considered the cradle of the Portuguese nation.

The scope of RCIS covers the thematic areas of information systems and their engineering, user-oriented approaches, data and information management, enterprise management and engineering, domain-specific information systems engineering, data science, information infrastructures, and reflective research and practice. RCIS 2024 focused on the special theme "Information Science: Evolution or Revolution?".

The 25 full papers presented in the first volume were carefully reviewed and selected from a total of 79 submissions to the main conference. Out of all the submissions, three were desk rejected because the Program Co-chairs found them to be outside the scope of the conference. All the remaining submissions were single-blind reviewed by at least three Program Committee (PC) members. That was followed by a discussion period moderated by Program Board (PB) members. A PB meeting was also held in Lisbon to discuss the final selection of papers for the conference program. The acceptance rate was 32%. The authors received recommendations and meta-reviews to be contemplated in the camera-ready versions.

The second volume includes 12 Forum papers and five Doctoral Consortium papers. The Forum track received 14 dedicated submissions, out of which seven were accepted. Six submissions to the main conference were recommended for presentation at the Forum. The Doctoral Consortium received seven submissions, of which five were accepted.

The contributions in the first volume have been organized in the following topical sections: Data and Information Management, Conceptual Modelling and Ontologies, Requirements and Architecture, Business Process Management, Data and Process Science, Security, Sustainability, and Evaluation and Experience Studies; the second volume contains the Forum papers, Doctoral Consortium papers, and Tutorials.

The conference program started with the workshops. The main conference included sessions on keynotes, research papers, tutorials, the Forum, the Doctoral Consortium, research projects, and journal-first presentations. The three invited keynote presentations were: "Information science research with large language models: between science and fiction" by Fabiano Dalpiaz, University of Utrecht, The Netherlands; "The power of Information Systems shaping the future of the Automotive Industry", by Carlos Ribas, Bosch, Portugal; and "BPM in the Era of AI and Generative AI: Opportunities and Challenges" by Barbara Weber, University of St. Gallen (HSG), Switzerland. The four accepted tutorials addressed relevant and well-timed topics at the core interest of the RCIS community.

We want to thank all authors who submitted their work to RCIS 2024, and also the PC and PB members for their hard work in reviewing and discussing the submitted papers. Finally, we want to thank all the Organization Committee members and the student volunteers for their valuable assistance.

May 2024

João Araújo
Jose Luis de la Vara
Maribel Yasmina Santos
Saïd Assar

Organization

Conference Chairs

General Chairs

Maribel Yasmina Santos	University of Minho, Portugal
Saïd Assar	Institut Mines-Télécom Business School, France

Program Chairs

João Araújo	NOVA University Lisbon, Portugal
Jose Luis de la Vara	University of Castilla-La Mancha, Spain

Local Organizing Team

Carina Andrade	University of Minho, Portugal
Victor Barros	University of Minho, Portugal
Pedro Guimarães	CCG/ZGDV, Portugal
Paula Monteiro	CCG/ZGDV, Portugal
Isabel Ramos	University of Minho, Portugal
António Vieira	University of Minho, Portugal

Doctoral Consortium Chairs

Selmin Nurcan	Université Paris 1 Panthéon-Sorbonne, France
Jaelson Castro	Universidade Federal de Pernambuco, Brazil

Forum Chairs

Dominik Bork	TU Wien, Austria
Jānis Grabis	Riga Technical University, Latvia

Workshop Chairs

Jean-Michel Bruel	Toulouse University, France
Nelly Condori-Fernandez	Universidad Santiago de Compostela, Spain

Tutorial Chairs

Ana Moreira	NOVA University Lisbon, Portugal
Renata Guizzardi	University of Twente, The Netherlands

Research Projects Chairs

Dimitris Karagiannis	University of Vienna, Austria
Tiago Prince Sales	University of Twente, The Netherlands
Camille Salinesi	Université Paris 1 Panthéon-Sorbonne, France

Journal First Chairs

Angelo Susi	Fondazione Bruno Kessler, Italy
Jolita Ralyté	University of Geneva, Switzerland

Proceedings Chairs

Miguel Goulão	NOVA University Lisbon, Portugal
Oliver Karras	Leibniz ISCT, Germany

Publicity Chairs

Ana León	Universidad Politécnica de Valencia, Spain
Carla Silva	Universidade Federal de Pernambuco, Brazil
Isabel Sofia Brito	Polytechnic Institute of Beja, Portugal

Program Board

Saïd Assar	Institut Mines-Télécom Business School, France
Marko Bajec	University of Ljubljana, Slovenia
Xavier Franch	Universitat Politècnica de Catalunya, Spain
Renata Guizzardi	University of Twente, The Netherlands
Evangelia Kavakli	University of the Aegean, Greece
Pericles Loucopoulos	Institute of Digital Innovation and Research, UK
Haralambos Mouratidis	University of Essex, UK
Selmin Nurcan	Université Paris 1 Panthéon-Sorbonne, France
Oscar Pastor	Universidad Politécnica de Valencia, Spain
Jolita Ralyté	University of Geneva, Switzerland
Maribel Yasmina Santos	University of Minho, Portugal
Jelena Zdravkovic	Stockholm University, Sweden

Program Committee

Ademar Aguiar	University of Porto, Portugal
Nour Ali	Brunel University, UK
Raian Ali	Hamad Bin Khalifa University, Qatar
Jose María Alvarez Rodríguez	Carlos III University of Madrid, Spain
Carina Alves	Universidade Federal de Pernambuco, Brazil
Vasco Amaral	NOVA University Lisbon, Portugal
Claudia P. Ayala	Universitat Politècnica de Catalunya, Spain
Clara Ayora	University of Castilla-La Mancha, Spain
Fatma Başak Aydemir	Boğaziçi University, Turkey
Dominik Bork	TU Wien, Austria
Carlos Cetina	San Jorge University, Spain
Mario Cortes-Cornax	Université Grenoble Alpes, France
Maya Daneva	University of Twente, The Netherlands
Andrea Delgado	Universidad de la República, Uruguay
Rebecca Deneckere	Université Paris 1 Panthéon-Sorbonne, France
Chiara Di Francescomarino	University of Trento, Italy
Sophie Dupuy-Chessa	Université Grenoble Alpes, France
Sergio España	Utrecht University, The Netherlands
Robson Fidalgo	Universidade Federal de Pernambuco, Brazil
Hans-Georg Fill	University of Fribourg, Switzerland
Andrew Fish	University of Liverpool, UK
Agnès Front	Université Grenoble Alpes, France
Frederik Gailly	Ghent University, Belgium
Arturo García	University of Castilla-La Mancha, Spain
Ignacio García	University of Castilla-La Mancha, Spain
Sepideh Ghanavati	University of Maine, USA
Giovanni Giachetti	Universidad Andrés Bello, Chile
Cesar Gonzalez-Perez	Spanish National Research Council (CSIC), Spain
Jaap Gordijn	Vrije Universiteit Amsterdam, The Netherlands
Miguel Goulão	NOVA University Lisbon, Portugal
Giancarlo Guizzardi	Federal University of Espirito Santo (UFES), Brazil
Jennifer Horkoff	Chalmers University of Technology, Sweden
Felix Härer	University of Fribourg, Switzerland
Mirjana Ivanovic	University of Novi Sad, Serbia
Christos Kalloniatis	University of the Aegean, Greece
Oliver Karras	Leibniz ISCT, Germany
Manuele Kirsch Pinheiro	Université Paris 1 Panthéon-Sorbonne, France
Elena Kornyshova	Conservatoire National des Arts et Métiers, France

Tong Li	Beijing University of Technology, China
Grischa Liebel	Reykjavik University, Iceland
Lidia Lopez	Universitat Politècnica de Catalunya, Spain
Andrea Marrella	Sapienza University of Rome, Italy
Beatriz Marín	Universidad Politécnica de Valencia, Spain
Raimundas Matulevicius	University of Tartu, Estonia
Nikolay Mehandjiev	University of Manchester, UK
Giovanni Meroni	Technical University of Denmark, Denmark
Denisse Muñante	ENSIIE, France
Elena Navarro	University of Castilla-La Mancha, Spain
Kathia Oliveira	Université Polytechnique Hauts-de-France, France
Jose Ignacio Panach Navarrete	Universitat de València, Spain
Henderik A. Proper	TU Vienna, Austria
Manfred Reichert	University of Ulm, Germany
Patricia Rogetzer	University of Twente, The Netherlands
Marcela Ruiz	Zurich University of Applied Sciences, Switzerland
Alberto Sardinha	Universidade de Lisboa, Portugal
Rainer Schmidt	Munich University of Applied Sciences, Germany
Florence Sedes	I.R.I.T. Univ. Toulouse III Paul Sabatier, France
Sagar Sen	Simula Research Laboratory, Norway
Samira Si-Said Cherfi	Conservatoire National des Arts et Métiers, France
Carla Silva	Universidade Federal de Pernambuco, Brazil
Denis Silveira	Universidade Federal de Pernambuco, Brazil
Anthony Simonofski	Université de Namur, Belgium
Pnina Soffer	University of Haifa, Israel
Isabel Sofia	Polytechnic Institute of Beja, Portugal
Erick Stattner	University of the French West Indies, France
Monika Steidl	University of Innsbruck, Austria
Eric-Oluf Svee	Stockholm University, Sweden
Ernest Teniente	Universitat Politècnica de Catalunya, Spain
Olivier Teste	IRIT, France
Nicolas Travers	Pôle Universitaire Léonard de Vinci, France
Juan Trujillo	University of Alicante, Spain
Aggeliki Tsohou	Ionian University, Greece
Tanja E. J. Vos	Universidad Politécnica de Valencia, Spain
Yves Wautelet	Katholieke Universiteit Leuven, Belgium
Hans Weigand	Tilburg University, The Netherlands

Program Committee, Forum

Syed Juned Ali	TU Wien, Austria
Clara Ayora	Universidad de Castilla-La Mancha, Spain
Judith Barrios Albornoz	University of Los Andes, Colombia
Cinzia Cappiello	Politecnico di Milano, Italy
Istvan David	McMaster University, Canada
Victoria Döller	University of Vienna, Austria
Anne Gutschmidt	University of Rostock, Germany
Simon Hacks	Stockholm University, Sweden
Abdelaziz Khadraoui	University of Geneva, Switzerland
Manuele Kirsch Pinheiro	Université Paris 1 Panthéon-Sorbonne, France
Elena Kornyshova	Conservatoire National des Arts et Métiers, France
Georgios Koutsopoulos	Stockholm University, Sweden
Emanuele Laurenzi	FHNW School of Business, Switzerland
Dejan Lavbič	University of Ljubljana, Slovenia
Beatriz Marín	Universidad Politécnica de Valencia, Spain
Patricia Martin-Rodilla	Spanish National Research Council, Spain
Giovanni Meroni	Technical University of Denmark, Denmark
João Moura-Pires	NOVA University Lisbon, Portugal
Mark Mulder	TEEC2 BV, The Netherlands
Christoforos Ntantogian	Ionian University, Greece
Michalis Pavlidis	University of Brighton, UK
Francisca Pérez	Universidad San Jorge, Spain
Iris Reinhartz-Berger	University of Haifa, Israel
Ben Roelens	Ghent University, Belgium
Marcela Ruiz	Zurich University of Applied Sciences, Switzerland
Natalia Stathakarou	Karolinska Institutet, Sweden
Gianluigi Viscusi	Linköping University, Sweden
Manuel Wimmer	Johannes Kepler University Linz, Austria

Program Committee, Doctoral Consortium

Ademar Aguiar	University of Porto, Portugal
José Borbinha	Universidade de Lisboa, Portugal
Nelly Condori-Fernández	Universidad Santiago de Compostela, Spain
Maya Daneva	University of Twente, The Netherlands
João Faria	University of Porto, Portugal
Xavier Franch	Universitat Politècnica de Catalunya, Spain

Renata Guizzardi	University of Twente, The Netherlands
Hugo Jonker	Open University of the Netherlands, The Netherlands
Manuele Kirsch Pinheiro	Université Paris 1 Panthéon-Sorbonne, France
Beatriz Marín	Universidad Politécnica de Valencia, Spain
Oscar Pastor	Universidad Politécnica de Valencia, Spain
Fethi Rabhi	University of New South Wales, Australia
Jolita Ralyté	University of Geneva, Switzerland
Rogier Van de Wetering	Open University of the Netherlands, The Netherlands

Program Committee, Research Projects

Alessandra Bagnato	Softeam, France
Xavier Boucher	École nationale supérieure des mines de Saint-Étienne, France
Robert Buchmann	Babes-Bolyai University, Romania
Nelly Condori-Fernández	Universidad Santiago de Compostela, Spain
Tolga Ensari	Arkansas Tech University, USA
Davide Fucci	Blekinge Tekniska Högskola, Sweden
Filippo Lanubile	University of Bari, Italy
Khaled Medini	École nationale supérieure des mines de Saint-Étienne, France
Marc Oriol Hilari	Universitat Politècnica de Catalunya, Spain
Dalila Tamzalit	University of Nantes, France
Wilfrid Utz	OMILAB NPO, Germany
Tanja E. J. Vos	Politécnica de Valencia, Spain
Robert Woitsch	BOC Group, Austria

Additional Reviewers

Renata Cruz
Nikolaos Marios Polymenakos
Maxwell Prybylo
Sara Haghighi
Vijanti Ramautar
Ioannis Paspatis
Roberto Sanchez Reolid
Katerina Soumelidou
Simone Agostinelli
Sara Haghighi
Stylianos Karagiannis
Simon Dechamps
Stylianos Karagiannis
Martin Eisenberg
Evita Roponena

Abstracts of Keynote Talks

Information Science Research with Large Language Models: Between Science and Fiction

Fabiano Dalpiaz

Utrecht University, The Netherlands
f.dalpiaz@uu.nl

Abstract. Large language models (LLMs) are in the spotlight. Laypeople are aware of and are using LLMs such as OpenAI's ChatGPT and Google's Gemini on a daily basis. While companies are exploring new business opportunities, researchers have gained access to an unprecedented scientific playground that allows for fast experimentation with limited resources and immediate results. In this talk, using concrete examples from requirements engineering, I am going to put forward several research opportunities that are enabled by the advent of LLMs. I will show how LLMs, as a key example of modern AI, unlock research topics that were deemed too challenging until recently. Then, I will critically discuss the perils that we face when it comes to planning, conducting, and reporting on credible research results following a rigorous scientific approach. This talk will stress the inherent tension between the exciting affordances offered by this new technology, which include the ability to generate non-factual outputs (fiction), and our role and societal responsibility as information scientists.

The Power of Information Systems Shaping the Future of the Automotive Industry

Carlos Ribas

Bosch, Portugal
carlos.ribas@pt.bosch.com

Abstract. The automotive industry is undergoing a seismic shift, fueled by the convergence of information systems and artificial intelligence. In this keynote, I will explore how these technologies are reshaping the whole automotive industry, from the idea to the innovative creation, development, and manufacturing through to the customer experience. Join me in my travel through data highways, decoding smart factories and steering towards intelligent, autonomous and safe vehicles as assistants and companions.

BPM in the Era of AI and Generative AI: Opportunities and Challenges

Barbara Weber

University of St. Gallen, Switzerland
barbara.weber@unisg.ch

Abstract. Artificial intelligence (AI) technologies, such as machine learning and natural language processing, empower Business Process Management (BPM) by enabling data-driven decision-making, predictive analytics, and automation. Leveraging AI algorithms, organizations can extract actionable insights from vast datasets, optimize processes, and enhance operational efficiency to boost productivity. Additionally, Generative AI introduces novel capabilities to BPM, facilitating creative problem-solving and innovation. Generative AI algorithms, exemplified by Large Language Models (LLMs), not only offer enhanced creativity but also excel in tasks such as text generation, context understanding, and natural language interaction. With generative AI algorithms, alternative solutions and scenarios can be generated, augmenting human creativity and driving organizational innovation. In this keynote presentation I will highlight how this impacts the field of BPM and discuss some of the challenges arising from that.

Contents – Part I

Sustainability

Evaluation and Experience Studies

Contents – Part II

Forum Papers

Doctoral Consortium Papers

Tutorials

Data and Information Management

Unified Models and Framework for Querying Distributed Data Across Polystores

Léa El Ahdab[1,4](✉), Imen Megdiche[2,4], André Peninou[3,4], and Olivier Teste[3,4]

[1] Université Toulouse III Paul Sabatier, Toulouse, France
[2] Université de Toulouse, INUC Champollion, Toulouse, France
[3] Université Toulouse Jean Jaurès UT2J, Toulouse, France
[4] IRIT, Toulouse, France
{lea.el-ahdab,imen.megdiche,andre.peninou,olivier.teste}@irit.fr

Abstract. Combining data sources from NoSQL and SQL systems leads to data distribution and complexifies user queries: data is distributed among different stores having different data models. This data implementation complexifies the writing of user queries. This work proposes a querying framework of a polystore with the use of unified models as a user vision of the polystore. Unified models hides the variety of data models and data distribution to the user. Our solution uses the Entity-Relationship model of the polystore to infer unified models and to identify intermediate required operations to fulfill querying on real polystore. Using these required transformations, a rewriting framework allows to automatically rewrite the user query (against the unified model) with respect to the real data distribution over the polystore. We apply this framework with one dataset (UniBench benchmark) between a relational, a document-oriented and a graph-oriented databases. We illustrate in this work performance and the low impact of our query rewriting solution when compared to query execution time.

Keywords: Polystore · NoSQL · SQL · Data distribution · Data fragmentation · Query rewriting

1 Introduction

With multi-store and polystore systems [1], the combination of schema systems and schema-less systems have led to distribution and querying issues [2]. One native language is not sufficient to interrogate data. This system heterogeneity emphasizes data complexity for polystore interrogation. Some hybrid languages were proposed in order to query a polystore [3,4] while others have chosen to adapt existing native languages with developed complementary functions [5]. The main stake is to create a link between different systems in order to have an integrated vision of data. Some works focus on adding an external algorithm to handle heterogeneity of systems [6]. User querying and the vision of the polystore

J. Araújo et al. (Eds.): RCIS 2024, LNBIP 513, pp. 3–18, 2024.
https://doi.org/10.1007/978-3-031-59465-6_1

need to be simplified. Logical views or models of each data system translate the physical implementation inside the polystore as if it is only stored in one system [7]. In this work, we present a framework able to query over heterogeneous polystores with data distribution and fragmentation based on the notion of unified modeling. It shows all data present in the polystore in a single model: relational, document or graph and hides the underlying data distribution, data models, and data fragmentation. This simplified vision is intended to provide transparency and simplicity for user querying. With the generation of a mapping dictionary, our solution is able to create links between the polystore and these unified models. The process may includes the adding of transformation and/or transfer of data. Our main advantage is the independence of our process despite data or structure updates on the polystore. This paper is organized as follows: Section 2 presents the main challenges with a motivation example on an e-commerce scenario. Section 3 details the prototype development with the explanation of the framework modules for querying the polystore. Section 4 shows results of the query rewriting and execution time using this framework on real data and over vertical data distribution. Finally we position our work in Sect. 5 and we conclude and give some perspectives about the future ones in Sect. 6.

2 Motivating Example

Prerequisite. A **polystore** PL is considered as a set of **databases** $\{DB_i\}$. Each database is composed **datasets** DS_j corresponding to the storage of entity classes and their data values. All datasets of one database are from the same family system but it is different for the databases of a polystore: data has different native forms. A *vertical distribution* is the distribution of attributes inside different databases. It can lead to data fragmentation where the distributed attributes belong to the same entity. The information linked to this entity is then fragmented between different datasets. In our context of polystores, it complexifies data distribution with the diversification of native forms. We consider data fragmentation where an entity class is distributed in several databases of the polystore. The entities distributed are linked according to a specific attribute called *distribution key.* This key is the unique value defining an instance of the entity class. It is used to identify all fragments of this entity.

A Motivating Example. Considering an E-commerce scenario from Unibench, data is distributed into three systems: relational (DB_1, DB_2), document-oriented (DB_3) and graph-oriented (DB_4). This example uses *Customers*, *Products*, *Orders*, *Reviews*, *Persons*, *Post* and *Tag* entity classes (Fig. 1). For vertical ***data distribution***, the entity class *Customers* is stored in one dataset DS_1 of one database DB_1 as shown in Fig. 1. Data shows ***fragmentation*** where one entity can have its attributes distributed in two databases. For example, the entity class *Products* is stored in two dataset DS_{2a} of database DB_2 and DS_{2b} of database DB_3. These databases are respectively from a relational storage system and document-oriented system (Fig. 1). One key attribute links the two fragments of *Products* and is called the attribute of fragmentation, *product_id*. Data fragmentation is illustrated with the vertical fragmentation of *Orders* and *Products* between the relational and the document-oriented system. Their primary key is also the fragmentation key for the distributed entity classes. To hide the complexity of four databases with three different modeling paradigms to the user, we introduce a *unified model.* It shows all data of the polystore "as if" it was a mono-store, thus hiding the real data distribution, data models, and data fragmentation. It is not implemented physically. Figure 2 shows the relational, document-oriented and graph-oriented unified models of the e-scenario data model used for this example.

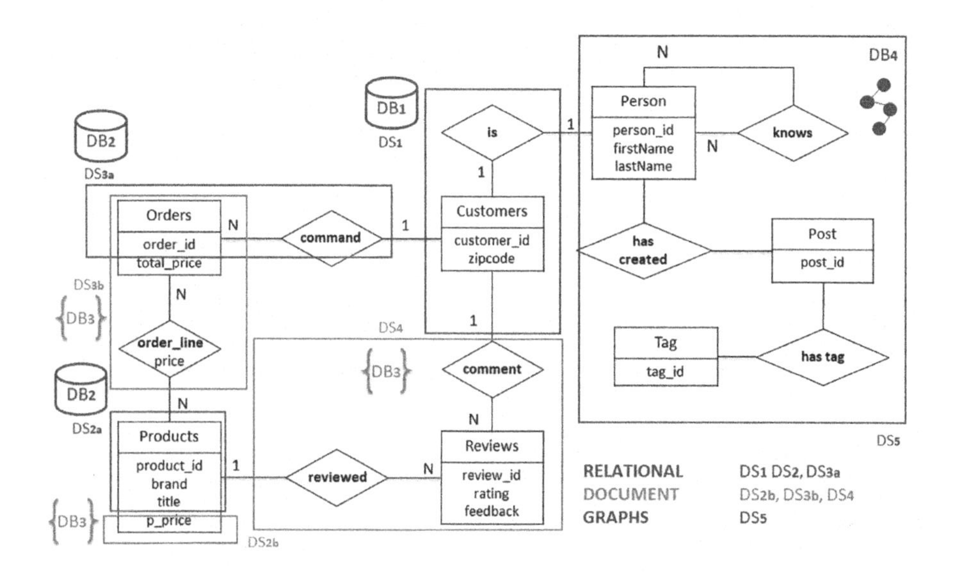

Fig. 1. Data model for the adapted e-commerce scenario

Query Use Case: "Do Customers Ordering the Same Products Have a Link Between Each Other?". Let's assume that this use case is formulated in SQL on the Relational Logical model. It implies to access to *Person*, *Customers*, *Orders* and *Products* entities stored in different *DB* systems. *Person* is stored in a graph database which is not compatible with the initial query language and system interrogated. Detailing entity location complexifies user query. *Orders* and *Products* are both fragmented between a relational database and a document-oriented database which implies query adaptation and potentially an entity rebuilding. In some works [6,9] attribute location is specified in the user query. It complexifies the rewriting of the query because the user should be aware of each system, each language, each database and each dataset of the polystore. Figure 3 shows the difference between a user query in CloudMdSQL and a user query on our system.

Relational Unified Model

```
Person [{person_id, firstName, lastName}]
Knows [{person_id1, person_id2}]
Customers [{customer_id, person_id, name, zipcode}]
Orders [{order_id, total_price, customer_id}]
Order_line [{order_id, product_id, price}]
Products [{product_id, brand, title, p_price}]
Reviews [{review_id, customer_id, product_id,
rating, feedback}]
Post [{post_id, post_label, customer_id]}
Has_tag [{post_id, tag_id}]
Tag [{tag_id, tag_label}]
```

Document Unified Model

```
Person [{person_id, firstName, lastName,
knows:[person_id2]}]
Customers [{customer_id, person_id, name,
zipcode}]
Orders [{order_id, total_price, customer_id,
order_line:[product_id, price]}]
Products [{product_id, brand, title, p_price}]
Reviews [{review_id, customer_id, product_id,
rating, feedback}]
Post [{post_id, post_label, customer_id,
has_tag:[tag_id]]}
Tag [{tag_id, tag_label, has_post:[post_id]}]
```

Graph Unified Model

```
Person [{person_id, firstName, lastName}]
Person→[Knows]→Person [{person_id1, person_id2}]
Person→[Knows]→Person [{person_id2, person_id1}]
Person→[Is]→Customers [{person_id, customer_id}]
Customers→[Is]→Person [{customer_id, person_id}]
Customers [{customer_id, person_id, name, zipcode}]
Customers→[Command]→Orders [{customer_id, order_id}
Orders→[Command]→Command [{order_id, customer_id}]
Orders [{order_id, total_price, customer_id}]
Orders→[Order_line]→Products [{order_id,
product_id, price}]
Products→[Order_line]→Orders [{product_id,
order_id, price}]
Products [{product_id, brand, title, p_price}]
Products→[Reviewed]→Reviews [{product_id, review_id}]
Reviews→[Reviewed]→Products [{review_id, product_id}]
Reviews [{review_id, customer_id, product_id, rating,
feedback}]
Post [{post_id, post_label, customer_id]}
Person→[Has_created]→Post [{person_id, post_id}]
Post→[Has_created]→Person [{post_id, person_id}]
Post→[Has_tag]→Tag [{post_id, tag_id}]
Tag→[Has_tag]→Post [{tag_id, post_id}]
Tag [{tag_id, tag_label}]
```

Fig. 2. Unified models of the data model showed in Fig. 1

USER QUERY ON CLOUDMDSQL

```
Customers(customer_id int, person_id int)@DB1 = (SELECT customer_id
FROM Customers)
Products (product_id string)@DB2 = (SELECT product_id FROM Products)
Orders1(order_id string, customer_id int)@DB2 = (SELECT order_id,
customer_id FROM Orders)
Orders2(order_id string, product_id string)@DB3 =
(db.orders.aggregate({$project:{_id:0, order_id:1, product_id:1}}))
Person (person_id1 int, person_id2 int)@DB4 = (MATCH (p1:Person)-
[k:KNOWS]->(p2:Person) RETURN p1.person_id as person_id1,
p2.person_id as person_id2)
SELECT C1.customer_id, C2.customer_id
FROM Customers C1, Customers C2, Person P1, Person P2, Orders1 O11,
Orders1 O12, Orders2 O21, Orders2 O22, Products P
WHERE P.product_id=O21.product_id AND P.product_id=O22.product_id
AND C1.customer_id = O11.customer_id AND
C2.customer_id=O12.customer_id
AND O11.order_id=O21.order_id AND O12.order_id=O22.order_id
AND (C1.person_id=P1.person_id OR C1.person_id=P2.person_id)
AND (C2.person_id=P1.person_id OR C2.person_id=P2.person_id);
```

USER QUERY WITH OUR SYSTEM

```
SELECT C1.customer_id, C2.customer_id
FROM Customers C1, Customers C2, Knows K,
Orders O1, Orders O2, Products P
WHERE P.product_id = O1.product_id AND
P.product_id=O2.product_id
AND C1.customer_id = O1.customer_id AND
C2.customer_id = O2.customer_id
AND (C1.person_id = K.person_id1 OR
C1.person_id = K.person_id2)
AND (C2.person_id = K.person_id1 OR
C2.person_id = K.person_id1);
```

Fig. 3. User query comparison between CloudMdsQL and our system for the use case

3 Our Proposed Framework

3.1 Problem Statement

The scope of our contribution considers a polystore PL with databases DB_i belonging to SQL and NoSQL paradigms. Data is distributed between relational, document-oriented and graph-oriented databases. Querying this heterogeneous polystore with a native language L is impossible without L being adapted to all paradigms. This shows two sub-problems: (i) How to provide a global view of

PL in one of DBMS type? (ii) How to execute the query written in an arbitrary chosen language L over PL?

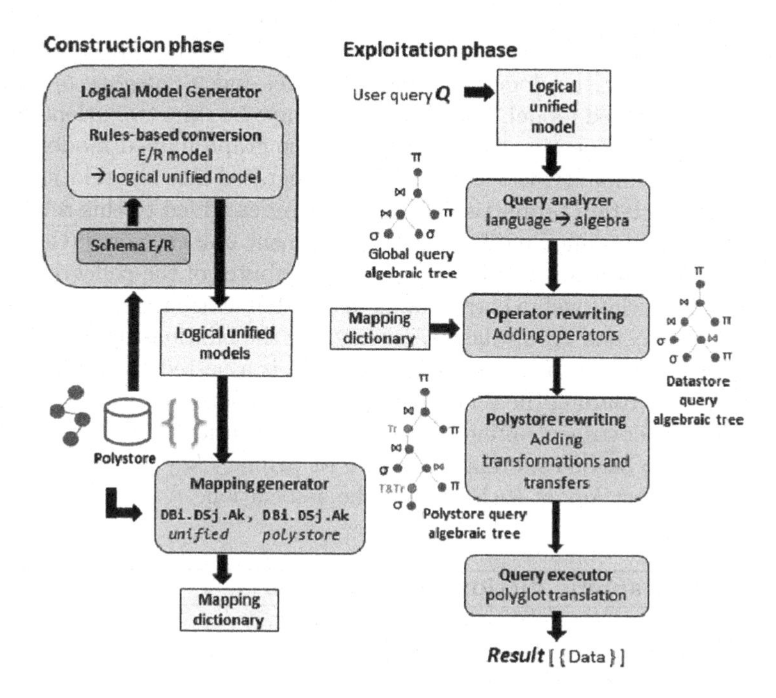

Fig. 4. Overview of our framework for the construction and the exploitation phases

3.2 An Overview of the Framework

Our solution aims to provide a transparency querying system over a heterogeneous polystore. A simpler representation of the polystore is used by the user to express his query in one language of his choice. It is analyzed and transformed part by part to take into account the real data location inside PL and to return results in the expected form. Our rewriting system considers data transfer and transformation and favors the use of DBMS operators and performance. Our solution is composed of two phases: the construction phase and the exploitation phase (Fig. 4).

1. **Construction phase:** composed of a module for logical unified model generation and a module for mapping dictionary generation:
   - *Logical Model generator:* gives a unified representation of the polystore for each included system (SQL, NoSQL). It hides the polystore complexity by representing it in a set of databases belonging to the same system (relational, document-oriented or graph-oriented). There is one U_M per

system of the polystore. The E/R model of the polystore is the input of this module. It includes all existing attributes of the polystore. It should be given to the module; the automatic extraction of such model is our of the scope of this paper. The U_M generation is based on specific rules: (i) Each entity class becomes a *relation* for the relational unified model, a *collection* for the document unified model and a *category of node* for the graph unified model; (ii) The relationships become two oriented edges linking their respective entity classes for the graph unified model. In case of (1, N) cardinality, the foreign key is integrated into the relationship to N and association properties are new attributes added to this relation for the relational unified model. For the document unified model, the foreign key and relationship attributes become attributes of the collection on the side 1. The relationship corresponds to a nested attribute for the collection on the side N and group the foreign key and the relationship attributes. In case of (N, M) cardinality, the relationship is a nested attribute for both collections grouping the relationship attributes and the respective foreign key for the document unified model. For the relational unified model, the relationship becomes a relation where its primary key is the composition of the primary key of each class of the association.

Algorithm 1. Mapping dictionary generation

```
1: input: U_V, PL                  → unified model, polystore
2: output: Mapp                    → mapping dictionary
3: Mapp ← {'Type' : U_V.type,' uv_tables' : [ ] }
4: for DS_i ∈ U_V do
5:     ds_infos ← {'name' : DS_i.name,' stores_infos : [ ] }
6:     for pl_db ∈ PL do
7:         stores ← {'name' : pl_db.name,' type' : pl_db.type}
8:         stores['columns'] ← [ ]
9:         for pl_ds ∈ pl_db do
10:            for attr ∈ pl_ds do
11:                column_map ← found(attr, PL, UV)
12:                stores['columns'] ← add(column_map)
13:            end for
14:        end for
15:        ds_infos['stores_infos'] ← stores
16:    end for
17:    Mapp['uv_tables'] ← ds_infos
18: end for
19: return Mapp
```

- *Mapping generator*: defines links between attributes inside the unified models and their actual location inside the polystore. It is visually represented as a table but it is implemented as a JSON file. The Algorithm 1 generates these links according to the unified model and the polystore.

For each property of the unified model, its equivalences are found in PL by browsing each dataset of each database of the polystore. Once, the correspondences are found with *found()* function, the result are added to the final mapping structure (with the function *add()*). Table 1 shows an example of the mapping dictionary for the relational unified model. Because the entity classes *Orders* and *Products* are fragmented between DB_2 and DB_3 this mapping dictionary allows our framework to link the real position of each attributes of these entity classes with their unified position presented to the user. $Order\_line$ is a relationship with (N, M) cardinality (Fig. 1), it is represented as a table in the U_{VR} when applying the *Logical Model generator.* The attributes of $Order\_line$ in the polystore are stored in the $Orders$ table for $DS_{3a}.order\_id$ and in the $Orders$ collection for $DS_{3b}.order\_id$ for the rest of the attributes.

Table 1. Extract of the dictionary showing the entity/relationship between *Orders* and *Products* for the relational unified model

U_{VR}	DB_2	DB_3
Orders.order_id	DS_{3a}.order_id	DS_{3b}.order_id
Orders.total_price		DS_{3b}.total_price
Order_line.order_id	DS_{3a}.order_id	DS_{3b}.order_id
Order_line.product_id		DS_{3b}.product_id DS_{3b}.details.product_id
Order_line.price		DS_{3b}.price DS_{3b}.details.price
Products.product_id	DS_{2a}.product_id	DS_{2b}.product_id
Products.brand	DS_{2a}.brand	
Products.title	DS_{2a}.title	
Products.p_price		DS_{2b}.p_price

2. **Exploitation phase:** This phase is composed of a module for query analysis, a module for operator rewriting, a module for optimization and the final module for query execution. For the relational unified model, we consider the operators of selection, projection and join in SQL, for the document unified model, we consider the operators of selection and projection in MongoDB.
   - *Query analyzer:* translates the user query into a global query algebraic tree G_T defined as $G_T = \{N_{node} \rightarrow (N_{left}, N_{right})\}$ where each node N is $N = (op, [E.A])$ and contains information about the associated operation op, the list of attributes accessed by this operation and the corresponding entity $E.A$. The last nodes are the entities interrogated and has no children nor operations;
   - *Operator rewriting:* works on the G_T and specifies data location. The entity classes are changed into their corresponding datasets. In case of

fragmentation, we introduce a rebuilding operator ρ to the new algebraic tree. This step adapts the query to the databases of the polystore and generates a multi-store query algebraic tree;
- *Polyglot rewriting:* works on the multi-store query algebraic tree and generates the final polystore query algebraic tree with the presence of transfers and transformations operations. These new nodes appear when there is a change of paradigms identified by the information given in the previous step;
- *Polyglot executor:* generates an execution plan according to the polystore query algebraic tree and translates the sub-trees of one algebraic tree into the respective languages (SQL for relational, MongoDB for document, Cypher for Graphs). This step follows the paradigms transformation rules presented in Table 2. The steps of transformation and transfers of sub-results are included between the execution of those sub-queries. The algebra for the document-oriented and the graph-oriented systems are proposed for the purpose of this work and are not a generalization.

Table 2. Operator equivalences for relational, document-oriented and graph-oriented systems

Operation System	selection	projection	join
Relational algebra - SQL	σ WHERE	π SELECT	$\bowtie$ FROM + WHERE
Document algebra - MongoDB	σ $match	π $project	$\lambda + \mu$ $lookup + $unwind
Graph algebra - Cypher	σ WHERE	π RETURN	$\rightarrow$ MATCH

4 Experiments

In this section, we evaluate with our framework according to two problems: **E1 (Adaptability):** Mapping dictionary generation time, query rewriting time and query execution time when there is a change in the polystore entity classes distribution between systems; **E2 (Volume):** Query execution time for rewritten queries according to volume variation.

4.1 Datasets

Benchmarks in the literature consider data that can be stored in multi-model polystores. *Unibench* is one of them and considers the following systems:

key/value, document-oriented, graph-oriented and relational. In our work, we use the relational, the document-oriented and the graph-oriented models. We extracted data from the University of Helsinki website [8] and we considered data from the JSON (document), the relational, the graph and the key/value systems. For the purpose of experimenting our framework, we decided to convert the key/value data into document-oriented data (because we do not support key/values systems for the moment). We work with vertical data distribution where one entity class is stored in one dataset of one database and on data fragmentation where one entity class is distributed in multiple datasets (in the same or in different databases).

4.2 Developed Framework Modules

The scope of our theoretical solution considers queries in SQL, MongoDB and Cypher corresponding respectively to relational, document-oriented and graph-oriented systems. We have developed the rewriting module that takes as an input a relational user query (with $\sigma, \pi, \bowtie$ operators) or a document user query (with σ, π operators). It generates its algebraic tree which is then rewritten with new operators. The graph user query is manually analyzed into its polystore algebraic tree. The unified model generation and the execution plan are manually produced. The obtained plan is manually rewritten into a python program to allow its execution over the databases. Their automation is a work perspective.

4.3 Experimental Setup and Protocol

The experiments were performed on a machine working with Intel i7 2.30 GHz and with 64 GB RAM. The polystore data is stored following the distribution scenarios in a version 5.1.1 of MySQL, in MongoDB 5.0.6-rc1 Enterprise of MongoDB Compass and in the version 1.5.9 Neo4J desktop. Our framework is implemented with Python 3.10.2 (jupyter lab) for the construction phase and the creation of the execution plan; and on IDE Netbeans 18 with JAVA 20 and Maven 3.9.2 for the query rewriting phase. We have implemented python files in order to generate relations, collections, type of nodes and type of relationships corresponding to data distribution for the experiments. They use Unibench files to generate new instances to increase the data volume. All new instances are then inserted into the different databases to fulfill the presented experiments. The considered use cases answer the following queries: (i) Q_1 *All products from the brand '54' with a price below 50$*, (ii) Q_2 *Every person that shares a rating of 3 on a product's review*, (iii) Q_3 *All orders for customers coming from the same geographical area*. With our framework, we have tested these queries expressed in SQL, MongoDB and Cypher applied on their respective unified views.

4.4 Evaluation of Our Framework Adaptability

We focus our experiments on our framework ability to adapt according to polystore dataset distribution. dd_1 is the reference distribution, dd_2 is a distribution without data fragmentation and dd_3 presents fragmentation in two datasets of the same database. These distributions are shown in Table 3. Unified models are based on the E/R model of data stored in the polystore, they are not impacted by the change of dataset distribution inside the polystore. Papping dictionaries depend on properties location inside polystore and they are regenerated following the new implementation. For each distribution, the time for mapping dictionary generation does not exceed 0.01 s.

Table 3. Data distribution scenarios for the E1

System DB Distribution	Relational DB_1	Relational DB_2	Document DB_3	Graph DB_4
dd_1	*Customers*	*Products, Orders*	*Products, Orders* *Reviews*	*Person, Post* *Tag*
dd_2	*Customers*	*Products, Orders*	*Reviews*	*Person, Post, Tag*
dd_3	*Customers*		$Reviews$, $Orders_1$ $Products_1$, $Orders_2$ $Products_2$	*Person, Post* *Tag*

Each query (Q_1, Q_2, Q_3) is tested using our framework to generate the final algebraic tree according to the polystore configuration. In these scenarios, the initial query was rewritten three times in order to get the average time of algebraic tree rewriting in our exploitation phase. Tables 4 and 5 show the average rewriting time for each dataset distribution and for each query when executed on the relational unified model and on the document unified model. It does not exceed 3 s. Depending on the distribution, the operators included can be different for the same query. Q_1 has one join, one transfer and one transformation when rewritten in dd_1 and does not have these operators when rewritten in dd_2 for the relational U_M. In comparison, for the document unified model, the operators of Q_1 in dd_1 are the same but in dd_2 the query has one transfer and one transformation. Q_2 works with two entities that keep the same place in the polystore whatever the considered distribution is, but they are in different systems.

Table 4. Results of algebraic tree generation for each query for each data distribution inside the polystore for the SQL interrogation of the relational unified model

Query Distribution	Q_1 dd_1	Q_1 dd_2	Q_1 dd_3	Q_2 dd_1	Q_2 dd_2	Q_2 dd_3	Q_3 dd_1	Q_3 dd_2	Q_3 dd_3
Average rewriting time (seconds)	**0.76**	**0.86**	**0.73**	**2.36**	**2.28**	**2.32**	**0.88**	**0.86**	**0.87**
Number of joins	1	0	1	2	2	2	2	1	2
Number of transfers	1	0	1	2	2	2	2	1	2
Number of transformations	1	0	1	2	2	2	1	0	2

Table 5. Results of algebraic tree generation for each query for each data distribution inside the polystore for the MongoDB interrogation of the document unified model

Query Distribution	Q_1 dd_1	Q_1 dd_2	Q_1 dd_3	Q_2 dd_1	Q_2 dd_2	Q_2 dd_3	Q_3 dd_1	Q_3 dd_2	Q_3 dd_3
Average rewriting time (seconds)	**1.28**	**1.46**	**1.32**	**1.58**	**1.55**	**1.82**	**1.54**	**1.57**	**1.62**
Number of joins	1	0	1	2	2	2	1	1	1
Number of transfers	1	1	0	2	2	2	2	2	1
Number of transformations	1	1	0	2	2	2	1	1	1

Based on the algebraic trees, execution plans of each query are obtained and their execution time was compared. We focus on Q_1 expressed on the relational unified model and over the three polystore configuration considered. The results returned show that the transfers and transformations from a paradigm to another are the more important factors for execution time increase. Using results of Table 4, the presence of transfer and transformation operators influence the execution time of the initial query. The high number of sub-results that will be converted and transferred in a specific distribution is the reason why the execution time is higher than for another distribution scenario.

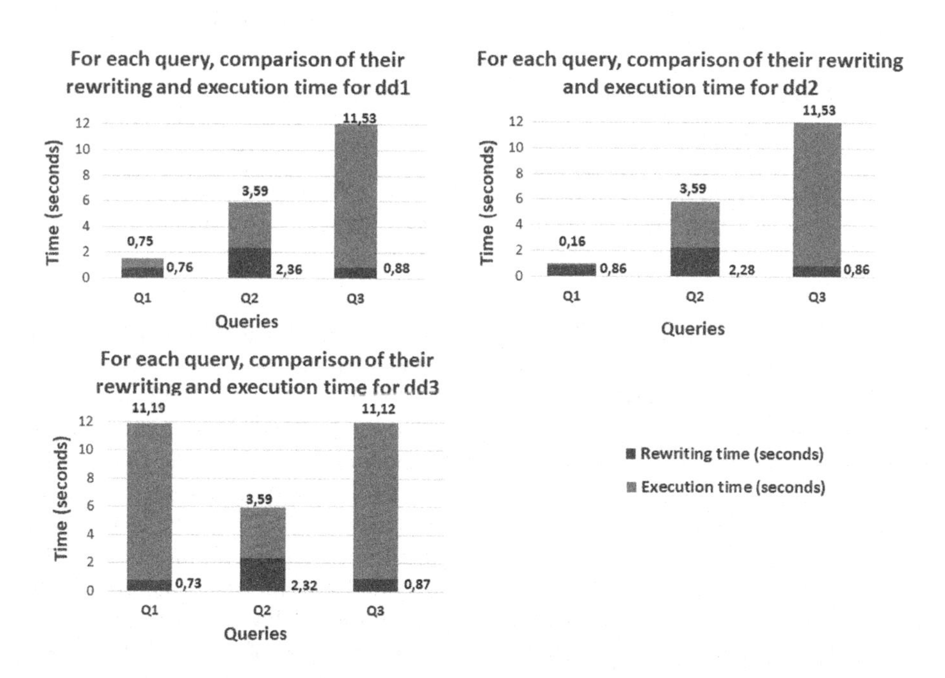

Fig. 5. Experiment result for query rewriting time and query execution time for the three polystore distribution considered in this paper considering the execution over the relational unified model

Figure 5 shows execution results and rewriting results for each query on each distribution inside the polystore. Q_2 presents the maximum of transfers and transformation needed to address data inside the polystore properly, whereas Q_1 only presents one transfer and one transformation and Q_3 is composed of two operations of transfer and only one of transformation. For Q_1, the change of distribution for the polystore impacts its execution time when the entity class is fragmented between two paradigms (0.75 s), is not fragmented (0.16 s), and fragmented in a different paradigm than the one interrogated (11.19 s). The two entity classes considered by Q_2 are in two different paradigms. One of the sub-results needs to be transformed and transferred into the relational one, which explains the execution time (3.59 s). Q_3 execution time (11.53 s) is not impacted by the change of dd_1 with dd_2. In these scenarios, the entity classes considered belong to the same paradigm even if they are found in two databases. In dd_3, one entity class changes its paradigm and needs a transformation operation before the transfer one. The execution time is still close to the one for dd_1 and dd_2 (11.12 s).

4.5 Evaluation of Our Framework with Data Volume

This experiment is based on the dd_1 dataset distribution where we added more instances to double the initial data volume as shown in Table 6.

Table 6. Data volumes and detail of instances for each DB_i of PL

Volume	DB_1	DB_2	DB_3	DB_4
V_1 (**103.09** Mb)	0.46 Mb	12.4 Mb	3.23 Mb	87 Mb
V_2 (**200** Mb)	15 Mb	55 Mb	15 Mb	115 Mb

New instances in V_2 are randomly generated from the initial Unibench dataset using a python algorithm and dependencies between entities are respected. We experiment how data volume impacts query execution. These modifications are linked to the number of instances, logical unified models and the mapping dictionary are not impacted. The rewriting time of each query is the same than in the previous experiment for dd_1 however the execution time is different. Q_1 shows a rewriting variation and the location of all attributes of *Products* is searched using the mapping dictionary. Depending on the original system chosen, the transfer and transformation operations are added to the algebraic tree. The increase of volume is expected to impact the number of lines returned for each sub-result and hence, the execution time. Q_1 is composed of two sub-queries: q_1 corresponding to the first selection on the *brand* attribute and q_2 corresponding to the second selection on the *price* attribute. The results obtained for Q_1 execution time comparing V_1 and V_2 are presented in Fig. 6. This experiment is based on an execution plan manually generated for each unified model interrogated. We expected an execution time twice higher for V_2 compared to V_1. We obtained an increase of execution time for V_2 that is almost 1,5 times higher than V_1. Q_1 considers *Products* entity which is fragmented between the relational database and the document-oriented database as considered in dd_1 data distribution. The execution time for the document-oriented unified model is explained by the low number of lines transformed and transferred from the relational database into the document database when compared to the execution time for the relational-oriented database which needs the transformation and transfer of thousands of lines. Finally for the graph oriented model, there are more transfers and transformations from the relational and document databases which explain the high value in Fig. 6.

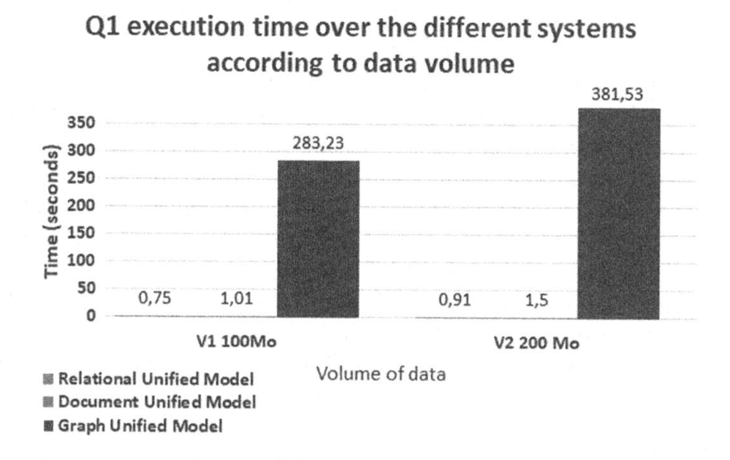

Fig. 6. Experiment result for query execution time for the two data volumes V_1 and V_2 considering the execution over the relational, document-oriented and graph oriented unified models for Q_1

5 Related Work

The appearance of polystores with the combination of SQL and NoSQL systems leads to different level of complexity: data distribution, different models and query language.

Data Distribution. The association of several systems leads to data distribution. Existing works focus on vertical distribution where one entity class is found in one dataset of one database of the polystore. Accessing data of each system is complicated when crossing systems. Operations are executed outside DBMS with an external function [6,14]. HydRa [5], a framework proposing physical possible models according to a specific conceptual model, mentions this type of vertical distribution for one entity but do not explain how to considerate it for querying purposes. Some solution [6] might apply to data fragmentation. Their approach uses an algorithm executing the join operation between datasets of different databases, we suppose they can handle this specific case of vertical data distribution but they do not experiment this in their work.

Data Model. Some works try to find a universal representation for a polystore to hide the complexity into one model. This schema inference helps to define the multiple systems as a simplified one [10]. It can be a graph representation [11–13] or a u-schema model [14] illustrating structural variations. Because "one size does not fit all" [15], it is question to transfer all data into this new model. Changing data representation impacts users and modifies the initial paradigm presented to them. They must adapt to one fixed vision of the polystore.

Query Language. Having different data storage systems brings the question of the user query language. Some works choose to rewrite the initial query according to user specifications or to one system of the polystore [16,17]. Some works [18] use parallel query methods outside data stores (map/reduce/filter). It is helpful to execute operators outside DBMS and to not consider paradigms conversion and they use specific languages limited to relational operators (CloudMdsQL language [6]). BigDAWG [9] uses the principle of islands which communicates with adapters. To query this polystore, the user needs to specifies the system interrogated (for example $RELATIONAL(\ SELECT\ *\ FROM...)$). This adds information to the user query. In this paper, we choose to provide unified models to allow the user to query with transparency because our systems deals with the sub-queries generated from the mono-language user query.

Overview. Table 7 illustrates the differences between our works and others working on vertical data distribution inside polystores. The comparison is for the relational R, document-oriented D, column-oriented C and graph G systems, the query language(s) considered and if it is question of entity class distribution in one or several system (fragmentation). In our work, the user has the choice of the unified model he wants to query without thinking about the distribution and possible fragmentation inside the polystore interrogated.

Table 7. A comparison of existing solutions on polystores

Authors	R	D	C	G	Query	Entity class fragmentation
El Ahdab et al.	●	●	○	●	*SQL - MongoDB - Cypher*	●
Barret et al. [12]	●	●	○	●	SparkQL	○
Candel et al. [14]	●	●	○	●	SQL	○
Ben Hamadou et al. [13]	●	●	●	○	SQL - MongoDB	○
Hai et al. [16]	●	●	○	●	SQL - JSONiq	○
Papakonstantinou [17]	●	●	○	○	SQL	○
Duggan et al. [9]	●	●	●	○	Declarative	○

6 Conclusion

In this paper, we focus on polystore systems with relational, document-oriented and graph-oriented systems. Data is distributed vertically. We define a framework composed of unified models intended to hide the polystore complexity to users. A mapping dictionary is generated to link these representations and the real data distribution. The user can query with a single query one unified model of his choice (relational, document, graph) whereas data is kept in its native form and in its specific location inside the polystore. Experiments were conducted on a Unibench dataset, showing the low impact of data distribution on the rewriting solution and the more important impact on the execution time. Unified models depend only of data structures, they need to be rebuilt only when some data structures change in the polystore (documents or graphs). Existing works on data model extraction may help to automatically produce E/R model. The execution plan optimization and automation will be a prospect of development. This part has an important cost due to data transfers and transformation. Considering our future work on polystore systems, we will expend our operators for each considered model (aggregation), data models (key-value) and we will experiment graph queries on graph unified model.

Acknowledgment. This work was supported by the French Government in the framework of the Territoire d'Innovation program, an action of the *Grand Plan d'Investissement* backed by France 2030, Toulouse Métropole and the GIS neOCampus.

References

1. Leclercq, É., Savonnet, M.: A tensor based data model for polystore: an application to social networks data. In: Proceedings of the 22nd International Database Engineering & Applications Symposium, pp. 110–118 (2018)
2. Forresi, C., Gallinucci, E., Golfarelli, M., et al.: A dataspace-based framework for OLAP analyses in a high-variety multistore. VLDB J. **30**(6), 1017–1040 (2021)
3. Ramadhan, H., Indikawati, F.I., Kwon, J., et al.: MusQ: a multi-store query system for IoT data using a datalog-like language. IEEE Access **8**, 58032–58056 (2020)
4. Misargopoulos, A., Papavassiliou, G., Gizelis, C.A., Nikolopoulos-Gkamatsis, F.: TYPHON: hybrid data lakes for real-time big data analytics – an evaluation framework in the telecom industry. In: Maglogiannis, I., Macintyre, J., Iliadis, L. (eds.) AIAI 2021. IAICT, vol. 628, pp. 128–137. Springer, Cham (2021). https://doi.org/10.1007/978-3-030-79157-5_12
5. Gobert, M., Meurice, L., Cleve, A.: HyDRa a framework for modeling, manipulating and evolving hybrid polystores. In: IEEE International Conference on Software Analysis, Evolution and Reengineering (SANER), pp. 652–656. IEEE (2022)
6. Kolev, B., Valduriez, P., Bondiombouy, C., et al.: CloudMdsQL: querying heterogeneous cloud data stores with a common language. Distrib. Parallel Databases **34**, 463–503 (2016)
7. El Ahdab, L., Teste, O., Megdiche, I., et al.: Unified views for querying heterogeneous multi-model polystores. In: Wrembel, R., Gamper, J., Kotsis, G., Tjoa, A.M., Khalil, I. (eds.) DaWaK 2023. LNCS, vol. 14148, pp. 319–324. Springer, Cham (2023). https://doi.org/10.1007/978-3-031-39831-5_29

8. Zhang, C., Lu, J., Xu, P., Chen, Y.: UniBench: a benchmark for multi-model database management systems. In: Nambiar, R., Poess, M. (eds.) TPCTC 2018. LNCS, vol. 11135, pp. 7–23. Springer, Cham (2019). https://doi.org/10.1007/978-3-030-11404-6_2
9. Duggan, J., Elmore, A.J., Stonebraker, M., et al.: The BigDAWG polystore system. ACM SIGMOD Rec. **44**(2), 11–16 (2015)
10. Koupil, P., Hricko, S., Holubová, I.: Schema inference for multi-model data. In: Proceedings of the 25th International Conference on Model Driven Engineering Languages and Systems, pp. 13–23 (2022)
11. Barret, N., Manolescu, I., Upadhyay, P.: Computing generic abstractions from application datasets. In: EDBT (2024)
12. Barret, N., Manolescu, I., Upadhyay, P.: Abstra: toward generic abstractions for data of any model. In: 31st ACM International Conference on Information & Knowledge Management, pp. 4803–4807 (2022)
13. Ben Hamadou, H., Gallinucci, E., Golfarelli, M.: Answering GPSJ queries in a polystore: a dataspace-based approach. In: Laender, A.H.F., Pernici, B., Lim, E.-P., de Oliveira, J.P.M. (eds.) ER 2019. LNCS, vol. 11788, pp. 189–203. Springer, Cham (2019). https://doi.org/10.1007/978-3-030-33223-5_16
14. Candel, C.F.J., Ruiz, D.S., García-Molina, J.J.: A unified metamodel for NoSQL and relational databases. Inf. Syst. **104**, 101898 (2022)
15. Khan, Y., Zimmermann, A., Jha, A., et al.: One size does not fit all: querying web polystores. IEEE Access **7**, 9598–9617 (2019)
16. Hai, R., Quix, C., Zhou, C.: Query rewriting for heterogeneous data lakes. In: Benczúr, A., Thalheim, B., Horváth, T. (eds.) ADBIS 2018. LNCS, vol. 11019, pp. 35–49. Springer, Cham (2018). https://doi.org/10.1007/978-3-319-98398-1_3
17. Papakonstantinou, Y.: Polystore query rewriting: the challenges of variety. In: EDBT/ICDT Workshops (2016)
18. Kranas, P., Kolev, B., Levchenko, O., et al.: Parallel query processing in a polystore. Distrib. Parallel Databases, 1–39 (2021)

Enabling Interdisciplinary Research in Open Science: Open Science Data Network

Vincent-Nam Dang[1(✉)], Nathalie Aussenac-Gilles[2], Imen Megdiche[3], and Franck Ravat[1]

[1] IRIT, CNRS (UMR 5505), Université Toulouse Capitole, Toulouse, France
vincent-nam.dang@irit.fr
[2] IRIT, CNRS (UMR 5505), Toulouse, France
[3] IRIT, CNRS (UMR 5505), INU Champollion, ISIS Castres, Université de Toulouse, Toulouse, France

Abstract. The aim of Open Science is to open up data to enrich knowledge creation processes. At present, Open Science actors face problems when trying to find and exchange data. Research data management platforms need to address the issue of interoperability to enable interdisciplinary research. Some solutions are available for specific communities, but none addresses the problem as a whole. Based on an extension of the theoretical model of interoperability, which enables us to define the criteria for an information exchange, we have quantitatively evaluated information exchange in Open Science. On the basis of this explorative study, we propose an inter-community and inter-disciplinary information exchange network solution enabling decentralised and federated data management as well as a unified search for datasets across all the entities registered in this network: the Open Science Data Network (OSDN). We carried out a proof of concept to assess the feasibility of the solution. We also evaluated this solution by applying it to a completed agronomy research project. This evaluation enabled us to measure a 7% increase in the volume of data, with an 80% reduction in the time needed to find this data. In addition, users have been able to design new intra- and interdisciplinary futures works whit data found.

Keywords: Information System · Interoperability · Data Integration · Metadata Management · Open Science

1 Introduction

Open Science is a global research movement aimed at opening up knowledge creation processes to enable collaboration and enrichment of the creative process. Open Science actors describe a need for a inter and intra-community coordination [8] to accelerate the adoption of Open Science movement. The FAIR principles define the direction that should be taken to improve information and data sharing, particularly in the context of Open Science [34]. Findability in the sense of the FAIR principles is defined as the ease of finding data for both

J. Araújo et al. (Eds.): RCIS 2024, LNBIP 513, pp. 19–34, 2024.
https://doi.org/10.1007/978-3-031-59465-6_2

humans and machines [12]. Researchers from different fields and communities agree to say that there is a problem of findability of datasets, whether raw data or datasets resulting from research work [4,5,11,15,19,20,22,25,27]. Open Science actors explain this problematic with 2 main reasons, mainly related with the dataset metadata models:

- the variety of the metadata models for these datasets alters findability. We find a large number of differents models, standardized [34] or specific to a platform [22]. Scientists have noticed a lack of interoperability between these models [20,25], therefore increasing the adoption cost of open data solutions.
- the lack of coordination between efforts [11] leads to a large number of research data management platforms, some of them being redundant [5,11,13,22], increasing the variety of used models. Users lack the resources – in time, manpower, knowledge and training – to get to grip with these models [19,20,25].

One way of addressing these issues is through a centralized data management platform in Open Science enabling harmonization of metadata models [30]. But several problems arise with centralisation: (i) impossibility to meet the specific needs of researchers from each different community [4,22]; (ii) too high volume for a single platform [30]; (iii) too costly to deploy such a platform that become a single point of failure [22]; (iv) no security, confidentiality or data control guarantee, whereas it is needed by researchers [13,19,20,25]. Centralisation is not viable to answer the Open Science information exchange problem. Several papers report that researchers claim the need for decentralized and federated data management platforms [22,30] with a unified access to metadata [22]. Several articles emphasize that to achieve an effective Open Science, the effort must be joint and community-based [20,25]. To put it another way, an appropriate solution should present the following features: (i) to manage the wide variety of metadata models from all Open Science communities and domains; (ii) to manage and coordinate with a decentralized solution the wide variety of pre-existing data management platforms, and take advantage of previous efforts; (iii) to provide an easy-to-adopt response to address users' lack of resources.

An intra- and inter-community coordination aims to exchange information to enable cooperation, which corresponds to interoperability [6,7]. However, before proposing a solution, it is needed to assess the type of required solution (infrastructure, user interface, communities, incentives or policy) [16] by getting a quantitative insight into the state of information exchange in Open Science. To the best of our knowledge, there is no quantitative evaluation available on the exchange of information.

We propose an extension of the theoretical model of interoperability [7]. This extension enables to explain the criteria for an information exchange in Open Science allowing us to propose a quantitative metric to assess the lack of information exchange. We then propose an architectural solution to address this lack of implementation of information exchange in Open Science: the Open Science Data Network (OSDN).

In Sect. 2, we explore the notion of interoperability and Open Science data management solutions. In Sect. 3, we extend the theoretical model of interoperability to apply this model to the quantitative evaluation of information exchange in Open Science. In Sect. 4, we describe how the interoperability of data management platforms is implemented in the OSDN and the network structuring of the OSDN. In Sect. 5, we developed a Proof of Concept of OSDN. We made an user experiment with a researcher in agronomy and evaluated the contribution of our solution in terms of time savings, data set enrichment and new future work for this researcher.

2 Related Works

Interoperability is a subject that regularly crops up in the scientific literature. A number of studies focused on various features of interoperability, leading to the distinction of several types of interoperability (technical, syntactical, semantic, platform, system, structural, conceptual, dynamic, ...) [17,24,29,35]. In state of art on interoperability, we find two main components of interoperability: (i) the technical interoperability, which comes from the field of software engineering domain [33], the networks and telecommunication domain [32] or the database domain [14] with the technical problem of exchanging information between several information systems, (ii) the semantic interoperability [10,35] bringing in the need to add value to the information exchanged. The link between the OSI model and interoperability is made [23] when describing interoperability as a layered characteristic. Interoperability also take an important role in the Semantic Web. The Semantic Web contributes to establish interoperability between systems and people on the Web [28]. This objective is shared in Open Science, which focuses on research community [31]. A major focus was placed on semantic interoperability with the issue of knowledge and data exchange [35], in particular when modelling machine-processable metadata with standard vocabularies. This exploration of the different approaches to interoperability reveals a wide variety of approaches to interoperability, which makes it difficult to understand in a comprehensive and generalizable way.

In Open Science, there are many data management platforms. But there are few solutions for interoperating these data management platforms. OAI-PMH[1] is an information exchange protocol based on a harvesting mechanism. The harvester retrieves the information proposed by a service provider. This protocol has a star-shaped architecture. But this solution has its limitations when it comes to setting up inter-community exchanges across the whole of Open Science, linked to the use of a pivotal metadata model and the star-shaped structure of the network created by the protocol. There are other intra-community solutions for data management. Open Science Framework [21] is a general-purpose data management solution. However, the centralisation of the solution does not allow us to respond to all the problems of Open Science. The Beacon Network[2] has a larger scope in its technical conception. However, this solution is designed for a specific field: genetic mutations.

[1] https://www.openarchives.org/pmh/.
[2] https://beacon-network.org/.

At present, we did not find a global solution for interdisciplinary and inter-community information exchange. To understand the challenges of such a solution, we proposed an unified theoritical framework [7] within which the many definitions of interoperability can be explained. In the following section, we complete this interoperability definition with additional characteristics of interoperability needed for an explanation of information exchange.

3 Open Science and Information Exchange

Interdisciplinary information exchange is one of the acknowledged challenge in Open Science. It requires metadata interoperability. In this section, we explore the link between interoperability and information exchange (see Fig. 1 as a guideline).

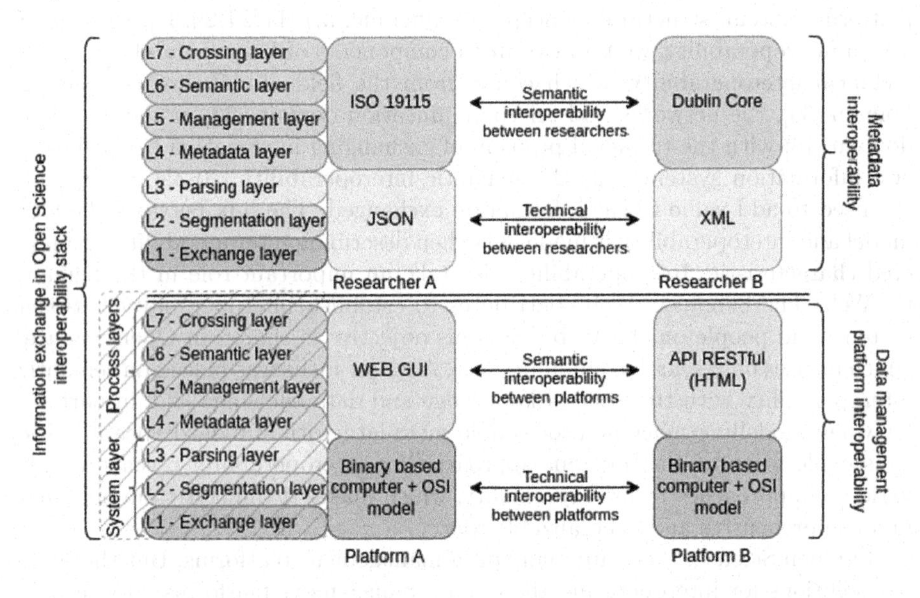

Fig. 1. Interoperability theoretical model applied to information exchange between researchers in Open Science

3.1 Formal Notation

We have chosen to formalise the concept of data using formal grammar. We use "information" and "data" interchangeably because data do not differ structurally from the information [1]. We use the definition of a model as the representation of a domain conceptualization [9]. In the following, we assume that models are faithful representations of domain concepts and properties, allowing us to associate semantic to models. We use "platform" to describe any data management solution that can be used in Open Science to manage research data (like data catalog, data repository, databases, etc..).

In the rest of the paper, we use the notions of graph, formal grammar, probability and set theory, with the following notations:

- $g^{\text{L}}_{e_i}$, a formal grammar of information managed by entity e_i. This grammar defines information management rules in e_i associated to interoperability layer L, depending on the layer purpose.
- $l^{\text{L}}_{e_1}$, a formal language generated by $g^{\text{L}}_{e_1}$ as the set of words that respects rules of layer L.
- $G = (V, E)$, a graph where V is the set of nodes in the graph and E is the set of edges in the graph.
- $\mathcal{P}(\mathcal{S})$, the powerset of a set S, that we can describe as the set of all subsets of S.
- $Pr(X)$ the probability of an event X
- $|S|$, the cardinality of a set S
- $\mathbb{N}$, the set of natural numbers
- API and MD, respectively the set of API implemented by a platform and the metadata model implemented by a platform

3.2 Interoperability and Information Exchange

To understand the mechanisms and requirements for setting up interoperability, it is necessary to have a complete understanding of interoperability. We defined interoperability as *the ability of two communicating entities to work cooperatively through an exchange of information to achieve an objective* [7]. We define interoperability as a stack of seven layers (see hatched area in Fig. 1) [7]. This stack can be divided into two groups of layers: system layers associated to **technical interoperability** and process layers associated to **semantic interoperability** (see blue braces in Fig. 1) [7]. When the entire interoperability stack is implemented, **global interoperability** is then achieved.

Each layer covers a group of mechanisms to be implemented, which vary according to the **context**, the **objective** and the involved **communicating entities** [7]. As an example, the mechanisms of the parsing layer (layer 3) to be implemented for interoperability between two humans will focus on finding the words in a sentence. In the context of interoperability between two computer servers, it is necessary to distinguish the header from the payload.

We distinguish two types of interoperability mechanisms [7]. The first type covers **standardisation** [7], with the aim to establish a common and formal vocabulary for communicating entities in digital format. The second one covers mechanisms of **gateways implementation** [7]. The objective is to set up an entity or a mechanism to act as a bridge between tools of a specific interoperability layer of the communicating entities involved. Let e_1 and e_2 be two communicating entities and f an application defining an interoperability mechanism. We distinguish several categories of such mechanisms: mechanisms applying between languages $f : l^{\text{L}}_{e_1} \rightarrow l^{\text{L}}_{e_2}$, corresponding to a **dictionary**; mechanisms applying on grammars $f : g^{\text{L}}_{e_1} \rightarrow g^{\text{L}}_{e_2}$, corresponding to **translator**. The literature contains other discrimination criteria, especially on translators [2]. The **level of interoperability** is defined as the percentage of domain elements to which it is possible to associate a codomain element using the interoperation mechanism. If these mechanisms are bijective, interoperability is **complete**. Completeness impacts what operations can be achieved out through cooperation. If an inverse mechanism f^{-1} is implemented, interoperability is **reciprocal**. Reciprocity impact the possible direction of information exchange.

To enable an **information exchange** between two communicating entities, it is necessary for these 2 entities to be globally interoperable and for an information exchange to be implemented. Reciprocal interoperability is needed when bidirectional information exchange is needed. To understand what is the state of the implementation of information exchange, a quantitative assesment is required. In the next section, we use interoperability model to propose a quantitative metric for assessing the state of information exchange in Open Science.

3.3 Open Science Information Exchange Quantitative Assessment

To assess the need for a solution and the type of solution needed (structural, incentive, ...) [16], this assessment is required. To the best of our knowledge, there is no quantitative evaluation available on the information exchange in Open Science. We propose to assess this quantity of information exchange through the percentage of data management platforms that exchange information in Open Science. We use data available on Re3Data, a catalog of data management platforms. This catalog contains information on 3117 research data management platforms (as of May 31, 2023) and is used by the European Union as an indicator of the state of open research[3]. We assume that an evaluation based on this catalog will provide a close-to-reality view of Open Science state. In the context of data exchange between Open Science platforms, two kinds of heterogeneity may prevent the platforms from being interoperable: the type of API provided by each platform to access to datasets (technical interoperability) and the kind and structure of the metadata used to describe the datasets (semantic interoperability). Metadata can be represented thanks to various schemas, vocabularies, thesauri and/or ontologies, that we will refer to as "metadata models" in the following.

Among the APIs used by the platforms described in Re3data, the only one that natively integrates an information exchange mechanism is OAI-PMH. This protocol enables metadata to be harvested on queried platform and to access the platform data from an external application, a catalog or another platform. However, OAI-PMH has a limitation when it comes to scaling up. Harvesting must be carried out by the harvester or the service provider, but it is necessary for these platforms to know each other beforehand. With the large number of existing platforms, it is not a viable approach. For the rest, we will assume that all platforms know each other, creating an overestimation of information exchange possibilities thanks to OAI-PMH. This hypothesis compensates for the lack of information on real exchanges between platforms.

[3] https://research-and-innovation.ec.europa.eu/strategy/strategy-2020-2024/our-digital-future/open-science/open-science-monitor/facts-and-figures-open-research-data_en.

To represent the state of Open Science, we define the undirected graph of information exchange in Open Science

$$G_{OSci} = (V_{OSci}, E_{OSci})$$

with V_{OSci} the set of vertices in the graph, where each vertex is a data management platform, and E_{OSci} the set of edges, where each edge defines an information exchange capability between 2 platforms. A platform $v \in V_{OSci}$ is defined with 2 components $v = (API, MD)$, based on the 2 heterogeneity issues previously mentioned. An edge exists between 2 platforms if both implement OAI-PMH and have at least 1 metadata model in common. The visualization of the graph in Fig. 2 clearly illustrates the problem of lack of information exchange in Open Science. A large number of data management platforms (2827 nodes, $\sim$ 90%), materialized as grey spots, are unable to exchange their data with other platform. A set of 290 interconnected nodes is visible ($\sim <$ 10% of total platforms), which materialize the platforms implementing OAI-PMH. Distinct communities can be observed based on node colors (automatically extracted using the Louvain method) with 2 distinct connected components, giving an indication of the lack of homogeneity in communities of Open Science regarding their ability to exchange information. The density of the graph is $\sim$ 0.0046, close to a set of unconnected nodes. This level of information exchange appears very low through this visualization. For a more precise interpretation, we define the event X_d "Find the desired data d on a platform". This event occurs when the search is carried out on a platform that provides access to its data. The empirical probability $Pr(X_d)$ is equal to the number of platforms on which a dataset is available $|avail(d)|$, with $avail(d)$ the function that returns the set of platforms where d is found, divided by the total number of platforms in Open Science $|V_{OSci}|$.

$$Pr(X_d) = \frac{|avail(d)|}{|V_{OSci}|}$$

OAI-PMH does not include a mechanism for harvesting data from indirect neighbours. We define h, a number of hops between two platforms. From a network point of view, counting hops may differ according to the protocols. In our context, we define the number of hops between 2 nodes as the distance between the two nodes on the path studied, i.e. the number of edges between these 2 nodes.

The maximum hop in OAI-PMH is 1. $avail(d)$ becomes $|\bigcup_{n=0}^{h} S^n(avail(d)|$, with $S : \mathbb{N} * \mathcal{P}(V_{OSci}) \rightarrow \mathcal{P}(V_{OSci})$ the successor function with hops such that $S^n(nodes) = \underbrace{S \circ ... \circ S(nodes)}_{\text{n times}}$, with $nodes \in \mathcal{P}(V_{OSci})$ and the successor function $S : \mathcal{P}(V_{OSci}) \rightarrow \mathcal{P}(V_{OSci})$ returning the set of neighbours of a set of nodes.

The probability of the event X_d when h hops are possible becomes

$$Pr^h(X_d) = \frac{|\bigcup_{n=0}^{h} S^n(avail(d)|}{|V_{OSci}|}$$

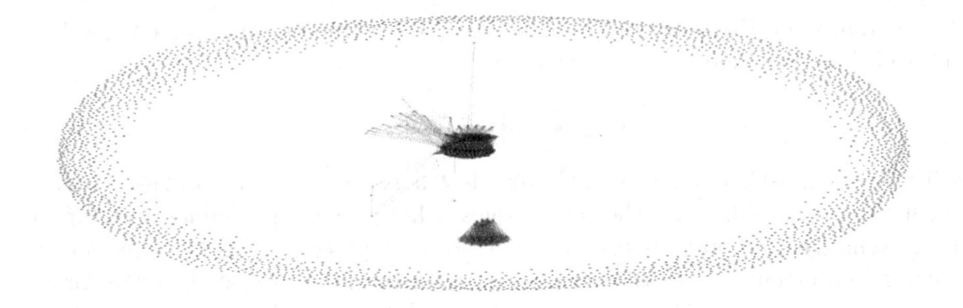

Fig. 2. Open Science information exchange graph visualization

OAI-PMH allows to request platforms 1 hop away from the initial one. *avail*(d) is equal to the average number of neighbours in Open Science plus the number of platforms managing that same dataset. We assume that data is not duplicated. The empirical probability of finding the desired data is equal to

$$Pr^1(X_d) = \frac{(1 + mean\_degree\_in\_OSci)}{total\_node\_number} \approx \frac{(1 + 14.24)}{3117} \approx 0.5\%$$

This probability is very low and confirms the description made by Open Science actors on the lack of cooperation between Open Science data management platforms and actors. This very low level of information exchange show the need to implement an information exchange in Open Science and not just improve it, **requiring architectural solutions** [16]. In the next section, we propose a network-based solution to implements information exchange in Open Science.

4 Proposition: The Open Science Data Network

Open Science contains many pre-existing data management platforms. Researchers are calling for decentralisation, unified research and control of open data by its owner. To interoperate existing systems and meet needs, we propose the Open Science Data Network (OSDN), a decentralised, federated and distributed interconnection network of data management platforms. We describe our solution in 2 parts. Firstly, we explore interoperability and information exchange within the OSDN. Then we explore a part of the scaling with the robustness of the solution, to ensure that the solution is sustainable.

4.1 Information Exchange and Interoperability in OSDN

This solution is based on a RESTful API using a registry shared among every platform (see Fig. 3a). This registry contains the information needed for the OSDN to operate (see Fig. 3b). A mechanism for propagating changes and queries is implemented by broadcasting them to all neighbours until the changes or queries have been propagated to the whole network. From a technical point of

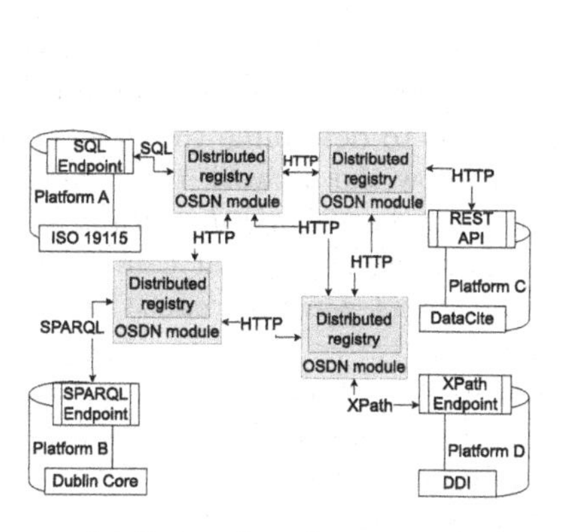

(a) Network architecture example

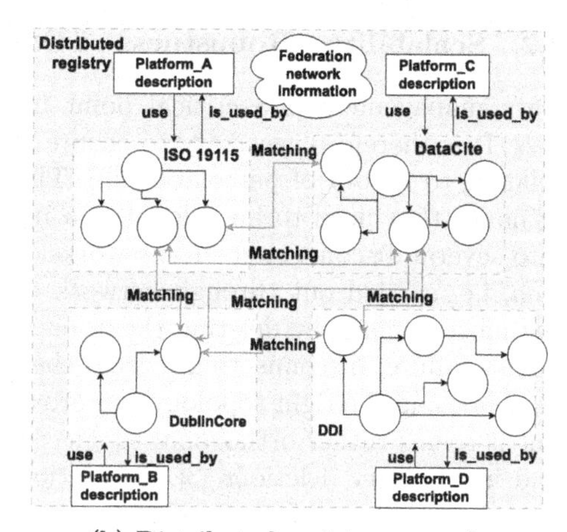

(b) Distributed registry example

Fig. 3. OSDN implementation example

view, this module takes the form of a Docker container with automatic deployment. The integration of this module requires a single operation: the implementation of the interoperation function between this module and the platform information retrieval mechanism, enabling the module to execute queries on the platform. **Reciprocal and complete technical interoperability** is achieved by standardising the exchange protocol between platforms based on the module. The registry contains information relating to the platforms (name, URL, interconnected platforms, etc.), the used metadata models (name, content, etc.) and the matchings between the metadata models.

We have observed a lack of adaptation of automatic matching solutions to the metadata models used in Open Science [7]. For the implementation of these matchings in OSDN, we propose a combined implementation of automatic and manual solutions. Firstly, the implementation of manual matchings is based on the establishment of community collaboration between the various actors in Open Science. Each platform manager sets up matches between its model and another model in the registry. Reducing the workload for each player goes some way to addressing the problem of scaling up manual matching. As it stands, the use of automatic matches can enable 2 solutions to be put in place: (i) a recommendation of matches reducing the cost of setting up manual matches, (ii) a crossing of the results of automatic matching solutions to improve the results benefiting from the collaborative structure of the OSDN. To address the problem of matching trustworthiness and the fact that they can contradict each other, we decided to keep all the matchings and associate them with a likelihood score, based on user feedback. The aim is to return query results based on the most likely matches. **Semantic interoperability** is achieved by setting up gateways thanks to the matchings between models, and so implementing a **global interoperability**.

4.2 Scalability - Robustness of OSDN

Data management is a critical point in the research knowledge creation process. It is therefore necessary to ensure that the network is robust and durable, relating to theory of percolation [3]. The objective is to determine the number of nodes that need to be deleted in a network to allow the network to be split into several disconnected sub-networks. Deletions may be either *voluntary deletion*, i.e. carried out through network attacks, targeting nodes that can cause the most damage by deleting them; or *random deletions*, generally caused by a node's failure, disconnecting it from the network.

Choosing the right topology can provide particular resistance to these events, but it seems rather orthogonal to succeed in protecting against both voluntary and involuntary deletions [3]. A solution that minimizes vulnerability to these two events is found for a scale-free network topology with a single hub and with the other nodes all having a degree of five [3]. To reduce adoption costs, we set the minimum degree of node equal to two. This network topology follows a power distribution in the node degree distribution. In the following, we denote γ the power law parameter followed by the network topology [3]. This topology offers an interesting feature for information exchange. Based on epidemiological approach applied to networks [3], the propagation time of a pathogen τ (which can be interpreted as an information) in a scale-free networks with a γ less than 3 tends towards 0, when the number of nodes increases [3]. Moreover, scale-free network topologies following a power law with $2 < \gamma < 3$ are in ultra-small world regime [3]. These networks have average distance between nodes growth equal to $lnlnN$. Linking a large number of nodes create a small distance between them [3], leading to a small increase in information propagation time.

Inscription Protocol: To be able to implement a specific topology, it is necessary to implement a control of the registration process. We chose a network parameter γ of 2.5 to avoid potential edge effects and to have the characteristics of networks with $2 < \gamma < 3$. This registration process is described by the Algorithm 1.

The *get_node_nearest_from_distribution* function returns a node of degree k_{opti} to connect to, such that k_{opti} minimizes the Kullback-Leibler divergence D_{KL}, giving a measure of dissimilarity between two distributions. In other words, it returns a degree k_{opti} node. We select the first node found in the list of degree k_{opti} nodes to reduce computation time in our implementation. To understand formally, let P be the initial degree distribution of the network graph with the node v_{chosen} added but not connected, P_n the degree distribution of the graph nodes after the connection of v_{chosen} to a node of degree n, Q a power law with $\gamma = 2.5$, E_k the set of nodes of degree k from the distribution P, and E the total set of nodes in the distribution P. We define that connecting v_{chosen} does not change its degree, assuming it has initially 2 self loops that we want to replace by connecting it to another node. We have $D_{KL}(P||Q) = \sum_{k \in K} q(k) \log(\frac{p(k)}{q(k)})$ with K the set of possible degrees in the graph, from 0 to k_{max} and $p(k)$ and $q(k)$,

Algorithm 1: Node inscription in network

```
Input  : A network graph G = (V, E), a node v_add, a node in the network to
         link to v_chosen
Output: A network graph G = (V', E')
1  if |V| = 0 then
2  |  add_node_to_network(G, v_add)
3  end if
4  if |V| = 1 then
5  |  add_node_to_network(G, v_add)
6  |  add_edge(v_add, v_0) /* v_0 is the only vertice in the network       */
7  end if
8  if |V| > 1 then
9  |  add_node_to_network(G, v_add)
10 |  add_edge(v_add, v_chosen)
11 |  v = get_node_nearest_from_distribution(v_add, V) /* Return the nearest
   |      node from set of nodes that should be connected to fit the
   |      most a power law with γ = 2.5                                    */
12 |  add_edge(v_add, v)
13 end if
14 return G
```

respectively the probability mass function of distribution P and Q. Connecting to a node of degree n increases the degree of the target node by 1. So we remove 1 node of degree n and add a node of degree $n+1$ to the distribution. We have

$$\begin{cases} p_n(k) = \frac{|E_k|-1}{|E|} = \frac{|E_n|-1}{|E|} \text{ for } k = n \\ p_n(k) = \frac{|E_k|+1}{|E|} = \frac{|E_{n+1}|+1}{|E|} \text{ for } k = n+1 \\ p_n(k) = P(k) = \frac{|E_k|}{|E|} \text{ otherwise} \end{cases}$$

where $p_n(k)$ is the empirical mass function of the distribution P_n. $p_n(k)$ is independent of n when $k \notin \{n, n+1\}$. We end up with the optimization problem in (1). We have $q(n) = \frac{n^{-\gamma}}{\zeta(\gamma)} = \frac{n^{-2.5}}{\zeta(2.5)}$, with ζ the Riemann zeta function [3]. Since the value of $\zeta(2.5)$ is independent of n, Eq. (1) can be simplified into (2).

$$k_{opti} = \arg\min_n (p_n(n) \log(\frac{p_n(n)}{q(n)}) + p_n(n+1) \log(\frac{p_n(n+1)}{q(n+1)})) \quad (1)$$

$$k_{opti} = \arg\min_n (\frac{|E_n|-1}{|E|} \log(\frac{\frac{|E_n|-1}{|E|}}{n^{-2.5}}) + \frac{|E_{n+1}|+1}{|E|} \log(\frac{\frac{|E_{n+1}|+1}{|E|}}{(n+1)^{-2.5}})) \quad (2)$$

Equation (2) gives the optimal degree of the node to be computed in the *get_node_nearest_from_distribution* function. Knowing that we are evaluating n from 2 to k_{max}, with k_{max} the highest degree in the distribution before adding a node, the registration function has a computational complexity in $O(k_{max})$.

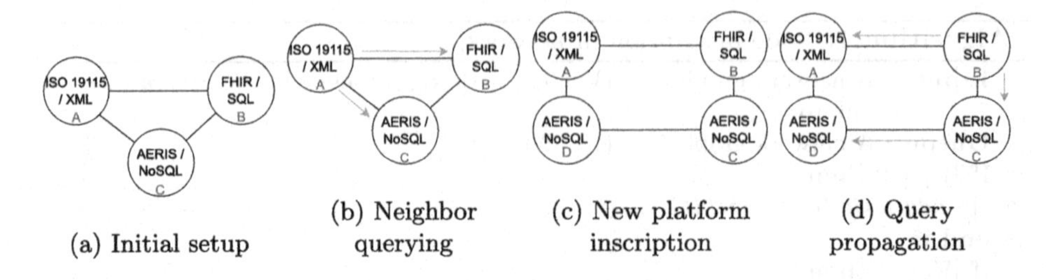

(a) Initial setup (b) Neighbor querying (c) New platform inscription (d) Query propagation

Fig. 4. POC implementation

5 Experiments

We experimented our solution in 2 distincts parts, based on the development of a proof of concept for OSDN for an assessment of our solution's ability to integrate platforms and metadata models from Open Science and then we developed a supplementary Web GUI to make an user experiment with agronomy researcher, in order to observe the benefits. All code is open, accessible and re-executable.

5.1 OSDN Network POC

Table 1. Classification of metadata models

Domain	Use case			
	Information retrieval	Data integration	Core data set	Secondary use
Humanities	**(14)**: (Struct) (SDMX) SDMX - Statistical Data and Metadata Exchange	**(11)**: (Struct) (OAI) OLAC	**(4)**: (Struct) (DDI) DDI - Data Documentation Initiative Metadata Standard	**(16)**: (Struct) (TEI) Text Encoding Initiative Guidelines
Life science	(Sem) (WHO) ICD-10	**(3)**: (Struct) (TDWG) Darwin Core		**(8)**: (Struct) (HL7) FHIR **(2)**: (Struct) (HL7) C-CDA
Natural science	**(12)**: (Struct) PDB		**(9)**: (Struct) (ISO) ISO-19115 **(1)**: (Struct) AERIS	
Engineering	**(7)**: (Struct) (RDA) EngMeta - Metadata for Computational Engineering		**(15)**: (Struct) (OGC) SensorML **(17)**: (Struct) (OGC) CoverageJSON	
General	(Sem) (ISO) ISO 639-2 **(6)**: (Struct) e-Government Metadata Standard	**(13)**: (Struct) (OCLC/RLG) PREMIS: Data Dictionary for Metadata Preservation	**(5)**: (Struct) Dublin Core **(10)**: (Struct) (DataCite) DataCite	

To assess the technical feasibility of OSDN and its mechanisms, we have developed a Proof of Concept[4]. We have integrated 3 platforms from Open Science with different metadata management technologies, with MySQL, MongoDB and an XML-based solution (see Fig. 4a). We executed a query on one platform and verified the propagation of this query to its neighbours (see Fig. 4b). Then, we evaluated the query propagation to indirect neighbours by adding a new node (Fig. 4c) and re-executing the query that also returns the dataset information from the new platform (Fig. 4d). To ensure that OSDN can integrate the metadata models from Open Science, we integrated and interoperate 19 metadata models in OSDN registry. To avoid biases, we selected them after a systematic

[4] https://github.com/vincentnam/Openscience_network_experiment.

review of metadata models [26] according to 3 criteria: domain, use case and type (structural (17) or semantic (2)). We added the model creator as supplementary criteria (see Table 1). We manually created matches between structural models to validate that models can be interoperated (see Table 2).

Table 2. Models interoperability : models matching

Concept	Model number (see Table 1)																
	1	2	3	4	5	6	7	8	9	10	11	12	13	14	15	16	17
Dataset Title	x	x	x	x	x	x	x	x	x	x	x	x	x	x	x	x	
Content localization	x		x					x		x							x
Content UOM	x	x					x								x		

5.2 Use Case - An Agronomic Research Project

To estimate the benefits of our solution, we developed a graphical interface to enable graphical information retrieval from a set of 11 data management platforms from different domains and communities[5], from which we downloaded a part of the data catalog. A single platform allowed us to download the whole catalog. We worked with Dr. Thomas Pressecq[0000-0003-0067-7903]. During the last 3 years, his research focused on the development of a decision support system to guide the use of biocontrol tools by farmers [18]. The main problem he met was the lack of data. Dr. Pressecq described a lateness in agronomy data management, with too few data and too few data management platforms either used or known by researchers. To carry out his research project, Dr. Pressecq extracted 381 row in his dataset, from 900 scientific publications on 41 strains of micro-organisms. Each row contains the combination of a biocontrol agent and its associated efficacy factor (propriety of biocontrol agent, environmental conditions, cropping practices, propriety of the pathogen). He estimates that processing a scientific article requires 20 min, including reading, analysing and extracting information. We asked him to reproduce his search for data, about a sub-part of his dataset (on "trichoderma harzianum T-22" strain) on OSDN. On this specific strain, his dataset contains 29 tabular rows extracted from a total of 115 scientific publications, for a mean time by line of 79 min. He performed a simple query applying only to the title or the description. We evaluated the time reduction gained using OSDN compared with his past experience to build the dataset. As side effect benefits, using OSDN allowed to get more datasets, to increase the volume of collected data and to generate new perspectives for future works.

[5] https://github.com/vincentnam/RCIS_userfeedback_experiment.

Time: Dr. Pressecq kept 17 datasets or publications from 3 different platforms out of the 11 platforms (13 results on NCBI[6], 3 on the Harvard generalist dataverse[7] and 2 on Figshare, a generalist platform[8]). The total time spent on this search was 2 h and 30 min. Of the 17 results selected, 12 were new datasets initially unknown to Dr. Pressecq. However, of the 29 tabular rows in the dataset, 3 rows could have been replaced by datasets containing extracted and usable data. With our solution, it took 45 min to Dr. Pressecq to find 3 new lines, which corresponds to a reduction in dataset construction time on this 3 lines of $\sim 80\%$ (compared to 3 times 79 min) thanks to datasets reuse.

Dataset Enrichment: OSDN provided 2 new lines that were not found by the initial method. This represents a **7%** increase in the dataset volume.

Interdisciplinarity: OSDN provided datasets from several domains (bioinformatics and health (NCBI), social science (Harvard dataverse)) and other communities (Figshare and Harvard). It shows an interdisciplinary and intercommunity data enrichment. But, this interdisciplinary and intercommunities information exchange also provided datasets that allow the creation of new futures works. Dr. Pressecq described several new futures works which we divide into 2 categories:

- **Intradisciplinary research works** with a study based on data crossing of soil characteristics (FORM@TER[9]) for microbial agents prospection in different types of soil.
- **Interdisciplinary research works** by crossing data taken from the European data platform[10], to assess the perception of biocontrol tools among farmers and consumers at European level as a collaboration of agronomy and social sciences domains.

6 Conclusion

Open Science still faces major obstacles to sharing and finding data, like the lack of coordinated solutions between data platforms [8]. We have proposed a quantitative assessment of the current state of information exchange in Open Science, based on percentage of platform that exchange information, estimated at 0.5%. To answer this lack of information exchange implementation in Open Science, we have proposed OSDN, a decentralized, distributed and federated network solution for research data management platforms. We assessed the technical feasibility and behaviour of this network. Finally, we identified the contributions of OSDN in the context of a real research project, saving time in building datasets, data enrichment and datasets reuses and research projects, and providing new interdisciplinary research perspectives. We observed that our solution allows

[6] www.ncbi.nlm.nih.gov.
[7] dataverse.harvard.edu.
[8] figshare.com.
[9] https://www.poleterresolide.fr/.
[10] data.europa.eu.

information exchange. But this solution faces a real challenge: adoption. Even if it has be thought to be as easy as possible to integrate OSDN, the development cost of the interoperability function of the OSDN module and the inscription process cost, especially model matching, may vary due to different contexts of data management platform (human resources, technologies used, etc.). Moreover, it may lead to an additional burden to process network queries. This could discourage platform manager without proper incentives. As a futures works, in addition to working on the adopting cost of this solution, there a 2 different axes. Firstly, an exploration of semantic, semantic interoperability concept and the problematic of automatic matching algorithms in the context of Open Science will be explored. Then, we plan to explore the scalability and several optimisation on resources consumption in the OSDN.

Acknowledgement. We thank Dr. Thomas Pressecq from INRAE (Institut national de la recherche agronomique) and APREL (Association Provençale de Recherche & d'Exprimentation Légumière) for his participation in our experiment and his feedback.

References

1. Ackoff, R.L.: From data to wisdom. J. Appl. Syst. Anal. **16**(1), 3–9 (1989)
2. Aho, A.V., Hopcroft, J.E., Ullman, J.D.: A general theory of translation. Math. Syst. Theory **3**, 193–221 (1969)
3. Barabási, A.L., Pósfai, M.: Network Science. Cambridge University Press, Cambridge (2016). http://barabasi.com/networksciencebook/
4. Beyan, O., et al.: Distributed analytics on sensitive medical data: the personal health train. Data Intell. **2**(1–2), 96–107 (2020)
5. Corpas, M., et al.: A fair guide for data providers to maximise sharing of human genomic data. PLoS Comput. Biol. **14**(3), e1005873 (2018)
6. Costin, A., Eastman, C.: Need for interoperability to enable seamless information exchanges in smart and sustainable urban systems. J. Comput. Civ. Eng. **33**(3), 04019008 (2019)
7. Dang, V.N., Aussenac-Gilles, N., Megdiche, I., Ravat, F.: Interoperability of open science metadata: what about the reality? In: Nurcan, S., Opdahl, A.L., Mouratidis, H., Tsohou, A. (eds.) RCIS 2023. LNBIP, vol. 476, pp. 467–482. Springer, Cham (2023). https://doi.org/10.1007/978-3-031-33080-3_28
8. Digital Science, Hahnel, M., Smith, G., Schoenenberger, H., Scaplehorn, N., Day, L.: The State of Open Data 2023 (2023). https://doi.org/10.6084/m9.figshare.24428194.v1
9. Guizzardi, G.: On ontology, ontologies, conceptualizations, modeling languages and (meta) models. In: Databases and Information Systems IV. IOS Press (2007)
10. Heiler, S.: Semantic interoperability. ACM CSUR **27**(2), 271–273 (1995)
11. Hughes, L.D., et al.: Addressing barriers in fair data practices for biomedical data. Sci. Data **10**(1), 98 (2023)
12. Jacobsen, A., et al.: Fair principles: interpretations and implementation considerations. Data Intell. **2**(1–2), 10–29 (2020)
13. Kathawalla, U.K., Silverstein, P., Syed, M.: Easing into open science: a guide for graduate students and their advisors. Collabra Psychol. **7**(1), 18684 (2021)

14. Litwin, W., Mark, L., Roussopoulos, N.: Interoperability of multiple autonomous databases. ACM Comput. Surv. (CSUR) **22**(3), 267–293 (1990)
15. National Academies of Sciences and Global Affairs and Board on Research Data and Information and Committee on Toward an Open Science Enterprise: Open science by design: Realizing a vision for 21st century research (2018)
16. Nosek, B.: Strategy for culture change. Center Open Sci. **11** (2019)
17. Noura, M., et al.: Interoperability in internet of things: taxonomies and open challenges. Mob. Netw. Appl. **24**(3), 796–809 (2019)
18. Pressecq, T., et al.: DeciControl: a participative tool to gather & share data regarding biocontrol efficacy. In: European Scientific Conference–Towards Pesticide Free Agriculture "What Research to Meet the Pesticides Reduction Objectives Embedded in the European Green Deal?" (2022)
19. Rainey, L., Lutomski, J.E., Broeders, M.J.: Fair data sharing: an international perspective on why medical researchers are lagging behind. Big Data Soc. **10**(1), 20539517231171052 (2023)
20. Sadeh, Y., et al.: Opportunities for improving data sharing and fair data practices to advance global mental health. Cambridge Prisms Glob. Ment. Health **10**, e14 (2023)
21. Spies, J.R.: The open science framework: improving science by making it open and accessible. University of Virginia (2013)
22. Tanhua, T., et al.: Ocean fair data services. Front. Mar. Sc. **6**, 440 (2019)
23. Thomesse, J.: Interoperability: an overview. IFAC Proc. **30**(7), 433–438 (1997)
24. Tolk, A., et al.: Applying the levels of conceptual interoperability model in support of integratability, interoperability, and composability for system-of-systems engineering. J. Syst. Cybern. Inf. **5**(5) (2007)
25. Top, J., et al.: Cultivating FAIR principles for agri-food data. Comput. Electron. Agric. **196**, 106909 (2022)
26. Ulrich, H., et al.: Understanding the nature of metadata: systematic review. J. Med. Internet Res. **24**(1), e25440 (2022)
27. Uribe, S.E., et al.: Dental research data availability and quality according to the fair principles. J. Dent. Res. **101**(11), 1307–1313 (2022)
28. Uschold, M.: Where are the semantics in the semantic web? AI Mag. **24**(3), 25 (2003)
29. Van Der Veer, H., Wiles, A.: Achieving technical interoperability. European Telecommunications Standards Institute (2008)
30. Vesteghem, C., et al.: Implementing the FAIR data principles in precision oncology: review of supporting initiatives. Brief. Bioinform. **21**(3), 936–945 (2020)
31. Vicente-Saez, R., Martinez-Fuentes, C.: Open science now: a systematic literature review for an integrated definition. J. Bus. Res. **88**, 428–436 (2018)
32. Wegner, P.: Interoperability. ACM CSUR **28**(1), 285–287 (1996)
33. Wileden, J.C.O.: Specification-level interoperability. Commun. ACM **34**(5), 72–87 (1991)
34. Wilkinson, M.D., et al.: The fair guiding principles for scientific data management and stewardship. Sci. Data **3**(1), 1–9 (2016)
35. Zeng, M.L.: Interoperability. KO Knowl. Organ. **46**(2), 122–146 (2019)

TD-CRESTS: Top-Down Chunk Retrieval Based on Entity, Section, and Topic Selection

Mohamed Yassine Landolsi(✉) and Lotfi Ben Romdhane

MARS Research Lab LR17ES05, SDM Research Group, ISITCom, University of Sousse, Hammam Sousse, Tunisia
{medyassine.landolsi,lotfi.benromdhane}@isitc.u-sousse.tn

Abstract. Retrieving specific information from extensive scientific documents presents a significant challenge. Existing information retrieval methods often focus on entire documents, even when only a small portion of a document is relevant. Also, achieving a balance between precise retrieval and optimal time complexity remains a persistent challenge. To address these issues, we propose TD-CRESTS (Top-Down Chunk Retrieval based on Entity, Section, and Topic Selection), a term-based document text chunk retrieval method. TD-CRESTS utilizes a hierarchical context architecture, indexing documents according to topics, named entities, sections, and individual text chunks. Key terms from each context guide a top-down search strategy across the index context levels, prioritizing the most relevant contexts based on their overlap with the query. Our method achieves chunk-level F1-measure of 71% and 77.14% on the SciREX and DrugSemantics benchmark datasets, respectively. It is able to handle diverse domains and languages with a balance between information retrieval efficiency and effectiveness.

Keywords: Text chunk retrieval · Domain-specific named entities · Document structure · Document clustering

1 Introduction

A large quantity of scientific online resources, particularly in textual document form, have become the largest source of information which is needed to be exploited by users in various domains, including medical and computer science, and even in several languages [3,22]. For example, the extensive number of scientific publications is expected to exceed 2.5 million annually, showing a continuous and rapid growth [7,25]. Thus, the user needs an efficient way to retrieve the desired information among these documents. There is many tasks in the fields of Information Retrieval (IR) and Natural Language Processing (NLP) that are able to deal with these documents, such as Named Entity Recognition (NER) [13,15], topic modeling and clustering [17], and document structure extraction [16].

The purpose of an IR system is to find and retrieve relevant information from documents to satisfy a user demand which is expressed by a query [3]. The

J. Araújo et al. (Eds.): RCIS 2024, LNBIP 513, pp. 35–46, 2024.
https://doi.org/10.1007/978-3-031-59465-6_3

system performance depends on three main tasks: (1) representing and indexing the documents, (2) representing the query, and (3) matching the query to the indexed documents. In the first stage, some words are usually extracted and selected to describe and index the document content, and they must be easily and efficiently matched later by a query during the IR [3,22]. The indexing is an important and challenging process to ensure a quick and efficient retrieval of relevant results from a large collection of documents.

Significant research on IR has been carried out on textual documents. Existing research studies are relied on term-based and machine learning based approaches [8]. Term-based methods [3,4,12] are able to select the most relevant terms for documents or queries and they can leverage external resources of terms. Machine learning based methods [2,17,21] are able to learn more important features from documents and queries and represent them in a reduced dimension using rich embedding vectors.

However, existing studies focus on extracting high-quality features and terms to represent texts while the retrieval complexity optimization is usually ignored or not sufficient. Thus, both efficiency and precision should be taken into account. Furthermore, studies tend to focus on retrieving whole documents instead the relevant parts of them.

In our work, we propose a hierarchical text chunk retrieval approach, reducing the search space across various context abstraction levels to enable efficient and precise searching from high to low level: (1) topic-level, (2) named entity-level, (3) Section and sub-section levels, and (4) chunk text level. For that, we perform document clustering to extract diverse topics and we create a named entity inverted index for selecting section contexts. Consequently, sub-section and text chunk context levels are incorporated. The topic, section, and sub-section contexts are represented by terms selected through TF-IDF. We utilize overlap similarity measure to compare query terms with the contexts, including the text chunks. For experiments, we have used two benchmark datasets annotated by domain specific named entities: SciREX [9] corpus with machine learning research papers in English and DrugSemantics [19] corpus with Spanish Summaries of medicinal Product Characteristics (SPC) in Spanish.

The remainder of this paper is organized as follows. Section 2 discusses state-of-the-art methods for information retrieval. Section 3 presents our method named TD-CRESTS; whereas Sect. 4 conducts an extensive experimentation of TD-CRESTS using standard criteria. The final section concludes this paper and discusses future research directions.

2 Related Work

In recent years, information retrieval has drawn worldwide attention as an important natural language processing task. In this section, we discuss some previous significant research conducted in this field. Based on the work of Hambarde and Proença [8], we can categorize information retrieval methods into two general categories based on the used techniques: term-based and machine learning based approaches.

The principle of term-based approach is to extract and assign relevant terms to documents or queries, then simply match them during retrieval. It usually selects terms in the document text, or use external knowledge base to enrich terms or to expand queries. As advantage, this approach is efficient and precise and can extract relevant key words. However, it faces challenges in effectively handling the polysemy and context of the terms [8]. For instance, Edinger et al. [6] have identified and normalized section titles to explicitly query specific sections. Brandsen et al. [4] have indexed document pages by domain specific named entities. Boukhari and Omri [3] have performed a partial matching between documents and external resource terms and generate a relevant knowledge representation. Kumar and Sharma [12] have expanded queries semantically through an optimization-based algorithm, incorporating external resources and utilizing a combined global and local word similarity ranking.

Machine learning based approach is able to leverage machine learning and deep learning models to represent a document or a query in a reduced dimension by dense or sparse vectors. Thus, it can use similarity metrics for ranking or extract key information from texts. As advantage, this approach is able to capture contextual information, learn high quality text embedding, and index documents. However, it usually introduces significant complexity, particularly during the searching process on a large scale, and information loss when reducing the dimension. For example, Khattab and Zaharia [11] have separately encoded a query and a document into contextual embedding sets using BERT, to efficiently predict the relevance between them. Lossio-Ventura et al. [17] have evaluated various topic modeling and clustering models, as they yield a high predictive performance on information retrieval tasks. Sarasu et al. [21] have used a CNN model with bag-of-word embedding to capture keyword relationships based on meaning and context, identifying polysemy and constructing a word structure. Bhopale and Tiwari [2] have generated document and query embeddings using BERT for document indexing and ranking, while summarizing lengthy documents and expanding queries by a phrase embedding based model.

In our work, we propose a precise and efficient term-based chunk retrieval method named TD-CRESTS (Top-Down Chunk Retrieval based on Entity, Section, and Topic Selection) using document clustering, named entities, and document structure to perform a hierarchical top-down query searching in different context levels. Thereby, search complexity should be significantly reduced at each level while retrieving relevant text chunks [8]. Although we use machine learning to cluster documents at the first context level, our method tends to be based on terms selection to leverage the simplicity and efficiency of this approach and to overcome the complexity and imprecision of machine learning [2,17]. For that, we leverage many NLP tasks so the results will be selected to be relevant in multiple perspectives [12]. TD-CRESTS is especially oriented to domain specific named entities which are the most important entities in the texts [4]. Also, it leverages the document structure to limit the search zone in multiple levels [6]. Even the results are retrieved in a minimized and concise form by retrieving text chunks instead of whole documents or pages [4].

In summary, the field of information retrieval faces several significant challenges that demand careful consideration and innovative solutions. Term-based methods offer simplicity and efficiency but grapple with understanding context and managing polysemy, while machine learning methods can introduce complexity and imprecision despite their ability to analyze context [2,17]. Thus, bridging the gap between precision and efficiency remains a primary concern [8]. Thus, bridging the gap between precision and efficiency remains a primary concern [8]. Furthermore, there is a need for multi-perspective searching to enhance precision [12]. Additionally, ensuring a more refined retrieval unit is crucial, particularly given the common practice of presenting results at the document or page level [4]. Addressing these challenges is pivotal for advancing information retrieval and ensuring users can efficiently access relevant information across various contexts. Our proposed solution are detailed in the next section.

3 Our Proposal

In our work, we propose a precise and efficient term-based chunk retrieval method named TD-CRESTS (Top-Down Chunk Retrieval based on Entity, Section, and Topic Selection) using document clustering, named entities, and document structure to perform a hierarchical top-down query searching in different context levels. Although we use machine learning to cluster documents at the first context level, our method tends to be based on terms selection to leverage the simplicity and efficiency of this approach and to overcome the complexity and imprecision of machine learning [2,17]. The goal of our method is to search for a query containing recognized domain-specific named entities within structured and indexed documents in a hierarchical way, aiming to quickly retrieve relevant text chunks. Note that even the results are retrieved in a minimized and concise form by retrieving text chunks instead of whole documents or pages [4]. For that, we propose an efficient approach by looking at contexts from high to low level. Thus, our approach should be able to reduce the searching zone after each level and evidently reduce the searching time complexity by avoiding the need to examine every single document [8]. Therefore, we leverage many NLP tasks so the results will be selected to be relevant in multiple perspectives [12]. We prepare context levels in the following order: topic, named entity, section, and sub-section. The topic-level contexts are determined after clustering the documents where each cluster represent a different topic [10,17]. Then, an inverted index is created for each topic to index text chunks by domain-specific named entities [4]. The text chunks collected from the documents are grouped by their sections and filtered according to each selected named entity in the index [18]. Thus, we can focus on the relevant sections during the search [5,6,14]. Finally, the same process is performed to the sub-section contexts which are assigned to their corresponding section contexts in order to select a more specific context [14]. To find the relevant chunk texts, the query is simply compared with each context starting from the top-level. Then, it is compared with the text chunks under the similar sub-section contexts. Figure 1 illustrates the key steps of our approach, which are detailed in the following subsections.

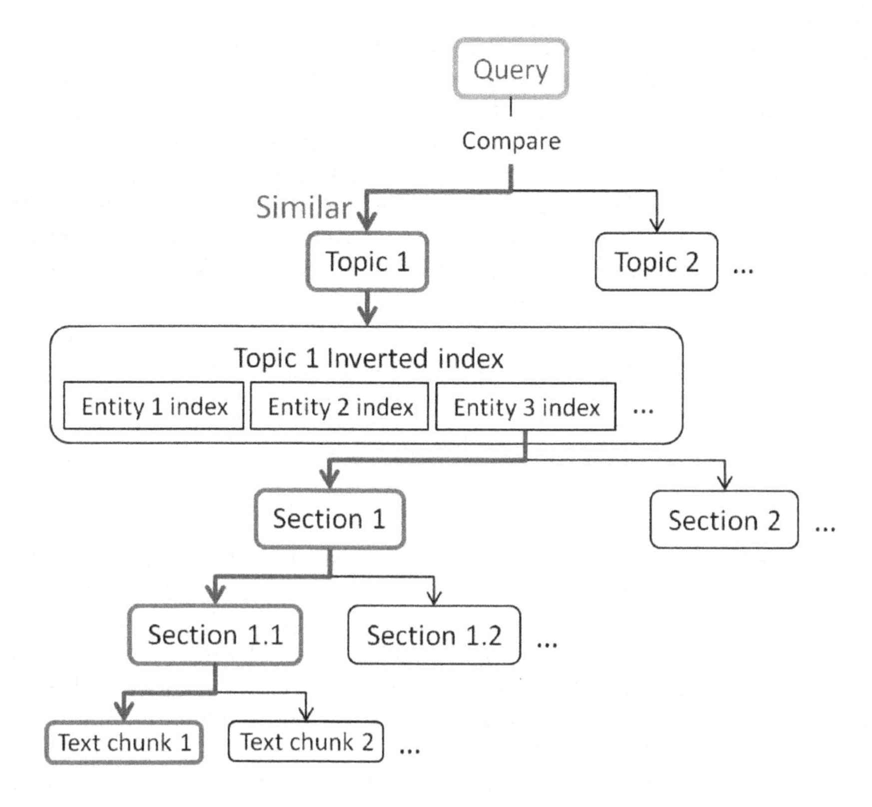

Fig. 1. The general architecture of our proposal.

3.1 Contexts Preparation

To construct and prepare the text chunk index, we generate a representation for each context and text chunk. For that, the documents are analysed and structured. In our work, we applied some preprocessing steps to any text in natural language to normalize the words and remove useless information in both the document text chunks and the query [17]. The preprocessing steps are the following:

1. Convert text to lowercase.
2. Remove URLs.
3. Replace numbers with a unique symbol.
4. Split text into words based on spaces.
5. Remove punctuation marks.
6. Remove stop words.
7. Reduce words to a root form using Snowball Stemmer.

In the following sub-sections, we describe the context construction at each level.

Topic Contexts. At the topic-level, we perform a document clustering task where each cluster represents a different topic [17]. At this stage, the named entities should be recognized and the document should be structured at a preliminary stage. For each document, we extract and preprocess the named entities and the title which represent its candidate key elements. Then, we use the Term Frequency-Inverse Document Frequency (TF-IDF) [24] measure to compute the weight of the tokens of the candidate elements which reflects their importance in the document among all the documents. We compute the TF-IDF of a token i in a document j by the following formula [20]:

$$tf_{ij} = 0.5 + 0.5 \cdot \frac{f_{i,j}}{\max_{\{t_k \in d_j\}} f_{k,j}} \tag{1}$$

$$idf_i = \log \frac{|D|}{|\{d_j : t_i \in d_j\}|} \tag{2}$$

$$tfidf_{ij} = tf_{ij} * idf_i \tag{3}$$

where $f_{i,j}$ is the frequency of the token i in the document j; t_i is the token i; d_j is the document j; and D is the set of all documents. Thereby, we select the most KD important key tokens to represent the document. Then, we apply the Fuzzy C-Means (FCM) [1] clustering on the documents based on their key elements since they might be overlapped. This technique uses Euclidean as distance measure while the key tokens of each document are embedded into a binary vector. The resulting clusters are considered as topic contexts, where each one is represented by a set of the most KC important tokens in all its documents. For that, we compute the TF-IDF of all preprocessed tokens in these documents to compute their importance to their context among all topic contexts. Therefore, we obtain topic contexts where each one is represented by some key tokens.

Named Entity Contexts. Since the domain specific named entities are the most important information in the text, we use inverted index to focus only on the text containing the target entities [4]. For that, named entities should be recognized in the text chunks where several automatic methods can be used [13,14]. Then, we create an inverted index for each topic context where we index the text chunks by their named entities. After passing the named entity level, all text chunks that do not contain the query entities are filtered. Thus, the named entity of a topic inverted index is used to select only text chunks where it appears, in the remaining context levels.

Section Contexts. We use the sections and sub-sections to reduce the searching zone by focusing on more specific and relevant context [5,14]. Sections are able to divide the document into different parts, or contexts, where each context often contains different type of information [14,15]. For that, we extract the structure of the documents to obtain their sections and sub-sections, where there is several automatic methods to obtain that structure [16]. To construct the context of a specific section, we collect text chunks that appear in this section from all the documents. Only documents and text chunks selected by the previous context levels are taken into account. For that, we can merge the sections by their preprocessed titles [18]. Note that we chose only the most top-level sections and sub-sections since they are usually the most homogenized in any type of documents, and merging them can't introduce numerous contexts. After collecting text chunks from documents for a section, its context can be represented by a set of its KC most important tokens among all section contexts in the same level based on the TF-IDF measure. Note that a dedicated section context is created for chunks that appear outside sections, and the same for sub-sections. Likewise, we can represent each context at the section level and then at the sub-section level.

Text Chunk. In our work, the text chunk represent the target unit to be retrieved instead of retrieving a whole document or a page [4]. Thereby, we are able to precisely retrieve a concise relevant part of text. Each sub-section context contains its text chunks that are selected by the higher levels. A text chunk can be a paragraph or a title and can be extracted by analyzing the document structure [16]. It is represented by its preprocessed tokens, original text, and position. The position references its document, section, sub-section, and number. The text chunks are considered to be the lowest-level contexts which are associated to the sub-section contexts. Also, they are used to construct all the contexts and serve as the retrievable units.

3.2 Text Chunk Retrieval

We apply a top-down searching on the constructed index to find relevant text chunks. At each level, the searching zone should be significantly reduced to focus on the relevant contexts. Given an input query, we transform it into preprocessed tokens and we extract named entities from it. Several methods can be applied to recognize named entities in a text [13,14]. Firstly, we compare the query tokens set with the key tokens set of each topic context using the Szymkiewicz-Simpson coefficient similarity measure to efficiently measure the overlap between the two sets [23]. The following formula is used to compute that overlap coefficient similarity between token sets X and Y:

$$overlap(X, Y) = \frac{|X \cap Y|}{\min(|X|, |Y|)} \tag{4}$$

A topic context is considered as similar when the similarity, ranging from 0 to 1, exceeds a threshold S. Then, we search the query named entities in the entity inverted index of each similar topic. After matching named entities in the index, we compare the query tokens set with the key tokens set of each section context where only text chunks that contain the matched named entity are considered. Therefore, we select similar section contexts. Likewise, we repeat the same process for their sub-section contexts. Finally, we compare the query with the preprocessed tokens of each text chunk under the similar sub-section contexts and we consider the chunks with high similarity to be relevant. Note that if we can't find a similarity which exceeds the threshold, we choose the most similar one. Thus, the relevant chunks are retrieved with their original texts and positions, and ranked based on their similarities.

4 Results and Discussion

4.1 Experimental Setup

For experiments, we have chosen $K = 6$ as the number of document clusters for the SciREX dataset and $K = 3$ for the DrugSemantics dataset. These values are selected according to the Silhouette measure. For clustering, we have selected the top $KD = 30$ keywords according to the TF-IDF weights to represent the documents. This value is sufficient to make a effective clustering since the results didn't change with a higher value. Likewise, we have chosen the top $KC = 100$ keywords to represent section and sub-section contexts by selecting all the important words in each context. We set the similarity threshold to $S = 70\%$ which is suitable for selecting the relevant contexts and text chunks by capturing high similarities. Also, we have limited the number of results by taking only the top 5 retrieved chunks.

4.2 Datasets

For experiments, we have chosen two benchmark datasets: SciREX [9] and DrugSemantics corpora [19]. Both of these datasets contain a pre-extracted document structure and are manually annotated by named entities. Thus, we have focused only on the main information retrieval task.

The statistics about these two datasets are represented in Table 1, including the number of preprocessed tokens, chunks, and unique sections and sub-sections. SciREX corpus contains 438 machine learning full-text articles in English. These documents are annotated by 4 types of machine learning related entities. DrugSemantics corpus contains 5 Spanish Summaries of medicinal Product Characteristics (SPC) in Spanish annotated by 10 types of medical named entities. Despite its limited document count, it has a suitable structure that can demonstrate the impact of our structure-based retrieval in obtaining relevant results. Also, we have manually crafted 75 queries for each dataset for evaluation which are sentences containing named entities.

4.3 Metrics

To evaluate the performance of TD-CRESTS, we calculated precision, recall, and F1-measure for each query and then averaged these scores across all queries. Precision is the ratio of the number of relevant text chunks retrieved to the number of chunks retrieved. Recall defines the ratio the ratio of the number of relevant chunks retrieved to the number of relevant chunks in the collection. F1-measure is the harmonic mean of the precision and recall. We have applied these

Table 1. Statistics about the used datasets.

SciREX		DrugSemantics	
Unit	Count	Unit	Count
Method	16 265	Disease	724
Task	5 356	Drug	657
Metric	2 294	Unit of Measurement	557
Material	1 642	Excipient	66
		Chemical Composition	62
		Pharmaceutical Form	45
		Route	42
		Medicament	37
		Food	31
		Therapeutic Action	20
Tokens	1 242 712	Tokens	11 792
Chunks	12 542	Chunks	195
Sub-sections	2 462	Sub-sections	20
Sections	1 644	Sections	8
Documents	438	Documents	5

metrics at both chunk and document levels. For document-level evaluation, we have considered the set of documents containing the query chunks rather than individual chunks.

4.4 Experiments

We evaluated TD-CRESTS using precision, recall, and F1-measure metrics on the SciREX and DrugSemantics datasets at both chunk and document levels, and the results are detailed in Table 2. In general, the results highlight the effectiveness of TD-CRESTS in processing datasets of various sizes at both chunk and document levels, achieving an F1-measure exceeding 71%. This reveals its capacity to handle complexity reduction while maintaining high performance across different document types and languages, given its language-independent nature and adaptability to documents with highly or lowly homogenized structures. It is worth noting that the average F1-measure difference between SciREX and DrugSemantics is only 5.8%. As anticipated, document-level results significantly outperform chunk-level results on both datasets, with an average F1-measure difference of 17.42%. This implies that even non-relevant chunks often originate from relevant documents. At the chunk level, TD-CRESTS exhibits substantially higher precision than recall, with an average difference of +8.78%, indicating the effectiveness of its multi-perspective selection technique in precisely identifying relevant text chunks. Note that TD-CRESTS achieves high performance on the large and structurally inconsistent SciREX dataset, containing 12 542 chunks and 1 644 unique section titles, which reflects its capability to handle complex data. However, many section titles are similar but written in various forms, which makes a reason for this data complexity. In contrast, the small and homogeneous DrugSemantics dataset, containing 5 documents and 195 chunks, is more beneficial for TD-CRESTS through its clear structure in terms of efficiency and precision in text chunk retrieval.

Table 2. Evaluation of TD-CRESTS on SciREX and DrugSemantics datasets.

Unit	SciREX			DrugSemantics		
	Precision	Recall	F1	Precision	Recall	F1
Chunk-level	77.22%	67.89%	69.67%	86.56%	91.33%	87.64%
Document-level	83.00%	74.78%	77.14%	93.78%	96.00%	94.44%

To better understand the efficacy of our method, we compared it with two state-of-the-art indexing methods at the document level: KMeansTFIDF [17], which relies on document clustering and topic modeling, and EntitySearch [4], based on named entity indexing. The results are presented in Table 3. As anticipated, our method retrieves more precise results by selecting them within specific contexts and outperforms the others by an average of 10.92% in terms of F1-measures across both datasets. KMeansTFIDF yields lower F1-measure results, being outperformed by more than 9% in the SciREX dataset and more than 19% in the DrugSemantics dataset. This can be attributed to the clustering of documents into a limited number of high-level contexts, which fails to precisely select relevant documents. Notably, the precision of KMeansTFIDF in the SciREX dataset is significantly lower than the recall with a difference of 12.02%, given that the dataset comprises 438 documents clustered into only 10 clusters, leading to imprecise document selection within clustered content. EntitySearch demonstrates enhanced precision compared to KMeansTFIDF, with an average improvement of 14.23% in F1-measures across both datasets, leveraging domain-specific named entities. However, its precision is significantly lower than its recall, with an average difference of −28.88%, resulting in numerous false negatives. This shows the inadequacy of relying solely on named entities for document selection. Notably, EntitySearch relies solely on named entities indexed by their types and words. However, our method demonstrates more stable results compared to both methods, maintaining an average difference of 5.22% between precision and recall, since it is able to accurately select results from multiple perspectives.

Table 3. Comparison of TD-CRESTS with state-of-the-art methods on SciREX and DrugSemantics datasets at document-level.

Unit	SciREX			DrugSemantics		
	Precision	Recall	F1	Precision	Recall	F1
KMeansTFIDF [17]	57.98%	70.00%	58.97%	62.44%	63.33%	62.31%
EntitySearch [4]	59.53%	93.56%	68.18%	74.93%	98.67%	81.56%
Document-level	83.00%	74.78%	77.14%	93.78%	96.00%	94.44%

5 Conclusion

In this paper, we have proposed TD-CRESTS (Top-Down Chunk Retrieval based on Entity, Section, and Topic Selection), a term-based method for efficient and precise text chunk retrieval. It indexes chunks by various context levels, represented by their key terms. This hierarchical structure enables a top-down search strategy, focusing retrieval on the most relevant contexts. Evaluation on the SciREX and DrugSemantics benchmark datasets demonstrates the high performance of TD-CRESTS in retrieving chunks from both English scientific articles in machine learning and Spanish summaries of medicinal product characteristics. Future directions include enhancing this method with document structure homogenization, named entity normalization, and adaptability to partially or incorrectly recognized named entities.

References

1. Bezdek, J.C.: Pattern Recognition with Fuzzy Objective Function Algorithms. AAPR, Springer, Boston, MA (1981). https://doi.org/10.1007/978-1-4757-0450-1
2. Bhopale, A.P., Tiwari, A.: Transformer based contextual text representation framework for intelligent information retrieval. Exp. Syst. Appl. **238**, 121629 (2024)
3. Boukhari, K., Omri, M.N.: DL-VSM based document indexing approach for information retrieval. J. Ambient. Intell. Humaniz. Comput. **14**(5), 5383–5394 (2023)
4. Brandsen, A., Verberne, S., Lambers, K., Wansleeben, M.: Can BERT dig it? Named entity recognition for information retrieval in the archaeology domain. J. Comput. Cult. Heritage (JOCCH) **15**(3), 1–18 (2022)
5. Deléger, L., Neveol, A.: Automatic identification of document sections for designing a French clinical corpus (identification automatique de zones dans des documents pour la constitution d'un corpus médical en français) [in french]. In: Proceedings of TALN 2014 (Volume 2: Short Papers), pp. 568–573 (2014)
6. Edinger, T., Demner-Fushman, D., Cohen, A.M., Bedrick, S., Hersh, W.: Evaluation of clinical text segmentation to facilitate cohort retrieval. In: AMIA Annual Symposium Proceedings, vol. 2017, p. 660. American Medical Informatics Association (2017)
7. Fortunato, S., et al.: Science of science. Science **359**(6379), eaao0185 (2018)
8. Hambarde, K.A., Proença, H.: Information retrieval: recent advances and beyond. IEEE Access **11**, 76581–76604 (2023)

9. Jain, S., van Zuylen, M., Hajishirzi, H., Beltagy, I.: SciREX: a challenge dataset for document-level information extraction. In: Proceedings of the 58th Annual Meeting of the Association for Computational Linguistics, pp. 7506–7516 (2020)
10. Karypis, G., Han, E.-H.: Fast supervised dimensionality reduction algorithm with applications to document categorization & retrieval. In: Proceedings of the Ninth International Conference on Information and Knowledge Management, pp. 12–19 (2000)
11. Khattab, O., Zaharia, M.: ColBERT: efficient and effective passage search via contextualized late interaction over BERT. In: Proceedings of the 43rd International ACM SIGIR Conference on Research and Development in Information Retrieval, pp. 39–48 (2020)
12. Kumar, R., Sharma, S.C.: Hybrid optimization and ontology-based semantic model for efficient text-based information retrieval. J. Supercomput. **79**(2), 2251–2280 (2023)
13. Landolsi, M.Y., Romdhane, L.B., Hlaoua, L.: Medical named entity recognition using surrounding sequences matching. In: 26th International Conference on Knowledge-Based and Intelligent Information & Engineering Systems. Elsevier (2022)
14. Landolsi, M.Y., Romdhane, L.B., Hlaoua, L.: Hybrid medical named entity recognition using document structure and surrounding context. J. Supercomput., 1–31 (2023)
15. Landolsi, M.Y., Hlaoua, L., Romdhane, L.B.: Extracting and structuring information from the electronic medical text: state of the art and trendy directions. Multimedia Tools Appl., 1–52 (2023)
16. Landolsi, M.Y., Hlaoua, L., Romdhane, L.B.: Hybrid method to automatically extract medical document tree structure. Eng. Appl. Artif. Intell. **120**, 105922 (2023)
17. Lossio-Ventura, J.A., Gonzales, S., Morzan, J., Alatrista-Salas, H., Hernandez-Boussard, T., Bian, J.: Evaluation of clustering and topic modeling methods over health-related tweets and emails. Artif. Intell. Med. **117**, 102096 (2021)
18. Lupşe, O.-S., Stoicu-Tivadar, L.: Supporting prescriptions with synonym matching of section names in prospectuses. In: Data, Informatics and Technology: An Inspiration for Improved Healthcare, pp. 153–156. IOS Press (2018)
19. Moreno, I., Boldrini, E., Moreda, P., Teresa Romá-Ferri, M.: DruGsemantics: a corpus for named entity recognition in Spanish summaries of product characteristics. J. Biomed. Inf. **72**, 8–22 (2017)
20. Salton, G., Buckley, C.: Term-weighting approaches in automatic text retrieval. Inf. Process. Manag. **24**(5), 513–523 (1988)
21. Sarasu, R., Thyagharajan, K.K., Shanker, N.R.: SF-CNN: deep text classification and retrieval for text documents. Intell. Autom. Soft Comput. **35**(2) (2023)
22. Sharma, A., Kumar, S.: Machine learning and ontology-based novel semantic document indexing for information retrieval. Comput. Ind. Eng. **176**, 108940 (2023)
23. Vijaymeena, M.K., Kavitha, K.: A survey on similarity measures in text mining. Mach. Learn. Appl. Int. J. **3**(2), 19–28 (2016)
24. Wan, Q., Xuanhua, X., Han, J.: A dimensionality reduction method for large-scale group decision-making using TF-IDF feature similarity and information loss entropy. Appl. Soft Comput. **150**, 111039 (2024)
25. Ware, M., Mabe, M.: The STM report: an overview of scientific and scholarly journal publishing. International Association of Scientific, Technical and Medical Publishers (2015)

Conceptual Modelling and Ontologies

An Ontology-Driven Solution for Capturing Spatial and Temporal Dynamics in Smart Agriculture

Laura Cornei[1](✉), Doru Cornei[2], and Cristian Foșalău[2]

[1] Faculty of Computer Science, Alexandru Ioan Cuza University, Iași, Romania
cornei.laura10@gmail.com

[2] Faculty of Electrical Engineering, Gheorghe Asachi Technical University, Iași, Romania
doru.cornei@student.tuiasi.ro, cfosalau@tuiasi.ro

Abstract. Semantic web technologies have been frequently used in smart agriculture systems to address integration and interoperability issues, while enhancing expressiveness through their automatic reasoning capabilities. Despite continuous advancements in the field, there is no existent ontology capable of providing a comprehensive representation of the spatial and temporal knowledge, essential in the context of smart farming. In this paper we propose such a model, pass it through a rigorous validation and verification process, and employ it as a representational layer in an IoT agricultural application. The resulting system utilizes real-world high-quality information to offer various functionalities, including crop health monitoring and disease risk forecasting. This work opens a new perspective regarding the development of smart farming applications, by enabling ontological models to fully exploit the spatial and temporal dimensions of agricultural information.

Keywords: Ontology · Knowledge Base · Time Series Forecasting · IoT

1 Introduction

Nowadays, the widespread adoption of IoT devices in agriculture has facilitated various tasks, such as monitoring the evolution of environmental factors, identifying and predicting the risk of crop disease.

State-of-the-art information systems in this field [1–3] employ IoT technologies, ML techniques, expert systems, and suitable knowledge representation models to enhance the decision-making processes and increase productivity.

Despite the progress in the domain, the persistent challenge of utilizing various heterogeneous data sources (e.g., observations of crop health, data from IoT devices, expert knowledge) continues to lead to integration and interoperability issues that reduce the potential of the adopted approaches. To overcome this problem, ontological models have been commonly employed to semantically interconnect data, offering additional benefits, such as increased flexibility and the possibility to automatically discover new knowledge through the reasoning process.

J. Araújo et al. (Eds.): RCIS 2024, LNBIP 513, pp. 49–65, 2024.
https://doi.org/10.1007/978-3-031-59465-6_4

Multiple ontologies have been proposed over the years to represent various agricultural concepts, including general terms and operations [4], as well as specific activities (e.g., disease identification and treatment recommendation [5]).

However, to the best of our knowledge, there is no existing ontology that can comprehensively represent agricultural spatial and time series data (e.g., related to plants' health and environmental factors), along with other information, necessary to facilitate tasks such as crop monitoring and disease prediction.

In this paper we introduce such a model, analyze its properties, and apply it in a real-world scenario to solve a series of crop-related challenges.

We summarize below the contributions brought in this work:

- Creating an ontology (AGROTS: Agricultural Temporal and Spatial Ontology) capable of representing agricultural temporal and spatial data, as well as related information, valuable for evaluating and forecasting crop health and disease risk. As far as we know, there is no current model able to comprehensively capture the temporal and spatial dynamics in the context of smart farming. The proposed ontology is carefully designed to take into account practical considerations specific to real-world use cases (e.g., crops and IoT devices have associated geolocations; the symptoms of a disease vary with the crop development stage and with the disease's severity; the ideal environmental conditions for crop development differ according to the type of plants, their growth stage, and the temporal dimension - day vs. night; the plant characteristics can be fuzzy associated to multiple diseases, etc.).
- Aggregating various real-world high-quality data sources (data collected by IoT sensors, domain knowledge, crop observations) to create a knowledge base that integrates information concerning crops, environmental factors, smart devices, plant diseases, pathogen agents, as well as optimal and disease development conditions.
- Performing a thorough evaluation of the proposed ontology and knowledge base.
- Developing a crop monitoring and disease prediction system that uses the constructed knowledge base as a representational layer. The created application integrates an ontology-based expert system and fully exploits the interoperability, expressiveness and reasoning capabilities offered by the ontological model.

The rest of the paper is organized as follows: Sect. 2 discusses related work; Sect. 3 presents the proposed ontological model; Sect. 4 describes the process of creating the knowledge base; Sect. 5 is dedicated to the ontology evaluation step; Sect. 6 introduces the architecture of developed ontology-based application, while the last section includes final discussions, conclusions and ideas for future improvements.

2 State of the Art

Semantic web technologies [6] have been previously used in the field of Smart Agriculture to solve data integration and interoperability challenges, enable automatic inference of new knowledge and improve the decision-making processes [7, 8].

In this section, we will review a series of relevant state-of-the-art agricultural ontologies, focusing on those that include IoT concepts.

The authors of [9] proposed a comprehensive ontology for the identification and management of plant diseases and pathogen agents, which encompassed information concerning pathologies, symptoms, pests and control methods. The ontology was well anchored in standard well-known vocabularies, such as AGROVOC [10].

Paper [11] introduced AgriOn, a systematic ontology for precision agriculture, which covered various concepts ranging from disease control to agricultural processes. In comparison to other recent knowledge representation models, AgriOn was designed to encode both spatial and temporal characteristics (e.g., crop locations, duration of a crop process).

[12] constructed an extension of ONTAgri, which incorporated local and global decision-making methods associated with various types of IoT devices. The model had the purpose to enhance the development of high-level agricultural business operations.

In paper [5], the authors presented an ontology (RiceDO v2), built upon previous work [13], that was created to identify rice disease and propose effective controls. An innovative aspect involved the integration of plant observations (human-detected symptoms), along with the associated location and timestamp, in the representation.

[14] introduced an ontology and a knowledge map model for encoding data mining results, used techniques, and algorithms in the context of smart farming. The ontology included general information from the domain, along with weather condition observations and crop locations.

[4] designed a methodology to map multiple agricultural ontologies in order to create a bigger representation containing core concepts. The final ontology integrated information concerning climate and various agriculture domains and operations.

Authors of [15] built an IoT ontology for classifying agricultural datasets based on given sensor metadata. The model included multiple sensor attributes, as well as semantic connections between IoT device information and several agricultural domains.

Despite continuous advancements in the field, very few existing knowledge representations incorporated spatial and/or temporal characteristics [7]. Moreover, as far as we are aware, there is no comprehensive ontology capable of modeling time series and geographical data, along with other relevant information, in order to support the crop monitoring and disease prediction tasks.

The spatial and temporal of features, along with information concerning IoT devices and plant condition diagnostics, are essential for tracking and forecasting crop health and disease risk, as they offer valuable insights for modeling real-world scenarios. For example, the spatial data related to plant observations and IoT devices could highlight the distribution of crop-related issues, as well as the geographical variation of the environmental factors' values; the temporal information could aid in analyzing the diseases' propagation behavior and the changes in the environmental conditions.

Table 1 summarizes the degree to which different types of data (temporal, spatial, related to IoT devices and disease diagnosis) are encoded in the reviewed state-of-the-art agricultural ontologies. The temporal dimension is divided in two, based on the nature of information included in the time series (observations concerning environmental factors or plants' health). Spatial data may include the locations of IoT devices and the areas of crops. As it can be observed, no current ontological model comprehensively integrates both spatial and temporal characteristics. Our current work aims to fill this gap in the literature by introducing AGROTS, an agricultural ontology aligned with the FAIR principles [16], which has the purpose to facilitate the tasks of crop monitoring and disease forecasting in the context of smart farming.

Table 1. Comparison between the proposed ontology (AGROTS) and state-of-the-art semantic models, based on the type of incorporated information

Ontology	Type of included information				
	IoT	Temporal		Disease diagnosis	Spatial
		Crop health	Environmental factors		
RiceDO v2 [5]	No	Yes	No	Yes	Partial (no representation for crop location)
AgriComO [14]	Yes	No	Yes	Partial (the disease symptoms are encoded as literals)	Partial (a crop location is represented as a point)
AgriSem [4]	Yes	No	No	No	No
ONTAgriX [12]	Yes	No	No	No	Yes
IoT agriculture taxonomy [15]	Yes	No	No	No	No
CropPestO [9]	No	No	No	Yes	No
AgriOn [11]	Yes	No	No	Partial (symptoms are not detailed by physical qualities)	Partial (a crop location is represented as a point)
AGROTS	Yes	Yes	Yes	Yes	Yes

3 AGROTS Ontology

This section introduces AGROTS (Agricultural Temporal and Spatial Ontology), which can be publicly accessed from: https://github.com/LauraCornei/AGROTS-ontology. The ontology was developed using the Protégé ontology editor.

To ensure semantic interoperability and promote the process of standardizing the domain knowledge, we reused (when possible) concepts from other well-known ontologies and vocabularies to create our model. Table 2 contains a list of 18 ontologies and vocabularies from which we reutilized various concepts.

Figure 2 describes the proposed ontological model, which can be divided into four comprehensive sections, based on the type of incorporated information. In the given diagram, leaf classes are depicted by yellow circles, parent classes by orange circles, object properties by blue arrows and data properties by green rectangles. According to Protégé's metrics, the final version of the ontology contained 146 classes, 51 object properties and 23 data properties.

Table 2. List of ontologies and vocabularies used to create AGROTS

Ontology/ Vocabulary	Reused concepts
Friend Of a Friend (FOAF) vocabulary [17]	person entity
Relations Ontology[1] (RO)	characteristics and localization properties
Sensor, Observation, Sample, and Actuator (SOSA) ontology [18]	observation, sensor, observable property, feature of interest entities
Unit of Measure ontology[2] (OM)	unit concept & associated properties
Extensible Observation Ontology [19]	observation collection entity
Agronomy Ontology[2] (AgrO)	crop concept
Agrontology [10]	properties concerning plant names
Biological Spatial Ontology[2] (BSPO)	anatomical region classes
Experimental Factor Ontology[2] (EFO)	environmental factor & symptom entities
NCBI taxonomy [20]	taxonomy of plant disease causal agents
Plant Trait Ontology[2] (TO)	plant trait concept
Plant Ontology[2] (PO)	plant development stage & plant part entities
Plant Disease Ontology[2] (PDO)	plant disease high-level taxonomy
Phenotype And Trait Ontology[2] (PATO)	physical qualities & environmental factors
Basic Formal Ontology[2] (BFO)	'part of' object properties
Geographical Entity Ontology[2] (GEO)	geodetic coordinates concepts
Interaction Network Ontology[2] (INO)	node entity
Semanticscience Integrated Ontology[2] (SIO)	taxonomy of intensities

Next, we will present and discuss the modeling decision used in each of the four components of AGROTS:

- **section A highlights general information related to crops and plant species, including the optimal & disease development conditions**. Plant species are identified via their scientific or common name, while crops are composed of plants in a specific growth stage. A plant development condition is characterized by the (minimum and maximum) optimal and extreme values of an environmental factor and appears during a particular plant growth stage. The range between the minimum and the maximum optimal bounds indicates favorable conditions for development, while everything outside of it signifies that the evolution (of the plant or of the disease) is hindered or not prone to happen at all. An environmental factor value outside the range of the minimum and maximum extreme limits means that the development is unlikely to occur. The properties indicating the values of an evolution condition are further categorized by the temporal aspect - daytime vs. nighttime, as it is necessary

[1] The ontologies can be accessed from: https://www.ebi.ac.uk/ols/ontologies.

[2] https://wiki.plantontology.org/index.php/plant_disease_ontology.

in some scenarios (depending on the plant type and the environmental factor) to make this difference. The ideal and the disease development conditions are essential in the task of assessing crop health.

- **section B encodes the spatial dimension of the ontology**, more exactly the crops' areas and the IoT devices' geolocations. This component of the ontology is useful for analyzing the geographical distribution of crop-related problems, as well as the fluctuations in the environmental factor values based on the sensors' positions.
- **section C describes the temporal dimension of the model.** It integrates information concerning crop and environmental factors observations made by observers, which can be either IoT sensors or individuals. The observations have associated sampling times and are organized in collections, based on their associated type. Each observation is associated with exactly one observable property (a plant trait in case of crop observations or an environmental condition in case of environmental factors observations). A crop observation collection contains only information related to a specific crop. The mentioned restrictions are also illustrated in Fig. 1. The temporal information encoded in this section can be used to monitor and predict the evolution of environmental factors' values and plant diseases.
- **section D introduces information regarding plant diseases, causal agents and symptoms**. A symptom is described by its intensity, the affected plant part and a set of physical qualities, which range from color and pattern attributes, to shape, texture and anatomical region features. The physical qualities form a class taxonomy, which includes base and compound classes. The compound classes, such as "Apical-SuperiorRegion" are formed by uniting multiple base classes (in this case "Apical-Region" and "Superior Side"). Plant diseases and symptoms can be fully associated (via the direct and inverse object properties: "hasSymptom" and "isSymptomAssociatedToDisease") or partially associated (using an intermediary class to express the degree of association). Direct matching is preferred when the symptom is constructed based on expert knowledge. In contrast, fuzzy matching becomes necessary when the symptom is part of a subjective observation made by a person or an IoT device, as its properties may correspond to different diseases in various degrees.

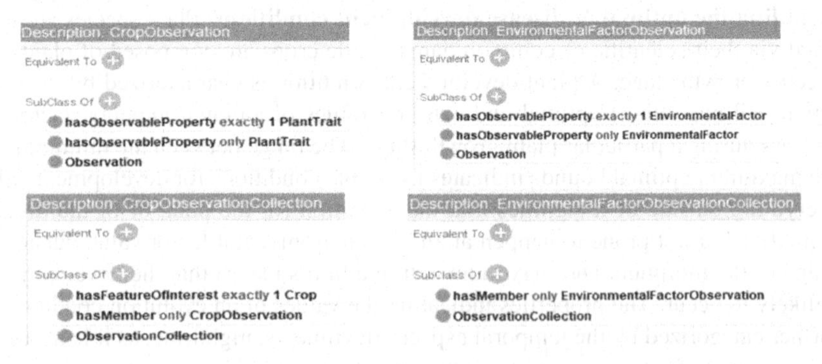

Fig. 1. Examples of Protégé class restrictions used in the proposed ontology

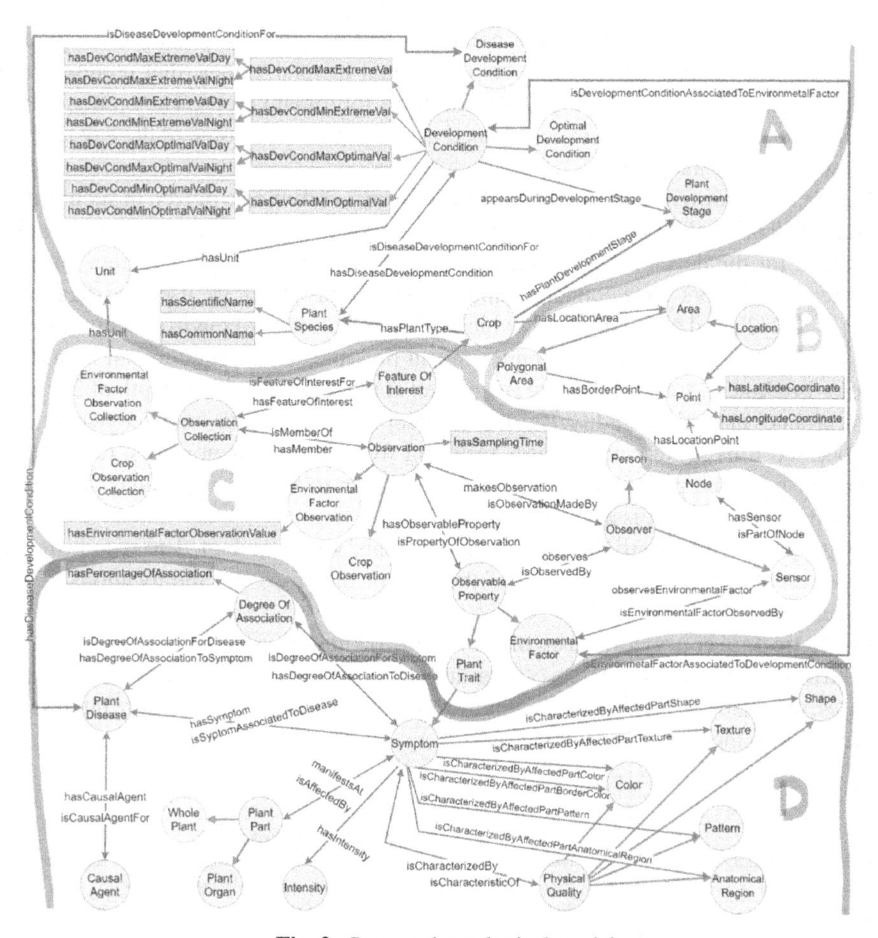

Fig. 2. Proposed ontological model

4 Creating the Knowledge Base

To create the knowledge base, we used the Cellfie Protégé plugin to semi-automatically aggregate different pre-processed data sources in the proposed ontological model, with the use of Manchester syntax rules. As the field of agriculture is very vast, we chose to integrate expert knowledge specifically focused on tomato cultivation. We detail the types of integrated information below:

- **expert knowledge related to tomato diseases, symptoms and pathogen agents, as well as the optimal and disease development conditions for tomato plants**. This knowledge was obtained by combining an agronomist's expertise with recommendations from tomato growth specialized reviews and technical guides [21, 22]. In case of diseases, we included information concerning five major well-known conditions: Early Blight, Late Blight, Gray Mold, Leaf Mold and Powdery Mildew. Figure 3 describes the symptoms of the mentioned diseases, with different characteristics, depending on the intensity of the condition and the affected plant part. Figure 4 contains the optimal and extreme environmental factors' values, that represent the ideal plant and disease development conditions for tomato crops. The optimal plant development conditions were determined in correlation to the time of the day and the growth stage of the plants.

- **hourly environmental data collected by IoT sensors during a period of six months** (second half of the year 2022). In previous work [23] we proposed an accessible method for the development of cheap IoT nodes and applied it to create a feasible system for sensors' data collection and storage. In this paper, we used the data from the N2a, N2b and N3 nodes [23], which measured a series of environmental factors relevant in the context of crop monitoring: air pressure, air temperature, air humidity, color brightness, soil temperature (measured at the surface and at a depth of 20cm in the ground) and soil humidity. Before introducing the data in the knowledge base, we processed it by: eliminating noises resulting from measurement errors, resampling data by hour and performing a linear interpolation to fill missing values. The final version of this dataset is accessible on Kaggle[3].
- **geographical information.** Field measurements were conducted to establish the approximate geodetic coordinates of the area of the crops and the locations of the nodes.
- **synthetic data concerning crop observation**. Taking into account that it would have been difficult to obtain real data concerning possible symptoms of various tomato diseases, we decided to generate it manually. This enabled us to simulate more interesting scenarios in the proposed application.

			Affected Plant Part Physical Quality					
Disease	**Affected Part**	**Intensity**	**Texture**	**Color**	**Pattern**	**Anatomical Region**	**Shape**	**Border Color**
Late Blight	Leaf	WeakToModerate	damp	lightGreen	blotchy	apicalSuperiorRegion	irregular	
Late Blight	Leaf	Strong	dampAndWrinkled	brown	blotchy	superiorSide	irregular	lightGreen
LateBlight	Leaf	SevereToFatal	dampAndWrinkled	purpleBrown	blotchy	superiorSide	irregular	greenBrown
LateBlight	Stem	StrongToFatal	damp	darkBrown	blotchy		irregular	
LateBlight	Fruit	SevereToFatal	dryAndRough	darkRedBrown	blotchy	basalToCentralRegion	irregular	darkRedBrown
EarlyBlight	Leaf	WeakToModerate	damp	yellowGreen	spotted		circular	yellowGreen
EarlyBlight	Leaf	StrongToFatal	damp	yellow	spotted		circular	yellow
EarlyBlight	Stem	StrongToFatal	damp	black	blotchy		elliptic	
EarlyBlight	Fruit	SevereToFatal	dryAndRough	darkBrown	blotchy	basalToCentralRegion	ringShape	
PowderyMilldew	Leaf	AllIntensities	chalky	yellowWhite	blotchy		irregular	
GrayMold	Leaf	AllIntensities	chalky	greenGrey	blotchy	apicalSuperiorRegion	vShape	
GrayMold	Stem	AllIntensities	wilted	greenGrey	blotchy		elliptic	
GrayMold	Fruit	WeakToModerate	smooth	white	spotted		ringShape	
GrayMold	Fruit	StrongToFatal		darkGrey	blotchy	basalToCentralRegion	irregular	
LeafMold	Leaf	AllIntensities		yellow	blotchy	superiorSide	irregular	
LeafMold	Leaf	AllIntensities		yellowBrown	blotchy	inferiorSide	irregular	

Fig. 3. Symptoms for the selected tomato diseases

[3] https://www.kaggle.com/datasets/lauramariacornei/agrots/.

	Day				Night			
	MinExt	MinOpt	MaxOpt	MaxExt	MinExt	MinOpt	MaxOpt	MaxExt
Growth Stage	Temperature [°C]							
Germination	10	22	24	35	10	22	24	35
Early Growth	18	20	26	32	16	18	22	26
Vegetative	14	18	22	32	12	14	18	24
Flowering	16	20	24	26	16	18	22	24
Fruit Formation	18	20	24	30	10	15	17	30
Fruit Ripening	10	20	24	30	10	20	24	30
Growth Stage	Light [μW/cm²]							
Germination	100	200	800	900	-	-	-	-
Early Growth	100	200	800	900	-	-	-	-
Vegetative	100	200	800	900	-	-	-	-
Flowering	100	200	800	900	-	-	-	-
Fruit Formation	150	250	900	1000	-	-	-	-
Fruit Ripening	150	250	900	1000	-	-	-	-
Growth Stage	Air Humidity [%rh]							
Germination	60	75	100	100	60	75	100	100
Early Growth	60	75	100	100	60	75	100	100
Vegetative	40	70	80	100	40	70	80	100
Flowering	40	60	80	100	40	60	80	100
Fruit Formation	40	60	80	100	40	60	80	100
Fruit Ripening	40	60	80	100	40	60	80	100
Growth Stage	Soil Humidity [mbar]							
Germination	-	-	-	-	-	-	-	-
Early Growth	100	150	250	300	150	200	300	350
Vegetative	200	250	350	400	250	300	400	450
Flowering	40	50	100	150	90	100	150	200
Fruit Formation	90	100	200	250	50	100	150	200
Fruit Ripening	200	250	300	350	250	300	350	400

	MinOpt	MaxOpt
Disease	Air Temperature [°C]	
Early Blight	24	29
Gray Mold	17	23
Late Blight	10	24
Leaf Mold	21	24
Powdery Mildew	22	30
Disease	Air Humidity [%rh]	
Early Blight	90	100
Gray Mold	90	100
Late Blight	90	100
Leaf Mold	85	100
Powdery Mildew	50	75

Fig. 4. Optimal (left) and disease (right) development conditions for tomato plants

5 Ontology Evaluation

This section discusses the evaluation methodology for the proposed model. Here, we will use the terms 'knowledge base' and 'ontology' interchangeably, assuming that the ontology contains both the ABox (Assertional Box) and the TBox (Terminological Box) components. The evaluation of the built knowledge model encompasses two phases: validation (checking whether the correct ontology has been developed) and verification (examining whether the ontology has been developed correctly).

5.1 Validation Step

To validate our model, we first used the OOPS! ontology pitfall scanner [24] to ensure that our proposed ontological model contained no important nor critical issues. Then, we analyzed if it satisfied six relevant specific criteria, widely utilized in the literature [25]. The results indicated that the constructed ontology fulfilled all the selected validation standards, as highlighted in Table 3. To verify the ontology's expressiveness, we checked that it can be used to answer a series of carefully elaborated 20 competency questions, listed in Table 4.

5.2 Verification Step

To verify the ontology, we first used the Pellet reasoner from Protégé to check that the model was free of inconsistencies. We then employed the Ontology Taxonomy Evaluation method [26] to manually assess whether the model contained inconsistent,

Table 3. Criteria utilized to validate the proposed AGROTS ontology

Validation criterion	Fulfilled?	Explanation
Accuracy	Yes	AGROTS accurately represents the real-world aspects of the covered field. It takes into account multiple practical dimensions (e.g., the optimal environmental conditions for crop growth vary based on factors such as plant type, growth stage and time)
Completeness	Yes	The model comprehensively encompasses the domain of interest, including information concerning IoT devices, crops, environmental factors, plant diseases, symptoms & pathogens, as well as optimal crop and disease development conditions. Both the geographical and the temporal dimensions are taken into account
Conciseness	Yes	AGROTS is free of any needless concepts or redundant properties. All incorporated knowledge is relevant for solving the given tasks
Expandability	Yes	The ontology is well anchored in important vocabularies and ontologies. The model can be easily extended with other knowledge related to the domain (e.g., disease control recommendations)
Expressiveness	Yes	AGROTS can be used to answer a series of competency questions, which validate the model's appropriate level of expressiveness
Practical usefulness	Yes	AGROTS represents the knowledge layer of the proposed system for crop health monitoring and disease risk prediction. This confirms that the introduced model has an increased practical usefulness

Table 4. Competency questions that can be answered with the use of the AGROTS model

Competency Question
Which sensors are part of a given IoT node?
What are the environmental factors (along with their units of measure) monitored by a specific sensor?
What are the geolocations of the IoT devices and crops?
Which IoT nodes monitor a certain crop?
What type of plant is cultivated on a given crop? What is the plants' growth stage?
Which crop observations are made by a certain person?
Which environmental factor observations are made by a specific IoT sensor?

(*continued*)

Table 4. (*continued*)

Competency Question
Which observations are related to a particular crop?
Which observations are related to a given environmental factor?
What are the observations made in a particular time range?
To what degree the observed plant symptoms can be associated with one or multiple diseases?
What are the crop observations associated with a particular disease?
What is the most probable disease that a crop can have if it shows one or multiple symptoms?
What are the diseases that a plant can have and which are the pathogen agents that cause them?
What are the symptoms associated with a given plant disease and how do they vary depending on the intensity of the disease?
What are the types of physical qualities that can be used to describe a symptom of a disease?
What are the ideal development conditions for a type of plant, depending on its growth stage and the time of the day (daytime vs. nighttime)?
Do the environmental factors values, corresponding to a specific crop and selected from a particular time range, satisfy the ideal plant development conditions?
What are the optimal development conditions for a disease, depending on the growth stage of the affected plant?
Do the environmental factors values, corresponding to a specific crop and selected from a particular time range, indicate a potential disease risk by satisfying the disease's ideal development conditions?

incomplete or redundant information. As indicated in Table 5, we found no issues while examining the ontology with respect to the mentioned criteria.

6 System Architecture

This section introduces the developed system for crop monitoring and disease forecasting, which uses the previously presented knowledge base. The application encompasses three major components:

- **The presentation layer**, consisting in a responsive web interface which exposes the various functionalities of the system.
- **The application layer**, which contains the business logic for developing the crop health diagnosis and disease risk analysis and forecasting tasks.
- **The knowledge layer**, established by loading the created knowledge base into Stardog, a platform for querying, analyzing and manipulating semantically interconnected data.

The presentation and application layers were developed with the use of Flask, a Python web microframework. Additional CSS and JS-based libraries (Chart.js, Bootstrap) were used to enhance the appearance and the features of the user interface.

Table 5. Criteria used to verify the proposed AGROTS model

Verification criterion	Found issues?	Explanation
1 Inconsistency		
1.1 Circulatory errors	No	no class is defined as a generalization or specialization of itself
1.2 Partition errors	No	no instance(/class) is a member(/subclass) of a reunion of multiple disjoint classes; no instance is a direct member of a class that has subclasses
1.3 Semantic errors	No	all instances belong to their corresponding classes from a semantic point of view; all class associations are semantically correct
2 Incompleteness		
2.1 Incomplete concept classification	No	all domain concepts are present in the ontology
2.2 Partition errors	No	all disjoint classes are defined; all the required specifications were encoded using OWL class restrictions
3 Redundancy		
3.1 Grammatical redundancy	No	all instances and classes have unique definitions
3.2 Identical formal definition	No	no instances or classes are redefined under a different name

Figure 5 illustrates a use-case diagram which depicts the system's functionalities, while Fig. 6 contains multiple print screens presenting the application's interface. In summary, a user is able to:

- **view and modify information about crops and IoT nodes** (e.g., the location area, the plant type and the growth stage of a crop)
- **add a crop observation**, consisting in one or multiple observed symptoms for a crop, where each symptom has associated a series of characteristics.
- **view and filter (by crop) disease diagnosis information** associated with previously added symptoms. Each symptom could be fuzzy matched to one or multiple diseases. To correlate a symptom S with a disease D, we first selected only the symptoms of D that corresponded to the intensity and affected plant part traits of S; then, we computed the final matching probability between S and D as being the average of the similarity scores between S and each retained symptom of D. The similarity score of S and a symptom S' of D was calculated as the percentage of *resembling* traits between S and S' over the total number of traits of S'. Two traits were considered to be *resembling* if the intersection of their associated classes in the ontology was

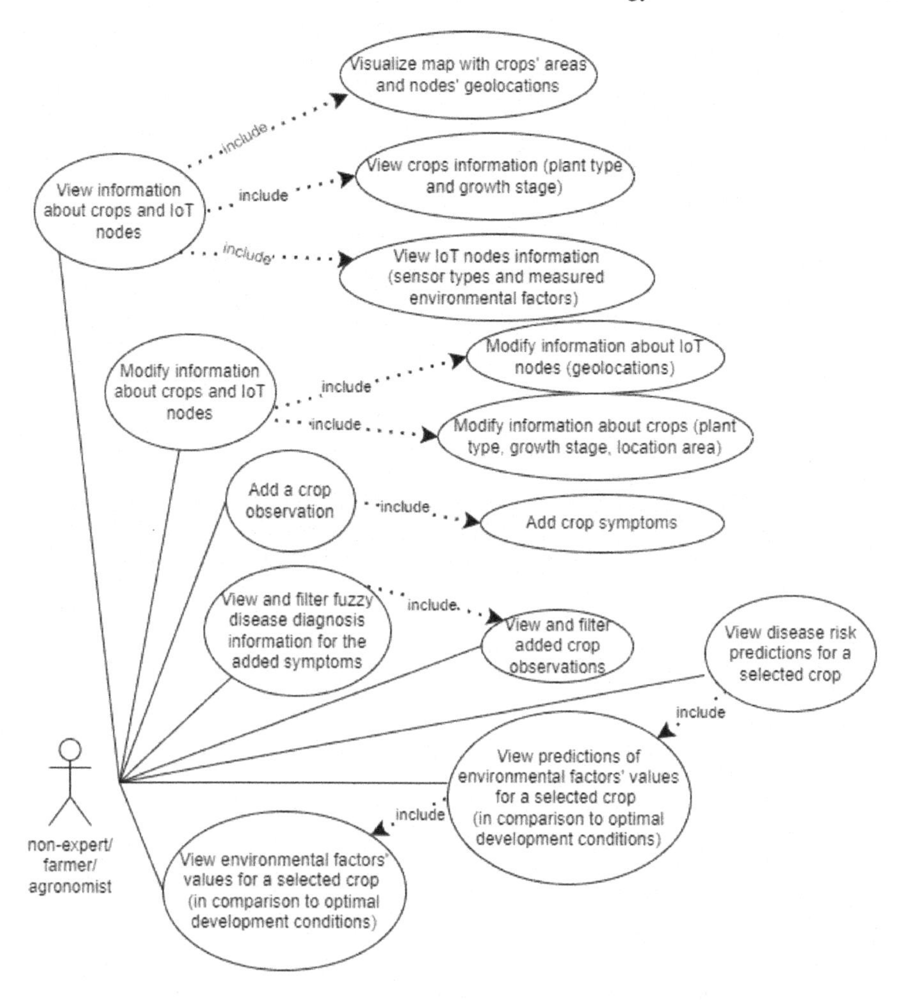

Fig. 5. Use-case diagram highlighting the functionalities of the developed system

non-empty (e.g., an instance belonging to 'dryAndRough', defined as the reunion between the 'dry' and 'rough' classes, *resembled* instances from 'dryAndWrinked'). With the use of ontological reasoning, we could efficiently identify properties like class intersection and disjointness.

- **view predictions of environmental factors' values for a chosen crop, in comparison to the optimal development conditions**. The user could visualize actual data spanning a period of five days, as well as forecasted values for two more days, starting from a selected initial date. To achieve this functionality, we first applied the Ray casting algorithm to determine which IoT nodes were located inside the area of the selected crop. Next, we obtained the sensors' data corresponding to the environmental conditions and then, the forecasted values using the Prophet model. We finally plotted the real and predicted values of the environmental factors along with the optimal crop

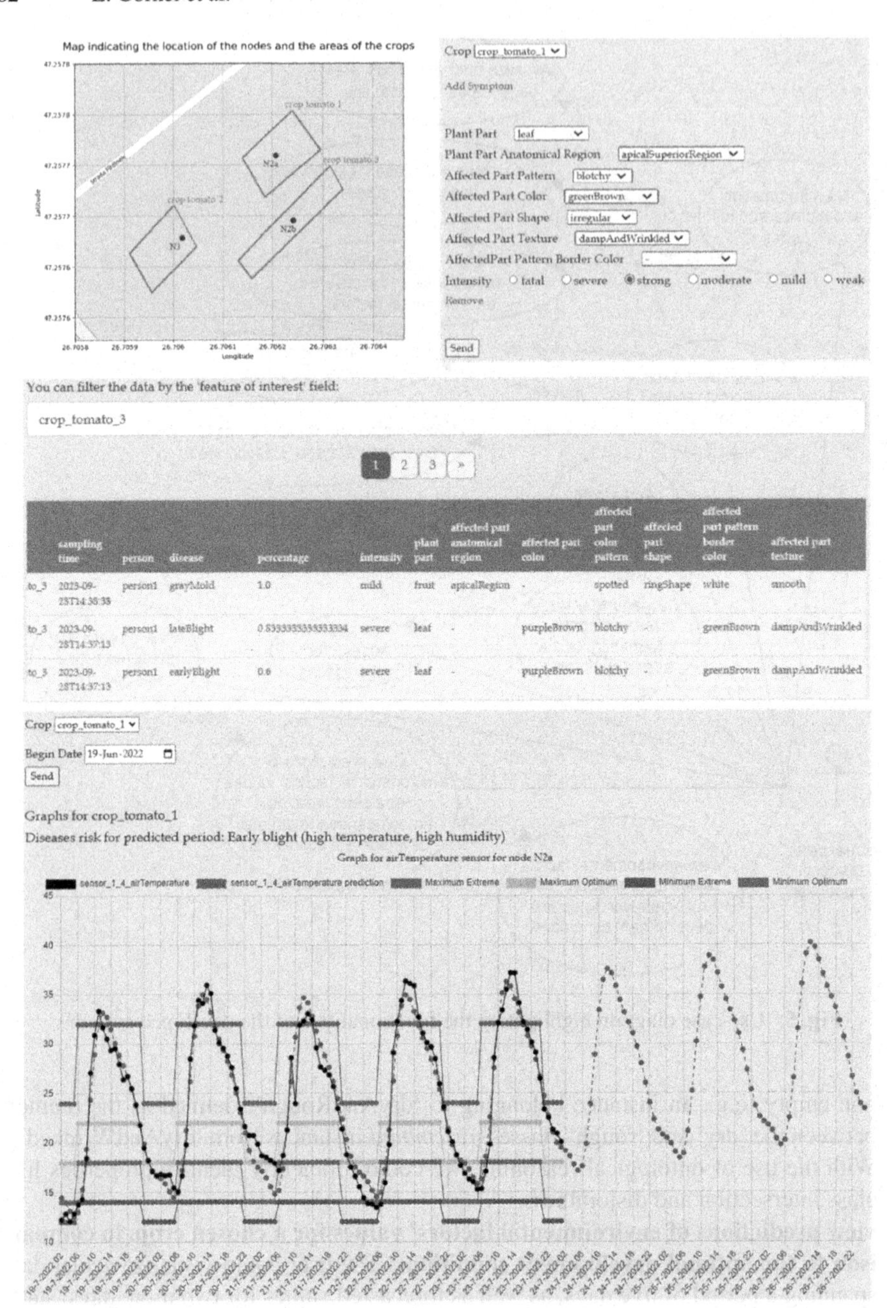

Fig. 6. Print screens depicting the interface of the proposed application

development conditions data, the latter being obtained given the crop's type and the growth stage. We used the Astral[4] library to associate the ideal plant development daytime and nighttime values to the actual time of the day.

- **view disease risk predictions for a selected crop**. To accomplish this functionality, we checked whether the forecasted values of the environmental factors fell within the optimal development range of any disease. If they did, that indicated a potential future evolution of that/those particular disease(s).

7 Conclusions and Future Improvements

In conclusion, this paper introduces AGROTS, an ontology for modeling agricultural temporal and spatial data, as well as other relevant information, that can be used for tasks such as crop health monitoring and disease risk forecasting.

The novel aspect of the proposed model lies in the fact that there is no existent semantic representation that can comprehensively encode temporal and spatial agricultural information, indispensable in the context of smart farming.

Starting from the proposed ontology, we created a knowledge base by aggregating various types of high-quality data sources (expert knowledge, geographical information, real sensor data, crop observations). After a careful evaluation, the final resulting model was applied in a real-world scenario by integrating it into a system for plant health monitoring and disease risk prediction.

One possible improvement would consist in extending the current ontology, for example by adding disease control methods and incorporating additional types of IoT devices (e.g., cameras and drones). The proposed model could be also integrated with other well-known ontologies capturing spatial/temporal aspects, such as OWLTime[5].

Another idea would be to use the ontology for solving new agricultural tasks (e.g., analyzing the diseases' transmission behavior). We also hope to attract the necessary funds in order to test the solution on a larger scale in the future.

References

1. Ahmad, A., Saraswat, D., El Gamal, A.: A survey on using deep learning techniques for plant disease diagnosis and recommendations for the development of appropriate tools. Smart Agric. Technol. **3**, 100083 (2023)
2. Attri, I., Awasthi, L.K., Sharma, T.P.: Machine learning in agriculture: a review of crop management applications. Multimedia Tools Appl. **83**(5), 12875–12915 (2023). https://doi.org/10.1007/s11042-023-16105-2
3. Fenu, G., Malloci, F.M.: Forecasting plant and crop disease: an explorative study on current algorithms. Big Data Cogn. Comput. **5**(1), 2 (2021)
4. Abdelmageed, A., et al.: A Core Ontology to Support Agricultural Data Interoperability. Gesellschaft für Informatik (2023)

[4] https://astral.readthedocs.io/en/latest/.

[5] https://www.w3.org/TR/owl-time/.

5. Jearanaiwongkul, W., Anutariya, C., Racharak, T., Andres, F.: An ontology-based expert system for rice disease identification and control recommendation. Appl. Sci. **11**(21), 10450 (2021). https://doi.org/10.3390/app112110450
6. Berners-Lee, T., Hendler, J., Lassila, O.: The semantic web: a new form of web content that is meaningful to computers will unleash a revolution of new possibilities. Sci. Am. (2001)
7. Bhuyan, B.P., Tomar, R., Cherif, A.R.: A systematic review of knowledge representation techniques in smart agriculture (urban). Sustainability **14**(22), 15249 (2022). https://doi.org/10.3390/su142215249
8. Drury, B., Fernandes, R., Moura, M.-F., de Andrade Lopes, A.: A survey of semantic web technology for agriculture. Inf. Process. Agric. **6**(4), 487–501 (2019)
9. Rodríguez-García, M.Á., García-Sánchez, F.: CropPestO: an ontology model for identifying and managing plant pests and diseases. In: Valencia-García, R., Alcaraz-Marmol, G., Del Cioppo-Morstadt, J., Vera-Lucio, N., Bucaram-Leverone, M. (eds.) CITI 2020. CCIS, vol. 1309, pp. 18–29. Springer, Cham (2020). https://doi.org/10.1007/978-3-030-62015-8_2
10. Subirats-Coll, I., et al.: AGROVOC: the linked data concept hub for food and agriculture. Comput. Electron. Agric. **196**, 105965 (2022)
11. Urkude, G., Pandey, D.: AgriOn: a comprehensive ontology for green IoT based agriculture. J. Green Eng. **10**, 7078–7101 (2020)
12. Fahad, M., Javid, T., Beenish, H., Siddiqui, A.A., Ahmed, G.: Extending ONTAgri with service-oriented architecture towards precision farming application. Sustainability **13**(17), 9801 (2021). https://doi.org/10.3390/su13179801
13. Jearanaiwongkul, W., Anutariya, C., Andres, F.: A semantic-based framework for rice plant disease management. New Gener. Comput. **37**(4), 499–523 (2019)
14. Ngo, Q.H., Kechadi, T., Le-Khac, N.-A.: Knowledge representation in digital agriculture: a step towards standardized model. Comput. Electron. Agric. **199**, 107127 (2022)
15. Lynda, D., Brahim, F., Hamid, S., Hamadoun, C.: Towards a semantic structure for classifying IoT agriculture sensor datasets: an approach based on machine learning and web semantic technologies. J. King Saud Univ. Comput. Inf. Sci. **35**(8), 101700 (2023)
16. Jacobsen, A., et al.: FAIR principles: interpretations and implementation considerations. Data Intell. **2**(1–2), 10–29 (2020)
17. Graves, M., Constabaris, A., Brickley, D.: FOAF: Connecting People on the Semantic Web. Knitting the Semantic Web **43**, 191–202 (2007)
18. Janowicz, K., Haller, A., Cox, S.J.D., Le Phuoc, D., Lefrançois, M.: SOSA: a lightweight ontology for sensors, observations, samples, and actuators. J. Web Semant. **56**, 1–10 (2019)
19. Madin, J., Bowers, S., Schildhauer, M., Krivov, S., Pennington, D., Villa, F.: An ontology for describing and synthesizing ecological observation data. Eco. Inform. **2**(3), 279–296 (2007)
20. Schoch, C.L., et al.: NCBI taxonomy: a comprehensive update on curation, resources and tools. Database **2020**, baaa062 (2020)
21. Shamshiri, R.R., Jones, J.W., Thorp, K.R., Ahmad, D., Man, H.C., Taheri, S.: Review of optimum temperature, humidity, and vapour pressure deficit for microclimate evaluation and control in greenhouse cultivation of tomato: a review. Int. Agrophysics **32**(2), 287–302 (2018)
22. Gleason, M.L., Edmunds, B.A.: Tomato Diseases and Disorders. Iowa State University, University Extension Ames, IA (2005)
23. Cornei, D., Foşalău, C.: Using IoT in smart agriculture: study about practical realizations and testing in a real environment. In: International Conference and Exposition on Electrical and Power Engineering (EPE), pp. 13–18 (2022)
24. Poveda-Villalón, M., Gómez-Pérez, A., Suárez-Figueroa, M.C.: OOPS! (OntOlogy Pitfall Scanner!): an on-line tool for ontology evaluation. Int. J. Semant. Web Inf. Syst. **10**(2), 7–34 (2014). https://doi.org/10.4018/ijswis.2014040102

25. Degbelo, A.: A snapshot of ontology evaluation criteria and strategies. In: Proceedings of the 13th International Conference on Semantic Systems, pp. 1–8 (2017)
26. Lovrencic, S., Cubrilo, M.: Ontology evaluation-comprising verification and validation. In: Central European Conference on Information and Intelligent Systems, p. 1 (2008)

A Knowledge Graph-Based Decision Support System for Resilient Supply Chain Networks

Wilhelm Düggelin and Emanuele Laurenzi(✉)

FHNW University of Applied Sciences and Arts Northwestern Switzerland, Riggenbachstrasse 16, 4600 Olten, Switzerland

wilhelm.dueggelin@alumni.fhnw.ch, emanuele.laurenzi@fhnw.ch

Abstract. Events in recent years such as the Russo-Ukrainian war of 2022 and the covid-19 pandemic have once again shown the importance of relying on resilient supply chain networks. The creation and maintenance of such networks is, however, a rather knowledge intensive task, which is still challenging. To tackle this, we introduce a first version of a knowledge graph-based decision support system aiming to help supply chain risk managers to make sourcing decisions. The system was designed by following the design science research methodology, which is supplemented with the Ontology Development 101 [25] for rigor in creation of the knowledge graph schema. Competency questions elicited with domain experts were used to evaluate the proposed system.

Keywords: Resilient Supply Chain Networks · Risk Aware Sourcing · Knowledge Graphs-based Decision-Support System

1 Introduction

Sources as old as thousands of years report from supply disruptions through criminal activity, accidents, and other events [1]. Some even go so far as partially linking the decline of the Roman empire to a lack of visibility and resilience of its– for those ancient times– highly globalized supply chains [2]. Today, dramatic, and highly disruptive events in our recent history such as the covid-19 pandemic, the Russo-Ukrainian war of 2022, global warming as well as political and economic tensions between superpowers such as the USA and China have once more highlighted the vulnerability of our supply chains [3].

The Association of Supply Chain Management (ASCM) stresses the importance of including risk as a factor in supplier selection [4]. Supplier selection methods of today consider multiple criteria including different types of risk [5] and companies commit enormous amounts of resources to the process [5].

Despite the existence of theoretical models (e.g., [6–8]) and commercially available supply chain risk management solutions like Coupa [9] or Sphera [10], there is still a lack of approaches that while coping with the complex and fast-changing underlying supply chain reality, also translate into tools that are intuitive and easy-to-use / adopt for managers to design and maintain resilient supply chain networks.

J. Araújo et al. (Eds.): RCIS 2024, LNBIP 513, pp. 66–81, 2024.
https://doi.org/10.1007/978-3-031-59465-6_5

To address this gap, we present a first version of a knowledge graph-based decision support system, which target supply chain risk managers. The knowledge graph conceptualizes the underlying complexity and leverages its reasoning capability to support risk-aware sourcing. The latter so far has only been viewed as a secondary concern [11]. The instantiation of the approach in the Metaphactory tool [30] ensures ease of use [34] at design time. In this work, the term 'knowledge graph' is referred to as a lightweight ontology, which is low in expressiveness, e.g., RDF(S).

This paper is structured as follows. First, Sect. 2 gives a summary of the related work. Next, Sect. 3 presents the followed design science research (DSR) methodology. In Sect. 4, the first step in the Design Science Research (DSR) process of this work is presented and awareness is given to the problem, summarizing interview results. In Sect. 5 the classes for the schema are defined. Section 6 documents the development of the knowledge graph, including instances and SPARQL queries. The findings are discussed in Sect. 7, followed by a conclusion and an outlook in Sect. 8.

2 Related Work

Early adoption of graph theory for supply chain risk management [13] can be found as far back as 2006. Since then, numerous researchers have devoted to the topic: Some developed an ontology based semantic model for supply chain management called Onto-SCM using the IDEF4 schematic language [14]. In another case, graph theory was applied to assess the vulnerability of supply chains with the goal of transforming fuzzy information into indices [15].

Emmenegger et al. [16, 17] developed an ontology-based early warning system for supply chain risk management. The system can identify early warning signals upstream (e.g., external events) and using reasoning capabilities to infer their impact on the focal firm and display them in the form of top 10 risks. This work follows a similar research approach, with respect to the use of ontology and reasoning capabilities but neglects the top 10 risks to give higher flexibility to the accommodation of new risk types by different audiences.

The work in [18] emphasize the importance of modelling company risks in supply graphs but is not intended to be a decision-support system.

Palmer et al. [19] propose ontologies of multiple levels for supply chain networks with the aim to facilitate the prediction of the risk impact on interoperability between different entities in a supply chain. Despite our work did not focus on risk prediction, findings of [19] could be used to extend our model.

An ontology-based decision support system for supply chain resilience, using semantic web rule language (SWRL) and mixed integer linear programming together with hybrid practical swarm optimization is described in [20]. The extension of the ontology, however, requires ontology expertise, which we aim to avoid in our work so that our conceptual model remains adoptable and extensible for a broad audience of non-ontology experts like supply chain risk managers.

The importance of relying on knowledge graphs in supply chain risk management has also been discussed [21].

Other research work deals with the creation of graph representation learning approaches for supply chain networks [22, 23]. An approach to detect dependency links which are potentially invisible to a focal firm has previously been presented [22]. The combination of neural networks and knowledge graph reasoning, to support the proactive discovery of invisible risks in supply chains has further been explored [23]. Specifically, this technique solves the problem of identifying hidden relationships in supply chains which are often not disclosed by suppliers. However, the work in [23] has no focus on external risk. Moreover, our approach relies on expert knowledge to design the supply chain network.

The reviewed literature led to identify the need of creating a new approach that accommodates the minimum number of relevant concepts about a resilient supply chain network, required to support the understanding and extension of the conceptual model by supply chain managers.

3 Methodology

This work follows the Design Science Research (DSR) strategy, which is supplemented with the approach in Ontology Development 101 [25], which builds on the previous ontology engineering approaches of [26, 27]. The research strategy allowed to answer the following main research question:

How can a knowledge graph-based decision support system support the design of resilient supply chain networks?

In the first DSR phase, we deepened the problem understanding by conducting semi structured expert interviews. Expert interviews have the advantage of delivering fast access to dense knowledge about a domain [24]. Given the impossibility of receiving a real-world case, the interviews were also used to create a realistic case. Three experts (two senior and one junior) with experience in managing global supply chains were interviewed. Two interviewees had more than 20 years of experience as business development executives, in COO as well as in CEO functions within the industrial sector and have later applied their knowledge as consultants, board members and investors. The less-experienced interviewee had about one year of project experience related to supply chain management within the railway industry as well as two years of operational experience in technology procurement at a cruise ship operator.

The suggestion phase accommodated the findings of the awareness of problem and was then instantiated into a running prototype.

Metaphactory was used for the prototypical implementation because of its user-friendliness towards business users and technology readiness to accommodate the conceptualized knowledge graphs schema and instances.

The approach has been evaluated by answering the competency questions, which according to the Ontology Development 101 [25], ensures the correctness and usefulness.

4 Challenge of Designing Resilient Supply Chain Networks

To deepen understanding on the relevant knowledge for designing resilient supply chain networks, findings from the literature were extended with the findings from the interviews (see Sect. 3 for the interviewee profiles). The key findings from the Interviews are displayed in Table 1. The semi-structured Interviews were conducted over 60-min video calls where the interviewer posed different questions from a prepared catalogue of 23 questions related to the interviewees, experience and opinion on supply chain risk management, resilience and approaches they encountered in their professional lives. Notes were directly taken on a prepared digital whiteboard, in addition to that the interviews were recorded and automatically transcribed. Notes and transcripts were then analyzed for contradictions and or agreements between the different interviewees.

Table 1. Key findings from the interviews

Nr	Interviews	Key finding
1	1–3	Supply chain risk management is generally underrated
2	1–3	Risks in the macro environment are especially underrated
3	2	Supplier Evaluation is often conducted through formal processes that do not address sourcing risk sufficiently
4	2	Companies often rely on a single supplier or on industrial clusters within a single country
5	1–2	Tradeoffs between acceptable risk and cost of diversification often lead to single sourcing
6	1–3	Supply chain networks of suppliers are often not transparent
7	1–3	Risk must be quantified as financial impact to support decision making

All three interviewees have experienced that companies underestimate the monitoring of external supply side risk, resulting in insufficient consideration of factors that create such risk. The three interviewees further perceived increasing instabilities in global supply chains and mentioned the covid 19 Pandemic as well as the Russo-Ukrainian War of 2020 and the ongoing tensions in US-China relations. One of the senior interviewees stressed the strong and thus often problematic dependency of European companies on Chinese manufacturing which was also mentioned by the other two interviewees but not stressed. One of the senior interviewees also stressed that a trade off between the acceptable risk of single-sourcing and the additional cost of multi-sourcing is mostly leading to risky sourcing decisions. The lack of transparency in supply chain networks is seen as another major problem by all the interviewees. There was a common acknowledgment on the increasing importance of political, economic, and environmental risks. All Interviewees mentioned the importance of a representation of potential negative impacts of risk in monetary terms to facilitate sourcing decisions with resilience in mind.

Adopted from financial management, in supply chain management, Value at Risk (VaR) has the purpose of quantifying supply chain risk in monetary terms [12]. VaR allows companies to consider all potential supply chain risks through one single metric, helping in the prisonization of actions for mitigation and resillience [12]. See (1) for VaR [12].

$$VaR = Probability\, of\; risk\; event * Monetary\, impact\, of\; risk\; event \quad (1)$$

Hence, in this work, we considered VaR as a value to be automatically inferred from the knowledge graph.

The competency questions agreed upon with the interviewees are the following:

1. What is the risk assigned to product *a*, produced by organizational unit *Z*?
2. What is the risk that propagates from organizational unit *Z* to organizational unit *Y* if organizational unit *y* integrates product *a* as a component in its product *b*?
3. What is the risk that propagates from organizational unit *y* to the focal firm *X* if the focal firm *X* integrates product *b* in its product *c*?
4. If product *b, d, e, and f* with their components products *a, j, k, l, m, n, o, p, …* are integrated in product *c, what* is the Value at Risk of the focal firm X given its annual turnover on sales of product *c*?

These questions define the scope of the knowledge graph.

5 Knowledge Graph Schema for Resilient Supply Chain Networks

This section describes the knowledge graph schema, which we name Risk Aware Sourcing Strategy Assistant (RASSA) schema. The latter was derived by extracting terms from the competency questions (see Sect. 4) and then comparing them to existing ontologies. The schema is meant to be used by supply chain risk managers in metal working, mechanical or electric engineering as well as COOs and CEOs of SMEs. Therefore, search was narrowed down to enterprise ontologies that have been applied or extended in real-world scenarios.

Table 2 shows the top three identified enterprise ontologies and related concepts.

Table 2. Existing ontologies considered

Ontology	Relevant Concepts	Standards	Reference
ArchiMeo	Location	RDFS	[28]
APPRIS	Location, Country, Product, Organizational Unit, Risk Factor	OWL, RDF, RDFS	[29]
FLEXINET	Risk Factor	Knowledge Framework Language (KFL)	[19]

These concepts were later adopted in our schema. The schema is purposefully kept lean and focused on addressing the competency questions. This helps the schema readability and future extension. Following the best practice of the Ontology Development 101 [25], we borrowed concepts and relations for existing and relevant ontologies wherever possible.

Table 3 shows the descriptions of some concepts. Each description is derived by combining the definition of the original ontology with one of the competency questions in this work.

Table 3. Concepts and their definitions

Term	Definition
Focal firm	The organizational unit that in our case sources components or materials
Turnover	Money that the focal firm receives for its sale of goods and services
Product	Outputs that organizational units produce for sale, products can be components or commodities required in other products
Component	Any material object or commodity that is built into a product or that a product is made off
Commodity	A raw material
Supplier	Company that could sell components to the focal firm
Country	A State in which a certain company is located
Risk factor	A factor that contributes to the likelihood of a negative event within a sourcing location / country
Value at Risk	Probability of an adverse event multiplied by financial impact of the event
Criticality	The relevance of a component for a product
Resilience	A company's ability to mitigate risk
Risk Propagation	The degree to which risk transcends from one node in the supply chain to another based on the resilience of each next node

To define the classes, the middle-out approach [27] was applied, meaning that to define the hierarchy, we neither started with the most generic nor the most specific terms. Table 4 displays the resulting class hierarchy.

Table 4. Hirerchy of the knowledge graph schema

Super Classes and their definition within the context of RASSA	Sub Classes and their definition within the context of RASSA
Business Object *Any concept related to a business and its activities*	Organizational Unit *A group of actors unified by a common goal e.g. to make a profit from the sale of goods or services*
	Product *A product which is made available by an organizational unit*
Location *Any physical location*	Country *A nation state*

(continued)

Table 4. (*continued*)

Super Classes and their definition within the context of RASSA	Sub Classes and their definition within the context of RASSA
Risk Factor *Information based on indices that quantify the probability of adverse events happening and indices indicating vulnerabilities towards such events in relation to a specific country*	Natural Risk [32] *Aggregated value of indices indicating the probability of different natural disasters*
	Conflict Risk [32] *Aggregated value of indices indicating the probability of violent conflict within a country*
	Socio Economic Vulnerability [32] *The aggregated value of indices indicating the vulnerability of a country's society towards natural disasters or violent conflict from a macro economic perspective*
	Infrastructure Vulnerability [32] *The aggregated value of indices indicating the vulnerability of a country's infrastructure towards natural disasters or violent conflict*

With the aim to minimize complexity, relationships from a class to itself were created. A product for example is offered by a company and a component is just another product offered by another company which supplies the latter company (see Tables 5 and 6).

Table 5. Classes and attributes

Classes	Attributes
Organizational Unit	Name / label (String) Risk Propagation Factor (decimal)
Product	Name / label (String) Identifier (String) Component (Boolean) Critical Type (Boolean) Commodity (Boolean) Price (Decimal) Annual Turnover (Decimal) Required Number of Components (Decimal)
Country	label (String)
Natural Risk	risk value (decimal) label (string)

(*continued*)

Table 5. (*continued*)

Classes	Attributes
Conflict Risk	risk value (decimal) label (string)
Socio Economic Vulnerability	risk value (decimal) label (string)
Infrastructure Vulnerability	risk value (decimal) label (string)

Table 6. Relationships

Domain	Predicate	Range
Organizational Unit	Is Located In	Country
Product	Is Manufactured By Is Shipped From Is Sold By Requires Component	Organizational Unit Organizational Unit Organizational Unit Product
Country	Is Exposed To Is Exposed To Is Exposed To Is Exposed To	Conflict Risk Natural Risk Socio Economic Vulnerabilities Infrastructure Vulnerabilities

6 Implementation of the Knowledge Graph Schema for RASSA

The implementation of the proposed schema was done in the enterprise knowledge graph software Metaphactory [30]. As mentioned in Sect. 3, this tool fulfilled the criteria for an intuitive and ease of use facilitating design as well as extension of knowledge graph schemas. This criterion has already been addressed by Haase et al. [31].

Figure 1 shows the implemented schema, which is consistent with the conceptualization presented in the previous section. Besides the sub-class assertion, the relations (object properties) automatically get the cardinality. This is an automatic functionality of Metaphactory, which did not compromise the conceptualized schema. The cardinalities are automatically translated by SHACL, but the business user only deals with the visualization, therefore compliant with the requirement of targeting supply chain risk managers and not ontology experts.

6.1 Instances

The implemented schema has been populated with instances. Risk and vulnerability factors are based on indices between 0 (no risk or vulnerability) and 10 (high risk or vulnerability). These indices are provided with open access by INFORM [32], a global, open-source risk assessment for humanitarian crisis and disasters, published by the European Commission.

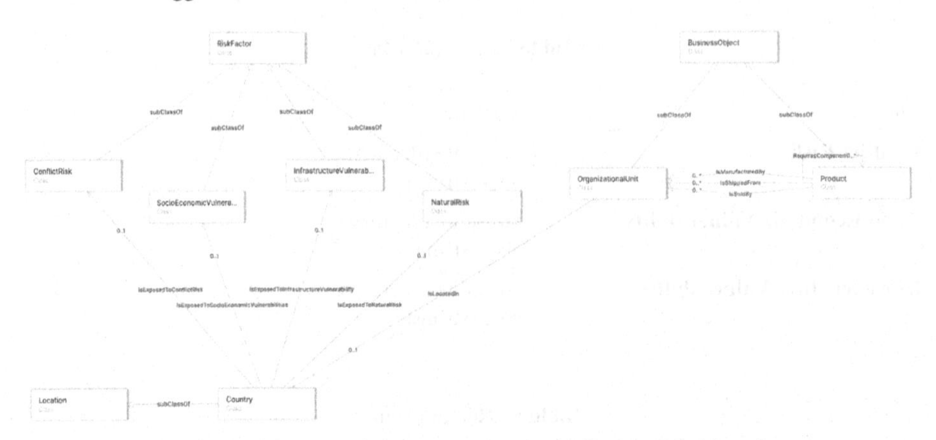

Fig. 1. The RASSA knowledge graph schema as displayed in Metaphactory

For the product data, a fictitious but realistic company was created. The company Velo AG is a Swiss SME that designs, engineers and assembles bicycles for sale to the end consumer. For its product Road Bike S, a Bill of Material with the main components could look as depicted in Table 7 which is based on the BOM structure presented in [33]. All product instances were entereed manually in Metaphactory.

Table 7. Bill of Material for the example product Road Bike S

Product	Assembly Group	Level-1-Component	Level-2-Component
Road Bike S	Frame (1)	Frame (1)	
		Fork (1)	
	Transmission (1)	Cranks (2)	
		Chain Wheels (3)	
		Pedal (2)	
		Chain (1)	
	Control (1)	Handlebar	
		Brakes	Brake Lever (2)
			Front Brake (1)
			Rear Brake (1)
		Shifter (1)	
	Wheel (2)	Rim (1)	
		Tire (1)	

Figure 2 depicts the resulting graph. The focal firm's product Road Bike S is on the left side of the graph, level 1 components (product instances) are directly linked to it as well as to the organizational unit that manufactures them. Level 1 components are linked to level 2 components which are then again linked to their manufacturers which are in different countries. The countries have direct relationships with their four individual risk scores.

Fig. 2. View of the RASSA knowledge graph (instance level) in Metaphactory

6.2 Automatic Reasoning for Manufacturing Risk

We call the risk that is directly related to each manufacturing node in the supply chain (i.e. an organizational unit) a manufacturing risk for the components produced by that unit.

$$\begin{aligned} &\textit{Manufacturing risk per Product} = ((\textit{Infrastructure Vulnerability} + \textit{Conflict Risk} + \\ &\textit{Natural Risk} + \textit{Socio Economic Vulnerability})/4)/10 \end{aligned} \tag{2}$$

The division by factor 10 in Eq. (2) allows us to create a probability value for a possible risk event from the index. This may be considered an unorthodox method; however, it serves the purpose of having a sample for risk data.

Figure 3 displays how the SPARQL query for Manufacturing Risk as well as the first ten rows of results. The results are shown in relation to components rather than the organizational units that produce them, this allows us to accommodate the propagation of risks to components that are used upstream in the supply chain. When using the INSERT rule instead of SELECT, the knowledge base is updated. It must be kept in mind that components are instances of the class product.

For space reasons the following Eqs. (3) and (4) in the subsequent sections are shown without screenshots.

6.3 Automatic Reasoning for Supply Risk

We refer to risk that is incoming from one node in the supply chain to another through components integrated upstream as Supply Risk. Equation (3) shows how Manufacturing Risk of different components that are integrated in another component are averaged. We consider the criticality of components by giving critical components a higher weight (1.5) than noncritical components, commodities that require no specific engineering are

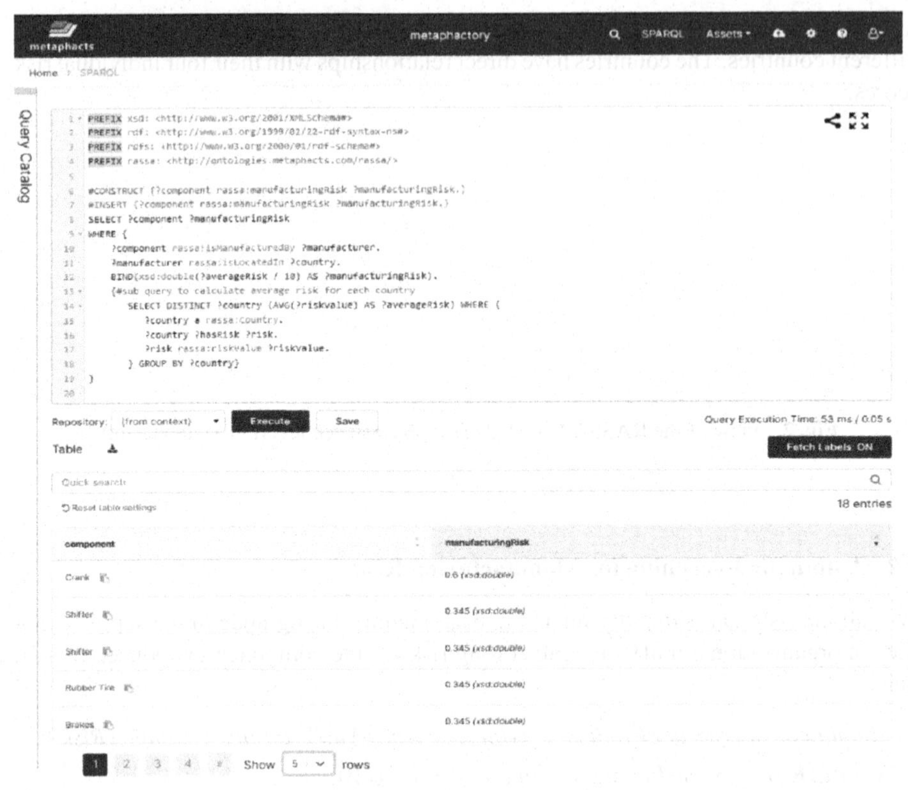

Fig. 3. Query execution manufacturing risk

weighted lower (factor 0.5). The weight assignment is arbitrary and may be defined differently by a user.

$$\begin{gathered}\textit{SupplyRisk to each node} = \\ ((\textit{ManufacturingRisk of all Components required} \\ \textit{where CriticalType equals ``true'' and Commodity equals ``false''} * 1.5) + \\ (\textit{ManufacturingRisk of all Components required where CriticalType equals ``false'' and} \\ \textit{Commodity equals ``false''}) + (\textit{ManufacturingRisk of all Components required} \\ \textit{where CriticalType equals ``false'' and Commodity equals ``true''} * 0.5)) / \\ \textit{Number of directly linked Components}\end{gathered} \tag{3}$$

6.4 Automatic Reasoning for Total Risk Propagated Per Supply Chain Node

The risk propagated by each node is the supply risk it inherits from upstream in the supply chain, averaged with its own direct manufacturing risk before the result is multiplied by the risk propagation factor of the respective organizational unit (4). The risk propagation factor is a decimal value between 0 and 1 that can be assigned to an organizational unit depending on the existing knowledge of an organizational unit's ability to mitigate risk.

A risk propagation factor below 1 would indicate some ability to mitigate risk, thus the risk propagated upstream is decreased.

$$\begin{aligned} \textit{Total Risk Propagated per Node} &= (\textit{Manufacturing Risk} + \textit{Supply Risk})/2 \\ &\quad *\textit{Risk Propagation Factor} \end{aligned} \tag{4}$$

6.5 Automatic Reasoning for Value at Risk

To retrieve the Value at Risk, the average total risk of all components in the focal firm's product of interest is multiplied by the firm's annual turnover in monetary units, yielding the amount of sales that is endangered by external supply chain risk (5).

$$\begin{aligned} &\textit{Focal Firm Product VaR} = \textit{AVG of Total Risks propagated through directly linked Components} \\ &*\textit{Annual Turnover in Monetary units of Product Road Bike S} \end{aligned} \tag{5}$$

In Fig. 4 the VaR for Velo AG's product Road Bike S is retrieved, the query returns the expected result of CHF 20′235.0 as VaR where the annual turnover that Velo AG realizes with Road Bike S amounts to CHF 200′000.00.

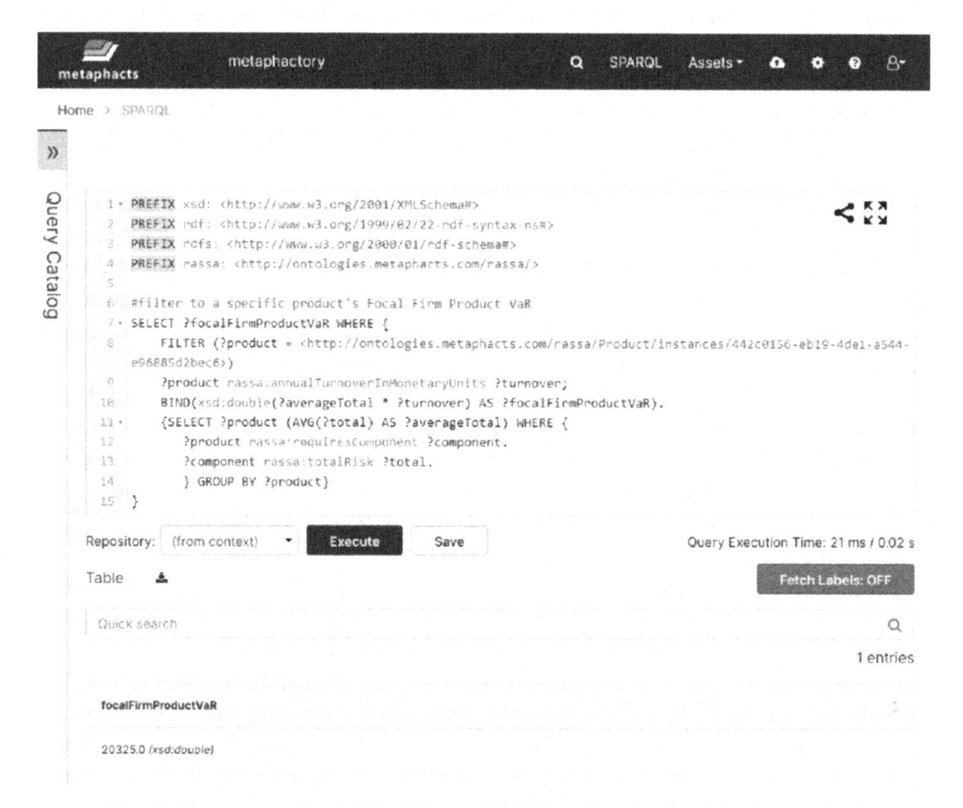

Fig. 4. Query execution of the Value at Risk (based on the focal firm's turnover)

7 Findings and Discussion

All competency questions were formalized in the standard SPARQL, and then executed against the knowledge base.

Namely, the first competency question, *What is the risk assigned to product a, produced by organizational unit Z?*, was successfully answered through a SPARQL query for the manufacturing risk formula (see Sect. 6.2). From the knowledge base, we were able to retrieve the risk values that each organizational unit is exposed to directly through its location.

The second competency question, *What is the risk that propagates from organizational unit Z to organizational unit Y if organizational unit Y integrates product a as a component in its product b?*, was answered by creating a SPARQL for the supply risk and product supply risk formula (see Sect. 6.3).

With the competency question, *What is the risk that propagates from organizational unit y to the focal firm X if the focal firm X integrates product b in its product c?*, the knowledge base is expected to deliver information about the total risk (manufacturing risk and supply risk from each component the vocal firm integrates in its product - see Sect. 6.4). With a SPARQL for total risk, we proved that it is possible to automate the calculation of the total risk values for all level 1 components of the product Road Bike S.

Lastly, the knowledge base was successful at providing an answer to our central question: *If product b, d, e, and f with their component's products a, j, k, l, m, n, o, p, … are integrated in product c, what is the Value at Risk of the focal firm X given its annual turnover on sales of product c?* (see Sect. 6.5).

All the query results were as expected and correct. Therefore, the proposed knowledge graph-decision support has proven its utility and correctness, which has been the core of investigation in this work. The ease of use [34] of the visualization has also been evaluated by interviewing the same three experts mentioned in Sect. 3. Each interviewer was introduced to the realistic company scenario. Next, the visual ontology was shown and explained. A discussion followed where the focal point was the ease of use of the approach. All participants agreed that the visual representation of the ontology helps to understand the supply chain network and to take decisions to make the supply chain more resilient. A customized user interface would be required to trigger and visualize the query results. The Metaphactory tool allows the creation of such user interfaces too, whose creation was not in scope of this work. The extension or change of the supply chain risk model was also out of scope.

The adoption of our system by companies would require integrating it to the existing IT architecture. The proposed ontology model should be constantly fed with data, e.g. from various data sources. The visualization of the ontology should require some admin rights to be accessed, whereas all supply chain risk managers and assistants with lower rights should see the results of the calculations through a separate interface where the ontology model cannot be adapted. This implies that query results should be refreshed automatically on a timely basis. Such an integration work is a daily business of the company Metaphacts.

8 Conclusion and Outlook

In this work, we presented RASSA, a knowledge graph-based decision support system, for the creation and maintenance of resilient supply chain networks. For this, we followed the Design Science Research methodology and the Ontology Development 101 [25] to ensure rigor in the creation of the ontology. The ontology model borrowed concepts of existing relevant research works. Expert interviews helped to deepen the problem understanding and to derive the primary requirements to address. The latter were presented in the form of competency questions. The ontology model was implemented in the tool Metaphactory, which offers ease of use in visualizing and adapting ontology models. The utility and correctness of the approach were evaluated by answering the competency questions, which have been formalized in SPARQL. The user-friendliness of the tool was also addressed, and major findings pointed to the need to create a customized user interface to simplify the visualization of the query results.

Future research could select real cases, empirically testing our approach's utility in businesses which could include extending our schema and knowledge graph with a broader coverage of county's risk landscape, including the propagation of risk in the event chain external to the supply chain itself. This would include the definition of company specific variables e.g. for risk propagation factors and component criticality.

Acknowledgements. This work has been supported by metaphacts GmbH. We would like to thank Jesse Lambert and Sebastian Schmidt. Further we thank all our interviewees for their willingness to share their knowledge and experiences with us. Especially J.M. for acting as a constant sparring partner throughout our research.

References

1. Zsidisin, G.A., Henke, M.: Revisiting Supply Chain Risk, vol. 7. Springer, Cham (2019). https://doi.org/10.1007/978-3-030-03813-7
2. Martus,C.: An empire-sized history lesson on supply chain security (2021). https://www.ismworld.org/supply-management-news-and-reports/news-publications/inside-supply-management-magazine/2021--mayjune-issue/insights/. Accessed 10 Dec 2022
3. Schneider-Petsinger, M.: Global trade in 2023: what's driving reglobalization? Chatham House Briefing (2023). https://doi.org/10.55317/9781784135560
4. ASCM: Develop Sourcing Strategy (2023). https://scor.ascm.org/processes/source/S1.3. Accessed 20 Jun 2023
5. Taherdoost, H., Brard, A.: Analyzing the process of supplier selection criteria and methods. Procedia Manuf **32**, 1024–1034 (2019). https://doi.org/10.1016/J.PROMFG.2019.02.317
6. Ravindran, A.R., Bilsel, R.U., Wadhwa, V., Yang, T.: Risk adjusted multicriteria supplier selection models with applications. Int. J. Prod. Res. **48**(2), 405–424 (2010). https://doi.org/10.1080/00207540903174940
7. Viswanadham, N., Samvedi, A.: Supplier selection based on supply chain ecosystem, performance and risk criteria. Int. J. Prod. Res. **51**(21), 6484–6498 (2013). https://doi.org/10.1080/00207543.2013.825056
8. Straube, F., Durach, C.F., Phung, J.: Developing and applying a supplier selection model to account for supplier risk impacts. Supply Chain Forum **17**(2), 68–77 (2016). https://doi.org/10.1080/16258312.2016.1171958

9. coupa: Supply chain risk | monitor spend at risk across suppliers | Coupa. https://www.coupa.com/products/supplier-management/supply-chain-risk. Accessed 11 Dec 2022
10. riskmethods: What is supply chain risk management? | riskmethods. https://www.riskmethods.net/scrm/what-is-supply-chain-risk-management. Accessed 11 Dec 2022
11. Lee, D., Glosserman, B.: How companies can navigate today's geopolitical risks. Harvard Business Review (2022). https://hbr.org/2022/11/how-companies-can-navigate-todays-geopolitical-risks. Accessed 29 May 2023
12. ASCM: Agility level-3 metrics overall value at risk (2023). https://scor.ascm.org/performance/agility/AG.3.1. Accessed 11 Jun 2023
13. Faisal, M.N., Banwet, D.K., Shankar, R.: Mapping supply chains on risk and customer sensitivity dimensions. Ind. Manag. Data Syst. **106**(6), 878–895 (2006). https://doi.org/10.1108/02635570610671533
14. Ye, Y., Yang, D., Jiang, Z., Tong, L.: Ontology-based semantic models for supply chain management. Int. J. Adv. Manuf. Technol. **37**(11–12), 1250–1260 (2007). https://doi.org/10.1007/s00170-007-1052-6
15. Wagner, S.M., Neshat, N.: Assessing the vulnerability of supply chains using graph theory. Int. J. Prod. Econ. **126**(1), 121–129 (2010). https://doi.org/10.1016/J.IJPE.2009.10.007
16. Emmenegger, S., Thönssen, B., Laurenzi, E.: Improving supply-chain-management based on semantically enriched risk descriptions (2012). https://www.researchgate.net/publication/236985852
17. Emmenegger, S., Hinkelmann, K., Laurenzi, E., Thönssen, B.: Towards a procedure for assessing supply chain risks using semantic technologies. In: Fred, A., Dietz, J.L.G., Liu, K., Filipe, J. (eds.) Knowledge Discovery, Knowledge Engineering and Knowledge Management, pp. 393–409. Springer Berlin Heidelberg, Berlin, Heidelberg (2013). https://doi.org/10.1007/978-3-642-54105-6_26
18. Carstens, L., Leidner, J.L., Szymanski, K., Howald, B.: Modeling company risk and importance in supply graphs. In: Lecture Notes in Computer Science (including subseries Lecture Notes in Artificial Intelligence and Lecture Notes in Bioinformatics), pp. 18–32. Springer Verlag (2017). https://doi.org/10.1007/978-3-319-58451-5_2/FIGURES/5
19. Palmer, C., et al.: An ontology supported risk assessment approach for the intelligent configuration of supply networks. J. Intell. Manuf. **29**, 1005–1030 (2018). https://doi.org/10.1007/s10845-016-1252-8
20. Singh, S., Ghosh, S., Jayaram, J., Tiwari, M.K.: Enhancing supply chain resilience using ontology-based decision support system. Int. J. Comput. Integr. Manuf. **32**(7), 642–657 (2019). https://doi.org/10.1080/0951192X.2019.1599443
21. Zhang, Q., Mu, W.: Research on search method of knowledge graph of supply chain risk based on cascading effect. In: Proceedings of the 2021 3rd International Conference on Management Science and Industrial Engineering, pp. 111–119 (2021). https://doi.org/10.1145/3460824.3460842
22. Aziz, A., Kosasih, E.E., Griffiths, R.-R., Brintrup, A.: Data considerations in graph representation learning for supply chain networks (2021). https://anonymous.4open.science/r/Link-. Accessed 18 Dec 2022
23. Kosasih, E.E., Margaroli, F., Gelli, S., Aziz, A., Wildgoose, N., Brintrup, A.: Towards knowledge graph reasoning for supply chain risk management using graph neural networks. Int. J. Prod. Res. **1–17** (2022). https://doi.org/10.1080/00207543.2022.2100841
24. Bogner, A., Littig, B., Menz, W. (eds.): Das Experteninterview: Theorie, Methode, Anwendung. Verlag für Sozialwissenschaften, Wiesbaden (2002). https://doi.org/10.1007/978-3-322-93270-9
25. Noy, N.F., McGuinness, D.L.: Ontology development 101 a guide to creating your first ontology (2001). https://protege.stanford.edu/publications/ontology_development/ontology101-noy-mcguinness.html. Accessed 29 May 2023

26. Grüninger, M., Fox, M.M.S.: Methodology for the design and evaluation of ontologies (1995). https://www.researchgate.net/publication/2288533_Methodology_for_the_Design_and_Evaluation_of_Ontologies. Accessed 10 Jun 2023
27. Uschold, M., Gruninger, M.: Ontologies: Principles, methods and applications. Knowl. Eng. Rev. **11**(2), 93–136 (1996). https://doi.org/10.1017/S0269888900007797
28. Hinkelmann, K., Laurenzi, E., Martin, A., Montecchiari, D., Spahic, M., Thönssen, B.: ArchiMEO: A standardized enterprise ontology based on the archimate conceptual model. In: Proceedings of the 8th International Conference on Model-Driven Engineering and Software Development, SCITEPRESS Science and Technology Publications, pp. 417–424 (2020). https://doi.org/10.5220/0009000204170424
29. Laurenzi, E.: An Ontology for the Assessment of Procurement Risk Management (2012)
30. metaphacts GmbH: Metaphactory (2023). https://help.metaphacts.com/resource/Help:Documentation. Accessed 21 Jun 2023
31. Haase, P., Herzig, D.M., Kozlov, A., Nikolov, A., Trame, J.: Metaphactory: a platform for knowledge graph management. Semant. Web **10**(6), 1109–1125 (2019). https://doi.org/10.3233/SW-190360
32. INFORM: INFORM risk index 2023 (2022). https://drmkc.jrc.ec.europa.eu/inform-index. Accessed 21 Jun 2023
33. Tung-Hung, L., Trappey, A.J.C.: Development of a web-based mass customization platform for bicycle customization services. In: Chou, S.Y., Trappey, A., Pokojski, J., Smith, S. (eds.) Global Perspective for Competitive Enterprise, Economy and Ecology, pp. 847–854. Springer London, London (2009). https://doi.org/10.1007/978-1-84882-762-2_80
34. Prat, N., Comyn-Wattiau, I., Akoka, J.: Artifact evaluation in information systems design-science research - a holistic view. In: Pacific Asia Conference on Information Systems (2014)

A Conceptual Model of Digital Immune System to Increase the Resilience of Technology Ecosystems

Beāte Krauze and Jānis Grabis[(✉)]

Riga Technical University, 6A Kipsalas Street, Riga 1048, Latvia
beate.krauze@edu.rtu.lv, grabis@rtu.lv

Abstract. In the light of technological advancements, disruptions and regulatory changes, various guidelines and standards emphasize the need for technology resilience. However, they often lack explicit evaluation methods, leaving organizations to determine their own implementation and assessment approaches. This absence of specific guidance amplifies the challenges organizations face in ensuring business continuity when critical systems fail. To address this, a "digital immune system" is proposed – a holistic approach to safeguard digital assets and mitigate IT-related risks. Digital immune system integrates processes, analytics and technologies to strengthen IT architecture, business operations and incorporates assessments to evaluate technology ecosystem resilience. Despite the acknowledged need for technology resilience, existing frameworks fall short in providing practical evaluation methods for the digital immune system. This paper confronts this challenge by focusing on the interconnected components of technologies, data and processes, considering emerging threats and compliance requirements. The goal of the research is to design an assessment framework for the digital immune system and establish a Digital Immune System Maturity Model. The model offers a structured path for organizations to measure and improve their IT risk assessment, resilience and business continuity plans.

Keywords: Digital Immune System · Resilience · Business Continuity Planning · IT Risk Management

1 Introduction

In the recent light of technology innovations, disruptions and geopolitical events, various regulations such as [1–3], guidelines as [4–8] and standards as [9, 10], require organizations to focus on technology resilience. These regulations and standards typically focus on the requirement to develop business continuity plans and ensure technology resilience, but often do not provide specific guidance on how to evaluate or measure technology resilience. While they establish the importance of having these measures in place, they leave the implementation, assessment and evaluation methods to the discretion of the organizations themselves. Therefore, the problem formulation is the following: How to

J. Araújo et al. (Eds.): RCIS 2024, LNBIP 513, pp. 82–96, 2024.
https://doi.org/10.1007/978-3-031-59465-6_6

perform an assessment of a digital immune system considering the interconnected components of software, infrastructure and people, as well as emerging threats and compliance requirements? Therefore, the goal of this paper is to design an analysis framework for evaluation of digital immune system and to develop Digital immune system maturity model (DIS MM).

The paper is organized as follows. Section 2 demonstrates the research methodology. It describes the connection between sections, including research methods and outcomes. Section 3 examines the regulatory landscape. In Sect. 4, authors explore the challenges faced by organizations in the current landscape characterized by unprecedented disruption, uncertainty and a convergence of external macro influences. Additionally, authors amalgamate techniques to address these challenges related to IT resilience and business continuity, identified through a literature review, emphasizing these as fundamental elements forming the core foundation of a digital immune system. Section 5 introduces a DIS MM for evaluation and showcases the practical application of this model within a case study for a professional services firm. In the last section, conclusions are drawn based on the findings and analyses conducted throughout the research. Moreover, the conclusions propose recommendations for future research directions and practical implementation strategies based on the identified gaps or areas for further exploration.

2 Approach

The development methodology of the model involves a structured approach that integrates various methods sourced from citations. It emphasizes the importance of a systematic method for synthesizing existing knowledge and creating a structured representation thereof.

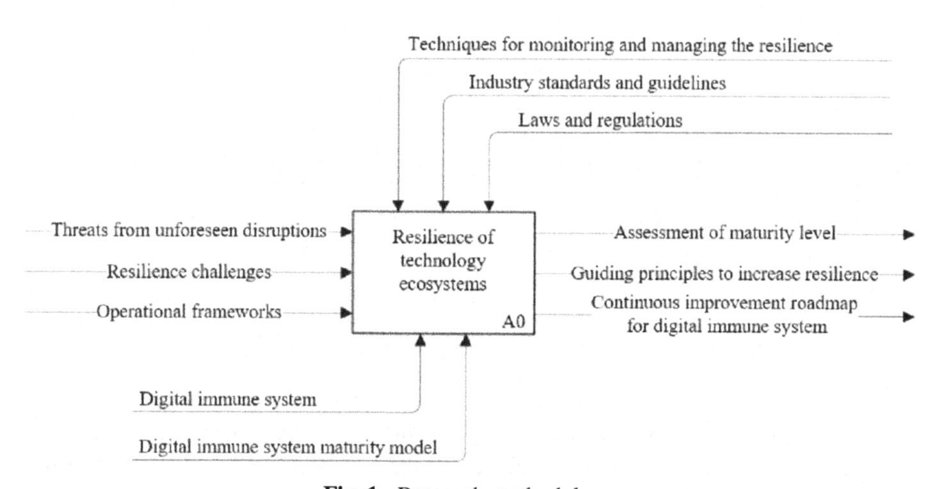

Fig. 1. Research methodology

The research methodology involves an ongoing literature review throughout the paper. To visualize the resilience of technology ecosystems, an IDEFØ method is used (see Fig. 1). The arrows are identified by the sides of the activity box. Thus, inputs are on the left, controls at the top, outputs on the right and mechanisms at the bottom.

Additionally, the methodology incorporates elements of Design Science. The design science principles applied in this context included problem identification, design creation, evaluation, demonstration, and communication. These principles guided the structured development of a resilience improvement plan tailored to address the real-world challenge of evaluating and enhancing technology resilience. This approach aligns with the systematic methodology described earlier; however, it extends beyond the mere synthesis of existing knowledge by actively contributing to the creation of novel solutions. The Digital Immune System Maturity Model introduced is a result of Design Science principles, providing organizations with a practical and comprehensive framework while addressing the gaps in existing regulatory guidance.

3 Background and Requirements Towards Digital Immune System

Given the recent advancements in technology, along with disruptions and geopolitical events, organizations are now obligated to prioritize technology resilience, as stipulated by regulations like [1–3], guidelines such as [4–8] and standards like [9, 10]. In this comprehensive examination of global standards and practices related to technology resilience, various international frameworks and initiatives are explored to provide insights into the challenges and opportunities faced by organizations striving to ensure the continuity of essential functions and services.

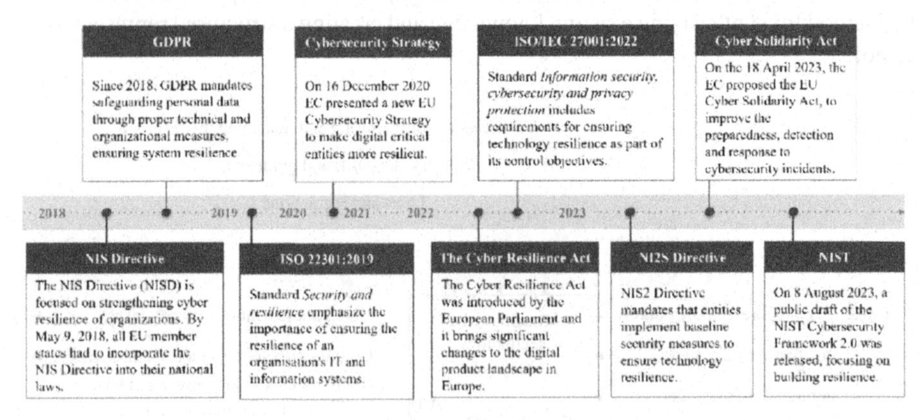

Fig. 2. Regulatory landscape towards digital immune system

The current regulatory landscape (see Fig. 2) emphasizes the critical need for organizations to enhance their technology resilience in the face of geopolitical events, natural disasters and cyber threats.

3.1 ISO 22301:2019

ISO 22301:2019 establishes a framework for business continuity management systems (BCMS), emphasizing the importance of ensuring the resilience of an organization's information technology (IT) and information systems. The standard addresses key aspects in clauses 6 and 8, encompassing general requirements for BCMS development and specific requirements related to incident response, business continuity planning and technology resilience. However, a notable challenge arises as ISO 22301:2019 lacks explicit, step-by-step guidance for assessing and implementing technology resilience measures. This deficiency requires organizations to invest significant time and resources in determining the most appropriate methods and controls.

3.2 ISO/IEC 27001:2022

ISO/IEC 27001:2022 integrates requirements for technology resilience within its control objectives. The standard lists 93 controls grouped into four themes: people, organizational, technological and physical. Controls such as ICT readiness, compliance with policies and secure system architecture contribute to business continuity and technology resilience. However, similar to ISO 22301:2019, ISO/IEC 27001:2022 lacks specific guidance on assessing and implementing technology resilience measures. This gap may lead to varying interpretations among organizations, potentially resulting in inconsistencies in technology resilience planning.

3.3 The NIST Cybersecurity Framework

The NIST Cybersecurity Framework, widely adopted as a best practice in the United States, emphasizes technology resilience within its core functions: identify, protect, respond and recover. While providing a risk-based framework for improving cybersecurity practices, the framework does not offer detailed assessments or specific actions for technology resilience.

3.4 European Union Initiatives

NIS2 Directive. The original NIS Directive was the first piece of EU-wide legislation on cybersecurity, aiming to provide legal measures to improve the overall level of cybersecurity in the EU. The NIS2 Directive focuses on enhancing overall cybersecurity and resilience for critical infrastructure. Mandating baseline security measures, it requires organizations to develop policies, incident response plans and business continuity plans. However, it does not provide specific actions for assessing technology resilience.

GDPR. GDPR, while not explicitly mentioning "technology resilience," emphasizes the need for appropriate security measures, including resilience, throughout its provisions. Articles such as 5, 25, 32 and 35 highlight the importance of ensuring the security and resilience of data processing systems.

Cybersecurity Strategy and Cyber Resilience Act. The EU's Cybersecurity Strategy aims at resilience, technological sovereignty and leadership. The Cyber Resilience Act mandates stringent cybersecurity requirements for products with digital elements. While the EU may not explicitly outline specific steps for self-assessment of technology resilience in every policy document, it's important to understand that self-assessment and cybersecurity readiness are often implied and encouraged within the broader framework of these initiatives.

Cyber Solidarity Act. Proposed to enhance preparedness for cybersecurity incidents, the Cyber Solidarity Act aligns with the recommendation to strengthen the resilience of critical infrastructure. The specific self-assessment steps required for technology resilience can vary widely depending on the nature of the organization, the sector and the technology in use. One-size-fits-all requirements may not be practical. Therefore, the EU often leaves room for organizations and member states to tailor their self-assessment processes to their specific circumstances.

The findings derived from the analyzed regulatory landscape is that while various EU directives and initiatives emphasize the importance resilience, there is no explicit, standardized set of actions for assessing technology resilience. Instead, these regulations, directives and initiatives emphasize the importance of tailored approaches to self-assessment, recognizing that one-size-fits-all requirements may not be practical due to variations in organizational circumstances, sectors, and technologies in use.

4 Challenges of IT Resilience, Business Continuity and Managing Techniques

4.1 Challenges of IT Resilience and Business Continuity

The challenges of IT resilience and business continuity are amalgamated from conducted literature review on the IT resilience and business continuity that are the core foundation of the digital immune system (see Table 1). Addressing these challenges is essential for maintaining IT resilience and ensuring effective business continuity in an increasingly complex and interconnected digital environment.

The first challenge involves managing the exponential surge in data volume, requiring investments in scalable solutions for efficient storage, backup and recovery methods. Dealing with diverse data structures and types is the second challenge, necessitating a decentralized, flexible enterprise architecture to handle both real-time and non-real-time analytics effectively. The third challenge centers on system integration and interoperability issues, particularly in dealing with legacy systems and complexities arising from mergers, emphasizing the need for strategic modernization. The unintended consequences of artificial intelligence constitute the fourth challenge, encompassing risks such as cybersecurity vulnerabilities and ethical concerns, prompting the need for regulatory frameworks. The fifth challenge revolves around the lack of disaster recovery abilities in complex IT environments, encompassing optimizing foundational elements, testing business continuity plans and meeting regulatory requirements.

Addressing these challenges is essential for maintaining IT resilience and ensuring effective business continuity in an increasingly complex and interconnected digital

Table 1. Challenges of IT resilience and business continuity.

ID	Description	Challenge is mentioned and analyzed	Challenge is mentioned, addressed and solution proposed
C1	Exponential increase in data volume	[11]	[12, 13]
C2	Ability to manage different data structures, types and formats	[14]	[13, 15]
C3	System integration and interoperability issues	[16–18]	[19, 20]
C4	Unintended consequences of artificial intelligence	[21, 22]	[23–26]
C5	Lack of disaster recovery abilities in complex IT environments	[27–29]	[30, 31]

environment. Organizations should proactively assess and mitigate these challenges to safeguard their operations against disruptions.

4.2 The Techniques for Monitoring and Managing the Resilience

The techniques for monitoring and managing the resilience of IT ecosystems are summarized in Table 2. These techniques are focused on monitoring and improving the resilience of IT ecosystems, ensuring that they can handle challenges, continue to function effectively and quickly recover from any disruptions or incidents.

The first technique is the Data Governance Framework. This involves performing a data asset inventory and developing a data management strategy throughout their lifecycle. The implementation process involves analyzing key processes, creating a data asset inventory, identifying system owners and assigning characteristics to each data asset. The second technique – Applied Observability – involves applying observable data in a highly orchestrated and integrated approach across business functions, applications and infrastructure. The implementation process includes core domains such as logging, tracing, visualization and monitoring. The third technique is the AI TRiSM Framework. This framework adopts an AI trust, risk and security management approach to identify, assess and mitigate the risks associated with acquiring, developing and deploying artificial intelligence systems. The implementation process introduces tasks performed by roles like Chief Information Security Officer, Chief Data Officer, Chief Compliance Officer, General Counsel and Internal Auditor. The fourth technique is BCP that involves conducting a business impact analysis and prioritizing recovery strategies based on critical functions. The implementation process emphasizes the Business Impact Analysis, Incident Response Plan and Crisis Management Plan.

Table 2. Resilience managing techniques.

Technique ID	Technique	Technique description	Challenge addressed (ID)	Adopted from
T1	Data governance framework	Perform data asset inventory and develop data management strategy throughout their lifecycle	C1, C2, C3, C4, C5	[32–34]
T2	Applied observability	Apply the use of observable data in a highly orchestrated and integrated approach across business functions, applications and infrastructure	C1, C3, C5	[35, 36]
T3	AI trust, risk and security management framework (AI TRiSM)	Adopt AI TRiSM to identify, assess and mitigate the risks associated with acquiring, developing and deploying artificial intelligence systems	C4	[24, 37–41]
T4	Business continuity planning (BCP)	Conduct business impact analysis and prioritize recovery strategies based on the criticality of these functions	C3, C4, C5	[42–46]

5 Framework and Evaluation

In this section, authors suggest the next steps organizations should perform to systematically improve their digital immune system's effectiveness, based on previously analyzed challenges and techniques. Adopting below mentioned approach helps organizations to maintain the reliability and availability of their IT systems, which is crucial for overall business operations and continuity.

5.1 Framework for the Evaluation of Digital Immune System

DIS MM amalgamates key components (see Fig. 3) from individual techniques, promoting improved collaboration and synergy in these critical aspects of technology ecosystem resilience.

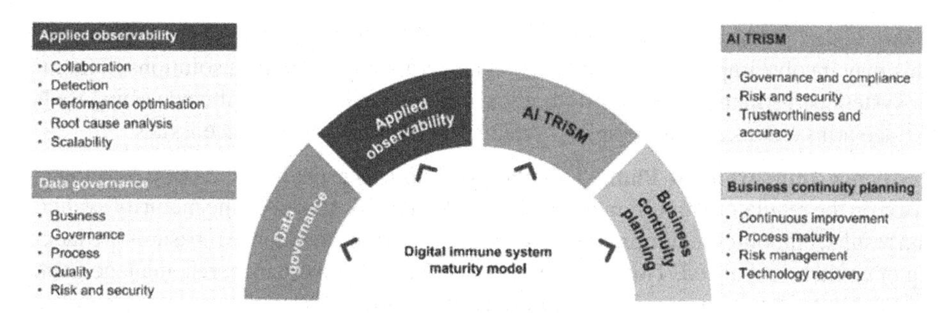

Fig. 3. DIS MM evaluation aspects

Implementation. For organizations that are on the initial stages of digital immune system implementation, the implementation shall begin sequentially. Starting with the data governance framework introduction, followed by observability practices and AI risk assurance application – AI trust, risk and security management (AI TRiSM) framework – and finally implementing comprehensive BCP procedures. For organizations that have already implemented a digital immune system, their focus might be on refining and optimizing existing processes rather than starting from scratch. DIS MM facilitates a culture of continuous improvement, encouraging organizations to regularly reassess their maturity and make iterative enhancements over time. Even for organizations at advanced stages of implementation, there's always room for refinement and adaptation in response to cybersecurity challenges and organizational needs.

Application. The proposed DIS MM offers a novel approach and a structured path for organizations to measure and improve the resilience as it includes evaluation criteria according to industry standards and best practices. This facilitates ongoing enhancements to the digital immune system. DIS MM amalgamates processes, analytics and technologies to strengthen IT architecture, business operations and mitigate business risks and practices, frameworks and assessments to evaluate technology ecosystems resilience.

Post-evaluation Recommendations. After conducting the maturity assessment, specific recommendations can be derived from identified weaknesses or gaps that led to low maturity levels. For example:

Processes. Implement a more structured process for managing data throughout its lifecycle, ensuring consistency and effectiveness across all data types and departments. Steps include developing standardized procedures for data creation, storage, usage, and archival and integrating data lifecycle management into existing business processes and workflows.

Analytics. Implement data quality tools and technologies to automate data profiling, cleansing, and validation processes. Introduce data stewardship program to ensure accountability and ownership of data quality issues across departments.

Technologies. Implement monitoring tools to track system performance in real-time. This may involve implementing application performance monitoring solutions to monitor system performance and identify bottlenecks and deploying distributed tracing tools to trace transactions across complex systems and identify performance issues.

Resilience Improvement Plan. The plan for improving resilience can be developed based on the regulatory requirements and DIS MM results. Based on the maturity evaluation results, pinpoint specific areas where the organization's resilience measures are lacking or could be enhanced. This includes aspects like data governance, risk management, cybersecurity, business continuity planning, and compliance with regulatory requirements. First, the organization must conduct an analysis comparing current resilience measures against regulatory requirements and DIS MM benchmarks (maturity level 5). Second, identify specific areas where the organization falls short and prioritize them based on risk exposure and potential impact.

5.2 Case Study

A case study was conducted to apply DIS MM to a case firm that is a global professional services network. The firm operates in 151 countries and its international operations fall under six regions: Europe, Africa, Americas, Asia/Asia Pacific, Eurasia, Middle East. The business operations are organized in several lines of services: Audit and Assurance, Advisory Deals, Tax, Legal, Consulting and Internal Firm Services.

The following section analyzes the results of the digital immune system evaluation using developed DIS MM. The evaluation of the firm's digital immune system was conducted by the authors, drawing upon expertise in cybersecurity, risk management, and organizational resilience. This assessment involved a comprehensive review of various aspects of the firm's digital infrastructure, including cybersecurity protocols, risk mitigation strategies, and resilience measures. The author's expert judgment was utilized to evaluate the effectiveness and maturity of the firm's digital immune system, identifying areas for improvement. The maturity degree was fixed based on the extent to which the firm met predefined criteria and demonstrated maturity in its digital immune system implementation.

The data presented in Fig. 4 indicate that the BCP, data governance and applied observability practices are managed on the business level, initiatives are planned, performed, measured and controlled. AI TRiSM is implemented to the initial phase, meaning that it is rather reactive than proactive, but work progresses until it achieves fulfilment or finalization.

Data Governance Maturity Evaluation. The data analyzed (see Fig. 5) indicate that from the data governance perspective, the processes and quality aspects are managed. There are documented guidelines and rules for data management, but their implementation varies across territories as the firm operates in multiple geographies. DIS MM acknowledges that organizations, especially those with global operations like the firm in question, face varying digital threats and regulatory environments across different territories. A basic data classification framework is defined, covering some aspects of sensitivity and criticality, but it lacks depth and consistency between territories in the regions.

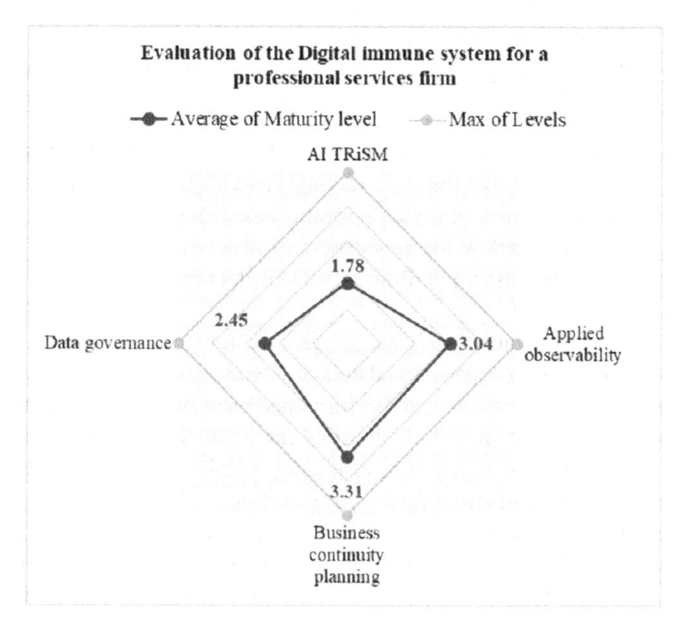

Fig. 4. Evaluation of the Digital immune system for a professional services firm

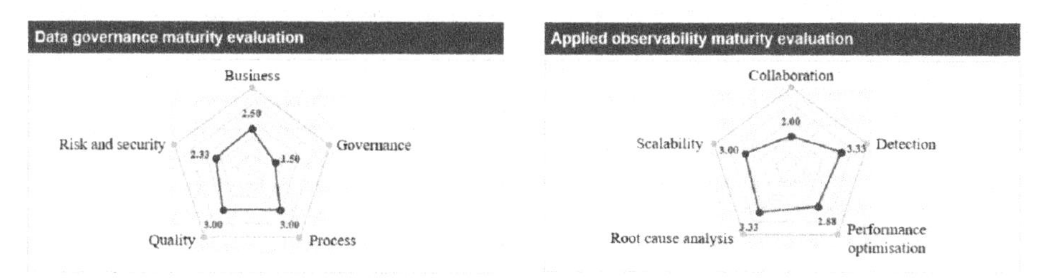

Fig. 5. Data governance (left) and applied observability (right) maturity evaluation

The business aspects are well defined and they include well-documented, comprehensive data governance policies and strategies that are consistently aligned with business goals and regulatory requirements. The case firm applies global ethics and independence principles and guidelines that limit services they can provide to specific clients. These guidelines are reinforced by regulatory restrictions. In addition to these restrictions, some member firms have further limited the services they provide to certain clients in response to local circumstances.

On the contrary, the data management costs are not transparently explained or analyzed, lacking optimization strategies. When risk and security is analyzed, authors identified that there is a structured backup and recovery plan in place, periodically tested, however no regular employee training is organized.

The major concern from the governance aspect includes monitoring mechanisms, feedback collection and ways for improvement in data governance processes that are sporadic, lacking a structured approach. The data governance quality is managed, there are structured and automated procedures for monitoring of data quality that include systematic examination and analysis of data to understand its structure, content, quality and relationships within a dataset or across multiple datasets.

Applied Observability Maturity Evaluation. The detection capabilities are discovered well managed, meaning that there are methods and tools employed for tracking and evaluating system performance, allowing periodic assessment and basic improvements. As major business systems allow the assessment of observability practices the insights into diverse data structures allow handling of various data formats and structures across the ecosystem.

When identifying root cause analysis, the firm demonstrates moderate capacity for error resolution, but encounters occasional delays. Some efforts for continuous improvements for overall efficiency and system performance exist but lack full systematic integration. The performance optimization attempts can be described as managed, as there are structured efforts to enhance resource efficiency, resulting in moderate cost savings, scalability improvements and streamlining of operations.

The scalability is well managed as there are initiatives to analyze diverse data sources, however this approach lacks implementation in all territories. The firm does not fully use the information on a shared basis across functions and business stakeholders, therefore limited sharing of information results in siloed data across functions is identified.

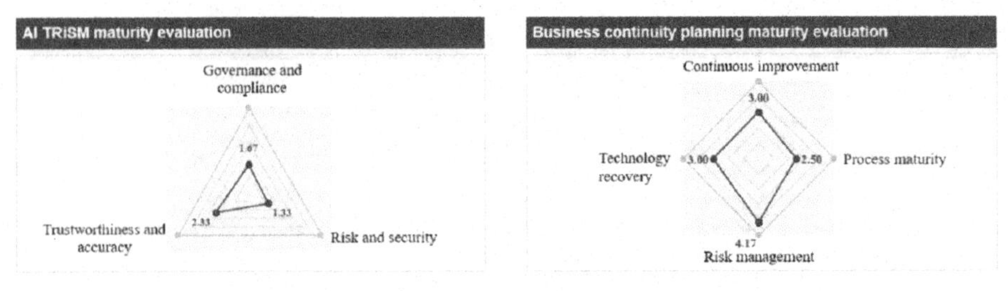

Fig. 6. AI TRiSM (left) and business continuity planning (right) maturity evaluation

AI TRiSM Maturity Evaluation. The data reveals significant lack of AI TRiSM maturity as presented in Fig. 6. The firm is moving quickly to develop and deploy machine learning and natural language processing capabilities across the global audit ecosystem.

A culture of innovation across the network of firms continues to drive the creation of tech-enabled solutions addressing both local and global needs.

Figure above identifies that the firm has procedures and frameworks implemented to assess and mitigate ethical risks associated with IT systems, however, it is not aligned in all territories and not all developed and available tools are accessed for proper governance and compliance.

As for risk and security of AI, there are only basic security measures implemented without comprehensive risk assessments, responding reactively to vulnerabilities or threats as they arise. There are security measures implemented during the development and deployment phases, but gaps exist in continuous security measures during ongoing operations.

There is a basic awareness of data privacy regulations, but lack of structured protocols for compliance in AI projects. The trustworthiness and accuracy are identified, there are structured measures implemented to mitigate biases, yet not fully integrated interpretability techniques universally across all AI systems.

BCP Maturity Evaluation. From the results, it is clear that, the process maturity is at core for BCP (see Fig. 6) meaning that Recovery Time Objective (RTO) and Recovery Point Objective (RPO) objectives are defined and included in BCP, however, no stakeholder involvement is organized to align the objectives with business priorities. The firm does not link these parameters to resource planning or technology decisions effectively. RTO and RPO are defined, based on technical abilities instead of business needs. The technology recovery is managed, the firm has established recovery procedures for critical systems or data. Superior results are seen for the risk management. Risk management is considered to be well defined as a cross-functional risk management team collaborates across departments, maintaining updated assessments closely tied to the BCP for a comprehensive and adaptable risk management strategy. The firm acknowledges and addresses third-party risks associated with vendors critical to the BCP. Contractual agreements define clear expectations and obligations regarding risk mitigation. The firm has developed comprehensive contingency plans specifically tailored to potential disruptions caused by third-party vendors. BCP incorporates and continuously updates various compliance aspects, including industry-specific guidelines, local legal requirements and global standards. As for continuous improvement, there are post-incident reviews and some documentation of lessons identified. However, the process lacks consistency and depth in analyzing lessons for broader improvements.

Proposed Guidance. The evaluation of the firm's digital immune system maturity indicates that its AI trust, risk and security framework currently operates in a reactive mode rather than a proactive one. The AI TRiSM framework is built and maintained by the following roles in the organization: Chief Information Security Officer, Chief Data Officer, Chief Compliance Officer, General Counsel and Internal Auditor. They must manage new and amplified risks as well as a deal with business, legal and regulatory challenges. While basic security measures are in place, there is a need for a more comprehensive approach, including proactive risk assessments. The firm addresses vulnerabilities and threats as they emerge, but lacks continuous security measures for sustained operations, indicating a crucial area for enhancement in AI risk management strategy.

Data governance framework's emphasis on data asset inventory and classification allows for a comprehensive risk assessment. Real-time risk detection mechanisms and adaptive controls could significantly enhance the framework's effectiveness.

While applied observability heavily relies on data, it's also about understanding human behavior within the system. Integrating user-centric metrics and behavior patterns into observability practices can lead to more tailored user experiences and better alignment of technological solutions with actual user needs.

6 Discussion and Conclusion

The emphasis on resilience in regulatory requirements highlights a shift towards a risk-centric approach, meaning that it requires to integrate resilience-building measures into core operational strategies. In order to enhance the resilience of technology ecosystems by detecting, mitigating, and recovering from various cyber threats and disruptions, Chief Technology Officers require an analysis framework for self-evaluation of a digital immune system for organizations. Therefore, the DIS MM serves as a living tool, enabling organizations to adapt to the evolving landscape of IT architectures. By offering actionable artefacts, the DIS MM is highly relevant for industry application and implementation in practice.

In order for CTO to enhance technology ecosystems resilience and therefore – digital immune system, they require a response from C-suite executives responsible for digital and/or IT strategy. This response should involve a decisive action or commitment to act within the next three years, considering the swift evolution of technology and ongoing threats.

Research also had certain practical limitations. Aspects such as continuous improvement and predictive learning, where DIS MM is designed to make predictions or forecasts based on patterns and relationships found in data, would have taken the research to the next level. Therefore, further research directions could be formulated as follows:

1. Investigate methodologies to create adaptive and self-improving DIS MM that evolves over time.
2. Research and develop frameworks that integrate security into the software development lifecycle, ensuring that digital immune system principles are embedded from the very beginning.

References

1. European Parliament, Council of the European Union: Directive (EU) 2022/2555 of the European Parliament and of the Council of 14 December 2022 on measures for a high common level of cybersecurity across the Union. Official Journal of the European Union (2023)
2. European Parliament, Council of the European Union: Regulation (EU) 2016/679 of the European Parliament and of the Council of 27 April 2016 on the protection of natural persons with regard to the processing of personal data and on the free movement of such data. Official Journal of the European Union (2018)
3. European Commission: Proposal for a REGULATION OF THE EUROPEAN PARLIAMENT AND OF THE COUNCIL laying down measures to strengthen solidarity and capacities in the Union to detect, prepare for and respond to cybersecurity threats and incidents. European Commission, Strasbourg (2023)
4. Council of the European Union: COUNCIL RECOMMENDATION of 8 December 2022 on a Union-wide coordinated approach to strengthen the resilience of critical infrastructure. Official Journal of the European Union, Brussels (2023)
5. European Commission and the High Representative of the Union for Foreign Affairs and Security Policy: The EU's Cybersecurity Strategy for the Digital Decade. European Commission, Brussels (2020)

6. European Commission, Secretariat-General: On the EU Security Union Strategy. European Commission, Brussels (2020)
7. National Institute of Standards and Technology: Framework for Improving Critical Infrastructure Cybersecurity (2023)
8. National Institute of Standards and Technology: The NIST Cybersecurity Framework 2.0, Gaithersburg (2023)
9. International Organization for Standardization: ISO/IEC 27001:2022, Information security, cybersecurity and privacy protection - Information security management systems – Requirements. Vernier (2022)
10. International Organization for Standardization: ISO 22301:2019. Security and resilience - Business continuity management systems – Requirements. Vernier (2019)
11. Technavio: Global Data Center Backup and Recovery Software Market 2023-2027. Infiniti Research Limited. (2020)
12. Smith, D.: Approaches to Avoid Common Cloud Strategy Pitfalls. Gartner (2022)
13. Raghavendar, K., Batra, I., Malik, A.: A robust resource allocation model for optimizing data skew and consumption rate in cloud-based IoT environments. Decis. Anal. J. **7**, 100200 (2023)
14. Chen, Z.: Observations and expectations on recent developments of data lakes. Procedia Comput. Sci. **214**, 405–411 (2022)
15. Cherradi, M., El Haddadi, A.: A scalable framework for data lakes ingestion. Procedia Comput. Sci. **215**, 809–814 (2022)
16. LaForce, E., Palmateer, S., Boone, J., Barnett, A.: 2020 Legacy Modernization Report. Levvel (2020)
17. Langer, A.: Legacy Systems and Integration (2011)
18. Harrell, H., Higgins, L.: IS integration: your most critical M&A challenge? (2002)
19. European Commission: EU eGovernment Action Plan 2016–2020, Brussels (2016)
20. Irani, Z., Abril, R.M., Weerakkody, V., Omar, A., Sivarajah, U.: The impact of legacy systems on digital transformation in European public administration: lesson learned from a multi case analysis (2023)
21. Glenn Cohen, I., Evgeniou, T., Husovec, M.: Navigating the New Risks and Regulatory Challenges of GenAI. Harvard Bus. Rev. (2023)
22. European Commission: EU AI Act: first regulation on artificial intelligence. European Commission (2023)
23. Kilian, K.A., Ventura, C.J., Bailey, M.M.: Examining the differential risk from high-level artificial intelligence and the question of control. Futures **151**, 103182 (2023)
24. Habbal, A., Ali, M.K., Abuzaraida, M.A.: Artificial intelligence trust, risk and security management (AI TRiSM): frameworks, applications, challenges and future research directions. Exp. Syst. Appl. **240**, 122442 (2024)
25. Malgieri, G., Pasquale, F.: Licensing high-risk artificial intelligence: toward ex ante justification for a disruptive technology. Comput. Law Secur. Rev. **52**, 105899 (2024)
26. Giudici, P., Centurelli, M., Turchetta, S.: Artificial Intelligence risk measurement. Exp. Syst. Appl. **235**, 121220 (2024)
27. PwC: PwC's Global Crisis Survey 2021 (2021)
28. Øverby, H., Audestad, J.A.: Lock-in and switching costs. In: Introduction to Digital Economics. CCB, pp. 177–192. Springer, Cham (2021). https://doi.org/10.1007/978-3-030-78237-5_12
29. Opara-Martins, J., Sahandi, R., Tian, F.: Critical analysis of vendor lock-in and its impact on cloud computing migration: a business perspective. J. Cloud Comput. **5**(1), 1–18 (2016). https://doi.org/10.1186/s13677-016-0054-z
30. Williams, T., Resto-Leon, M.: Cracking the code: the keys to a successful business impact analysis. J. Bus. Continuity Emerg. Plann. **16**, 313–319 (2023)

31. Liu, S., Keil, M., Wang, L., Lu, Y.: Understanding critical risks of business process outsourcing from the vendor perspective: a dyadic comparison Delphi study. Inf. Manag. **60**, 103837 (2023)
32. Davidson, E., Wessel, L., Winter, J.S., Winter, S.: Future directions for scholarship on data governance, digital innovation, and grand challenges. Inf. Organ. **33**(1), 100454 (2023)
33. Zhang, Q., Sun, X., Zhang, M.: Data matters: a strategic action framework for data governance. Inf. Manage. **59**, 103642 (2022)
34. Jarvenpaa, S.L., Essén, A.: Data sustainability: data governance in data infrastructures across technological and human generations. Inf. Organ. **33**(1), 100449 (2023)
35. Perri, L.: What Is a Digital Immune System and Why Does It Matter? Gartner (2022)
36. Marie-Magdelaine, N., Ahmed, T., Astruc-Amato, G.: Demonstration of an observability framework for cloud native microservices. In: IFIP/IEEE Symposium on Integrated Network and Service Management (IM) (2019)
37. World Economic Forum: Adopting AI Responsibly: Guidelines for Procurement of AI Solutions by the Private Sector. World Economic Forum (2023)
38. PwC: Managing the risks of generative AI. PwC (2023)
39. U.S. Department of Commerce, U.S. National Institute of Standards and Technology: Artificial Intelligence Risk Management Framework (AI RMF 1.0). U.S. National Institute of Standards and Technology (2023)
40. Microsoft Corporation: Microsoft Responsible AI Standard, v2. Microsoft Corporation (2022)
41. NIST Trustworthy & Responsible Artificial Intelligence Resource Center (AIRC): NIST AI RMF Playbook (2023)
42. Russo, N., Reis, L.: Methodological approach to systematization of business continuity in organizations. In: Research Anthology on Business Continuity and Navigating Times of Crisis (2022)
43. Lincke, S.: Addressing business impact analysis and business continuity. In: Security Planning, pp. 85–102. Springer, Cham (2015). https://doi.org/10.1007/978-3-319-16027-6_5
44. Surianarayanan, C., Chelliah, P.R.: Disaster recovery. In: Essentials of Cloud Computing. TCS, pp. 291–304. Springer, Cham (2019). https://doi.org/10.1007/978-3-030-13134-0_12
45. Demin, P.: Service continuity management: ITIL 4 Practice Guide. AXELOS Ltd. (2020)
46. Kadam, A.: Evaluating Business Service Continuity and Availability Using COBIT 2019. ISACA (2020)

Requirements and Architecture

Dealing with Emotional Requirements for Software Ecosystems: Findings and Lessons Learned in the PHArA-ON Project

Mohamad Gharib[1](✉), Mariana Falco[1], Femke Nijboer[2], Angelica M. Tinga[2], Stefania D'Agostini[3], Erika Rovini[4], Laura Fiorini[4], Filippo Cavallo[4], and Kuldar Taveter[1]

[1] University of Tartu, Tartu, Estonia
{mohamad.gharib,mariana.falco,kuldar.taveter}@ut.ee
[2] University of Twente, Enschede, The Netherlands
{femke.nijboer,a.m.tinga}@utwente.nl
[3] Engineering Ingegneria Informatica, Rome, Italy
stefania.dagostini@eng.it
[4] University of Florence, Florence, Italy
{erika.rovini,laura.fiorini,filippo.cavallo}@unifi.it

Abstract. Requirements Engineering (RE) stands as the cornerstone in ensuring that a system comprehensively captures and analyzes the needs and expectations of its users and stakeholders. Despite the numerous approaches designed for dealing with functional and quality (non-functional) requirements, approaches for dealing with emotional requirements still lag. Emotional requirements capture how users should feel when using a system, and inadequate consideration of such requirements results in end-users reluctance to use the system. In this paper, we report on our experience in dealing with emotional requirements as part of an H2020 European Project, namely PHArA-ON (Pilots for Healthy and Active Ageing in Europe) for the development of the PHArA-ON ecosystem that aims at improving the well-being and active aging of older adults. Specifically, we present the process we followed for dealing with emotional requirements, and we summarize the findings and lessons learned from this experience.

Keywords: Emotional Requirements · Emotions · Goal Modeling · Ecosystems · Requirements Engineering

1 Introduction

The success of any software system relies on its ability to fulfill the needs and expectations of its stakeholders [1]. These needs and expectations serve as a key source for specifying the requirements for the system [2]. In the Requirements Engineering (RE) community, requirements traditionally fall into two categories: functional and non-functional (quality) requirements, both of which a successful system must satisfactorily address. Engineering methodologies have evolved

J. Araújo et al. (Eds.): RCIS 2024, LNBIP 513, pp. 99–114, 2024.
https://doi.org/10.1007/978-3-031-59465-6_7

to adeptly manage functional and various types of quality requirements (e.g., usability, reliability [3]). However, emotional requirements, despite their significance, have remained relatively underexplored within the RE community [4].

Emotions are inherently complex, involving cognitive appraisals, physiological responses, and behavioral tendencies [5]. Although emotions can be broadly classified under positive (e.g., happiness and relaxation) and negative (e.g., anger and stress) emotions [6], recent studies have suggested that most emotions encompass elements of both positivity and negativity [7], which results in ambiguity and conflicts in interpreting emotions [8], rendering the task of capturing and understanding emotions highly challenging. On the other hand, emotional requirements capture how users should feel when using the system [6,9]. For example, a user might want to feel motivated and empowered while avoiding feelings of anxiety, stress, or frustration during system interaction. Accordingly, inadequate consideration of such requirements can result in end-users reluctance to use the system [3]. Specifically, even a flawlessly functional system might face rejection if it fails to resonate with the emotional needs of its users [3].

Emotional requirements/needs can be highly relevant to other functional and quality requirements, i.e., they can influence and/or can be influenced by several functional and quality requirements. In this context, capturing and addressing the emotional requirements of end users during the early phases of a system design is essential for a successful system. This also improves system acceptance and usability on the users' side [3]. However, this is not an easy task as existing approaches for dealing with emotional requirements (e.g., [3,4]) are relatively scarce, they capture emotional requirements a high level of abstraction and predominantly focus on elicitation, as revealed by a recent mapping study [6]. Moreover, available approaches were not designed to fit the needs of ecosystems, which are large software systems/platforms that consist of various, constantly interacting, and partly autonomous subsystems [10]. Ecosystems have a considerable number of stakeholders such as internal and external developers, domain experts, and various types of end-users, which makes specifying their requirements a very challenging task [11].

In this paper, we report on our experience in dealing with emotional requirements within the PHArA-ON ecosystem under the larger context of the PHArA-ON project. Our primary contribution lies in presenting a comprehensive process tailored specifically for handling emotional requirements. The process is composed of several activities accompanied by rules and suggestions that need or are advised to be followed respectively. This guarantees that the activities will be conducted while leaving sufficient flexibility to cope with any emergent situations. We aimed to create an easily replicable process, providing detailed descriptions for each activity to facilitate seamless adoption by others. The process has been followed across six different pilot[1] sites for capturing[2], prioritizing, and validating/consolidating the emotional requirements for several types of users including older adults, formal and informal caregivers. Additionally, we describe

[1] A pilot study is a small-scale test of a method to be used on a larger scale.
[2] Information on how each pilot has elicited its requirements can be found in [12].

how the process has been used to analyze the requirements for the PHArA-ON ecosystem, and we briefly discuss how such requirements were used to derive design solutions and the proposed methods for their evaluation.

The rest of the paper is organized as follows; we briefly describe the PHArA-ON project in Sect. 2. Section 3 describes the process for dealing with emotional requirements, and Sect. 4 discusses how the process has been applied to deal with the requirements for the PHArA-ON ecosystem with special emphasis on emotional requirements. In Sect. 5, we present our findings and lessons learned, and related work is presented in Sect. 6. Finally, we conclude the paper and discuss future work in Sect. 7.

2 The PHArA-ON Project

Given the preference of older adults to remain in their own homes, compounded by the scarcity of caregivers and the expenses associated with home care nursing, there is an increasing demand for smart solutions to assist the well-being and active aging of older adults. With this in mind, the main aim of PHArA-ON[3] (Pilots for Healthy and Active Ageing in Europe), an H2020 European project, is to actualize smart and active living for Europe's aging population, and make such smart and active living a reality [12]. To achieve this ambitious goal, the PHArA-ON consortium has pinpointed several critical challenges (presented in Table 1) that need to be addressed. Addressing these challenges entails leveraging advanced services, devices, and tools to forge an integrated and highly customizable interoperable open platform-the PHArA-ON ecosystem. This ecosystem will be crafted by harnessing platforms, technologies, and tools provided by consortium members or developed by third parties through open calls.

The primary users of the PHArA-ON ecosystem encompass older adults, healthcare professionals, and both formal and informal caregivers. Accordingly, a user-centric approach [13] has been followed for developing the PHArA-ON ecosystem, which will equip its users with the right tools and technologies that enable them to utilize different services from different providers focusing on well-being, socialization, training, etc. In this context, a pivotal success determinant for the PHArA-ON ecosystem lies in capturing the requirements of its users and stakeholders. These requirements will be used as a foundation for defining the core functionalities and qualities of the ecosystem. Emphasis is particularly placed on emotional requirements, recognizing their critical role in ensuring the usability and acceptance of the PHArA-ON ecosystem. These requirements have been specified, mainly, by six different pilots (Spain (Murcia and Andalusia), Portugal, the Netherlands, Slovenia, and Italy), where each pilot concentrates on a subset of the identified challenges. The PHArA-ON ecosystem will be tested and validated in two stages: pre-validation and large-scale pilots (LSPs) validation within these six pilots. These comprehensive validation phases aim to ensure the efficacy and readiness of the ecosystem for broader implementation.

[3] https://www.pharaon.eu/.

Table 1. PHArA-ON key Challenges (PCH)

ID	Description
PCH1	The behavior and the approach of elderly to friendly technological devices
PCH2	Health status definition and its progress over time
PCH3	Non-intrusive monitoring and alarm triggering
PCH4	Promote social cohesion
PCH5	Define specific personalized care plan on the basis of user's needs
PCH6	Reduce loneliness and enhance autonomy through connectivity and digital tools
PCH7	Promote accessibility and provision of proximity services through IT platforms
PCH8	Promote capacity building and awareness on citizenship and cultural traditions
PCH9	Indoor environmental quality
PCH10	Support to caregivers toward more efficient and personalized care services

3 A Process for Engineering Emotional Requirements

The process for engineering the emotional requirements (depicted in Fig. 1) consists of four main interrelated activities that have been constructed based on the human-centered design approach proposed in the ISO 9241 [13]:

1. Specifying the context of use: aims at identifying relevant users[4] and stakeholders of the ecosystem; their characteristics; their goals and tasks as well as the characteristics of tasks that can influence usability and accessibility. Consequently, two sub-activities must be conducted: *1.1 Users & stakeholders analysis:* aims at understanding the overall scope of the ecosystem by identifying all users and stakeholders that may influence or be influenced by the ecosystem, and classifying them into coherent groups to better identify their needs and expectations concerning the ecosystem. *1.2 Users & stakeholders goals, tasks and expectations analysis:* aims to identify the goals and expectations of the identified groups concerning the services to be provided by the ecosystem. A pre-analysis of the target ecosystem and an analysis of existing/similar systems/ecosystems along with other sources can be used as input to specify key user and stakeholder groups as well as their goals. These sub-activities produce the *context-of-use description*, which should be described in sufficient detail to support the requirements, design, and evaluation activities.
2. Specifying users and stakeholders requirements (USRs): In a software ecosystem, the integration of disparate software components or subsystems, each developed by distinct entities, presents a series of interconnected challenges, herein termed as *problems*. The entities tasked with eliciting the requirements to address these challenges are referred to as *Consortium Members (CMs)*. This activity comprises several sub-activities:

2.1 Identifying CMs relevant problems: aims at assigning key ecosystem *problems* to different *CMs* based on their characteristics and the *problem's* nature. Specifically, each CM should identify the stakeholders and users relevant

[4] Users are a subset of stakeholders, we differentiate between the two as users are the main source for eliciting emotional requirements.

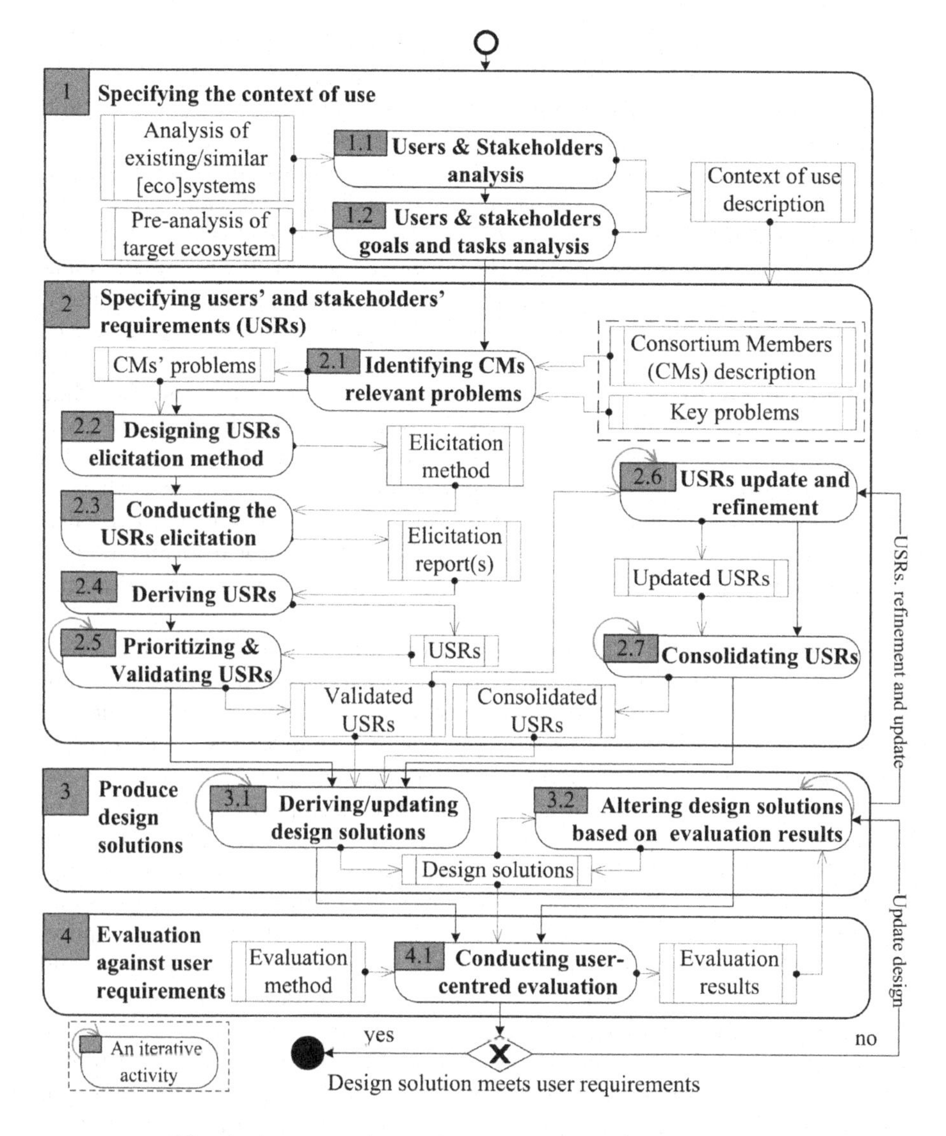

Fig. 1. A process for engineering emotional requirements

to its assigned problems, then, elicit requirements related to these problems. This activity should ensure that all identified problems are assigned to CMs.

2.2 Designing user and stakeholders requirements (USRs) elicitation method: Following the identification of their respective problems, each CM is tasked with devising an appropriate method for eliciting USRs. We recommend co-creation/co-design workshops/sessions aligned with the motivational goal modeling technique [3], which produces the following four lists: *Do* (functional goals), *Be* (quality goals), *Feel* (emotional goals), and *Who* (stakeholder roles). However, any traditional elicitation methods (e.g., interviews, surveys) can be used if adapted to produce the aforementioned four lists.

2.3 Conducting the USRs elicitation: Once the USRs elicitation method is established, the method should be strictly followed during the elicitation of USRs. While each CM takes charge of designing and executing the USRs elicitation specific to their designated problem(s), the resulting elicitation report must adopt a standardized format across all CMs. This standardization facilitates the derivation of comprehensive USRs for the entire ecosystem. It also aids in identifying and reconciling potential conflicts among these requirements. Ultimately, this uniformity streamlines the process of deriving design solutions that can be easily integrated.

2.4 Deriving USRs: The elicitation reports obtained from the preceding activity contain the initial version of the USRs in terms of three sets of goals (functional, quality, and emotional), which might lack consistency, or missing inter-relationships among these goals. Consequently, these issues are dealt with in this activity, and the reports should be carefully checked and used to elicit USRs for the overall ecosystem and present them in a uniform format (e.g., a well-structured unified template is highly advised) that allows their review and validation.

2.5 Prioritizing and Validating USRs: Aims to gather user and stakeholder feedback on the compiled list of USRs to effectively prioritize and validate them. The selection of prioritization techniques can vary based on the system/ecosystem's characteristics. Techniques such as Analytic Hierarchy Process (AHP), Cumulative Voting, Ranking, Ten Requirements, or Numerical Assignment (Grouping) can be chosen to suit the system's specific attributes. The requirements validation process encompasses three key activities outlined in [1]: 1. *Completeness checks:* Ensure that the requirements comprehensively capture all system functionalities, properties, constraints expected by the users and stakeholders; 2. *Consistency checks:* Verify the absence of conflicting requirements, ensuring there are no contradictions or discrepancies among them; and 3. *Realism checks:* Verify that the requirements can be implemented.

2.6 USRs refinement and update: Occurs after activity *3* and aims at refining and updating the USRs list based on the received feedback, which may suggest relaxing or refining some USRs to better fit the users' needs, modified to better match the technology chosen to realize them. It is important to note that *2.6* is an iterative activity that might be repeated several times until the USRs list attains a level where further refinement is unnecessary.

2.7 Consolidating the final list of USRs: It takes the updated USRs list that resulted from the *2.6* activity as input, then, applies the three checks we used in *2.5* (*completeness, consistency, and realism check*) to them, which will result in the final consolidated USRs list.

3. Producing design solutions: It is comprises of two sub-activities: *3.1 Deriving design solutions:* that takes the validated/consolidated USRs and derives the reference architecture of the overall ecosystem in terms of its main components. *3.2 Altering design solutions based on evaluation results:* is needed when the

evaluation results show that design solutions do not meet USRs. In such cases, this sub-activity tries to refine/alternate design solutions to meet USRs.
4. Evaluating the design: In ISO 9241 [13], two widely used methods of user-centered evaluation can be employed: 1- *user-based testing* can be carried out to assess whether usability objectives, including measurable usability performance and satisfaction criteria, have been met in the intended context(s) of use. 2- *inspection-based evaluation* is ideally performed by usability experts who put themselves into the role of the user. Regardless of the adopted evaluation method, a clear evaluation and acceptance criteria for the evaluation should be defined.

4 Applying the Process for Engineering the Emotional Requirements for the PHArA-ON Ecosystem

This section provides a detailed description of how the process has been used to engineer the requirements for the PHArA-ON ecosystem with special emphasis on emotional requirements.

1. Specifying the context of use: started with *1.1 Users & stakeholders analysis,* in which, we depend on the PHArA-ON proposal and analysis of existing/similar [eco]systems as input to identify key users and stakeholders groups. These encompassed older adults, healthcare professionals, formal and informal caregivers. Following this, we conducted *1.2 Users & stakeholders goals, tasks & expectations analysis*, in which, we analyzed each of these groups to identify their goals and expectations concerning the services to be provided by the PHArA-ON ecosystem. Both of these sub-activities contributed to the *context-of-use description*, which identifies, besides user and stakeholder groups and objectives, the physical, technical, and social aspects of the overall ecosystem. This detailed description aids in delineating the ecosystem boundary and defining its 'sub-systems', along with their respective boundaries and interdependencies. Such clarity is essential to facilitate seamless integration among these 'sub-systems'.
2. Specifying users and stakeholders requirements (USRs): consists of the following sub-activities:

2.1 Identifying pilots' relevant challenges: Drawing from the *context-of-use description*, the PHArA-ON CHallenges (PCH), representing the *problems*, have been allocated to different pilots, which represent the *Consortium Members (CMs)* responsible for eliciting requirements relevant to these PCHs. Due to space constraints, our focus will be directed solely toward the activities of one pilot (Andalusia), which has been assigned two PCHs: *PCH4* and *PCH6.*

2.2 Designing User Requirements (USRs) Elicitation Method: The intended approach for eliciting user requirements within the PHArA-ON project involved co-creation/co-design workshops/sessions [14], conducted in the format of motivational goal model workshops [3]. However, due to the prevalence of the COVID-19 virus, only the Netherlands pilot conducted a physical co-design workshop. The remaining pilots resorted to virtual co-design/co-creation sessions, alongside phone and online interviews. Still, all these elicitation methods were adapted to produce the same output prescribed in [3], consisting of Do/Be/Feel/Who lists, which are then structured as motivational goal models.

Regarding the Andalusian pilot, initially, three co-design workshops were scheduled, involving care professionals, older adults with varying levels of technology experience, and those without such experience. However, due to the COVID-19 outbreak, these workshops had to be transformed into online sessions. They were restructured as individual semi-structured phone interviews conducted with the same three groups, all on a voluntary participation basis. The interviews with older adults were structured into several sections: 1- Obtaining informed consent and providing general project information, and the specific challenges within the pilot, 2- Collecting socio-demographic information, 3- Exploring their typical daily routine, 4- Discussing their interaction with technology, 5- Discussing how technology could enhance their everyday experiences, and 6- a pool of questions (a subset is shown in Table 2) for eliciting their requirements (*functional*, *quality* and *emotional* goals). Interviews with professionals consist of the same sections but are adapted to professionals' points of view.

2.3 Conducting the USRs elicitation: As previously mentioned, regardless of the selected elicitation method, it should produce the following four lists: *Do* (functional goals); *Be* (quality goals); *Feel* (emotional goals); and *Who* (stakeholder roles). Followingly, Do/Be/Feel/Who lists, representing the *elicitation report*, were hierarchically structured as motivational goal models [3].

Table 2. Partial phone interview guide - Goals Model section - Andalusia

TO DO Goals - (Functional goal) - *What it should do?*
What hobbies would you like to share with other people?
What kind of activity reminders would you like the solution to have?
What kind of entertainment would you like the system would offer you?
What kind of people would you like to be connected with?
TO BE Goals - (Quality Goal) - *How it should be?*
How should the information be represented in the system (e.g., text, voice)?
Which type of messages best fit your interest?
Would you like to see other people? Or just hear them?
Would you like the system to take your preferences into account?
TO FEEL Goals - (Emotional goal) - *How it should feel?*
How would you feel if you could talk to people you've never seen/heard before?
How would you feel if the system informed you of social/cultural events?
How would you feel with a device that allows you to connect with similar people and propose meetings/appointments?
Are there any negative emotions that might be related to using a technology to meet people? What fears do you have?

In motivational goal model, *goals* represent the objectives/aims of [eco]system users, and it can be *functional*, *quality*, or *emotional*. *Functional goals* are based on motives, and describe an intended state of affairs a performer of a particular role wants to/can achieve. A *functional goal* can consist of sub-goals, where these sub-goals contribute to achieving the higher-level goal. *Quality goals* and *emotional goals* are attached to *functional goals*, where the first describe required *qualities* that should be maintained, and the last describe desired emotions of relevant *role*, which represents a user (group of users) of the system that can be characterized by capacities and responsibilities that are needed for the achievement of functional goals.

Concerning the Andalusian pilot, the result of the interviews carried out with different users and stakeholders groups to specify their needs have been constructed in the form of a motivational goal model (Fig. 2). Specifically, the model is developed starting with specifying the top-level functional goal, which represents the main objective/goal the system needs to achieve. Then, the top-level goal is refined into sub-goals, which in turn, can be also refined into more refined sub-goals. After that, the roles, quality goals, and emotional goals are attached to relevant functional goals. The main functional goal of the motivational goal model, representing the purpose of the sociotechnical system - Support wellbeing - has been elaborated into three sub-goal: *Improving digital skills*, *Participating in the community*, and *Providing cognitive stimulation*, which are characterized by several quality and emotional goals. Each of the three functional goals has been further refined into three additional motivational goal models, which we could not include in this paper.

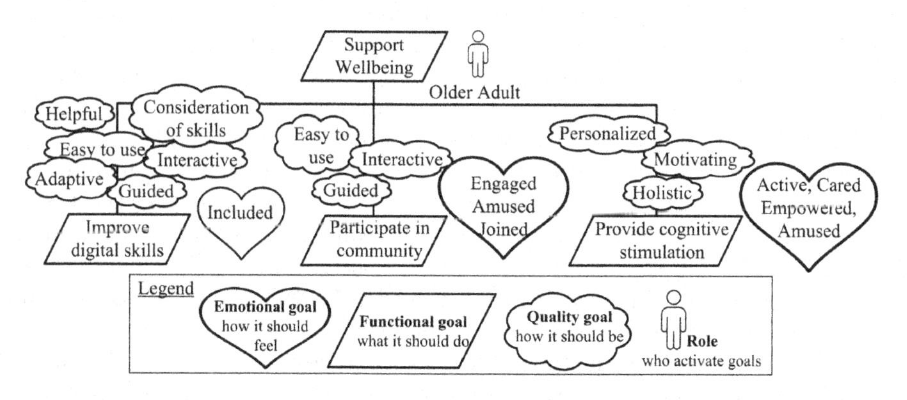

Fig. 2. The motivational goal model for the Andalusian pilot of supporting well-being of older adults

2.4 Deriving USRs: The motivational goal models generated in the preceding stage have been carefully checked and used to elicit the USRs and represent them as use case scenarios. A standardized format in the form of tables was adopted for uniformity across all pilots. An example of these scenarios related to the Andalusian pilot is presented in Table 3. Each scenario is described by

Table 3. Improving digital skills Scenario - Andalusian Pilot

Scenario 1					
Goal	Improve digital skills				
Initiator	Volunteer				
Trigger	An action by the volunteer				
Description	allows older adults to train their digital skills to promote the use of technology to facilitate digital inclusion and eradicate digital divide in Andalusia				
Condition	Step	Activity	Roles	Quality goals	Emotional goals
Sequence	1	The older adult provides information about his or her digital skills and education level for assessing the digital skills	Older Adult, Volunteer	Interactive, Easy to use, Guiding, Helpful	Included, Amused, Empowered
	2	A volunteer schedules offline classes for the older adult			
-	-	-	-	-	-

a *number* for identification purposes, a *goal* corresponding to one of the key functional goals identified in the motivational goal model, an *initiator* that acts to trigger the scenario, and a *description*. Each scenario is composed of several *activities*, described in the *activity* field, and may have a *condition* that can be: *interleaved* means a set of activities may be performed in any order; *sequence* means a set of activities need to be performed one after another, and *optional* when performing an activity is not mandatory. *Steps* play a role when activities have a sequence to follow. The fifth and sixth columns respectively list *quality goal(s)* and *emotional goals(s)* represented by the motivational goal model.

2.5 Prioritizing and Validating USRs: Following, the numerical assignment prioritization technique, the USRs was structured around three priority levels: *priority 1* indicates a requirement that needs to be implemented, but at the moment it has a low priority for the given pilot; *priority 2* indicates a requirement that needs to be implemented, even if it is not the first priority of the given pilot; and *priority 3* indicates a requirement that has the maximum priority for the given pilot. While various methods were employed across pilots to prioritize their USRs, adherence to these predefined priority levels was mandatory.

Regarding the *validation of USRs*, this process was overseen by a member of the PHArA-ON consortium, the University of Tartu. After collating the USRs lists from all pilots in the specified format, the validation followed a structured approach described in [1], comprising three key activities: *1. completeness check* was conducted by checking whether the lists of USRs identified by pilots capture all the functionalities and characteristics (qualities and emotions) the ecosystem is expected to deliver. During this check, we contacted several pilots for additional information and clarifications when there were issues concerning missing USRs. After that, we asked each pilot to check the elaborated list of USRs again. Some pilots asked to add new USRs that were not included in the list, refine some existing USRs, and even delete some USRs as they were considered out of the ecosystem scope or covered by more refined USRs. When the overall USRs list was considered complete concerning the users' and stakeholders' needs, we conducted the *2. consistency checks* the objective of which was to identify and resolve conflicts among the USRs, where a conflict is a situation when two or more requirements cannot be achieved together. This check was able to detect some conflicts, which we managed to resolve by revising the conflicting USRs with the help of corresponding pilot(s). After the overall USRs list was verified as consistent, we conducted the *3. realism checks* mainly with pilots together with technology providers. As previously mentioned, USRs were accompanied by potential implementation technologies, which facilitated this check. No significant issues were identified due to the nature of the PHArA-ON project - being an Innovation Action - to combine and integrate technological solutions rather than developing new solutions from scratch.

2.6 USRs update and refinement: occurred after activity *3*, where the project made several progresses. Notably, a comprehensive analysis of the overall PHArA-ON ecosystem was conducted, concretizing its constituent components, and defining its reference architecture. Additionally, the pilots significantly enhanced their understanding of the anticipated services and technologies. Consequently, the initially specified USRs had to be updated to reflect these advancements. Therefore, each pilot was tasked with identifying any modifications (e.g., additions, alterations, removals of requirements) to their initial USRs. Primarily, the changes centered around technology specifications-either due to certain technologies failing to meet project requirements or the consideration of newer, more suitable alternatives. Furthermore, some modifications involved refining USRs with more specific, detailed information, alongside updates in the requirement priorities. Additionally, we transformed USRs into the user stories format [15], as such format is easy to specify, understand, revise, and most importantly, it facilitates the communication between technical and non-technical stakeholders, i.e., it allows cross-team clarity on what to develop, for whom, and why. We suggest the following format of user stories[5]:

> As a ⟨performer of some Role⟩, I want/shall be able to perform ⟨some action⟩, using ⟨some technology⟩, to achieve ⟨some goal⟩[, with the consideration of ⟨some quality goal⟩, so that achieving ⟨some emotional goal⟩]

Table 4 presents an updated USR in the user stories format. To facilitate traceability, all IDs of the USRs use the following format: PUCS_-

Table 4. An example of an updated USR as a user story - Andalusian Pilot

Req. ID	Requirements as a user story	Priority
PUCS_A01 .1.001	As an ⟨older adult⟩, I ⟨should be able to provide information about my digital skills and education level⟩, using ⟨an App deployed in a Tablet⟩, to achieve ⟨improve digital skills⟩ with the consideration of ⟨ guided and ease of use ⟩ so that achieving ⟨ amused and empowered ⟩	2

[5] The inclusion of quality and emotional goals in user stories is optional (indicated by square brackets [..]) since not all user goals have associated quality/emotional goals.

PilotIDScenarioID.ActivityNumber.USRID, where the PUCS prefix stands for the initials of PHArA-ON Use Case Scenario, and PUCS_A01.1.001 means that the user story was elicited by the Andalusian pilot (A), from the first Scenario (01) and the first activity in that scenario (1), with ID 001.

2.7 Consolidating the final list of USRs: We took the list of USRs as user stories as an input, then, we rigorously applied the same three validation checks (*completeness, consistency, and realism checks*) utilized in *2.5*, where pilots were asked to check and verify the USRs again to ensure that there were no missing or wrong details introduced during the transformation process. The received feedback contains suggestions to revise several USRs, which were corrected accordingly. The outcome of this activity has been used to update the design solutions that have been initially derived based on the validated USRs resulting from activity *2.5*.

3. Producing design solutions: is composed of two sub-activities: *3.1 Deriving design solutions:* in the initial iteration following Activity *2.5*, we outlined the reference architecture (high-level design solutions) of the PHArA-ON ecosystem based on the validated USRs. As the PHArA-ON ecosystem will, mostly, consist of a series of software components that will be delivered by the PHArA-ON partners or via open calls, this activity mapped USRs to these components, and when there is no available component to realize a USR, such USR was marked as to be realized by solution/technology to be provided by third-parties via an open call. Subsequently, in the second iteration after Activity *2.7*, the reference architecture was updated based on the refined and consolidated USRs. Here, the open calls resulted in solutions/technologies sourced from third-party providers. An illustrative example of USR-to-design solution mapping is detailed in Table 5. Please not that the mapping was done for the functional requirements, and relevant quality and emotional requirements were derived from the related scenario using the Scenario ID. *3.2 Altering design solutions based on evaluation results:* will refine design solutions based on the received evaluation results.

4. Evaluating the design: following ISO 9241 [13], *user-based testing* has been adopted as the evaluation method, where developed solutions are evaluated against USRs. This activity covered all USRs, yet we focus on emotional requirements due to space limitations. The following evaluation method has been drafted, and is followed when assessing the satisfaction of emotional requirements. *Evaluation method:* pilots will analyze the list of the emotional goals identified in the consolidated USRs, then, each pilot will decide whether a *quantitative* and/or *qualitative evaluation method* is appropriate for collecting feedback

Table 5. An example of mapping a USR to design solutions - Andalusian Pilot

Scenario ID	Activity	Functional Requirement	Technologies
PUCS_A01	Participant's digital skills profile	Older adult should be able to fill in a template user profile on the tablet	SentabTV, OneSait
		Interactions of older adult with the tablet and the platform should be facilitated by "text to speech"	

from end-users to allow the assessment of the satisfaction of emotional requirements. Concerning the *quantitative evaluation*, two types of questionnaires have been designed: 1- *Standard questionnaire:* that will include the most appropriate tools for the assessment of emotional requirements.

The identification of these tools was done by investigating various domains to select the most relevant tools/measures for the assessment of emotional requirements depending on their nature. For example, PSS (Perceived Stress Scale) [16] was selected to assess *Relaxed & Less stressed*, Item 9 from the System Usability Scale (SUS) [17] was selected to assess *Confidence*, UCLA Loneliness Scale [18] was selected to assess *Connected & Involved*, MSC (Most Significant Change) [19] was selected to assess *Engaged, Integrated & Stimulated*, and Lubben Social Network Scale [20] was selected to assess *Engaged & Involved.* 2- *Ad-hoc questionnaire:* designed to collect feedback from end-users taking into consideration the emotional requirements included in the USRs. A partial ad-hoc questionnaire is shown in Table 6. Concerning the *qualitative evaluation*, focus groups/individual interviews that represent target user groups are used to verify how the satisfaction of emotional requirements were perceived.

5 Findings and Lessons Learned

Through using our process for dealing with the requirements for the PHArA-ON ecosystem with special emphasis on emotional requirements, we have faced several challenges related to the different activities we performed. In what follows, we summarize our findings, suggestions, and lessons learned:

Considering co-creation workshops (or co-design sessions) for the elicitation of emotional requirements: co-creation approaches empower users and stakeholders while specifying their needs/requirements, especially, the emotional ones. Specifically, they give users a voice and treat them as active

Table 6. A partial ad-hoc questionnaire for assessing emotional requirements

Domain		User Code:				
Open Question		How do you feel using the PHArA-ON System?				
After using the PHArA-ON system, do you feel (answer from 1 to 5):						
Emotion	Description	S. disagree	Disagree	Neutral	Agree	S. agree
Stressed	Less stressed by the usage of the Technology?	1	2	3	4	5
Informed	More aware of your health situation?	1	2	3	4	5
Empowered	Do you have increased opportunities for socialization?	1	2	3	4	5

contributors. This increases their involvement in the requirements identification process, which is a key success factor for the adoption and usage of the resulting ecosystem as it will shape it based on their needs/requirements. However, successful co-creation workshops require the involvement of all key relevant users and stakeholders.

Capturing emotional requirements: is usually done relying on traditional psychological methods such as self-reporting and observation-based, which heavily rely on participant cooperation and are subject to human biases [8]. However, recent technologies such as wearable technologies, if available, could overcome such limitations allowing direct collection of affective states representing emotions.

Considering motivational goal models as the artifact resulting from co-creation workshops: a co-creation workshop results, usually, in visual artifacts and documentation concerning the [eco]system to be developed. Such artifacts/documentation should facilitate capturing knowledge generated during the workshop and documenting it in a way that allows its understanding and review by involved entities. Accordingly, using motivational goal models as an artifact to capture, present, and document users' (and stakeholders') needs is highly advised. Using motivational goal model in the PHArA-ON project has proven that it is easy to develop, understood, and very suitable for facilitating communications between technical and non-technical stakeholders.

Considering dependencies between functional, quality, and emotional requirements: motivational goal models can explicitly capture such dependencies, yet most other goal-modeling languages do not. Therefore, if a language other than motivational goal models was adopted, it is strongly advised to extend it with such dependencies.

Considering user stories (US) as a [final] format for requirements: US mostly capture user-visible functional requirements, therefore, they face various challenges when dealing with quality and emotional requirements, as emotional requirements are first-class citizens and not subsumed under NFRs. We solve this problem by proposing a flexible template that allows for explicitly capturing the relationship between functional requirements and their associated emotional and quality requirements, which was essential to be captured in many cases. Specifically, this will ensure that emotional requirements will not be overlooked in the final solution.

Mapping requirements to implementation technologies: Most ecosystems are built by modifying/extending software solutions, then, integrating them rather than building new solutions from scratch. The PHArA-ON ecosystem is no exception, thus, we have mapped the requirements to existing software solutions provided by the consortium members or by third parties. This facilitate performing the *realism check* while validating/consolidating the requirements and, in turn, deriving/updating the reference architecture for the PHArA-ON ecosystem. Concerning emotional requirements, they were not mapped directly to implementation technologies since most of them were not expected to be realized by a single technology but a set of them.

Managing the validation of emotional requirements: Unlike functional and many types of quality requirements, there are no agreed-upon methods for validating emotional requirements [6]. Consequently, we have to devise a method for the validation of emotional requirements that considers both *quantitative* and/or *qualitative evaluation methods*, which have been constructed based on the pilots' experience as well as adopting various well-known tools to be used for the assessment of emotional requirements.

6 Related Work

There are relatively few approaches for dealing with emotional requirements. For instance, Miller et al. [3] propose preliminary work on including emotions in requirements models using emotional goals. Their work was evaluated in a controlled user study, and by applying it to an emergency system for older adults case study. Grundy et al. [9] proposed a model-driven approach to model various human-centric aspects/requirements including emotions. The authors plan to extend their work to cover code generation relying on these models in their future work. Iqbal et al. [4] propose a methodology compatible with the theory of constructed emotion for designing and developing emotive software artifacts. Apart from requirements, Seligman [21] proposes the PERMA (Positive emotions, Engagement, positive Relationships, Meaning, and Accomplishments) framework that offers elements required to facilitate well-being. According to the author, PERMA is not an exhaustive framework, and it could be expanded.

7 Conclusions and Future Work

We introduce an RE process specifically tailored for handling emotional requirements. The process has been developed within the context of the PHArA-ON project and has been successfully used for dealing with the emotional requirements of the PHArA-ON ecosystem. We aimed to make the process easy to follow, and we accompanied each of its activities with a detailed description of how it can be conducted. On the other hand, this research has uncovered several critical research gaps, laying the groundwork for potential future work. For instance, motivational goal models capture emotional requirements at a high abstract level, and lack a systematic process for their refinement. They also overlook various types of dependencies among requirements (e.g., requires), which we consider essential for appropriately dealing with emotional requirements. Finally, dealing with emotional states might have ethical implications, which we aim to investigate in future research. An area of concern in our process involves the potential threat to validity due to the *generalization of findings* validity since it was applied to one ecosystem. However, we intend to validate the process further by applying it to various case studies spanning different domains.

Acknowledgment. This research received funding from the EC via the Horizon 2020 project 'Pilots for Healthy and Active Ageing' (Pharaon), Grant agreement no. 857188

References

1. Sommerville, I.: Software Engineering. Pearson Education Limited (2007)
2. Loucopoulos, P., Karakostas, V.: System Requirements Engineering. McGraw-Hill, Inc. (1995)
3. Miller, T., Pedell, S., Lopez-Lorca, A.A., Mendoza, A., Sterling, L., Keirnan, A.: Emotion-led modelling for people-oriented requirements engineering: the case study of emergency systems. J. Syst. Softw. **105**, 54–71 (2015)
4. Iqbal, T., Marshall, J.G., Taveter, K., Schmidt, A.: Theory of constructed emotion meets RE: an industrial case study. J. Syst. Softw. **197**, 111544 (2023)
5. Frijda, N.H., Kuipers, P., Schure, E.: Relations among emotion, appraisal, and emotional action readiness. J. Pers. Soc. Psychol. **57**(2), 212–228 (1989)
6. Iqbal, T., Anwar, H., Filzah, S., Gharib, M., Mooses, K., Taveter, K.: Emotions in requirements engineering: a systematic mapping study. In: International Conference on Cooperative and Human Aspects of Software Engineering, pp. 111–120. IEEE (2023)
7. An, S., Ji, L.J., Marks, M., Zhang, Z.: Two sides of emotion: exploring positivity and negativity in six basic emotions across cultures. Front. Psychol. **8**, 1–14 (2017)
8. Gharib, M.: Emotions, readiness for act, and safe/unsafe acts in safety critical systems: a position paper. In Proceedings of the 31st IEEE International Requirements Engineering Conference Workshops, REW 2023, pp. 12–15 (2023)
9. Grundy, J., Khalajzadeh, H., McIntosh, J.: Towards human-centric model-driven software engineering. In: the International Conference on Evaluation of Novel Approaches to Software Engineering, pp 229–238 (2020)
10. Knauss, A., Borici, A., Knauss, E., Damian, D.: Towards understanding requirements engineering in IT ecosystems. In: International Workshop on Empirical Requirements Engineering, EmpiRE, pp 33–36. IEEE (2012)
11. Bosch, J.: From software product lines to software ecosystems. In: Proceedings of the International Software Product Line Conference, pp. 111–119 (2009)
12. Mooses, et al.: Involving older adults during COVID-19 restrictions in developing an ecosystem supporting active aging: overview of alternative elicitation methods and common requirements from five European countries. Front. Psychol. **13**, 676 (2022)
13. ISO: Ergonomics of human-system interaction - Part 210: Human-centred design for interactive systems (2019)
14. Nielsen, L.: Personas in co-creation and co-design. In: Proceedings of the Danish Human-Computer Interaction Research Symposium, pp. 38–40 (2011)
15. Fallis, A.G.: User Stories Applied for Agile Software Development, vol. 53. Addison-Wesley Professional (2013)
16. Cohen, S., Kamarck, T., Mermelstein, R.: A global measure of perceived stress. J. Health Soc. Behav. **24**(4), 385–396 (1983)
17. Brooke, J.: SUS: a 'Quick and Dirty' usability scale. Usab. Eval. Industry **189**(194), 207–212 (1996)
18. Russell, D.W.: UCLA loneliness scale (version 3): reliability, validity, and factor structure. J. Pers. Assess. **66**(1), 20–40 (1996)
19. Davies, J., Dart, R.: The 'most significant change' (MSC) technique. TR (2005)
20. Lubben, M., Gironda, J.E.: Social support networks. Comprehensive geriatric assessment, pp. 121–137 (2000)
21. Seligman, M.: PERMA and the building blocks of well-being. J. Positive Psychol. **13**(4), 333–335 (2018)

A Tertiary Study on Quality in Use Evaluation of Smart Environment Applications

Maria Paula Corrêa Angeloni[1](✉), Rafael Duque[2], Káthia Marçal de Oliveira[1], Emmanuelle Grislin-Le Strugeon[1,3], and Cristina Tirnauca[2]

[1] UPHF, CNRS, UMR 8201 - LAMIH, Valenciennes, France
mariapaula.correaangeloni@uphf.fr
[2] Depart. de Matemáticas, Estadística y Computación, Universidad de Cantabria, Santander, Spain
[3] INSA Hauts-de-France, Valenciennes, France

Abstract. As the population grows older, the need for special assistance increases, and a modern alternative to mitigate the absence of face-to-face caregivers (which is expensive) is to take advantage of technological devices in so called smart environments, which can be an economical and practical solution. Guaranteeing the software quality of applications in these spaces before providing it to end users is essential, especially in situations involving senior citizens or people with motor disabilities. In order to investigate how the quality evaluation of smart environment applications has been performed, we carried out a tertiary study. From a total of 1,028 studies, 21 were carefully selected for analysis. The results confirmed that classical questionnaires and interviews are the techniques that are still used the most for evaluation, but that simulation appears as a new trend to that end.

Keywords: Quality in Use · Smart environments · Literature review

1 Introduction

Smart environments have been considered as an appropriate living alternative for both ageing safely at home and also receiving care as needed [16]. A smart home can be described as a "residence wired with technology features that monitor the well-being and activities of their residents to improve overall quality of life, increase independence and prevent emergencies" [13]. Ensuring the quality of these systems is essential if they are to be effectively utilized: without a solid reference on user evaluation, a system evaluation cannot be tackled well [30].

The quality of a system application is defined as "the degree to which the system satisfies the stated and implied needs of its various stakeholders and thus provides value" [26]. These needs are represented in the SQuaRE International Standards by two quality models: (i) *a product quality model* made

J. Araújo et al. (Eds.): RCIS 2024, LNBIP 513, pp. 115–130, 2024.
https://doi.org/10.1007/978-3-031-59465-6_8

up of characteristics (such as functional suitability, performance efficiency and maintainability) related to the static properties of the software and the dynamic properties of the computer system; and (ii) a *Quality-in-Use (QinU) model* made up of characteristics (such as effectiveness and efficiency) related to the outcome of the human-computer interaction when a product is used in a specific context of use. While the first model assesses the quality of the product itself, the QinU perspective assesses the effect of the interaction between the user and the software, taking into consideration the user experience (UX). We focused this study on QinU, as our main intention is to evaluate the quality of software used in smart environment contexts.

Given our main interest in evaluating QinU for software applications in smart homes that are usually inhabited by elderly people, we wondered how these evaluations have been carried out and whether the classic user evaluation sessions have been applied. The QinU evaluation of smart environment applications involves placing users in the smart environment (in this case, smart homes) for evaluation sessions due to the array of sensors that usually capture information for this type of application. Evaluations like that may be difficult or even unsafe for the elderly and/or people with reduced mobility; therefore, we should secure the QinU even before carrying them out. Aiming to find out how researchers are addressing this problem and which approaches have been applied for evaluating smart home applications that could address this situation properly, we decided to carry out a tertiary study [12] by rapid review [45].

The rest of this paper is organized as follows: Sect. 2 briefly presents basic concepts of QinU. Section 3 describes the research protocol and execution procedures of the study. Then, Sect. 4 discusses the results of the performed review and Sect. 5 the threats to the validity of this study. Finally, Sect. 6 presents some final remarks and our ongoing work.

2 Background

QinU considers how much a software can address the user needs in a specific context of use and it is divided into five categories [26]: *effectiveness*, for how accurately users can perform their intended tasks; *efficiency*, for how easily and fast such tasks can be accomplished; *satisfaction*, for how pleased the users are with the product; *freedom from risk*, for how much such product lessens potential risks related to either the environment, economic status or humans' health; and *context coverage*, for the degree to which the product can be used with the four previous characteristics in specified contexts of use (context completeness) and also in contexts beyond those initially explicitly identified (flexibility).

When we talk about evaluating interactive systems, we immediately turn to usability issues, as presented in the ISO 9241-11 standard [21]. Usability is defined as "the degree to which a product can be used, by identified users, to achieve defined objectives with effectiveness, efficiency and satisfaction, in a specified context of use". Effectiveness, efficiency and satisfaction, which are the tripod of the ISO/IEC 9241-11 definition, are present in the QinU model of the SQUaRE standard [26].

The evaluation of QinU and usability issues has been largely explored in literature with methods and applications. We can quote, for instance: mathematical simulations (e.g. [10]), where the authors establish values and apply them to mathematical formulas for validation, or agent-based simulations (e.g. [8]), to emulate humans' behaviour. One of the most common ways of evaluating QinU and usability is through questionnaires that can be answered by end users (e.g. [38]) or by domain experts (e.g. [40]). In such situations, users must either interact with the evaluated application or have it demonstrated to them before answering a set of questions. Some authors decide to apply ad-hoc questionnaires, that is, a group of questions created for assessing a specific application with no intentions of recreating the process, or published questionnaires that are already established with a pattern of interrogations. Among the latter, there may be standardized questionnaires [5] or non-standardized ones.

3 The Tertiary Study

A tertiary study is described as a "systematic review of systematic reviews" that can be performed in a field where a number of reviews related to the same research questions has already been made [28]. In this section, we describe the planning (Sect. 3.1) and execution (Sect. 3.2) of our tertiary study on QinU of smart environment applications[1].

3.1 Planning: Research Protocol

From the main issues presented previously, the research questions (RQs) were defined to guide the information that should be extracted from the studies:

- RQ1 (main RQ): What are the most common evaluation approaches for QinU in smart environments?
- RQ2: What are the most evaluated types of systems regarding smart environments?
- RQ3: Which quality characteristics were the most evaluated in smart environments?

To find the secondary studies, the databases Scopus and Web of Science were selected considering that they are the most widely used databases for analysis [44]. Besides, they are largely used for systematic studies and gather most of the Computer Science references including ACM, Elsevier, IEEE and Springer.

The search string to be ran in such databases was defined using the PICOC strategy [42]: *P*opulation, for all studies related to QinU[2], considering all its subcharacteristics; *I*ntervention, for finding evaluations and assessments for the quality previously mentioned; *C*omparison, which was not included since we

[1] The detailed process and replication packages are available on [1,2].
[2] The acronym QinU was not included, as we tried it on its own in the search string and no studies came up.

did not know of any other reviews with the same goal; *O*utcome, to establish that the results should include secondary studies; *C*ontext, to include intelligent environments, pervasive systems, and so on.

The search string was defined as follows, and each part was then linked with an "and" connector:

- **Population**: ("usability" OR "quality in use" OR "effectiveness" OR "efficiency" OR "satisfaction" OR "freedom from risk" OR "context*" OR "usefulness" OR "trust" OR "pleasure" OR "comfort" OR "flexibility")
- **Intervention**: ("evaluation" OR "assessment" OR "quality evaluation" OR "quality assessment")
- **Outcome**: ("*systematic literature review" OR "systematic* review*" OR "mapping study" OR "systematic mapping" OR "structured review" OR "secondary study" OR "literature survey" OR "review of survey*" OR "state of research" OR "state of art" OR "rapid review" OR "SLR" OR "scoping review")
- **Context**: ("smart*" OR "intelligent environment" OR "pervasive" OR "ubiquitous" OR "AAL" OR "ambient intelligence" OR "assisted living" OR "systems of systems" OR "internet of things" OR "Cyber-Physical Systems" OR "Industry 4" OR "fourth industrial revolution" OR "web of things" OR "Internet of Everything" OR "IoT" OR "CPS")

To be included in this tertiary study, each one of the reviewed studies had to comply with the following aspects of the defined inclusion criteria (IC):

- IC1: Published in the Computer Science or Engineering area;
- IC2: Published in the proceedings of a conference, journal, or as a book chapter;
- IC3: Published as a secondary study;
- IC4: Studies concerning the evaluation of the QinU of smart environment applications.

The exclusion criteria (EC) was established as well:

- EC1: Studies not published in English;
- EC2: Editorials, books or erratums;
- EC3: Studies duplicated in the results;
- EC4: Studies that have not been found fully available.

For extracting data, it was established that the following information would be collected from each one of the studies: (i) goal; (ii) number of papers reviewed; (iii) approaches used for evaluation; (iv) type of the system that was evaluated; (v) quality characteristics that were evaluated; (vi) type of secondary study; (vii) type of publication; and (viii) year of publication. A spreadsheet file and a Google form (that stores the data in a spreadsheet) were used, respectively, for the review process and extraction of data. Both spreadsheets were then accessed by all reviewers via Google Sheets.

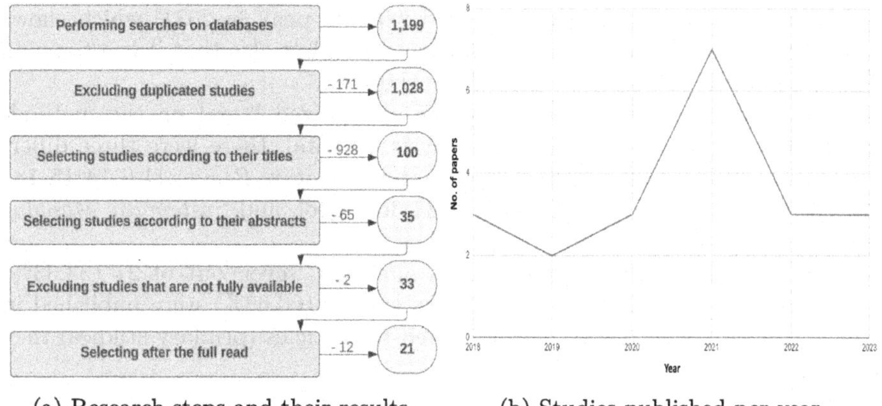

(a) Research steps and their results. (b) Studies published per year.

Fig. 1. Research protocol and year distribution in the final selected set.

3.2 Execution

One PhD student and four professors expert on HCI and IoT applications participated in this phase. The primary selection was performed by one of the reviewers and peer-reviewed by another. In case of disagreements between these two, a third reviewer provides an opinion regarding the respective study to resolve the tie. In case of persistent doubt, a discussion with another reviewer was done to make a decision.

All selection criteria (IC, EC) was applied from the beginning of the research steps (Fig. 1a). The search string was executed in both selected databases on July 19th, 2023. At this moment, thanks to the engine tools of the databases, IC1, IC2, EC1, and EC2 were already included in the search string automatically filtering the results: 338 from Scopus and 861 from Web of Science. Combining both results (1,199 papers), we could therefore apply EC3 to exclude duplicated studies before the manual review, which resulted in a total of 1028 studies. Then, all the titles were reviewed by one reviewer to include only secondary studies that worked to measure the QinU of software (IC3, IC4), which resulted in 100 studies. Then, all the abstracts were analyzed for the same purpose, which resulted in 35 studies. The next step implies downloading all papers for data extraction. However, 2 papers were not available (EC4) which resulted in 33 papers. After reading the full publications, we concluded that 12 articles concerned mobile applications and were not effectively applied in smart environments (IC4), which was not clear from the abstract, so they were also excluded. Finally, 21 studies remained for data extraction.

4 Discussion of the Results

We did not establish any date limits. We noticed that the studies brought in the search are very recent, ranging from 2018 to 2023, as shown in Fig. 1b, with at least two articles published each year and with its peak in 2021, which shows that not only it is a relatively new field of research, but also that it is a topic on which researchers are currently actively working on.

The secondary studies selected can be categorized based on the methodology used to choose the studies for analysis. In total, there were three different categories of secondary studies: *Systematic Literature Review* [4,6,14,15,18–20,29,33], *Scoping Review* [7,9,11,31,32,34,36,37,46], and *Systematic Mapping* [17,39,43].

Considering all the secondary studies analyzed, seven out of 21 (33,33%) were published in international conferences and 14 (66,66%) were published in journals. Not all the secondary studies listed the articles (primary studies) they analyzed.

4.1 RQ1. What Are the Most Common Evaluation Approaches for QinU?

Different types of evaluation were found by the secondary studies (Table 1), with the most common evaluation approach being the questionnaire, whether the authors created it themselves for their study [4,6,7,9,15,17,20,29,31,33,43,46] or applied an already published questionnaire [6,7,14,17,29,31,33,34,37,39,46], followed by interviews [4,6,7,9,17,20,29,31,33,34,36,37,43,46], focus groups [9,17,20,29,31,33,34,46], observation [4,7,17,18,20,29,34,46], log data [11,17,33,34,46], the think aloud protocol [7,17,29,31], surveys [18,32,46], user feedback [9,29], heuristic evaluations [29,31], and others. Table 1 presents the reference for each secondary study, the number of primary studies they analyzed (#), and the methods and techniques identified (e.g. [4] analyzed 24 primary studies).

More than half of the secondary studies analyzed (12) reported the use of questionnaires that were created by the authors with the sole purpose of answering their research questions and without any plan for repetition [4,6,7,9,15,17,20,29,31,33,43,46], while other authors used established questionnaire forms to reach their goals. Between those, the secondary studies indicated a total of 44 different questionnaires, including well-known and standardized ones, according to [5] and [14], that are presented in Table 2 (which shows not only the standardized questionnaires' names, but also how many secondary studies found them and how many primary studies were found applying such questionnaires in each secondary study). The most common used questionnaire was System Usability Scale (SUS), which was found by 9 secondary studies applied in its original form [4,6,7,15,17,29,31,37,46] and was also in an adapted way in three secondary studies [7,17,46]: one of them adapted the questionnaire according to the context [17] and the others [7,46] customized it by putting it together with other established questionnaires. Between the latter, one of them [46] found it adapted with the NASA Task Load Index (NASA-TLX), while another secondary study [7] spotted SUS adapted in two different primary studies: one combined it with PSSUQ (Post-Study System Usability Questionnaire) and another combined it with CSUQ (Computer System Usability Questionnaire). Three different secondary studies [4,17,37] also found the application of

Table 1. Methods and techniques found in the primary studies (#).

Ref.	#	Ad-hoc question-naire	Established question-naire	Interview	Focus group	Observation	Log data	Think aloud	Survey	User feedback	Simulation	Heuristics	Automated test	Real testbed	Prototype	Experimental protocol	Others
[4]	24	13	–	2	–	7	–	–	–	–	–	–	–	–	–	–	–
[6]	44	16	19	13	–	–	–	–	–	–	–	–	–	–	–	–	–
[7]	35	12	26	6	–	4	–	2	–	–	–	–	–	–	–	–	–
[9]	21	8	–	8	3	–	–	–	–	2	–	–	–	–	–	–	1
[11]	63	–	–	–	–	–	7	–	–	–	–	–	–	–	–	–	56
[14]	553	–	553	–	–	–	–	–	–	–	–	–	–	–	–	–	–
[15]	21	21	–	–	–	–	–	–	–	–	–	–	–	–	–	–	–
[17]	65	29	41	8	1	2	17	8	–	–	–	–	–	–	–	–	–
[18]	15	–	–	–	–	10	–	–	9	–	–	–	–	–	–	–	–
[19]	146	–	–	–	–	–	–	–	–	–	76	–	–	37	11	–	22
[20]	29	1	–	6	4	1	–	–	–	–	–	–	–	–	–	–	17
[29]	22	2	2	2	2	1	–	1	–	2	–	11	5	–	–	–	–
[31]	133	39	66	37	13	–	–	45	–	–	18	–	–	–	–	–	57
[32]	9	–	–	–	–	–	–	–	6	–	–	–	–	–	–	3	1
[33]	34	2	2	1	1	–	9	–	–	–	–	–	–	–	–	19	–
[34]	111	–	9	4	2	1	101	–	–	–	–	–	–	–	–	–	–
[36]	8	–	–	4	–	–	–	–	–	–	–	–	–	–	–	–	4
[37]	51	–	29	24	–	–	–	–	–	–	–	–	–	–	–	–	–
[39]	12	–	10	–	–	–	–	–	–	–	–	–	–	–	–	2	–
[43]	10	7	–	5	–	–	–	–	–	–	–	–	–	–	–	–	–
[46]	31	14	17	16	2	4	3	–	3	–	–	–	–	–	–	–	–

Table 2. Standardized questionnaires found in the secondary studies.

Questionnaire	#Sec. studies	#Pri. studies per sec. studies
System Usability Scale (SUS)	9	1 [4], 16 [6], 16 [7], 1 [15], 11 [17], 2 [29], X [37], 44 [31], 3 [46]
System Usability Scale (SUS) adapted	3	2 [7], 1 [17], 1 [46]
Technology Acceptance Model (TAM)	4	4 [7], 5 [31], X [37], 2 [43]
Post-Study System Usability Questionnaire (PSSUQ)	3	1 [6], 1 [7], 12 [31]
Attrakkdiff	2	341 [14], 2 [17]
Modular evaluation of key Components of User Experience (meCUE)	1	12 [14]
Usefulness, Satisfaction and Ease of use questionnaire (USE)	2	1 [6], 1 [17]
User Experience Questionnaire (UEQ)	2	200 [14], 1 [17]
Computer System Usability Questionnaire (CSUQ)	3	1 [4], 2 [6], 1 [7]
Software Usability Measurement Inventory (SUMI)	1	1 [46]
Usability Metric for User Experience (UMUX)	1	1 [17]
Unified Theory of Acceptance and Use of Technology (UTAUT)	1	2 [7]

the NASA-TLX in its original form, being the non-standardized questionnaire to be applied the most.

One study [14] focused particularly on three different standardized questionnaires: AttrakDiff, UEQ (User Experience Questionnaire) and meCUE (Modular Evaluation of key Components of User Experience), with the purpose of determining how they were applied in the past for Ambient Intelligence (AmI) and ubiquitous computing. Apart from the standardized questionnaires, one published non-standardized questionnaire was found applied by primary studies three times [31], seven were found applied twice in the primary studies [4,17,43,46], and eighteen were found applied only once in the primary studies [4,7,17,34,46].

Plenty of secondary studies encountered methods and techniques that involved the researchers having direct contact with end user for evaluating the QinU. Some of them performed direct interviews to collect the information they wanted to report on, asking predefined questions regarding the software they wanted to evaluate [4,6,7,9,17,20,29,31,33,34,36,37,43,46], while others took into consideration people's opinion through user feedback [9,29], listening to the users' comments and thoughts after they interacted with the applications in a free manner, and some applied the think aloud protocol [7,17,29,31], where the people involved use the software that is being evaluated while describing, out loud, their actions (and expectations regarding what will come from it).

A number of secondary studies found authors that used focus groups to evaluate the software quality [9,17,20,29,31,33,34,46], while some also indicated the observation method [4,7,17,18,20,29,34,46].

Four secondary studies found the use of log data in primary studies [11,17, 33,34,46], and in such cases, they were used for calculating metrics, even though they were evaluating the QinU in different systems. Two of these studies focused on the assessment of the QinU of smart environments for the elderly [33,46], one [34] aimed to evaluate QinU characteristics in wearable devices, such as smartwatches, and another [17] had the goal of analyzing multi-touch systems in general. In all these cases, the log data came from the interaction of the end users themselves with the application that was being evaluated.

We noticed that only one study [19] pointed out the use of simulations applied in different platforms for different contexts of use. Most of them provided mathematical simulations, simulating "energy cost, hardware frequency rate, and computation time" of the scheme to be validated, but some other examples were of simulating "a resource description framework for heterogeneous IoT devices". There was no mention of any primary studies performing agent-based simulations for replacing human-computer interaction.

Some secondary studies pointed out evaluation methods and techniques based on different testing performances, which included real testbeds and prototypes [19], automated tests [29], and experimental protocols [32,33,39].

Finally, we defined a category "others" in Table 1 to represent some specific findings. First, to represent some methods and techniques for evaluation found only once in a secondary study: card sort [9], which is a method to help the researcher perceive how others categorize information (guaranteeing a data architecture that matches the users' expectations), and randomized controlled trials [36], which are techniques that balance the characteristics of the participants between the trial groups to allow differences in the intervention results. Similar to this last technique, four studies are identified in [32] as applying randomized clinical trials. Another study [19] found the use of formal techniques, employed to model complex systems as mathematical entities, and used what the authors called "design techniques" for evaluation, in which the primary studies aim to provide new approaches or frameworks. Task completion was also quoted [31] as another method for evaluation that considers whether or not the users were able to perform (with adequate standards) certain tasks within a defined period of time. Finally, several other approaches were quoted in the studies, not really precising which technique or method was applied [11,19,20]. For instance, [11] mentioned that the identified approaches in the primary studies worked on visual data anonymization methods to try "to retain all the informative richness of visual data acquired with RGB cameras" and on security based issues for user authentication and data encryption. Another one [20] mentioned the use of experimentation field tests in general to collect data.

It is noticeable that for most of the secondary studies there was a higher number for methods and techniques than for primary studies, which is caused by many of them combining different ways of evaluating the QinU of the application under analysis. One example is when a secondary study [29] pointed out that two primary studies applied focus groups and interviews together.

We cannot quantify all the information from the primary studies found in the secondary ones, as some of them did not provide all the references and it is likely that there is an overlap of information. Besides, even though some secondary studies went into detail about their findings, some of them did not disclose specific numbers for the data taken from primary studies (for example, how many of them applied the same method or technique for assessing software quality or how many measured a specific QinU characteristic).

4.2 RQ2: What Are the Most Evaluated Types of Systems?

The most evaluated type of systems were smart environments/AmI [6,7,9,11,14, 15,18,33,36,43], with ten studies (47,61%), as seen in Fig. 2, followed by IoT [19, 20,32,34,37,39,46] with seven studies (33,33%). The other software categories were evaluated by only one study each, those being cyber-physical systems [4], multi-touch systems [17], eGovernment [29], and eHealth [31].

Most of the secondary studies about smart environments/AmI focused on Ambient Assisted Living (AAL) [6,7,9,11,14,15,33,36]. However, one researched specifically ubiquitous healthcare [43] and another one smart learning environments [18]. Among the seven studies that included QinU evaluations for IoT, two of them dealt with navigation apps: one of them [20] involved mobile safety alarms with GPS, RFID tags and readers, product design assessment for safe navigation and more. The other study [37] focused on indoor navigation apps for people with mobility disabilities. Three other studies about IoT investigated wearable devices (such as smartwatches, wristbands or neckwear) for different reasons: tracing people and objects regarding hospital-like scenarios involving medical teams and patients [19], improving medication adherence [32], or analyzing physical activities [34]. One study focused on assistive technology for older adults [46] and another on IoT in general [39].

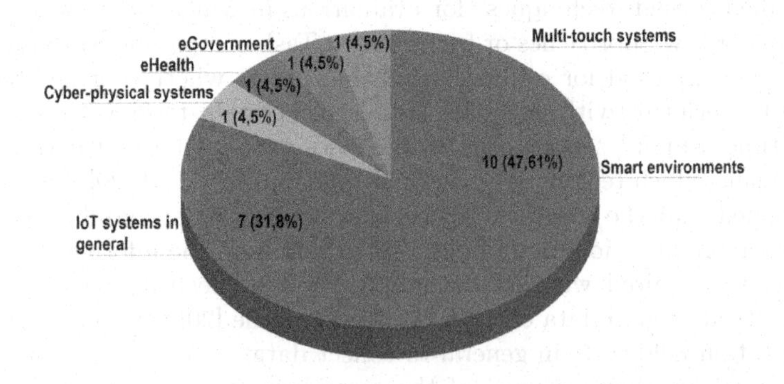

Fig. 2. Types of systems evaluated in the secondary studies.

Table 3. QinU characteristics found in the primary studies (#).

Ref.	#	Usability	Effectiveness	Efficiency	Satisfaction	Freedom from risk	Context coverage
[4]	24	12	5	3	3	1	–
[6]	44	44	–	–	–	–	–
[7]	35	35	–	–	–	–	–
[9]	21	21	–	–	–	–	–
[11]	63	40	–	–	6	17	–
[14]	553	553	–	–	–	–	–
[15]	21	15	17	17	17	14	13
[17]	65	65	13	11	2	–	–
[18]	15	7	–	–	3	–	–
[19]	146	–	–	X	–	X	–
[20]	29	29	–	–	–	–	–
[29]	22	22	–	–	–	–	–
[31]	133	133	–	–	–	–	–
[32]	9	–	6	2	1	3	–
[33]	34	–	34	–	–	–	–
[34]	111	16	107	–	20	–	–
[36]	8	–	8	–	–	–	–
[37]	51	51	–	–	–	–	–
[39]	12	4	2	2	2	5	–
[43]	10	10	–	–	–	–	–
[46]	31	–	31	–	–	–	–

4.3 RQ3: Which Quality Characteristics Were the Most Evaluated?

All QinU characteristics as well as usability were found to be evaluated in the secondary studies, as seen in Table 3. It is usual for a study to evaluate more than one QinU characteristic at once. Usability was evaluated the most, being found in sixteen secondary studies (76,19%) [4,6,7,9,11,14,15,17,18,20,29,31, 34,37,39,43]. This result is expected, considering that usability is the focus of questionnaires, which were the most commonly used method for evaluation (see Sect. 4.1). Effectiveness was the second most mentioned, with nine studies [4, 15,17,32–34,36,39,46], followed by satisfaction, with eight studies [4,11,15,17, 18,32,34,39]. Then, tied with six studies each, efficiency [4,15,17,19,32,39] and freedom from risk [4,11,15,19,32,39]. Lastly, context coverage was found to be assessed by only one of the secondary studies [15].

It is noticeable that many studies apply questionnaires for evaluating usability (and, as a consequence, effectiveness, efficiency and satisfaction as defined by ISO 9241-11 [21]) since the Human-Computer Interaction domain has been investigating usability issues for a long time.

Most of the studies aimed at evaluating the main QinU characteristics, but some had the goal of evaluating ones that were considered as subcharacteristics, according to the ISO/IEC 25010 standard [26], such as usefulness, acceptability [11,18,46], which fall under the satisfaction umbrella, and reliability [39], safety [11], security [11], which are considered a subcharacteristic of freedom from risk. It is important to follow a pattern to guarantee that studies have the goal of evaluating the same characteristics, as "it is evident that the concepts of UX and usability are not the same among the authors" [17]. However, few of the secondary studies extracted information about alignment with quality standards. One of them [15] pointed out that five primary studies followed ISO/IEC 25010 [26], four followed ISO/IEC 9126 [23], two followed ISO/IEC 25012 [24], and lastly, one followed ISO/IEC 14598 [22]. Another [39] classified the selected primary studies according to the characteristics and subcharacteristics they evaluated, splitting them between system/software product quality and QinU, according to ISO/IEC 25000 [25]. Moreover, none of the analyzed studies followed a standard such as ISO/IEC 25022 [27], which brings measures for the said characteristics to be assessed. One of the secondary studies [29] suggested that future researches should take ISO/IEC 25010:2011 [26] into consideration, as it defines usability in a context of use.

5 Threats to Validity

The analyses of threats to the validity of this study have been based on Petersen [41], as follows:

- *Descriptive validity*, related to the gathering of data, which includes recording, storing, and analyzing the information that is being reviewed. To mitigate this threat, a Google Form was created for extracting data according to the established RQs, and that information was then accessed in a spreadsheet on Google Sheets. Moreover, all quantitative analysis was double checked by one or two peers;
- *Theoretical validity*, which involves the identification of the studies to be analyzed and also the data extraction. It is possible that some studies have not been included in this review because the search was performed on two databases, which might provide incomplete results. However, Scopus and Web of Science are recognized for indexing different conferences and journals, and both have been widely used in reviews [35]. To mitigate the risk of bias interpretation, all data extraction was peer-reviewed. However, since this step involves human judgment, the threat cannot be completely eliminated;
- *Generalizability validity* worries about how generalizable the obtained results can be in a wider scenario (within an institution or between several different groups or organizations). As a tertiary study, this factor depends on the generalizability of the 21 secondary studies performed. Another issue that might impact our work is that since not all of the secondary studies mention their references, it is very likely that there are overlaps of information regarding the analyzed QinU evaluations. Therefore, we prefer to be cautious and say that we cannot guarantee the generalizability of the results in different organizations, but since we used the databases available for our institution, it can be considered generalizable in our institution;

- *Repeatability validity* requires that all methods are established and reproducible. We consider this threat under control, since the protocol and all the analysis procedures that took place were well documented, with the detailed process [2] and replication package [1] being available on HAL[3].

6 Final Remarks

This paper presents a tertiary review about evaluating the Quality in Use of smart environments, pervasive systems and intelligent applications. Our goal was to identify how this evaluation has been done considering, in particular, the case of smart environments inhabited by elderly or disabled people.

By carrying out this study, contrary to our expectation, we found that most of the assessment approaches to evaluate the QinU of smart environment applications use classic methods and techniques requiring that users interact directly with the application and answer usability questionnaires/interviews. That means the user should have a live experience in these environments before answering questionnaires or attending interviews to offer their feedback. We believe that carrying out these experiences with elderly or people with disabilities is not adequate. There is, therefore, an urgent need to develop tools that can carry out these evaluations in an automated way. Our ongoing work looks to address this gap with the development of an approach to evaluate the QinU of smart environment applications without having to involve end users, and instead involve artificial intelligence and agent-based modeling and simulation as a step that would come before any end user evaluation [3].

Acknowledgments. The present research work is partially supported by the Hauts-de-France Region and the REUNICE (EUNICE Research) project funded by the European Union and the CNRS. C. Tirnauca's work is partially supported by the project PID2022-139237NB-I00 financed by MICIU/AEI/10.13039/501100011033 and FEDER, UE.

References

1. Angeloni, M.P.C., Duque Medina, R., Marçal de Oliveira, K., Grislin-Le Strugeon, E., Tirnauca, C.: Initial results for a tertiary study on quality in use evaluation of smart environment applications (2024). https://hal.science/hal-04506543
2. Angeloni, M.P.C., Duque Medina, R., Marçal de Oliveira, K., Grislin-Le Strugeon, E., Tirnauca, C.: Research protocol and results for a tertiary study on quality in use evaluation of smart environment applications (2024). https://hal.science/hal-04507278
3. Angeloni, M.P.C., Marçal de Oliveira, K., Grislin-Le Strugeon, E., Duque, R., Tirnauca, C.: A review on quality in use evaluation of smart environment applications: what's next? In: Companion Proceedings of the 2023 ACM SIGCHI Symposium on Engineering Interactive Computing Systems, pp. 9–15 (2023). https://doi.org/10.1145/3596454.3597177

[3] https://hal.science/.

4. Apraiz, A., Lasa, G., Mazmela, M.: Evaluation of user experience in human-robot interaction: a systematic literature review. Int. J. Soc. Robotics **15**(2), 187–210 (2023)
5. Assila, A., Marçal de Oliveira, K., Ezzedine, H.: Standardized usability questionnaires: features and quality focus. Elec. J. of Comp. Sc. and Inf. Tech. **6**(1), 15–31 (2016)
6. Bastardo, R., Martins, A.I., Pavão, J., Silva, A.G., Rocha, N.P.: Methodological quality of user-centered usability evaluation of ambient assisted living solutions: a systematic literature review. Int. J. Environment. Res. Public Health **18**(21), 11507 (2021)
7. Bastardo, R., Pavão, J., Rocha, N.P.: A scoping review of the inquiry instruments being used to evaluate the usability of ambient assisted living solutions. In: Proceedings of the 15th INSTICC International Joint Conference on Biomedical Engineering Systems and Technology - vol. 5: HEALTHINF, pp. 320–327 (2021)
8. Carbo, J., Sanchez-Pi, N., Molina, J.M.: Agent-based simulation with netlogo to evaluate ambient intelligence scenarios. J. Simulation **12**(1), 42–52 (2018). https://doi.org/10.1057/jos.2016.10
9. Choukou, M.A., et al.: Evaluating the acceptance of ambient assisted living technology (AALT) in rehabilitation: a scoping review. Int. J. Med. Inform. **150**, 104461 (2021)
10. Cioroaica, E., Buhnova, B., Kuhn, T.: Predictive simulation within the process of building trust. In: 2022 IEEE 19th International Conference on Software Architecture Companion, ICSA-C 2022, pp. 47–48 (2022). https://doi.org/10.1109/ICSA-C54293.2022.00017
11. Colantonio, S., et al.: Are active and assisted living applications addressing the main acceptance concerns of their beneficiaries? Preliminary insights from a scoping review. In: Proceedings of the 15th ACM International Conference on Pervasive Tech. Related to Assistive Environments, pp. 414–421 (2022)
12. Costal, D., Farré, C., Franch, X., Quer, C.: How tertiary studies perform quality assessment of secondary studies in software engineering. In: XXIV Iberoamerican Conference on Software Engineering (ESELAW@CIbSE) (2021). https://doi.org/10.48550/arXiv.2110.03820
13. Demiris, G., Hensel, B.K., Skubic, M., Rantz, M.: Senior residents' perceived need of and preferences for "smart home" sensor technologies. Int. J. Technol. Assess. Health Care **24**(1), 120–124 (2008). https://doi.org/10.1017/S0266462307080154
14. Díaz-Oreiro, I., López, G., Quesada, L., Guerrero, L.A.: UX evaluation with standardized questionnaires in ubiquitous computing and ambient intelligence: a systematic literature review. Adv. Hum. Comput. Interact. **2021** (2021)
15. Erazo-Garzon, L., Erraez, J., Cedillo, P., Illescas-Peña, L.: Quality assessment approaches for ambient assisted living systems: a systematic review. In: Botto-Tobar, M., Zambrano Vizuete, M., Torres-Carrión, P., Montes León, S., Pizarro Vásquez, G., Durakovic, B. (eds.) ICAT 2019. CCIS, vol. 1193, pp. 421–439. Springer, Cham (2020). https://doi.org/10.1007/978-3-030-42517-3_32
16. Felber, N.A., Tian, Y.J.A., Pageau, F., Elger, B.S., Wangmo, T.: Mapping ethical issues in the use of smart home health technologies to care for older persons: a systematic review. BMC Med. Ethics **24**(1) (2023). https://doi.org/10.1186/s12910-023-00898-w
17. Filho, G.E.K., Guerino, G.C., Valentim, N.M.C.: A systematic mapping study on usability and user experience evaluation of multi-touch systems. In: Proceedings ACM 21st Brazilian Symposium on Human Factors in Computing Systems (2022)

18. Gambo, Y., Shakir, M.Z.: Review on self-regulated learning in smart learning environment. Smart Learn. Environ. **8**(1), 12 (2021)
19. Haghi Kashani, M., Madanipour, M., Nikravan, M., Asghari, P., Mahdipour, E.: A systematic review of IoT in healthcare: applications, techniques, and trends. J. Netw. Comput. Appl. **192**, 103164 (2021)
20. Holthe, T., Halvorsrud, L., Karterud, D., Hoel, K.A., Lund, A.: Usability and acceptability of technology for community-dwelling older adults with mild cognitive impairment and dementia: a systematic literature review. Clin. Interv. Aging **13**, 863–886 (2018). https://doi.org/10.2147/CIA.S154717
21. ISO 9241-11: Ergonomic requirements for office work with visual display terminals (VDT) s- Part 11 Guidance on usability (1998)
22. ISO/IEC: 14598-5:1998: information technology—software product evaluation—part 5: Process for evaluators (1998). https://www.iso.org/standard/24906.html
23. ISO/IEC: 9126. software engineering - product quality (2001). https://www.iso.org/standard/22749.html
24. ISO/IEC: 25012:2008. software engineering—software product quality requirements and evaluation (square)—data quality model (2008). https://www.iso.org/standard/35736.html
25. ISO/IEC: 25000: Systems and software quality requirements and evaluation (square)—guide to square (2011). https://www.iso.org/standard/64764.html
26. ISO/IEC: 25010:2011. systems and software engineering—systems and software quality requirements and evaluation (square)—system and software quality models (2011). https://www.iso.org/standard/35733.html
27. ISO/IEC: 25022:2016. systems and software engineering—systems and software quality requirements and evaluation (square)—measurement of quality in use (2016). https://www.iso.org/fr/standard/35746.html
28. Kitchenham, B., Charters, S.: Guidelines for performing structural literature reviews in software engineering (version 2.3). Tech. Report, Keele University and University of Durham (2007)
29. Lyzara, R., Purwandari, B., Zulfikar, M.F., Santoso, H.B., Solichah, I.: E-government usability evaluation: insights from a systematic literature review. In: Proceedings ACM 2nd International Conference on Software Engineering and Information Management, pp. 249–253 (2019)
30. Mao, C., Chang, D.: Review of cross-device interaction for facilitating digital transformation in smart home context: a user-centric perspective. Adv. Eng. Inform. **57**, 102087 (2023). https://doi.org/10.1016/j.aei.2023.102087
31. Maramba, I., Chatterjee, A., Newman, C.: Methods of usability testing in the development of ehealth applications: a scoping review. Int. J. Med. Inform. **126**, 95–104 (2019). https://doi.org/10.1016/j.ijmedinf.2019.03.018
32. Marengo, L.L., Barberato-Filho, S.: Involvement of human volunteers in the development and evaluation of wearable devices designed to improve medication adherence: a scoping review. Sensors **23**(7), 3597 (2023)
33. Maresova, P., et al.: Health-related ICT solutions of smart environments for elderly-systematic review. IEEE Access **8**, 54574–54600 (2020)
34. McCallum, C., Rooksby, J., Gray, C.: Evaluating the impact of physical activity apps and wearables: interdisciplinary review. JMIR mHealth uHealth **6**, e58 (2018). https://doi.org/10.2196/mhealth.9054
35. Motta, R.C., de Oliveira, K.M., Travassos, G.H.: A conceptual perspective on interoperability in context-aware software systems. Inf. Softw. Technol. **114**, 231–257 (2019). https://doi.org/10.1016/j.infsof.2019.07.001

36. Moyle, W., Murfield, J., Lion, K.: The effectiveness of smart home technologies to support the health outcomes of community-dwelling older adults living with dementia: a scoping review. Int. J. Med. Inform. **153** (2021). https://doi.org/10.1016/j.ijmedinf.2021.104513
37. Nasr, V., Zahabi, M.: Usability evaluation methods of indoor navigation apps for people with disabilities: a scoping review. In: Proceedings of the 2022 IEEE International Conference on Human-Machine Systems, pp. 1–6 (2022)
38. Ntoa, S., Margetis, G., Antona, M., Stephanidis, C.: Uxami observer: an automated user experience evaluation tool for ambient intelligence environments. Adv. Intell. Syst. Comput. **868**, 1350–1370 (2018). https://doi.org/10.1007/978-3-030-01054-6_94
39. Paiva, J.O., Andrade, R.M., Carvalho, R.M.: Evaluation of non-functional requirements for IoT applications. In: International Conference on Enterprise Information Systems, ICEIS - Proceedings. vol. 2, p. 111–119 (2021)
40. Pavlovic, M., Kotsopoulos, S., Lim, Y., Penman, S., Colombo, S., Casalegno, F.: Determining a framework for the generation and evaluation of ambient intelligent agent system designs. Adv. Intell. Syst. Comput. **1069**, 318–333 (2020). https://doi.org/10.1007/978-3-030-32520-6_26
41. Petersen, K., Gencel, C.: Worldviews, research methods, and their relationship to validity in empirical software engineering research. In: 23rd International Workshop on Software Measurement and the 8th International Conference on Software Process and Product Measurement, pp. 81–89 (2013). https://doi.org/10.1109/IWSM-Mensura.2013.22
42. Petticrew, M., Roberts, H.: Systematic Reviews in the Social Sciences: A Practical Guide. Blackwell Publishing Ltd, Oxford (2006). https://doi.org/10.1016/B978-0-12-374708-2.00014-0
43. Saleemi, M., Anjum, M., Rehman, M.: Ubiquitous healthcare: a systematic mapping study. J. Ambient Intell. Humanized Comput. **14**(5), 5021–5046 (2023)
44. Singh, V.K., Singh, P., Karmakar, M., Leta, J., Mayr, P.: The journal coverage of web of science, scopus and dimensions: a comparative analysis. Scientometrics **126**(6), 5113–5142 (2021)
45. Tricco, A., Langlois, E., Straus, S., Alliance for Health Policy and Systems Research SCI, WHO Organization: Rapid reviews to strengthen health policy and systems: a practical guide. World Health Organization (2017)
46. Tónay, G., Pilissy, T., Tóth, A., Fazekas, G.: Methods to assess the effectiveness and acceptance of information and communication technology-based assistive technology for older adults: a scoping review. Int. J. Rehabil. Res. **46**(2), 113–125 (2023). https://doi.org/10.1097/MRR.0000000000000571

A Reference Architecture for Dry Port Digital Twins: Preliminary Assessment Using ArchiMate

Joana Antunes[1(✉)], João Barata[1(✉)], Paulo Rupino da Cunha[1(✉)], Jacinto Estima[1(✉)], and José Tavares[2(✉)]

[1] CISUC, DEI, University of Coimbra, Coimbra, Portugal
joanantunes@student.dei.uc.pt, {barata,rupino,estima}@dei.uc.pt
[2] Fordesi, Lisboa, Portugal
jose.tavares@fordesi.pt

Abstract. Dry Ports are critical infrastructures in the logistics chain, optimizing seaport operations by providing customs services and container storage on land. Some of their main challenges include traffic congestion, environmental impact, high dependence on regulations and paper documentation, and the need to provide reliable information to government authorities and road and rail freight industries. More recently, Digital Twins have emerged as a promising solution to monitor real-time information and optimize Dry Port operations simultaneously. This paper presents the initial proposal of a reference architecture for developing Dry Port's Digital Twins. It results from a design science research project conducted in cooperation with a major IT company in the logistics sector and a Dry Port operator in Portugal. ArchiMate was the selected language for architecture modeling. Our theoretical contribution is in the form of a high-level reference architecture, confirming the suitability of the ArchiMate language. For practitioners, this work is part of the Portuguese NEXUS agenda: Innovation Pact for Green and Digital Transition for Transport, Logistics, and Mobility, assisting in adopting Blockchain, Optical Recognition, and Artificial Intelligence as crucial technological enablers.

Keywords: Dry Port · Digital Twin · Enterprise Architecture · Archimate

1 Introduction

Seaports are pillars of global supply chains. However, many locations that produce and consume goods lack access to advantageous coastal areas or have limited storage capacity [1] and customs authority services. These problems ultimately led to the creation of Dry Ports. The logistic links between seaports and Dry Ports enable import and export supply chains at regional, country, and cross-board levels and are usually supported by railway transport, which is more efficient and greener than trucks.

These logistics infrastructures extend port operations inshore, also called the port's hinterland, by providing customs authority services, managing larger container flows,

J. Araújo et al. (Eds.): RCIS 2024, LNBIP 513, pp. 131–145, 2024.
https://doi.org/10.1007/978-3-031-59465-6_9

and expanding storage capacity. The digital transformation of Dry Ports is a priority due to some of their challenges [2]: high dependence on regulations and paper documents, traffic congestion requiring optimization of all movements in the facility, environmental impacts that need to be minimized (e.g., reduce energy resources), and the necessity of providing reliable information to all stakeholders (e.g., government, railroad and truck operators, insurance companies, among others).

Digital Twins are virtual replicas of physical objects or systems, supporting bidirectional communication between both spaces through which data is exchanged [3]. The information, captured in real-time using Internet of Things (IoT) devices, can be used for multiple purposes, from preventing and predicting faults to simulating future behaviors. As Digital Twins architectures increase in complexity, shifting from representing single objects to broader contexts like factories or cities, the need for a reference architecture for Dry Ports ultimately justifies this preliminary assessment.

Architectures are crucial not only to better understand the elements that describe a system but also to adapt it more quickly to changes. Research showed a few examples of Digital Twin architectures, modeled with ArchiMate, of complex systems such as smart cities and smart factories. Still, none focus on Dry Ports.

This research is part of NEXUS[1], a project whose stakeholders include Sines' Port and the Portuguese Government, which intends to digitalize and decarbonize ports and value chains through technological solutions. This particular work is part of WP3-Smart Gates And Smart Terminal, and intends to model the architecture of a Dry Port Digital Twin using the modeling language ArchiMate.

This paper is structured as follows: Sect. 2 explains the methodology used to conduct the research, and Sect. 3, Literature Review, defines the foundations of Enterprise Architecture (EA), explores the adoption of Digital Twins in Dry Ports, and presents how EA has been used to model Smart Spaces. Section 4 analyzes the key technologies of the Dry Port, presents a high-level view of the TO-BE architecture, and identifies the technological additions. Section 5 presents both the conclusion and future work.

2 Methodology

Considering the lack of studies about Dry Port's Digital Twins and the insufficient guidance on creating the correspondent architecture, the chosen methodology was Design Science Research (DSR) [4].

DSR aims to create knowledge by designing artifacts (e.g., models, frameworks, or digital solutions) in real settings. It considers six iterative steps, starting with the identification of both the problem and the motivation, followed by the objectives' definition, design and development, demonstration, evaluation, and, lastly, the conclusion. It is not mandatory to realize each step consecutively, as the research can start in any of the steps and move outward. To better describe the research steps, Table 1 shows the adapted version of the DSR method, one iteration completed, and relates each step with the specific case of the Dry Port considered.

Overall, the steps taken to conduct this research were the analysis of existing Digital Twin architectures and the high-level blueprint of a Dry Port. After modeling the AS-IS

[1] https://nexuslab.pt/ [Online] [Accessed 18th of January 2024].

Table 1. DSR Method (adapted from [4]).

Problem	Objectives	Design
Lack of existing architectures of Dry Port's Digital Twin, using ArchiMate	Model a Dry Port's Digital Twin Architecture using ArchiMate	Design the planned architecture
Development	**Evaluation**	**Conclusion**
Implement the envisioned architecture, with all requirements and technological changes identified	Evaluation of the proposed architecture by a major IT company in the logistics sector	Application of changes to the architecture and a possible representation of a Dry Port's Digital Twin is made available

architecture using ArchiMate and comparing both representations, it was concluded that this modeling language could accurately represent all architectural elements. Then, the TO-BE architecture was modeled, with all the new technologies, and evaluated by an IT company, specialized in developing technological solutions for the logistics sector.

3 Literature Review

3.1 Digital Twins in Ports

The increasing research on Digital Twins [5] has justified the need for accurate ways to describe the elements involved, the data flow, and the capabilities and interactions with the user. Most of their benefits, such as the increase in productivity, efficiency, and scalability, are achieved by the implementation of autonomous systems and IoT devices, mixed with machine learning capabilities. Digital Twins are crucial components in an IoT ecosystem, "*...bringing many vendors and technologies together*" [6] and allowing a flexible integration and configuration of other applications and devices.

In [7], it is explained that the conceptualization and implementation of a Digital Twin involves four phases that are essential to consider when modeling its architecture: (1) the collection of data, (2) the addition of sensors to gather information in real-time, (3) the creation of simulation tools, and (4) the implementation of the platform to be accessed by the user. From virtually representing a unit to an entire system, the increase in complexity and flexibility [8] has further complicated maintenance and development operations, justifying the need for methods to describe each component more comprehensively. Creating an adequate architecture is challenging and can have long-lasting impacts on the organization that will consider it [8].

Dry Ports have multiple processes that can be improved with the adoption of Digital Twins, ranging from automatization to asset optimization, sustainability, safety, regulatory compliance, and stakeholder engagement [9]. The three core capabilities of this technology that can potentially improve these infrastructures are increased situational awareness, smarter decision-making, and collaboration [10].

A study considering Qingdao Port [11] proposed a Digital Twin that integrated data from different sources and enhanced its visualization, allowing the early warning of risks and the optimization of business processes. Singapore's Port [12], through real-time traffic tracking, increased awareness about the procedures by the time the trucks arrived, reducing traffic congestion. In China, Mawan's Port [12] visually represents the entire port, facilitating access to information in different formats and from various sources and creating simulations. Oulu's Port, in Finland [13], adopted Digital Twins to formulate better environmental plans, and, in Livorno's Port [12], Italy, data from sensors and cameras is used to simulate the best placement for the cargo and the most efficient order for the tasks.

These examples indicate a rising interest in adopting Digital Twins within the context of ports. Their advantages range from the capacity to model various elements of ports to collecting data from multiple sources, which can then be used to simulate potential issues. However, despite the benefits of implementing this technology, there are still some challenges [11] related to their adoption in ports, specifically regarding modeling, data acquisition, complexity, and changes adaptability [10]:

- Each port is unique, keeping a wide array of information systems with limited or no interconnection between them, due to different actors and processes. As a result, deploying and integrating real-time data with Digital Twins poses additional challenges [13].
- The large scale and high volume of data produced by enterprises [10–12] results in increased complexity and the need for more sophisticated prediction algorithms to anticipate issues and the system's response to changes.
- Considering the complexity of ports, with various dynamic processes and actors, the need to define standards for their Digital Twin has been recognized but slowly accomplished [13]. For example, ISO 23247-2 Automation Systems and Integration – Digital Twin Framework for Manufacturing (Part 2: Reference Architecture) [14] outlines a reference architecture comprising four domains for modeling a manufacturing Digital Twin: (1) the user domain, which hosts applications, (2) the simulation and analysis operations, (3) the device communication domain, responsible for data collection and monitoring, and (4) the observable manufacturing domain, which consists of observable resources and equipment. These domains inspire key building blocks for Digital Twin architectures in manufacturing and also other sectors of the economy.

Some of the most relevant challenges to make the digital transformation of ports a reality are data-related. For example, managing and integrating external data efficiently ensures trustworthiness and consistency of information [12], protecting it from loss and attacks during transmission [15]. Privacy concerns and regulatory compliance can't be ignored when using drones [6] or other devices to capture data in ports, requiring a focus on trustworthiness in the data lifecycle management. Along with these challenges, protecting and managing cargo documentation has increased the interest in adopting Blockchain [12] to minimize the likelihood of data tampering.

3.2 Foundations of Enterprise Architecture

Architecture is the design of every structure that makes a product, device, system, and enterprise [8], from its properties and interfaces to relationships. A proper architecture enables the early analysis of potential trade-offs related to certain decisions and is easy to update and control over time [16].

Enterprise Architecture (EA) addresses the very specific design concerns related to large Information Technology (IT) systems and the company dependent upon them, documenting the human interactions with the technology used to conduct business operations [8]. A well-defined architecture is crucial for identifying potential changes and integrating new applications, allowing the enterprise to grow and innovate in a controlled way [17]. Several frameworks for Enterprise Architecture [5] define the tasks and artifacts needed for the architectural process and the views that answer different stakeholders' needs, easing the discussion among them [17].

Some examples of prominent frameworks include:

- Zachman Framework [8], created in the 1980s is a two-dimensional table with thirty cells, classified by many as the fundamental structure for Enterprise Architecture. Easy to create [18], it is challenging to read and does not allow a strategic analysis of possible new technological solutions to add value to the enterprise.
- The Federal Enterprise Architecture Framework (FEAF) [18] was defined specifically for the U.S. Federal Government in the 1990s to guide the integration of technology, business, and strategic processes across all U.S. Federal Agencies. Heavily influenced by the Zachman Framework, it shares its disadvantages and is challenging to apply to enterprises that do not follow the U.S. Federal Agencies' standards.

The Open Group Architecture Framework (TOGAF) [18], created in the 1990s, is one of the most widely used frameworks nowadays. It provides the steps required to model an Enterprise Architecture, helping to identify the business and IT goals, and their alignment with the enterprise. This framework intends to define a shared language, understood by everyone involved. Additionally, it is possible to align TOGAF with popular EA languages like ArchiMate, allowing the creation of multiple views, layers, and strategic justifications for changes. ArchiMate is a modeling language that provides detailed data about the architectures' structure and allows the definition of viewpoints to convey information that addresses the concerns of specific stakeholders, isolating certain aspects of the model [31]. Examples provided by the ArchiMate [32] specification are split into four viewpoint categories, from more basic aspects (e.g., organization viewpoint), motivation, strategy, to implementation/migration.

However, when considering the complexity of concepts such as cities and ports addressed in our research, most of the EA frameworks have limitations [19], as they were created to operate within an enterprise with a complete overview of its data and information systems [20]. For instance, a larger number of stakeholders guided by different rules, multiple data sources with various restrictions of access, and the service-oriented nature of these enterprises, are some of the challenges yet to be solved when modeling in smart spaces [19].

3.3 How EA Shapes the Current Advances of Smart Spaces

We found no studies modeling Dry Port architectures or addressing Digital Twins developments in this context. Therefore, our review explored the adoption of Enterprise Architecture approaches in complex spaces like cities, ports, or rail stations.

Two of the driving forces for cities are sustainability and digital transformation [21]. Smart Cities have evolved into systems that now provide citizens with services that fit their needs and improve their quality of life [19, 20]. The data captured with IoT devices and from other sources is crucial for creating valuable services [20].

Two of the problems repeatedly identified in some of the EA frameworks were the lack of focus on the user perspective and the unclear description of the data relevant to improving services [20]. To answer that challenge, the authors of [20] established a framework to model a Smart City architecture with two perspectives and seven layers. The two perspectives are those of (1) stakeholders, which focus on all entities involved and their relationships with data (from regulations to policies, privacy, and even access limitations), and (2) data, enhancing the need for its governance and security. Regarding the layers, from top to bottom, *context* describes the drive of each service, *services* meet the needs of the citizens, *business* depicts the flow of the enterprises' procedures, *application* identifies the systems that manage the heterogeneity of data sources, *data* provides storage for the information, and both *technological* and *physical* layers regard the devices that produce the real-time data and the infrastructure to run the applications [20].

The framework proposed by [20] has been adopted by other authors, for instance, in the work of [19], where the focus is on designing services according to citizens' needs. Based on the context, service, information, and technology layers, guidelines to model both the context and service layers are shown, from the identification of the requirements to the definition of the scope and goals, as well as the explanation of service functions and the actors' resources. Other examples regarding, for example, the adoption of electric mobility services and focusing on data management and governance are [21] and [22], and to model an evaluation and monitoring platform for Smart Cities [23] was conducted. However, a few limitations [24] have already been identified when adopting ArchiMate to model Smart Cities, leading to extensions of its elements, as presented in [25]. These authors present new elements such as, for example, the *domain* (which describes a critical field of urban development), the *objectives* (which shows the progress towards a city goal), and the *decision* (which presents an option based on data collected to support the decision-making process).

Overall, the research regarding Smart Cities is significant, and their modeling using ArchiMate has led to the creation of specific frameworks that could inspire other complex smart spaces, for example, Smart Ports. Generally referred to as intelligent ports that can operate autonomously, they use real-time information to improve operations and ensure harmonious communication among all devices, actors, and elements of the Port [26, 27]. With the need to remain competitive and with processes whose tasks are repetitive and monotonous, their automatization can be achieved with adequate data processing and management [26]. Their key components include [27] (1) the smart infrastructure, which describes the technological solutions to increase productivity; (2) the smart traffic flows, responsible for the seamless flow of moving assets; and (3) the smart logistics, which supports the automatic movement and handling of containers, as well as the communication between stakeholders.

In contrast to Smart Cities, EA adoption in the development of Smart Ports is still fragmented and underexplored [27], especially when considering the impact those have on response time, asset utilization, and the environment. Existing contributions focus on the operational aspects of a Port. For instance, the authors of [26] used ArchiMate to model a platform that allows stakeholders to access the various applications and port information, focusing on the application and business layers. In [28], the processes were identified and split into administrative and operational levels, followed by the description of technological services that could improve each one.

Seaports can combine transport options like vessels, trucks, or trains that are particularly relevant in their configurations. The work presented by [29] studied the adoption of a Digital Twin within the rail freight in the Netherlands. Firstly, all operations, actors, applications, and their interfaces were identified. The main problems were the lack of transparency, data accuracy, and liability. To answer these questions, a Digital Twin prototype was created, and its architecture was developed using ArchiMate. The high-level architecture includes six layers chosen based on modularity, scalability, and considering the Model-View-Controller (MVC) architectural pattern [30]. It splits a system into three main logical components: (1) the model represents all the actions related to data, functioning as the intermediate between the controller's data requests and the database access, handling the data logic; (2) the controller enables the connection between the model and the view, sending requests to the model and rendering the final output through the view; (3) the view is dynamically rendered and generates the user interface, to be accessed by the user.

In the work of [29], the models to be used by the Digital Twin are stored in the database included in layer 5, while the views to be interacted with by the user are described in layer 6. Lastly, layer 4 represents the controller. Considering this modulation, it is possible to add more models and views to the Digital Twin, based on future changes. To effectively connect the existing technology, REST APIs are used to integrate data from different sources, and an event hub ensures the asynchronous data acquisition from the various sensors installed and the easier integration of new event-triggered solutions. Regarding the remaining layers, layer 3 represents the data server, where all data gathered by the IoT devices will be stored and shared with the event hub, layer 2 includes all the device controllers, and Layer 1 represents the devices themselves. This more generalist architecture can be adapted to other industries.

Our literature review confirmed the importance of EA approaches to digital transformation in Smart Spaces, the potential of the ArchiMate language, and the lack of research in Dry Ports Digital Twins. However, it was also possible to identify frameworks to guide new developments in this field and relevant layers to model a Dry Port architecture. The following section presents the results of our design science research on developing an innovative reference architecture for Dry Port Digital Twins.

4 Dry Port Digital Twin Architecture

The main applications already implemented in the Dry Port are identified in Sect. 4.1, based on a high-level blueprint of this infrastructure. The high-level view of the TO-BE architecture considers the seven layers introduced in [20] and, in Sect. 4.2, the new technological solutions are explained. In Sect. 4.3, the preliminary modeling of the Dry Port Digital Twin with Archimate is described. In Sect. 4.4, a brief evaluation of the architecture of the Dry Port Digital Twin is presented.

4.1 Key Applications

To model the architecture of the Dry Port, it is crucial to first identify all applications and APIs accessed. The PHC Enterprise Resource Planning (ERP) is an internal financial management and payroll system and MSC is an API that provides route information.

- JUL[2] is an application implemented in 2019 by APP (Portuguese Ports Association). It is installed on all Portuguese ports' administrations with a strong connection with the national tax authority (AT), to improve information exchange between stakeholders and to centralize access to all information regarding Seaports and hinterland Dry Ports. JUL acts as a hub, facilitating the message exchange, increasing data flow efficiency, and amplifying the Port's capabilities, reducing the need for physical documentation, administration costs, and waiting times.
- MedTOS[3] is a Multimodal Terminal Operating System with a strong and complete API with JUL. It manages information regarding yard operations and shares real-time data with all freighters and cargo agents through JUL, such as the gate in and gate out of trucks and trains.

4.2 Requirements Identification

Table 2 presents the technological additions to the Dry Port, previously defined within the project's scope, meant to be modeled in the correspondent architecture. The solutions intend to optimize the dry port's operations, ensure information reliability, reduce the dependency on paper documentation, and decrease environmental impact.

The companies participating in this research identified the architecture's business requirements. They aim to address different Dry Ports' challenges, including the need to reduce environmental impact (e.g., movements of yard vehicles), optimize operations' time (e.g., automatic gate control), or ensure information reliability. With the fundamental elements of a Digital Twin being the modeling, connectivity, and data analysis capabilities, each can be associated with a specific ArchiMate layer, as discussed in [33]. To model the services provided by the Digital Twin, the business layer is the most adequate, the acquisition and flow of data via sensors can be modeled using the technology layer and, lastly, its capability to process information for decision-making and behavior prediction can be represented with the application layer.

4.3 Preliminary TO-BE Architecture

Figure 1 shows the proposed architecture for the Dry Port Digital Twin.

Based on the framework in [20], seven layers were considered to model the Dry Port's Digital Twin. The (1) context layer (on the left of Fig. 1) captures the information regarding the strategies, goals, drivers of each stakeholder, and the port's capabilities. The main stakeholders of the Dry Port are the port's administration and the national government, which share the same drivers: performance, environmental impact, information reliability, and competitiveness. The assessments and how they influence each

[2] https://www.projeto-jul.pt/ [Online] [Accessed 17th of January 2024]

[3] https://www.fordesi.pt/solucoes/gestao-para-terminais-multimodais-de-mercadorias/ /[Online] [Accessed 19th of January 2024].

Table 2. Technological Requirements for the Dry Port Architecture.

Technology	Description
License Plate Recognition (LPR) cameras	Cameras that capture trucks' license plates and confirm if each is allowed to enter (or exit) the Port. These remove the need for someone to control the flow of traffic manually
Optical Character Recognition (OCR) cameras	Cameras that will capture images of all containers that enter and or exit the yard
AI to confirm containers' state	Based on the images captured of each container, the model is meant to be able to detect damage
Public Blockchain	A public Blockchain will be created to store crucial documentation for the Port's operations. For instance, bill landing documentation, CMR documents, dispatch orders and pictures of containers that arrived damaged have to be stored securely
AI to optimize the interior movements in the port and container organization	Based on previous data regarding the arrival and exit times of containers and their characteristics, the model is meant to predict and present the best route for the container's movement within the Port
MedTOS ENT	Application to be used by the Dry Port's operators which, based on the vehicle selected, shows the tasks to be done
Trucker App	Application to be used by the truck drivers which allows a remote check-in. When done, the information is stored and, on arrival, the camera can confirm the operation and allow the entrance into the port. This removes, as well, the need for someone to control the flow of traffic manually

other allowed the comprehension that, for instance, because "yard vehicles' movements are not optimized," the "level of environmental impact is high." The outcomes expected for the Dry Port's digitalization ("neutralized carbon emissions", "insured information reliability", and "reduced time in manual containers' assessment, gates control and containers' handling") realize each one of the goals identified, which include operations' automatization and heavier data control. The capabilities of the port are identified and linked to the defined courses of action: the definition of AI models to optimize the yard vehicles' movements and containers' organization and handling, among others, and the priority to adopt Blockchain for specific information flows.

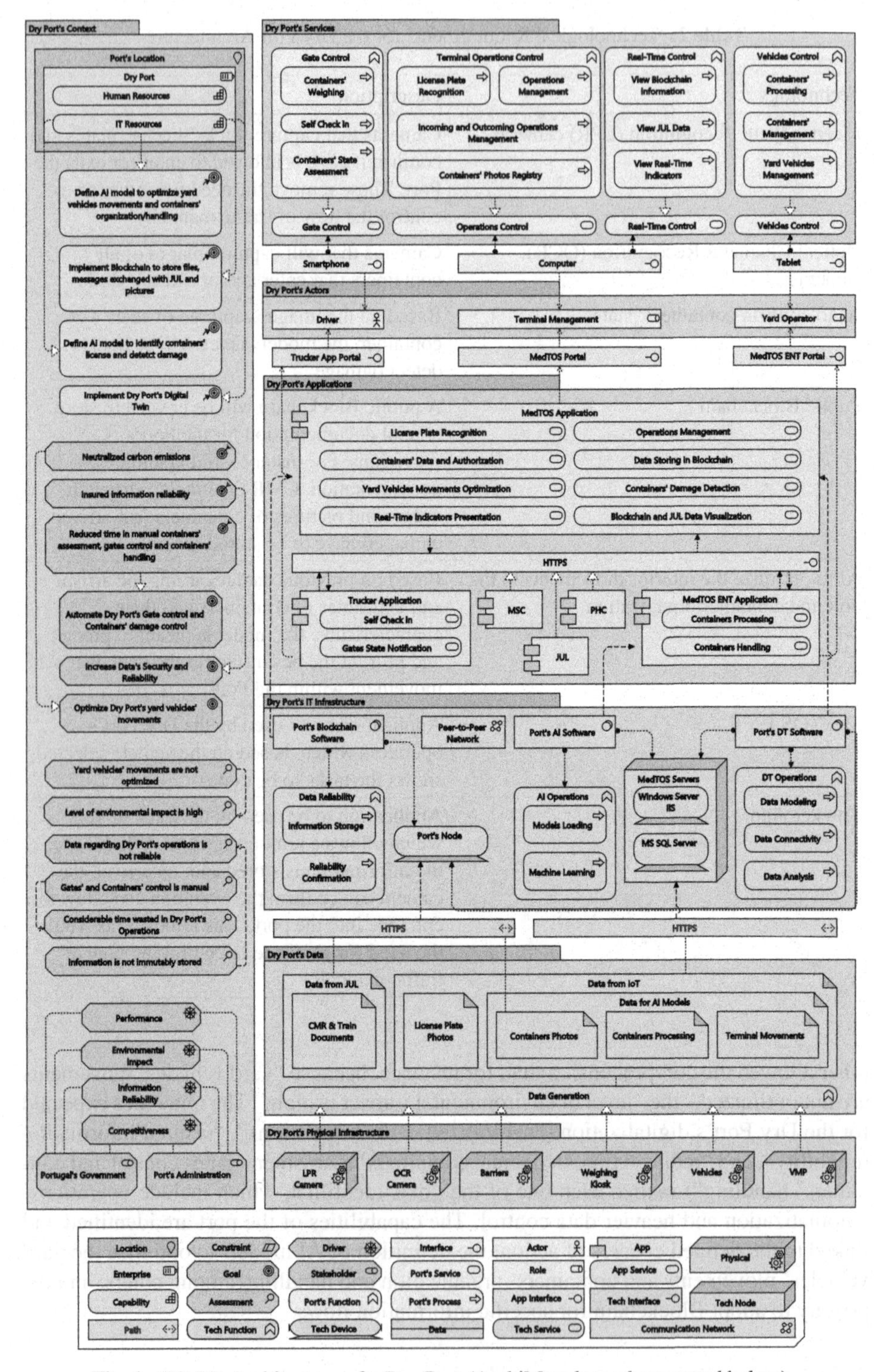

Fig. 1. TO-BE Architecture of a Dry Port (ArchiMate legend presented below).

In (2), the service layer (top-right), identifies the actual services that will be accessed by the actors that make up the Dry Port: the functions of this infrastructure are presented, with the correspondent processes, considering four key services: gates, terminal operations, vehicle, and real-time control. Between this layer and (3), the actors' layer (which presents actors that interact with the Dry Port, including yard operators, terminal managers, and drivers), the interfaces, through which the actors interact with the port's services, are identified.

The application layer (4) (blue elements) presents two new applications, the Trucker Application and the MedTOS ENT Application (explained in detail in Table 2 and which exchange information with MedTOS), the new services added to the MedTOS Application, which include the capability to store data in the Blockchain, detect containers' damage based on the correspondent photos taken at arrival, visualize JUL's messages exchanged, present real-time indicators, and optimize the yard vehicles' movements. The MedTOS application uses data from the applications/APIs presented (JUL, MSC, and PHC Enterprise Resource Planning) to implement the required services.

The IT infrastructure layer (5) identifies the servers and software responsible for the AI, Digital Twin, and Blockchain capabilities, as well as the functions of each. Regarding the AI capabilities of the Dry Port, the main functions are the models' loading and machine learning, the Digital Twin operations include data modeling, connectivity, and analysis and, lastly, to explain the Blockchain component, a peer-to-peer network and the port's node were modeled, as well as the information storage and data reliability confirmation processes. The data layer (6) describes information used by the Dry Port applications and connects to the IT infrastructure layer through https. The data is split into two categories: data from JUL and data realized by the IoT devices (which includes the license plate pictures and the containers processing, photos, and terminal movements, the last three considered as models to train the AI capabilities). The data from JUL, which includes information about truck arrivals and departures, and the containers' photos when damaged, are stored in the Blockchain, while the remaining information is kept in the MedTOS servers. Lastly, (7) the physical infrastructure layer identifies all devices that capture information in real-time, from cameras to the yard vehicles, the weighing kiosk, barriers, and the Variable Message Panel (VMP). Blockchain [34] is essential to ensure trust and reliability in specific Dry Port's data, namely, photos collected during container inspections (e.g., to certify the source of potential damages – before arrival or at the port facilities) and the data changed in JUL, which is critical to ensure traceability of editing records about container movements.

Compared with the existing architecture, the interactions of the drivers with the port's services are now achieved through a mobile application, the Trucker Application, which allows self-check-in before arrival at the port. MedTOS ENT Application shows yard operators the route of each container based on the vehicle selected and the least number of movements considered. The other noticeable change is the addition of Blockchain, which will be considered to store electronic consignment note documents (eCMR, one of the essential documents in logistic operations) and pictures of the containers (if they are detected damaged at arrival).

The main Digital Twin operations include data modeling, connectivity, and analysis, which are represented as functions and use data from both the MedTOS servers and the

Blockchain. Our framework will enable the creation of trustworthy Digital Twins in three key dimensions. First, implementing real-time monitoring capabilities of all movements from/to the Dry Port (trucks, trains, and yard vehicles) and the inspection procedures required. Second, adopting Blockchain to ensure trustworthy data exchange between parties, contributing to implementing the requirements of the Portuguese regulation for Dry Ports, Decree-Law n.° 53/2019, 17th April. Third, the "intelligent part" is supported by artificial intelligence (e.g., computer vision, machine learning), enabling simulation to define the best yard vehicle route to organize the containers and load trains, minimizing movements and wasted resources, and contributing to the reduction of carbon emissions.

4.4 Summative Evaluation of the Dry Port Digital Twin Architecture

A summative evaluation was conducted using the FEDS framework [35], created explicitly for DSR projects. This form of evaluation is "*...more often used to measure the results of a completed development or to appraise a situation before development begins*" [35].

According to the project partners responsible for the digital transformation of the Dry Port, the ArchiMate model is an advance compared to the traditional (informal) architectural representations. First, providing a standard representation language for different building blocks of a Digital Twin. The mere adoption of IoT is not enough to create a digital replica of a complex system, requiring the inclusion of (1) the business impact, (2) the applications supporting the key processes at the port, and (3) the elements responsible for Digital Twin intelligence. In the NEXUS vision, artificial intelligence is crucial to ensure resilient, human-centric, and sustainable port operations, aligned with the emerging priority of Industry 5.0 [36]. However, a more detailed analysis will include multiple views, for example, an *Application Usage* viewpoint, focused on describing dependencies between applications and business processes, and an *Information Structure* viewpoint, presenting and relating the data processed by the Dry Port's applications and gathered from various sources. As discussed in [25] and [20], the relevance of services in the context of smart spaces justifies the modeling of the *Service Realization* viewpoint of the Dry Port as well.

The proposed architecture for Dry Port Digital Twin will be the leading guide for multiple teams developing specific IT components at the port (e.g., computer vision for container identification and damage analysis and Blockchain-based file management system are implemented by different teams), ensuring alignment between architectural elements, assisting in hardware identification requirements (green elements supporting the application portfolio), and providing a simple blueprint of business – IT alignment (on the top of Fig. 1, representing business layer interactions).

5 Conclusion

One of the intentions of this research was to model a Dry Port Digital Twin using ArchiMate, as there is still a considerable lack of guidelines to do so, even though this infrastructure has been recognized as fundamental for logistics chain optimization. Research showed a considerable increase in the modeling, using ArchiMate, of Smart Spaces, the most easily comparable to a Smart Port. Based on the modeling of these complex smart systems, the TO-BE architecture of a Dry Port Digital Twin was built, identifying the elements mandatory for each of the seven layers. Our work contributes to a systemic approach to Digital Twin design and provides a real example for Dry Port Digital Twins.

There are important limitations that need to be stated. First, the proposed architecture modeled in ArchiMate was considered an advance to the current practice of Digital Twin modeling in Dry Ports, but the system has not yet been fully deployed. In the case presented, the TO-BE architecture is classified as preliminary because a lot of the digital elements to be added (the Blockchain, the Artificial Intelligence Model to optimize the movements within the port, among others) are still being conceptualized and developed, which limits the level of detail possible to model. Second, only the high-level architecture is presented at this stage. The following steps include the detailed modeling of all the necessary views and an extension proposal for ArchiMate. One possibility is to improve the reference architecture using approaches proposed in the software engineering field [37]. Third, although the experts participating in this project confirmed the applicability of the proposed architecture to other Dry Ports and, with some adaptations, the use in traditional seaports (an extension would be required for vessel interoperability, including berth planning features to optimize the port) is feasible. However, our summative evaluation conducted in the scope of Work Package 3 of Nexus was restricted to the company managers and software developers, which does not allow us to confirm generalizability. A detailed evaluation is necessary, for instance, as the one presented in [25], which assesses utility and quality based on real case studies and validation of concepts from experts.

Acknowledgments. This project was funded by Project "Agenda Mobilizadora Sines Nexus". Ref. No. 7113, supported by the Recovery and Resilience Plan (PRR) and by the European Funds Next Generation EU, following Notice No. 02/C05-i01/2022, Component 5 - Capitalization and Business Innovation - Mobilizing Agendas for Business Innovation. The work was also funded by national funds through the FCT-Foundation for Science and Technology, I.P., within the scope of the project CISUC-UID/CEC/00326/2020 and by the European Social Fund, through the Regional Operational Program Centro 2020.

References

1. Khaslavskaya, A., Roso, V.: Dry ports: research outcomes, trends, and future implications. Marit. Econ. Logist. **22**(2), 265–292 (2020). https://doi.org/10.1057/s41278-020-00152-9
2. Mohd Zain, R., Salleh, N.H.M., Mohd Zaideen, I.M., Menhat, M., Jeevan, J.: Dry ports: redefining the concept of seaport-city integrations. Transp. Eng. **8**, 100112 (2022). https://doi.org/10.1016/j.treng.2022.100112
3. Singh, M., Fuenmayor, E., Hinchy, E., Qiao, Y., Murray, N., Devine, D.: Digital twin: origin to future. Appl. Syst. Innov. **4**, 36 (2021). https://doi.org/10.3390/asi4020036
4. Peffers, K., Tuunanen, T., Rothenberger, M., Chatterjee, S.: A design science research methodology for information systems research. J. Manag. Inf. Syst. **24**, 45–77 (2007)
5. Liu, Z., Zhang, A., Wang, W.: A framework for an indoor safety management system based on digital twin. Sensors **20**(20) (2020). Art. No. 20. https://doi.org/10.3390/s20205771

6. Yigit, Y., et al.: TwinPort: 5G drone-assisted data collection with digital twin for smart seaports. Sci. Rep. **13**(1) (2023). Art. No. 1. https://doi.org/10.1038/s41598-023-39366-1
7. Hadjidemetriou, L., et al.: A digital twin architecture for real-time and offline high granularity analysis in smart buildings. Sustain. Cities Soc. **98**, 104795 (2023). https://doi.org/10.1016/j.scs.2023.104795
8. Lattanze, A.: Architecting Software Intensive Systems - A Practitioners Guide, 1st edn. CRC Press. Auerbach Publications (2008)
9. Ports Australia Digital Twins Working Group - Findings Document (Final) November 2022.pdf. https://assets-global.website-files.com/5b503e0a8411dabd7a173eb7/638d87369bf1382aadf303bb_Ports%20Australia%20Digital%20Twins%20Working%20Group%20-%20Findings%20Document%20(Final)%20November%202022.pdf. Accessed 14 Jan 2024
10. Klar, R., Fredriksson, A., Angelakis, V.: Digital twins for ports: derived from smart city and supply chain twinning experience. IEEE Access **11**, 71777–71799 (2023). https://doi.org/10.1109/ACCESS.2023.3295495
11. Wenqiang, Y., et al.: A digital twin framework for large comprehensive ports and a case study of Qingdao Port. Int. J. Adv. Manuf. Tech. (2022). https://doi.org/10.1007/s00170-022-10625-1
12. Wang, K., Hu, Q., Zhou, M., Zun, Z., Qian, X.: Multi-aspect applications and development challenges of digital twin-driven management in global smart ports. Case Stud. Transp. Policy **9**(3), 1298–1312 (2021). https://doi.org/10.1016/j.cstp.2021.06.014
13. Klar, R., Fredriksson, A., Angelakis, V.: Assessing the maturity of digital twinning solutions for ports. In: 2023 IEEE International Conference on Pervasive Computing and Communications Workshops and Other Affiliated Events (PerCom Workshops), pp. 552–557. IEEE, March 2023
14. International Standard, ISO 23247-2:2021: Automation Systems and Integration - Digital Twin Framework for Manufacturing, 1st edn. (2021)
15. Wang, K., Hu, Q., Liu, J.: Digital twin-driven approach for process management and traceability towards ship industry. Processes **10**(6) (2022). Art. No. 6. https://doi.org/10.3390/pr10061083
16. Bass, L., Clements, P., Kazman, R.: Software Architecture in Practice. Software Engineering, 2nd edn. Addison-Wesley Professional (2003)
17. Jonkers, H., Lankhorst, M., ter Doest, H., Arbab, F., Bosma, H., Wieringa, R.: Enterprise architecture: Management tool and blueprint for the organization. Inf. Syst. Front. **8**, 63–66 (2006). https://doi.org/10.1007/s10796-006-7970-2
18. Kotusev, S.: A Comparison of the Top Four Enterprise Architecture Frameworks. BCS - The Chartered Institute for IT. https://www.bcs.org/articles-opinion-and-research/a-comparison-of-the-top-four-enterprise-architecture-frameworks/. Accessed 12 Jan 2024
19. Pourzolfaghar, Z., Bastidas, V., Helfert, M.: Standardisation of enterprise architecture development for smart cities. J. Knowl. Econ. **11**(4), 1336–1357 (2020). https://doi.org/10.1007/s13132-019-00601-8
20. Petersen, S.A., Pourzolfaghar, Z., Alloush, I., Ahlers, D., Krogstie, J., Helfert, M.: Value-added services, virtual enterprises and data spaces inspired enterprise architecture for smart cities. In: Camarinha-Matos, L.M., Afsarmanesh, H., Antonelli, D. (eds.) PRO-VE 2019. IAICT, vol. 568, pp. 393–402. Springer, Cham (2019). https://doi.org/10.1007/978-3-030-28464-0_34
21. Anthony, B., Petersen, S.A.: A practice based exploration on electric mobility as a service in smart cities. In: Themistocleous, M., Papadaki, M. (eds.) EMCIS 2019. LNBIP, vol. 381, pp. 3–17. Springer, Cham (2020). https://doi.org/10.1007/978-3-030-44322-1_1

22. Bokolo, A.J.: Examining the adoption of sustainable emobility-sharing in smart communities: diffusion of innovation theory perspective. Smart Cities **6**(4) (2023). Art. No. 4. https://doi.org/10.3390/smartcities6040095
23. Jnr, B.A., Petersen, S.A.: Validation of a developed enterprise architecture framework for digitalisation of smart cities: a mixed-mode approach. J. Knowl. Econ. **14**(2), 1702–1733 (2023). https://doi.org/10.1007/s13132-022-00969-0
24. Bastidas, V., Bezbradica, M., Bilauca, M., Healy, M., Helfert, M.: Enterprise architecture in smart cities: developing an empirical grounded research agenda. J. Urban Technol. **30**(1), 47–70 (2023). https://doi.org/10.1080/10630732.2022.2122681
25. Bastidas, V., Reychav, I., Ofir, A., Bezbradica, M., Helfert, M.: Concepts for modeling smart cities: an ArchiMate extension. Bus. Inf. Syst. Eng. **64**(3), 359–373 (2022). https://doi.org/10.1007/s12599-021-00724-w
26. Kapkaeva, N., Gurzhiy, A., Maydanova, S., Levina, A.: Digital platform for maritime port ecosystem: port of Hamburg case. Transp. Res. Procedia **54**, 909–917 (2021). https://doi.org/10.1016/j.trpro.2021.02.146
27. Min, H.: Developing a smart port architecture and essential elements in the era of Industry 4.0. Maritime Econ. Logist. **24**(2), 189–207 (2022). https://doi.org/10.1057/s41278-022-00211-3
28. Lepekhin, A.A., Levina, A.I., Dubgorn, A.S., Weigell, J., Kalyazina, S.E.: Digitalization of seaports based on enterprise architecture approach. IOP Conf. Ser. Mater. Sci. Eng. **940**(1), 012023 (2020). https://doi.org/10.1088/1757-899X/940/1/012023
29. Pool, A.: Digital Twin in Rail Freight - The Foundations of Future Innovation. University of Twente (2021)
30. Harper, K.E., Ganz, C., Malakuti, S.: Digital Twin Architecture and Standards. Industrial IoT Consortium, November 2019
31. Bogea Gomes, S., Santoro, F.M., da Silva, M.M., Iacob, M.-E.: Visualization of digital transformation initiatives elements through ArchiMate viewpoints. Inf. Syst. Front. (2024). https://doi.org/10.1007/s10796-023-10469-4
32. The Open Group: Archimate 3.2 Specification. https://pubs.opengroup.org/architecture/archimate3-doc/index.html. Accessed 17 Mar 2024
33. Bandeira, C., Barata, J., Roque, N.: Architecting digital twin-driven transformation in the refrigeration and air conditioning sector. In: Kumar, V., Leng, J., Akberdina, V., Kuzmin, E. (eds.) Digital Transformation in Industry: Digital Twins and New Business Models, pp. 13–28. Springer, Cham (2022). https://doi.org/10.1007/978-3-030-94617-3_2
34. Curty, S., Härer, F., Fill, H.-G.: Towards the comparison of blockchain-based applications using enterprise modeling. ER Demos/Posters, October 2021
35. Venable, J., Pries-Heje, J., Baskerville, R.: FEDS: a framework for evaluation in design science research. Eur. J. Inf. Syst. **25**(1), 77–89 (2016). https://doi.org/10.1057/ejis.2014.36
36. Barata, J., Kayser, I.: Industry 5.0 – past, present, and near future. Procedia Comput. Sci. **219**, 778–788 (2023). https://doi.org/10.1016/j.procs.2023.01.351
37. DeBaud, J.M., Flege, O., Knauber, P.: PuLSE-DSSA—a method for the development of software reference architectures. In: Proceedings of the Third International workshop on Software architecture, pp. 25–28 (November 1998)

Business Process Management

Enhancing the Accuracy of Predictors of Activity Sequences of Business Processes

Muhammad Awais Ali(✉), Marlon Dumas, and Fredrik Milani

University of Tartu, Tartu, Estonia
{muhammad.awais.ali,marlon.dumas,fredrik.milani}@ut.ee

Abstract. Predictive process monitoring is an evolving research field that studies how to train and use predictive models for operational decision-making. One of the problems studied in this field is that of predicting the sequence of upcoming activities in a case up to its completion, a.k.a. the case suffix. The prediction of case suffixes provides input to estimate short-term workloads and execution times under different resource schedules. Existing methods to address this problem often generate suffixes wherein some activities are repeated many times, whereas this pattern is not observed in the data. Closer examination shows that this shortcoming stems from the approach used to sample the successive activity instances to generate a case suffix. Accordingly, the paper introduces a sampling approach aimed at reducing repetitions of activities in the predicted case suffixes. The approach, namely Daemon Action, strikes a balance between exploration and exploitation when generating the successive activity instances. We enhance a deep learning approach for case suffix predictions using this sampling approach, and experimentally show that the enhanced approach outperforms the unenhanced ones on event logs that exhibits a high frequency of repeated activities with respect to both control-flow and temporal accuracy measures.

Keywords: Process Mining · Predictive Process Monitoring · Sequence Prediction · Deep Learning

1 Introduction

Predictive process monitoring [19] is a set of techniques that predict future states or properties of ongoing cases of a business process. These techniques are based on event logs extracted from information systems [8]. These predictions [13] help operational managers make decisions that improve the performance of the process by, for instance, reallocating resources, increase the probability of achieving positive case outcomes, or reduce delays [22].

Existing predictive process monitoring techniques address a range of prediction targets [3], such as predicting the outcome of a case [12,21], its remaining cycle time [23], the next activity instance [19], and the sequence of activity

Work funded by the European Research Council (PIX Project).

J. Araújo et al. (Eds.): RCIS 2024, LNBIP 513, pp. 149–165, 2024.
https://doi.org/10.1007/978-3-031-59465-6_10

instances until case completion [1,19,20]. In this paper, we focus on the latter prediction target. Given an ongoing case of a process and the sequence of activity instances that have occurred up to that point in time (a *case prefix*), the goal is to predict the sequence of activity instances until the completion of the ongoing case (herein called the *case suffix* for short). For each activity instance in this sequence, we predict its activity type and its completion time. Thus, we also tackle the problem of predicting the remaining time until completion of the case as the remaining time is equal to the difference between the completion time of the last activity instance in the case suffix and the completion time of the last activity instance in the case prefix.

Various methods for case suffix prediction have been proposed [1,4,14,19]. These methods generally rely on deep learning models trained to predict the next activity instance in a case. These methods predict a case suffix by predicting the next activity instance after a given case prefix, then the following activity instance, and so on, until the next-activity model predicts the end of the case.

A shortcoming of these approaches is their tendency to generate suffixes that contain an excessive number of consecutive (repeated) instances of the same activity, even when such repetitions do not occur or rarely occur in the historical execution data [1,17,19]. These self-repetition patterns negatively impact the performance of these techniques, both with respect to their ability to replicate the control-flow of the process (sequence of activities) and the temporal behavior (the completion times). To reduce or avoid activity repetitions not present in the historical process execution data, it has been proposed to use random selection [1] of the next activity, instead of always selecting the next activity with the highest probability [19]. This approach, however, does not consider historical activity patterns in a case when predicting the next activity. As a result, these models end up in an "exploitative cycle", where they favor familiar patterns (an activity is repeated), instead of exploring a more diverse set of possibilities. In this setting, this paper addresses the following problem:

> Given an ongoing (incomplete) case of a process, of which we know the sequence of activity instances up to a point in time (herein called a *case prefix*), predict the sequence of activity instances until completion of the case and the end timestamp of each activity instance (herein called the *case suffix*) while avoiding generating activity repetitions that are not present in the historical process execution data.

The contribution of this paper is a sampling method to predict the sequence of activities in a case suffix, given a model that predicts the next activity instance in an ongoing case. The proposed method seeks to strike a balance between exploring less common activities and exploiting frequent ones, enabling predictions that incorporate diverse activities. We evaluate our method against baselines using real-life event logs, demonstrating its effectiveness in generating accurate case suffixes. The proposed method can be used to predict how many activity instances, of each activity type, will be generated by ongoing cases of a business process. These predictions, in turn, enable operational managers to estimate the expected near-term workload for different types of resources, and thus to make data-driven resource allocation decisions.

2 Related Work

Existing methods for predicting activity sequences of business processes often use neural networks, categorized by their architectures: Long Short-Term Memory (LSTM), Recurrent Neural Networks (RNN), Convolutional Neural Networks (CNN), and Generative Adversarial Networks (GAN).

Tax et al. [19] demonstrated the effectiveness LSTM architectures in predicting the next activity, the case suffix, and the remaining time of ongoing cases in a process. Di Francescomarino et al. [4] built on this, using process structures and pre-existing knowledge to enhance LSTM accuracy. Others have developed LSTM variants focusing on more accurate remaining time predictions [10,23]. Lin et al. [14] highlighted event attribute's role in enhancing predictions with their RNN model, which incorporates a component modulator to integrate multiple event attributes, improving performance on real-life datasets.

Mauro et al. [15] found that CNNs, particularly using stacked inception modules, could be more efficient and accurate in predicting activity sequences than some RNN methods. Similarly, Di Pasquadibisceglie et al. [16] used CNNs to forecast activities by converting the event log data into 2D images for CNN training, showing promise in activity prediction but struggling with timestamp prediction. Taymouri et al. [20] critique LSTM and CNN methods for next-activity and case suffix prediction, suggesting a GAN-based adversarial framework that enhances case suffix generation by pitting two neural networks against each other for more accurate outputs.

The discussed neural network models predict the next activity in a case autoregressively: starting with a case prefix, they predict the next activity, then use the updated prefix to predict subsequent activities until the case ends. These models internally consider multiple potential next activities with varying probabilities, selecting the next one based on established sampling methods. Argmax [19] sampling methodically selects the next sequence item with the highest probability. Top-K [9] sampling narrows the selection to the K most probable choices, with the model randomly choosing from these top options. In contrast, Nucleus sampling [11] selects from a dynamically sized pool of top predictions, choosing a subset that collectively reaches a specific probability threshold (p). This approach allows for a more flexible selection range compared to the fixed number in Top-K sampling. These methods tend to predict frequent activities, leading to repetitive predictions. Camargo et al. [1] found that random selection based on probability enhances prediction accuracy and quality by considering both less and more probable options, an explorative approach. Building on this, the present paper aims to propose an approach that strikes a balance between exploration and exploitation for selecting the next activity in a case suffix.

3 Daemon Action Approach

In line with the previous discussion, we propose an approach for selecting the next activity in a case suffix from the most likely activities suggested by the next activity predictors. This approach is designed to strike a balance between exploration and exploitation. To achieve this, we introduce a 'Daemon Action' approach, which aims to refine the process of sampling the next activity for a case suffix. This method focuses on precisely choosing from the most likely activities, ensuring a more targeted and effective balance between the two strategies.

Daemon actions are a component of ant colony optimization algorithms, which serve to direct the search process towards higher-utility results [7]. Given a set of possible directions in a search process (in our context, the possible next-activities), the goal of a Deamon action is to select the most promising one, taking into account both the estimated goodness of each direction as well as the choices that were made previously in the search process (in out context, the activities in the case prefix). A Daemon action is inherently heuristic and does not come with any soundness guarantees. However, the efficacy of Daemon Action heuristics has been extensively demonstrated via empirical studies in the field of Ant Colony Optimization [5–7] and also in the context of prediction problems [18].

Equation 1 shows the proposed Daemon Action approach formulated as a mathematical equation and can be integrated with existing deep learning approaches for case suffix and remaining time prediction.

$$F(a) = \frac{P(a) \cdot \frac{1}{\text{count}(a)}}{\sum_{i=1}^{N} (P(i) \cdot \text{count}(i))} \tag{1}$$

Here, $F(a)$ represents the function that determines the selection of the next activity in a suffix from a sample. The term $P(a)$ signifies the probability of selecting activity a among possible next activities. The term $\text{count}(a)$ represents the count of historical occurrence of activity a. The summation term $\sum_{i=1}^{N} (P(i) \cdot \text{count}(i))$ calculates the overall preference or likelihood for any activity based on their respective probabilities and historical occurrences in a case prefix of length N.

The purpose of the Daemon Action approach is to balance between exploration and exploitation in selecting next activity in a suffix. Exploration refers to the attempt to discover new or less frequent activities, rather than relying solely on what has been successful in the past. In Eq. 1 the term $(1/count(a))$ encourages exploration. The $count(a)$ represents the number of times an activity a has been selected in the past. If an activity has been selected less frequently (count(a) is low), the reciprocal $1/count(a)$ will be higher. By multiplying the probability $P(a)$ by this reciprocal, activities historically less frequent are given more weight in the calculation of $F(a)$. This means that the equation will tend to favor less common activities, encouraging the model to explore new or less frequent activities.

Exploitation refers to using strategies that were successful in the past. It leverages known paths or activities that previously showed optimal results, termed as the *most promising areas* in the search space. The algorithm minimizes redundant exploration by building on past insights. In Eq. 1, the term $P(a)$ encourages exploitation. This term represents the likelihood of selecting an activity among all available activities in a sample. If an activity has a higher probability, it indicates that the model has found this activity to be more frequent in the past. Hence, activities with higher probabilities (or higher historical success rates) are given more weight in the calculation of $F(a)$.

Consider a predictive task in a manufacturing process with activities: material procurement (A), design (B), assembly (C), quality check (D), and shipping (E). The goal is to predict the sequence of activities following the design phase, with historical data indicating that while some repetitions of activities (like assembly due to rework) occur, excessive repetitions are not typical. Given a case prefix up to the design stage (A-B), and using a trained next activity predictor, we obtain the next activity probability distribution: P(A) = 0.1, P(B) = 0.2, P(C) = 0.5, P(D) = 0.1, and P(E) = 0.1. Traditional sampling methods i.e. Argmax [19] overly favor activity C (assembly) due to its highest probability, while Random Choice [1] will randomly select the next activity leading to unrealistic predictions of repeated activity C. Hence, the problem lies in the sampling approach for the selection of the next activity in a case suffix, which leads to excessive repetitions of a particular activity.

Our approach recalibrates the selection process by considering both the probability distribution and the historical occurrence frequencies of activities. This method promotes a realistic balance by possibly adjusting the likelihood of choosing 'C' again, based on its past frequency, and increases exploration of other activities like 'D' (quality check) that logically follow assembly in the historical process flow. By applying Daemon Action, we predict a balanced and realistic sequence A-B-C-D-E, which includes a necessary iteration of 'C' but prevents unrealistic repetitions, aligning predictions closely with historical execution data.

The Eq. 1 offers a dynamic trade-off between exploration and exploitation by balancing the two conflicting goals. It achieves this by multiplying the probability $P(a)$ by the reciprocal of the count $1/count(a)$, and then normalizing by the sum over all activities. By combining these values, the Daemon Action approach allows the model to explore new or less common activities while still taking advantage of what has been successful in the past and considering the probabilities to guide decisions about which activity to include in a case suffix. Hence, this balance can lead to more effective and robust predictive modeling.

4 Experiment Design

This section reports on an evaluation of the proposed Daemon Action approach via computational experiments on real-life event logs. The objective of this evaluation is to compare the accuracy of the Daemon Action approach, layered on top of different LSTM architectures for next-activity prediction, relative to other sampling approaches.

4.1 Questions

Given that the targeted problem is that of predicting sequences of activity instances, such that each activity instance consists of the activity type and the timestamp, we are interested in assessing the relative accuracy of the proposed approach both with respect to the control-flow perspective (the order in which the activity types appear in the predicted sequence) and with respect to the temporal perspective (the accuracy of the timestamps, particularly the last timestamp in the predicted sequence). Additionally, since the Daemon Action approach is designed to prevent the occurrence of spurious activity repetitions in the predicted suffix, we are also interested in assessing if the repetition patterns produced by Daemon Action are more reflective of the ground truth than those produced by baseline approaches. Accordingly, the experiments presented herein address the following Experimental Questions (EQs):

- EQ1 Does the proposed Daemon Action approach, combined with different LSTM architectures for next-activity prediction, predicts sequences of activity types that are closer to the ground truth, compared to other sampling approaches?
- EQ2 Does the proposed Daemon Action approach, combined with different LSTM architectures for next-activity prediction, predicts case suffixes with a distribution of activity repetitions closer to the ground truth, compared to other sampling approaches?
- EQ3 Does the proposed Daemon Action approach, combined with different LSTM architectures for next-activity prediction, predicts case suffixes with an overall temporal accuracy closer to the ground truth, compared to other sampling approaches?

4.2 Datasets

We conducted our experiments using 15 real-life event logs, specifically chosen to validate the effectiveness of the daemon action approach in reducing redundant spurious activities in case suffixes. To ensure the reproducibility of our experiments, the logs employed are publicly accessible at the *4TU Centre for Research Data*[1] as of November 2023. Notably, several of these logs originate from various years of the Business Process Intelligence Challenges.

In our evaluation, logs with extreme outliers were omitted. For instance, BPIC 2011 was excluded due to its daily timestamp granularity, causing many activities to be indistinguishable in chronological order. Logs lacking both case and event attributes, like BPIC 2014, were also removed. BPIC 2016, being a click-stream dataset, and BPIC 2018, with its strong seasonal patterns making predictions straightforward (76% of cases ending on specific dates), were excluded. BPIC 2019 was omitted due to the brevity of its cases, with 75% having a trace length of up to 6.

[1] https://data.4tu.nl/search?search=bpi.

Table 1 outlines the characteristics of each dataset, including *#Cases* (total process executions), *#ActivityInstances* (total activity executions), and *#Activity* (unique activities). It also lists *Avg.Act/Trace* (average activities per trace) and *Rep.Index(%)* (activity repetition proportion). Additionally, *Avg.Dur.* and *MaxDur.* indicate average and maximum cycle times, respectively. Further details of each dataset are as follows:

- BPI12[2]: This dataset comprises sequences from a loan application process at a Dutch financial institution, captured through an online system between October 1, 2011, and March 14, 2012. Within this process, there are three distinct sub-processes, one of which is labeled as W. Consequently, we isolate this sub-process from the dataset, termed BPI12w.
- BPI13[3]: These event logs feature real-life data from Volvo IT's VINST system, focusing on incident and problem management processes. we utilized the closed problem set of event log from the BPI Challenge 2013.
- BPI15[4]: This dataset comprises event logs from five Dutch municipalities related to the building permit application process, with each municipality's dataset treated as a separate event log.
- BPI17[5]: The dataset comprises sequences from a loan application process at the same Dutch financial institute as BPI12, covering 2016 and extending up to February 2, 2017.
- BPI20[6]: The dataset includes two years of travel expense claims from two university departments, covering documents like declarations and payment requests, all following a similar process. Each document has been treated as a separate event log.

Table 1. Descriptive Statistics of Datasets

Eventlog	#Cases	#Activity Instances	#Act.	Avg. Act/Trace	Max. Act/Trace	Rep. Index(%)	Avg. Dur.	Max Dur.
BPI2012	13087	164505	23	12.57	96	34.99%	8.61 days	91.32 days
BPI12w	9658	72413	6	7.5	74	69.48%	11.41 days	91.02 days
BPI2013	1487	6660	7	4.48	35	34.53%	178.99 days	6 years, 62 days
BPI2015_1	1199	27409	38	22.8	61	53.35%	95.9 days	4 years, 26 days
BPI2015_2	832	25344	44	30.4	78	52.65%	160.3 days	3 years, 230 days
BPI2015_3	1409	31574	40	23.4	69	51.11%	62.2 days	4 years, 52 days
BPI2015_4	1053	27679	43	26.2	83	48.47%	116.9 days	2 years, 196 days
BPI2015_5	1156	36234	41	31.3	109	52.0%	98 days	3 years, 248 days
BPI2017	31509	561671	26	17.83	61	16.19%	21.84 days	168.94 days
BPI2017(W)	31500	128227	8	4.07	20	20.53%	11.27 days	167.72 days
BPI20 (Domestic_Declaration)	10500	56437	17	5.37	24	3.12%	11.55 days	1 year, 102 days
BPI20 (International_Declaration)	6449	72151	34	11.19	27	4.21%	86.45 days	2 years, 11 days
BPI20 (Permit_Log)	5417	64107	51	11.83	90	12.45%	86.75 days	3 years, 95 days
BPI20 (Prepaid_Travel_Cost)	2099	18246	29	8.69	21	2.37%	36.83 days	325.10 days
BPI20 (Request_For_Payment)	6886	36796	19	5.34	20	2.89%	12.04 days	1 years, 40 days

[2] doi: 10.4121/uuid:3926db30-f712-4394-aebc-75976070e91f.
[3] doi: 10.4121/uuid:500573e6-accc-4b0c-9576-aa5468b10cee.
[4] doi: 10.4121/uuid:31a308ef-c844-48da-948c-305d167a0ec1.
[5] doi: 10.4121/uuid:5f3067df-f10b-45da-b98b-86ae4c7a310b.
[6] doi: 10.4121/uuid:52fb97d4-4588-43c9-9d04-3604d4613b51.

4.3 Experiment Setup

We implemented the proposed approach in Python 3.8, by leveraging a previous implementation of an LSTM approach for next activity and case suffix prediction reported in [1]. We extended this existing implementation with an implementation of the proposed Daemon Action approach as well as other baseline sampling approaches discussed above in Sect. 2. The resulting Python scripts as well as instructions on how to reproduce the reported experiments can be found at GitHub Repository[7]

Data Splitting and Hyperparameter Optimization. To simulate real-life scenarios where models are trained on historical data and tested on current cases, we treat each training and test instance as a pair consisting of one prefix and one suffix event, denoted as $(\sigma \leq k, \sigma > k)$, where the prefix length k is at least 1. To prevent any data leakage, we utilized a strict temporal split [23] for dividing the event log into training and testing sets. This process orders the cases by their start time, allocating the first 80% of cases up to a specific point in time, for training and the remaining 20% for evaluating predictive accuracy. It's important to note that cases which began before this cutoff point and are still ongoing should not be included in either set to maintain the integrity of the data.

Hyperparameter optimization is a crucial step for enhancing model performance. We achieve this by dividing the training set further into 80% for training and 20% for validation. This allows us to fit the predictors using different hyperparameter combinations and evaluate their performance on the validation set. The Mean Absolute Error (MAE) serves as the metric for assessing each combination on the validation data. The best-performing combination, as determined by the MAE, is then used to retrain the model on the full training set. The hyperparameter oprimization ranges have been shown in Table 2. Moreover, to comprehensively explore the hyperparameter space for each dataset, we implement a random search with 50 iterations, focusing on different hyperparameter combinations. This approach ensures a thorough and effective training process for the predictor on each dataset.

Table 2. Hyperparameter Configuration Space

Parameter - LSTM	Explanation	Search Space
Batch size	No of Samples to be Propagated	[32,64,128]
Normalization method	Preprocessing - Scaling	[lognorm, max]
Epochs	Number of training epochs	100
N_size	Size of the ngram	[5, 10, 15, 20, 25, 30, 35, 40, 45, 50]
L_size	LSTM layer sizes	[50,100,150,200,250]
Activation	Activation function to use	[selu, tanh, relu, sigmoid, linear, softmax]
Optimizer	Weight Optimizer	[Nadam, Adam, SGD, Adagrad]

[7] https://github.com/AwaisAli37405/LSTM---Daemon-Action.git.

Baselines. To integrate the proposed Daemon Action approach into a Deep Learning framework, we considered three LSTM-based neural network architectures [1] for predicting case suffixes and remaining time in business processes. These architectures are: Specialized, Shared Categorical, and Full Shared. The Specialized architecture functions as three independent models without sharing information. The Shared Categorical architecture merges inputs related to activities and roles, sharing the first LSTM layer, aiming to minimize noise from mixing different attribute types (categorical or continuous). The Full Shared architecture combines all inputs and fully shares the first LSTM layer. The evaluation explores which LSTM-architecture best fits the nature of each event log.

To further validate the effectiveness of our approach, we compared it against two existing LSTM-based approaches for case suffix prediction. Specifically, we considered the approach proposed by [19], which utilizes the Argmax selection heuristic, and the approach proposed by [1], which employs the Random Choice approach. In addition to comparing our method with these two prior LSTM-based approaches for suffix prediction, our evaluation also includes the same LSTM-based approach coupled with two other sampling approaches: Top-K [9] and Nucleus sampling [11].

We aim to validate our approach's effectiveness through a comprehensive evaluation of its performance in predicting suffixes and remaining time. This analysis will focus on the rate of activity repetitions within the log, emphasizing our method's superiority in high-repetition scenarios compared to alternative sampling techniques.

Measures of Goodness. The purpose of *EQ2* is to evaluate the closeness of the sequences of activity types generated by different sampling approaches to the ground truth at the control flow level. To evaluate the proposed Daemon Action approach from this perspective, we applied the established Damerau-Levenshtein distance (DL) metric [2].

DL metric enhances the Levenshtein distance by adding a swapping operation, making it more appropriate for evaluating the quality of predicted suffixes. As an illustration, when comparing the sequences $(a1, a2, a3)$ and $(a1, a3, a2)$, the metric assigns a cost of 1.0 for the swap between $a2$ and $a3$.

For two given activity sequences $s_1 = fa(\sigma_1)$ and $s_2 = fa(\sigma_2)$, the similarity is calculated as:

$$\mathrm{SDL}(s_1, s_2) = 1 - \frac{\mathrm{DL}(s_1, s_2)}{\max(\mathrm{len}(s_1), \mathrm{len}(s_2))} \tag{2}$$

Here, $\mathrm{len}(s)$ denotes the length of sequence s, that is, the count of its elements. The value of SDL ranges between $[0, 1]$. It reaches 1.0 when sequences are identical and touches 0.0 when the sequences have completely different elements.

For *EQ3*, to evaluate the closeness of the activity repetitions generated by different sampling approaches to the ground truth at the control flow level, we introduce the Repetitive Activity Similarity (RAS) metric defined as:

$$RAS = 1 - \frac{\sum_{a \in A} |count_G(a) - count_P(a)|}{\sum_{a \in A}(|count_G(a)| + |count_P(a)|)} \tag{3}$$

This metric quantifies the variance in activity frequency between a *ground truth* set(G) and a *predicted suffix* set(P). The activity occurrences are counted in both sets to determine the penalty. The process is repeated for each activity. The summation of the penalties is reported as absolute count difference. It is then normalized by the sum of the maximum possible penalties for all pairs in both sets, P and G. A *RAS* score approaching 1 signifies a high predictive accuracy, suggesting that the predicted activities closely mirror the actual ground truth activities. A score near 0 denotes maximum variance in the frequency of activities in both the predicted and the ground truth set.

The *EQ4* examines the temporal accuracy of predicted case suffixes versus the ground truth and other methods, focusing on the duration between the last activity in the case suffix and case prefix. Temporal accuracy is evaluated using Absolute Error (AE) to measure the time difference between these activities, with Mean Absolute Error (MAE) representing the average discrepancy across cases. The MAE for Remaining Time can be defined as:

$$\text{MAE} = \frac{1}{N}\sum_{i=1}^{N} |t_{\text{actual},i} - t_{\text{predicted},i}| \tag{4}$$

5 Results

This section summarizes the experimental results, addressing the key questions outlined in Sects. 4 and 6 delves into the implications derived from these experiments. For an in-depth review, the complete set of experimental data is available in the supplementary material, which can be accessed online[8].

The experiments compare the proposed approach against three other sampling approaches across three LSTM architectures described in [1]: Specialized, Shared-Categorical, and Full-Shared. For brevity, we report only the top-performing results (e.g., best SDL value) for each sampling approach-dataset combination, rather than all possible combinations with LSTM architectures. The supplementary material contain detailed results, confirming consistent findings across LSTM architectures.

The results across the three experimental questions are summarized in Figs. 1, 2 and 3 and Tables 3, 4 and 5. Figures 1, 2 and 3 are grouped bar charts plotting the accuracy of each of the five approaches (the Daemon Action approach and the four baselines) on each of the datasets. Each group in the chart corresponds to one dataset, and each data series corresponds to one of the approaches. Figure 1 plots the SDL metric, Fig. 2 the RAS metric, and Fig. 3 the MAE metric. All the reported values are means across all cases in the test set (e.g. Fig. 1 reports mean SDL values across the cases in the test set).

Tables 3, 4 and 5 rank the sampling techniques in descending order by repetition index for each dataset. Table 3 showcases SDL rankings, Table 4 for RAS, and Table 5 for MAE metrics. For instance, a '1' in Table 3 for BPI12W under D-Action indicates it scored the highest SDL value for that dataset. When sampling

[8] https://figshare.com/s/90177dbff5f4786a11cd.

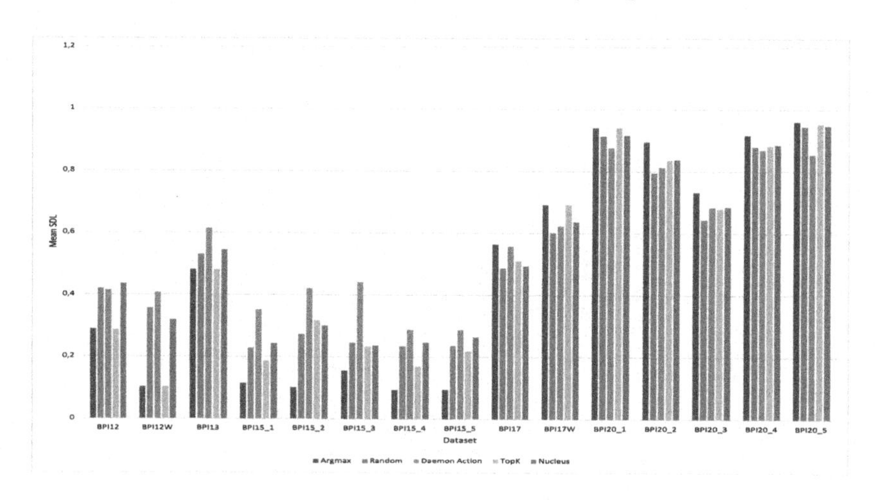

Fig. 1. Mean SDL

approaches tie, they share same rankings, as seen in Table 5 where all methods scored equally on BPI12, each receiving rank 1.

Table 3. Mean SDL Ranks

Dataset	Sampling Approaches				
	ArgMax [19]	Random [1]	D-Action	Top-K	Nucleus
BPI12w	4	2	1	4	3
BPI15-1	5	3	1	4	2
BPI15-2	5	4	1	2	3
BPI15-5	5	3	1	4	2
BPI15-3	5	2	1	4	3
BPI15-4	5	3	1	4	2
BPI12	4	2	3	5	1
BPI13	4	3	1	4	2
BPI17w	1	5	4	1	3
BPI17	1	5	2	3	4
BPI20-3	1	5	3	4	2
BPI20-2	1	5	4	3	2
BPI20-1	1	4	5	1	3
BPI20-5	1	4	5	2	3
BPI20-4	1	4	5	3	2

Regarding *EQ1*, we observe from the Fig. 1 that SDL Score for the Daemon Action approach generally outperforms other sampling approaches for logs with larger vocabularies (#Act's), longer trace lengths and higher repetition index (*Rep.Index*(%)) percentage, such as BPI12w, BPI13 and BPI15. Similarly, the performance of Daemon Action ranked as the second best for BPI2017, closely followed by a marginally different yet high Mean SDL value for BPI2012. Likewise, Table 3 highlights Daemon Action robust performance, leading in 7 out of 15 datasets. Additionally, it's notable that 10 out of these 15 datasets exhibit the highest percentage of activity repetitions. Meanwhile, the Argmax approach, matches this consistency, secures the top position in a different set of 7 out of

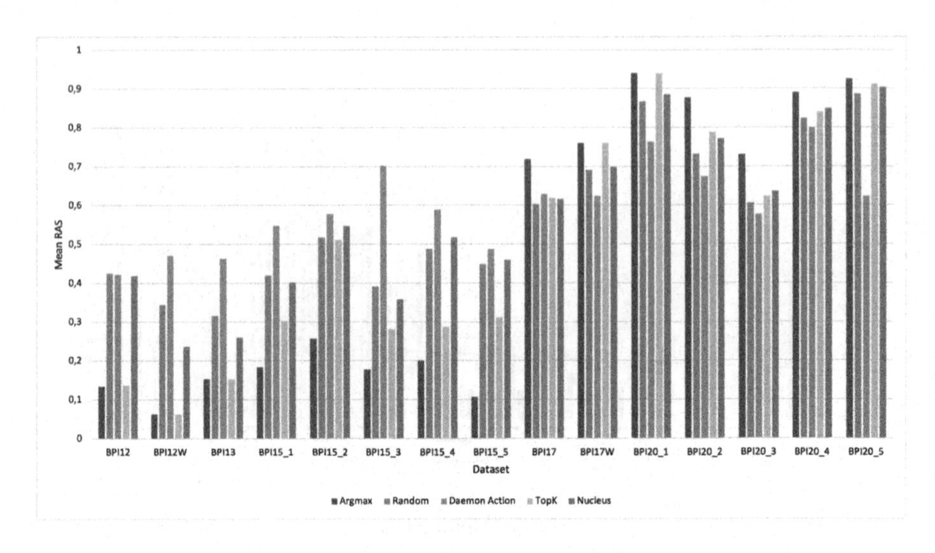

Fig. 2. Mean RAS

the 15 datasets having least number of repetitions. In contrast, approaches like Nucleus, Top-K, and Random Choice show limited efficacy, with Nucleus ranking first only in a dataset like BPI12.

Regarding *EQ2*, Fig. 2 and Table 4 reveal Daemon Action as the most effective at minimizing repetitions, especially in high-repetition logs (e.g., BPI12, BPI12w, BPI13, BPI15, BPI17), with Random Choice also performing well. Argmax stands out in logs with fewer repetitions, like BPI17w and BPI20. Nucleus and Top-K show limited success across the logs, though Top-K slightly performs better in BPI17W and BPI20-1 for Mean RAS scores.

Table 4. Mean RAS Ranks

Dataset	Sampling Approaches				
	ArgMax [19]	Random [1]	D-Action	Top-K	Nucleus
BPI12w	4	2	1	4	3
BPI15-1	5	2	1	4	3
BPI15-2	5	3	1	4	2
BPI15-5	5	3	1	4	2
BPI15-3	5	2	1	4	3
BPI15-4	5	3	1	4	2
BPI12	5	1	2	4	3
BPI13	4	2	1	4	3
BPI17w	1	4	5	1	3
BPI17	1	5	2	3	4
BPI20-3	1	4	5	3	2
BPI20-2	1	4	5	2	3
BPI20-1	1	4	5	1	3
BPI20-5	1	4	5	2	3
BPI20-4	1	4	5	3	2

As for *EQ3*, Fig. 3 shows that the performance of the Daemon Action approach is comparable, but arguably not superior, to that of other sampling approaches. This observation is confirmed in Table 5, which shows many tie breaks between approaches. While the Daemon Action approach ranks highest in 7 out of 15 datasets in this category, so does the Top-K approach (often ex aequo). The Random Choice, Argmax, and Nucleus methods also have a top rank in several datasets. At first glance, one would expect that higher control-flow accuracy (SDL and RAS metrics) would translate into higher temporal accuracy (MAE of remaining time). However, the results suggest that the predictors are able to correctly predict the remaining time of the case, even if there are errors in the prediction of the sequence of activity labels.

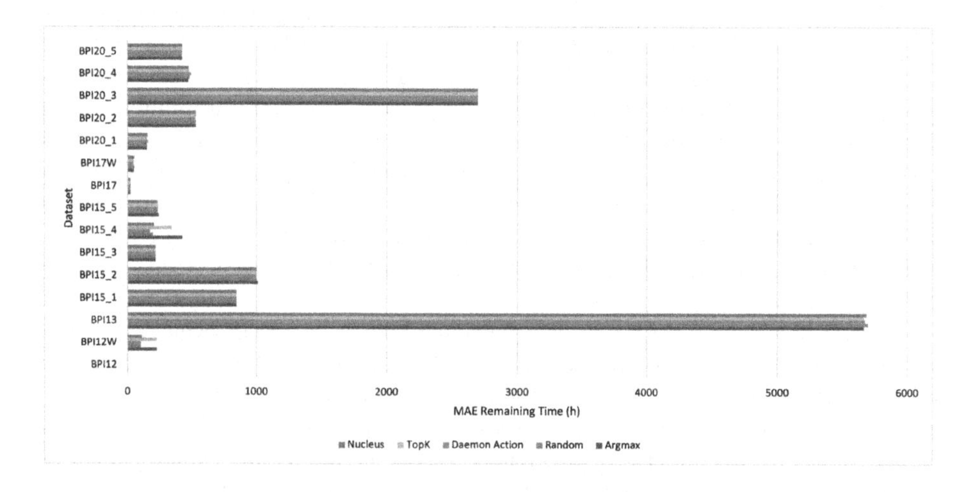

Fig. 3. Remaining Time MAE (h)

Table 5. MAE Ranks

Dataset	Sampling Approaches				
	ArgMax [19]	Random [1]	D-Action	Top-K	Nucleus
BPI12w	4	**1**	2	4	3
BPI15-1	5	4	2	**1**	3
BPI15-2	5	4	2	**1**	3
BPI15-5	5	2	3	4	**1**
BPI15-3	5	2	**1**	4	3
BPI15-4	5	2	**1**	4	3
BPI12	**1**	**1**	**1**	**1**	**1**
BPI13	**1**	5	3	**1**	4
BPI17w	2	5	**1**	2	4
BPI17	4	2	**1**	5	3
BPI20-3	**1**	**1**	**1**	**1**	**1**
BPI20-2	5	2	**1**	3	4
BPI20-1	**1**	3	5	**1**	4
BPI20-5	4	2	5	**1**	3
BPI20-4	**1**	4	5	3	2

6 Discussion

This paper addresses the issue of auto-regressive prediction within deep learning frameworks by proposing a new sampling technique aimed at minimizing looped sequences in predictions. By addressing three key evaluation questions (EQs) mentioned in Sect. 4, it focuses on enhancing the accuracy of sequence predictions made by LSTM models. LSTM's are typically structured at predicting sequential activities but struggle with parallel or recurrent sequences. This directly relates to our first EQ about the accuracy of predicted sequences compared to ground truth, and the second EQ that examines the frequency of repeated activities within these predicted sequences. These EQs are important for assessing the quality of predictions. According to Figs. 1 and 2, current sampling techniques, including Argmax and Top-K, tend to aggravate looping behavior, especially noticeable in event logs with frequent repetitions (e.g., BPI12W, BPI13, BPI15). These method's preference for the most probable next activity often leads to poor performance by promoting repetitive sequences. Random sampling emerges as a relatively better method by selecting the next activity from the LSTM's output distribution randomly, thus reducing activity repetition. However, its disregard for the historical occurrence of activities, a flaw also seen in nucleus sampling, illustrates a limitation. Such biases in sampling strategies contribute to LSTM's generating redundant sequences, highlighting the necessity for our novel approach.

The LSTM-Daemon Action combination underperforms in event logs with limited parallel or repetitive activities, notably in simpler logs such as BPI20, as evidenced by Figs. 1 and 2. Its difficulty lies in accurately predicting activity suffixes, particularly for logs with extensive vocabularies and brief trace lengths like BPI20. This results from its ineffectiveness in balancing exploration and exploitation during the next activity selection, causing suboptimal exploration loops. In these cases, the argmax method often outperforms Daemon Action.

For *EQ3*, concerning temporal performance, Fig. 3 and Table 5 show that the Daemon Action method performs comparably to other sampling strategies, though not always superior, with numerous ties observed. Despite differences in sequence accuracy (SDL and RAS), predictors consistently achieve high precision in predicting remaining time (MAE metric), suggesting that accurate sequence prediction does not guarantee improved temporal accuracy.

In summary, Daemon Action surpasses Argmax, Nucleus, Random, and Top-K in Mean SDL and RAS scores, particularly in logs with high parallelism or repetitions. While Argmax enhanced LSTM's perform well in simpler logs like BPI20, they struggle with more complex ones, as seen in BPI15. The results regarding MAE scores remain consistent, indicating the need for further research to enhance temporal accuracy alongside improved control-flow predictions.

Threats to Validity. We acknowledge threats to our study's external and internal validity. Externally, the generalizability of our findings may be limited by our selection of specific datasets, despite efforts to choose diverse datasets to mitigate this concern. Internally, the study's reliance on LSTM model for sequence prediction, while appropriate for such tasks [17], might bias our results and overlook the potential of other architectures. Future research should consider a broader range of architectures to enrich insights.

7 Conclusion and Future Work

Existing LSTM-based approaches to case suffix prediction often produce suffixes with spurious activity repetitions. This phenomenon can be traced down to the fact that these approaches are trained to predict the next-activity given a case suffix. When an activity occurs frequently, multiple times in a case, these approaches tend to repeatedly predict it as the most likely next activity.

This paper presented an approach that addresses this limitation by using a Daemon Action heuristic to select the next activity in a case suffix, given the distribution of possible next activities produced by an LSTM model. This approach balances between exploitation steps (predicting the most likely activity) and exploration steps (exploring a diverse set of options). The results show that, when layered on top of different LSTM-based predictors, the Daemon Action approach leads to less spurious activity repetitions in the generated case suffixes. The improvement is notable when the approach is applied to business processes with a larger number of parallel or recurrent activity sequences per case.

The experimental evaluation reported in this paper focuses on applying the Daemon Action approach on top of next-activity predictors with LSTM architectures. Future work will aim at evaluating the benefits of the proposed approach when applied to other architectures, such as CNNs and GANs.

While the Daemon Action approach enhances the control-flow accuracy (measured via SDL and RAS scores) relative to baseline sampling techniques, we did not observe any notable enhancements in the temporal accuracy (MAE metric). A direction for future work is to design case suffix prediction approaches that explicitly take into account waiting times between activity instances during the generation of the predicted suffixes. This future work would require exploring different types of neural network architectures in combination with different sampling approaches.

References

1. Camargo, M., Dumas, M., González-Rojas, O.: Learning accurate LSTM models of business processes. In: Hildebrandt, T., van Dongen, B.F., Röglinger, M., Mendling, J. (eds.) BPM 2019. LNCS, vol. 11675, pp. 286–302. Springer, Cham (2019). https://doi.org/10.1007/978-3-030-26619-6_19
2. Damerau, F.: A technique for computer detection and correction of spelling errors. Commun. ACM **7**(3), 171–176 (1964)
3. Di Francescomarino, C., Ghidini, C., Maggi, F.M., Milani, F.: Predictive process monitoring methods: which one suits me best? In: Weske, M., Montali, M., Weber, I., vom Brocke, J. (eds.) BPM 2018. LNCS, vol. 11080, pp. 462–479. Springer, Cham (2018). https://doi.org/10.1007/978-3-319-98648-7_27

4. Di Francescomarino, C., Ghidini, C., Maggi, F.M., Petrucci, G., Yeshchenko, A.: An eye into the future: leveraging a-priori knowledge in predictive business process monitoring. In: Carmona, J., Engels, G., Kumar, A. (eds.) BPM 2017. LNCS, vol. 10445, pp. 252–268. Springer, Cham (2017). https://doi.org/10.1007/978-3-319-65000-5_15
5. Dorigo, M., Birattari, M., Stutzle, T.: Ant colony optimization. IEEE Comput. Intell. Mag. **1**(4), 28–39 (2006)
6. Dorigo, M., Caro, G.D.: Ant colony optimization: a new meta-heuristic. In: CEC, pp. 1470–1477. IEEE (1999)
7. Dorigo, M., Maniezzo, V., Colorni, A.: Ant system: optimization by a colony of cooperating agents. IEEE Trans. Syst. Man Cybern. Part B **26**(1), 29–41 (1996)
8. Dumas, M., Rosa, M.L., Mendling, J., Reijers, H.A.: Fundamentals of Business Process Management. Springer, Heidelberg (2018). https://doi.org/10.1007/978-3-662-56509-4
9. Fan, A., Lewis, M., Dauphin, Y.N.: Hierarchical neural story generation. In: ACL (1), pp. 889–898. Association for Computational Linguistics (2018)
10. Gunnarsson, B.R., vanden Broucke, S., De Weerdt, J.: A direct data aware LSTM neural network architecture for complete remaining trace and runtime prediction. IEEE Trans. Serv. Comput. **16**(4), 2330–2342 (2023)
11. Holtzman, A., Buys, J., Du, L., Forbes, M., Choi, Y.: The curious case of neural text degeneration. In: ICLR. OpenReview.net (2020)
12. Kratsch, W., Manderscheid, J., Röglinger, M., Seyfried, J.: Machine learning in business process monitoring: a comparison of deep learning and classical approaches used for outcome prediction. Bus. Inf. Syst. Eng. **63**(3), 261–276 (2021)
13. Kubrak, K., Milani, F., Nolte, A., Dumas, M.: Design and evaluation of a user interface concept for prescriptive process monitoring. In: Indulska, M., Reinhartz-Berger, I., Cetina, C., Pastor, O. (eds.) Advanced Information Systems Engineering. CAiSE 2023. LNCS, vol. 13901, pp. 347–363. Springer, Cham (2023). https://doi.org/10.1007/978-3-031-34560-9_21
14. Lin, L., Wen, L., Wang, J.: MM-PRED: a deep predictive model for multi-attribute event sequence. In: SDM, pp. 118–126. SIAM (2019)
15. Di Mauro, N., Appice, A., Basile, T.M.A.: Activity prediction of business process instances with inception CNN models. In: Alviano, M., Greco, G., Scarcello, F. (eds.) AI*IA 2019. LNCS (LNAI), vol. 11946, pp. 348–361. Springer, Cham (2019). https://doi.org/10.1007/978-3-030-35166-3_25
16. Pasquadibisceglie, V., Appice, A., Castellano, G., Malerba, D.: Predictive process mining meets computer vision. In: Fahland, D., Ghidini, C., Becker, J., Dumas, M. (eds.) BPM 2020. LNBIP, vol. 392, pp. 176–192. Springer, Cham (2020). https://doi.org/10.1007/978-3-030-58638-6_11
17. Rama-Maneiro, E., Vidal, J.C., Lama, M.: Deep learning for predictive business process monitoring: review and benchmark. IEEE Trans. Serv. Comput. **16**(1), 739–756 (2023)
18. Scianna, M.: The AddACO: a bio-inspired modified version of the ant colony optimization algorithm to solve travel salesman problems. Math. Comput. Simul. **218**, 357–382 (2024)
19. Tax, N., Verenich, I., La Rosa, M., Dumas, M.: Predictive business process monitoring with LSTM neural networks. In: Dubois, E., Pohl, K. (eds.) CAiSE 2017. LNCS, vol. 10253, pp. 477–492. Springer, Cham (2017). https://doi.org/10.1007/978-3-319-59536-8_30

20. Taymouri, F., Rosa, M.L., Erfani, S., Bozorgi, Z.D., Verenich, I.: Predictive business process monitoring via generative adversarial nets: the case of next event prediction. In: Fahland, D., Ghidini, C., Becker, J., Dumas, M. (eds.) BPM 2020. LNCS, vol. 12168, pp. 237–256. Springer, Cham (2020). https://doi.org/10.1007/978-3-030-58666-9_14
21. Teinemaa, I., Dumas, M., La Rosa, M., Maggi, F.M.: Outcome-oriented predictive process monitoring: Review and benchmark. ACM Trans. Knowl. Discov. Data **13**(2), 17:1–17:57 (2019)
22. Toosinezhad, Z., Fahland, D., Köroglu, Ö., van der Aalst, W.M.P.: Detecting system-level behavior leading to dynamic bottlenecks. In: ICPM, pp. 17–24. IEEE (2020)
23. Verenich, I., Dumas, M., La Rosa, M., Maggi, F.M., Teinemaa, I.: Survey and cross-benchmark comparison of remaining time prediction methods in business process monitoring. ACM Trans. Intell. Syst. Technol. **10**(4), 34:1–34:34 (2019)

Which Legal Requirements are Relevant to a Business Process? Comparing AI-Driven Methods as Expert Aid

Catherine Sai[1], Shazia Sadiq[2], Lei Han[2], Gianluca Demartini[2], and Stefanie Rinderle-Ma[1](✉)

[1] TUM School of Computation, Information and Technology, Technical University of Munich, Garching, Germany
{catherine.sai,stefanie.rinderle-ma}@tum.de

[2] School of Information Technology and Electrical Engineering, University of Queensland, Brisbane, Australia
shazia@itee.uq.edu.au, {l.han,g.demartini}@uq.edu.au

Abstract. Organizations are obliged to ensure compliance with an increasing amount of regulatory requirements stemming from laws, regulations, directives, and policies. As a first step, it is to be determined which of the requirements are relevant in a certain context, depending on factors such as location of the organization and the business processes. For the processes, the identification of relevant requirements can be detailed by an assessment of which parts of each document are relevant for which step of a given process. Nowadays the identification of process-relevant regulatory requirements is mostly done manually by domain and legal experts, posing a tremendous workload due to the extensive number of regulatory documents and their frequent changes. Hence, this work examines how organizations can be assisted in the identification of relevant requirements for their processes based on embedding-based NLP ranking and generative AI. The evaluation highlights strengths and weaknesses of both methods regarding applicability, automation, transparency, and reproducibility. The evaluation results lead to guidelines on which method combinations will maximize benefits for given characteristics such as process usage, impact, and dynamics of an application scenario.

Keywords: Business Process Compliance · Regulatory Relevance · Legal Information Retrieval · Large Language Models

1 Motivation

In our globalized world, organizations are faced with keeping track of an increasing amount of requirements from various sources to stay compliant. Yet, identifying relevant requirements for a business process from vast amounts of documents requires extensive manual work by highly qualified legal and domain experts [8,31]. Recent work [29] defines 13 Regulatory Compliance Assessment Solution Requirements (RCASR) of which this study investigates RCASR 1 (regulatory document relevance) by identifying which regulatory documents a company

J. Araújo et al. (Eds.): RCIS 2024, LNBIP 513, pp. 166–182, 2024.
https://doi.org/10.1007/978-3-031-59465-6_11

needs to comply with and RCASR 2 (content relevance) by identifying which part of the document is relevant. As businesses become increasingly complex, we further analyze the relevance within the business by not only identifying if a regulatory text is relevant for the entire organization but for a specific process, the sub-processes within the process, and the tasks and throwing events (as these are the active elements) within each sub-process (cf. Fig. 1).

Our work aims to develop novel technological solutions to help reduce compliance burdens and breaches. Relevance identification is a challenging and high-stakes task that is currently performed manually by experts [8].

We evaluate the capabilities of two automated methods: an embedding-based NLP ranking and generative AI as a possible aid for the experts to better cope with the vast amounts of regulatory texts. Specifically, the study focuses on the design of a hybrid system to aid the regulation implementation experts in their compliance assessments. We analyze for which process level granularity (Fig. 1) relevance can be identified in sufficient quality and how the selected methods can be used and possibly combined to best support the human experts.

Consequently, this work offers the following contributions:

- **feasibility analysis of state-of-the-art NLP and generative AI methods** to retrieve regulatory relevant texts in a complex legal and business setting for multiple levels of business process detail
- **novel approach** to identify relevant regulatory requirements for specific process aspects, resulting in scenario-based application recommendations for the analyzed methods and their combinations
- **publicly available data set** with two real-world use cases, their models, process descriptions on all three levels (Fig. 1), and the mapped regulatory text passages, aiming to facilitate further research in the area

The paper is structured as follows: The analyzed aspects, approach and methods are presented in Sect. 2, while Sect. 3 provides the study implementation details. Evaluation and discussion of the findings are presented in Sects. 4 and 5. Section 6 discusses related work, followed by Sect. 7 concluding the paper.

2 Approach

2.1 Analyzed Aspects

Regulatory documents can i.a. vary in their structure, level of detail, geographic region and domain of applicability. The diversity of regulatory documents [31] should be reflected in the data set, therefore, this study includes regulatory documents from multiple countries and domains as well as domain-independent documents. Additionally, regulatory documents can stem from business external sources (i.e., laws, regulations, directives) or business internal sources. All internal documents are automatically business-relevant, but might not be relevant for the process at hand. Additionally, not all internal documents are publicly available. This study includes external documents and publicly available internal documents, representing the following 3 groups of regulatory documents, reflecting different combinations of relevance:

- **Group A** contains internal and external documents that are both, business-relevant and process-relevant.
- **Group B** contains external documents that are business-relevant and process-irrelevant.
- **Group C** contains external documents that are business-irrelevant and process-irrelevant.

Furthermore, for the **relevance judgement**, the process relevance is further detailed into 3 levels (cf. Fig. 1). We define 4 levels of relevance, level 0 being the most general and level 3 the most specific. After level 0 identifies relevance for the business as a whole, level 1 assesses if a regulatory requirement is relevant for a specific process within the business. Even more detail is provided by level 2, which identifies if a requirement is relevant for a sub-process within a process. Finally, level 3 displays the deepest aspect of a process by identifying relevance for a specific task or throwing event. Through this, we want to analyze if the methods are feasible for all or only certain levels of relevance identification and identify the necessary circumstances for applicability.

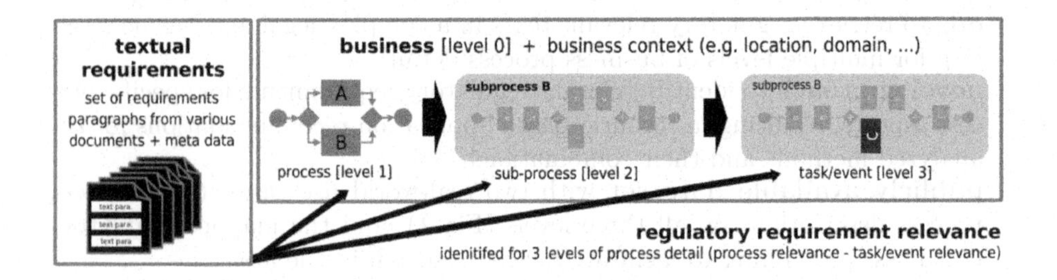

Fig. 1. Overview of study aspects analyzed

Moreover, interviews with three regulation implementation experts led us to distinguish two kinds of relevance: *Compliance relevance* is a description of an action an organization has to fulfill to be compliant with a regulation (relevance in the stricter sense). *Informative relevance* applies when texts contain information related to the process but do not require a clear action by the organization (e.g. customer obligations to fulfill policy requirements).

For the **business** side, **processes** with multiple relevant regulatory documents should be chosen. The business (level 0) is characterized through meta information, similar to the regulatory documents meta information (location, domain, etc.). The processes are visualized as business process models and contain textual descriptions at all three process levels (level 1–3, cf. Fig. 1), as *"textual process descriptions are widely used in organizations"* and hence *"provide a valuable source for process analysis, such as compliance checking"* [1]. The presented design ensures that the approach is validated by sources from various origins.

2.2 Method Analysis

The study aims at improving the relevance identification of regulatory texts for business processes. This task is currently performed manually by an extensive expert analysis of potentially relevant regulatory texts. One dimension for improvement that the study is capturing is therefore the direction towards a (semi-)automated approach, shown in the horizontal axis of Fig. 2. Furthermore, the improvement deals with the regulatory compliance of businesses. This is a critical field where violations are quickly resulting in a big negative impact on a business's finances and reputation. Therefore, it needs to be understandable for the stakeholders how the method reached its relevance judgment and the relevance judgment needs to be consistent, meaning the method should reach the same result when given the same input. This dimension is measured by the vertical axis of transparency and reproducibility (cf. Fig. 2).

The **Expert Analysis** is the basis for the approach as on the one hand it is the current standard way to perform the relevance judgment and on the other hand, it is necessary to create the gold standard to evaluate the other methods quantitatively. This method is purely manual and also transparent and reproducible as the experts come to reasoned relevance judgments and document these.

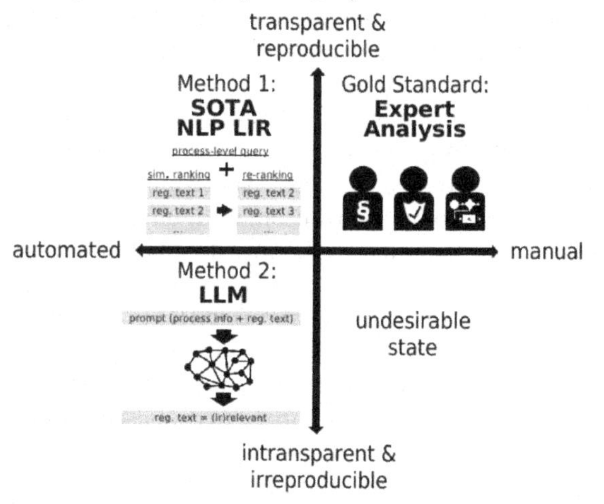

Fig. 2. Overview of methods applied for process relevance identification

[8] identified that legal experts, followed by domain experts perform best in interpreting legal requirements and definitions. An assessment with five business auditors conducted by [31], reveals that it is not clear at the beginning of an audit which and how many regulatory documents need to be considered for compliance. The collection of documents also increases iteratively as one document refers to another document. Thus, the *"search for [...] regulatory documents is often performed exploratively and can take some time"* [31]. For this study, we interviewed three experts, responsible for ensuring that the regulatory requirements are fulfilled within their business. Their position can be considered as a combination of domain experts and legal experts as they work closely together with both legal and domain teams and need to understand both sides to ensure regulatory compliance. The combined challenges **[C]** for the expert analysis elicited from [8,31] and our interviews are:

- high degree of **manual** work/ lack of automation approaches **[C1]**
- lack of **context** as the search for relevant documents is often only based on a few keywords [31] **[C2]**
- need for **highly skilled workers** to correctly interpret legal requirements, which are costly and hard to obtain [8] **[C3]**
- **decentralized** data (e.g. multiple tools, excel sheets, documents) **[C4]**
- **variety of regulatory documents** which *"differ greatly in terms of subject matter as well as structure, vocabulary, and level of detail"* [31] **[C5]**

Based on these challenges and the dimensions *automation* and *transparency/reproducibility* (cf. Fig. 2), we select the other methods to be compared in the study. The aim is a semi-automated approach that supports the experts, knowing that full automation is neither technologically feasible nor desired from a responsibility and reliability point of view **[C1+C3]**. The identification input includes as much context information as feasible by each methods nature **[C2]** and we aim at a generalizable approach that can deal with various regulatory documents **[C4+C5]**.

As a second method, our approach analyses the applicability of state-of-the-art natural language processing legal information retrieval (**SOTA NLP LIR**), ranking the regulatory texts according to their relevance for the process query. This can be completely automated, offers full reproducibility, and is also explainable and transparent in the way it calculates the relevance ranking based on word embeddings.

Legal Information Retrieval (LIR) is concerned with, e.g., retrieving supporting case laws based on a query case (cf. COLIEE 2023 [7]) or key phrases [31]. Yet, no approach exists, that analyses the applicability of these methods for the retrieval of relevant regulatory texts based on detailed business process information. Our application scenario poses new challenges compared to the existing use cases as e.g. with the complex business process context, we have more information to consider for the query design than a 2–3 word key phrase. The similarity computation in LIR can be lexical, which needs an exact match, or semantic, which can identify semantically similar terms [12,19]. [3,31] identifies the combi-

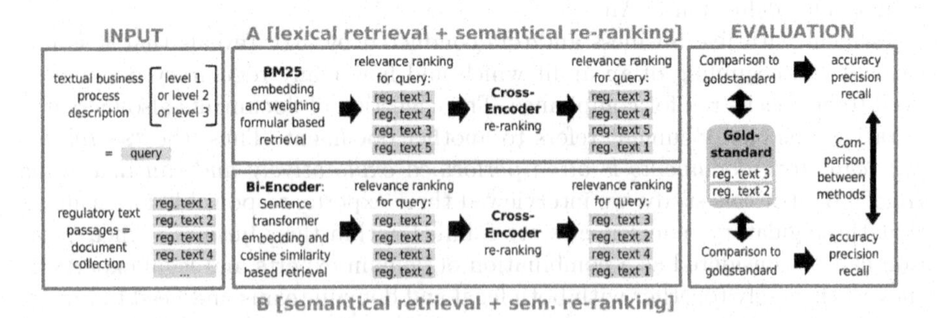

Fig. 3. SOTA NLP LIR design inspired by [26,31]

nation of a lexical and semantic search with a cross encoder re-ranking as most promising to retrieve potentially relevant text passages based on an input query.

Figure 3 illustrates the steps of our SOTA NLP LIR method for relevance judgment concerning regulatory business process requirements. We implemented two state-of-the-art NLP LIR methods, one utilizing lexical and semantical retrieval and the other purely semantical, to use the better-performing method as a baseline automation approach. **BM25** is a lexical ranking function based on the "bag-of-words" approach in NLP, where words are considered independent of their position in the text [27]. Using measures like term frequency (TF) and inverse document frequency (IDF), the words are transformed into a mathematical representation (also called embedding) and weighed in a certain formula to evaluate their ranking. In contrast to BM25, both semantic components (Bi-Encoder and Cross-Encoder) rely on sentence-transformer models. **Sentence Transformer** is a transformer-based embedding framework that, takes the context of the word into consideration [26]. The **Bi-Encoder** [26] passes the texts independently to Sentence-Transformer and compares the resulting embeddings based on cosine similarity. This set-up is less computationally intensive (than, e.g., Cross-Encoder) and thus recommended for information retrieval tasks as an initial step to retrieve a pre-selection of potentially relevant texts from a larger collection.

Cross-Encoder [26], like Bi-Encoder is a semantics-based encoder, but *"Cross-Encoder does not produce a sentence embedding"*[1]. Both texts are passed together to the Sentence-Transformer and the result is a similarity classification between 0 and 1. Cross-Encoders deliver better results in relevancy ranking. However, they are too computationally intensive to use directly on large document collections.

Method A is based on a lexical search with BM25, re-ranking the retrieved passages with a semantic Cross-Encoder. Method B already bases the initial retrieval stage on semantic search with Bi-Encoder and also re-ranks all retrieved passages with the semantic-based Cross-Encoder.

Large language models (**LLMs**) are currently considered the best supportive system for various tasks that were thus far performed manually and have also shown great results for classifying texts (cf. Sect. 6). This study analyses their capability to support the relevance identification for business processes. LLMs only need a well-designed prompt to perform their relevance judgment which can be automatically generated from the process information, they are thus a completely automated solution. However, the LLM relevance judgment can differ if given the same prompt multiple times and it is unclear how the relevance is calculated (low transparency/reproducibility).

For the generative AI Study a zero-shot approach is chosen, as they have performed well for multiple tasks [18] and for the alternative of few-shot learning, use case-specific examples would need to be created which would be against our design principle of creating an approach as general and automated as possible. In this kind of setup, the prompt used for the generation needs to be carefully

[1] https://www.sbert.net/examples/applications/cross-encoder/README.html.

designed. [32] propose a "legal prompt stack", where the prompt should consist of the information needed to answer the task (in their case the legal document text), in our case the business context and requirements text. Additionally, a description of what the model is expected to do is required. In [4] prompt engineering for BPM applications is discussed, stating that a task description is needed for zero-shot design and that prompt templates are task-specific. They also state that it is an open challenge to create prompt templates for business processes and their complex representations.

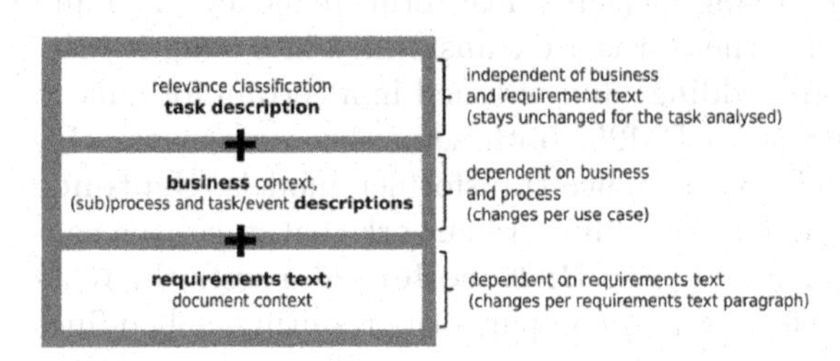

Fig. 4. Prompt design components in this study

Based on these existing prompt proposals and our application setting, we design the prompt consisting of 3 parts, as shown in Fig. 4: 1) a description of the task to be performed, 2) information about the business process: organizational context, the process description, sub-process descriptions and event descriptions, 3) information about the regulation: one paragraph from a regulatory text and regulatory document context information. Part 1) stays the same for all runs, 2) is the same per process and 3) is changed for each run with a different regulatory document text.

3 Implementation

LLMs are designed to handle large amounts of context data while the existing LIR methods are based on comparably very short queries. Due to this, while the same input data (cf. Table 1) is used for both methods, the information is provided in different forms (cf. Sect. 2.2). The created data sets, prompts for the generative AI and the SOTA NLP implementation are publicly available on Github[2].

3.1 Expert Analysis

Following the described criteria (cf. Sect. 2.1) and procedure (cf. Sect. 2.2) two suitable business processes with a composition of relevant and irrelevant regulatory documents for these are identified and a gold standard is created. The procedure is, that for each of the regulatory text paragraphs, the relevancy type (irrelevant, compliance relevant, or informative relevant) is decided for the process (level 1). The processes identified as relevant are then further labeled at process levels 2 and 3 (cf. Fig. 1). For use case 1 the gold standard is created and

[2] https://anonymous.4open.science/r/regulatory_relevance4process-D73C.

reviewed with three experts responsible for the regulation implementation in the business, for use case 2 the gold standard is created and reviewed independently by two business process experts.

Regulatory Requirements. This study analyzes text paragraphs from 7 regulatory documents. 5 of these originate from Australia, 1 from the UK and one is a business internal document that is publicly available. The later document is anonymized for the study (e.g., company name replaced by a placeholder, company address deleted) and is excluded from the published data set to preserve the anonymity of our industry cooperation partner. A random selection of all textual content paragraphs from the described documents is extracted, excluding table of contents, tables, and definitions. As the regulatory text passages are extracted from different documents, some context about the regulatory document (title, section title, subsection) is included together with the document meta-information (i.e. applicable country).

Business Process. In 2021, [35] identified 404 primary studies for their survey in the field of Natural Language Processing for Requirements Engineering. However, only 7% of these were evaluated in an industrial setting, which *"highlights a general lack of industrial evaluation of NLP4RE research results"* [35]. Therefore, for this study, we cooperate with an industry partner from the insurance domain for use case 1.

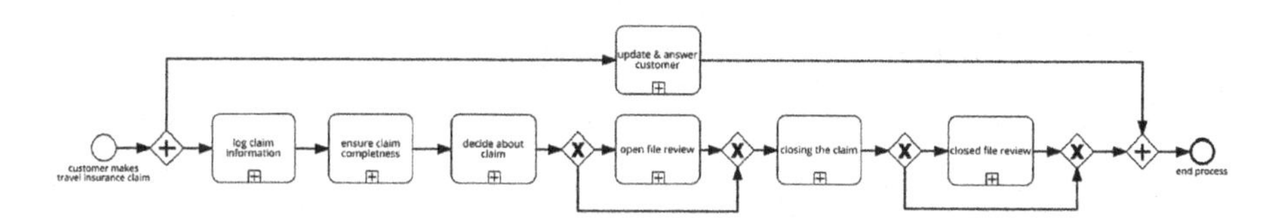

Fig. 5. Travel insurance claim process modeled with SAP Signavio

The *use case of travel insurance claims* needs to comply with multiple regulatory documents. As neither a process model nor textual process descriptions

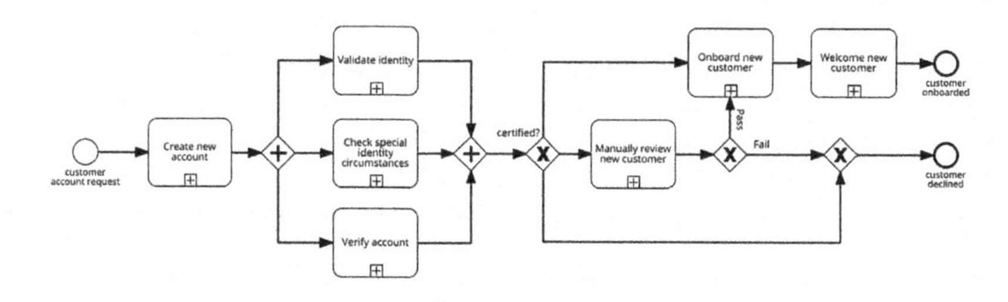

Fig. 6. KYC process model based on SAP Signavio workflow example

existed at our industry partner, we create them together with domain experts in an iterative manner. The process deals with the handling of travel claims, customers might make against their travel insurance company. Figure 5 displays the root process with 7 underlying sub-processes.

The *use case Know Your Customer (KYC)* from the banking industry is chosen as it is an international standard that needs to comply with different regulatory documents depending on the application country which made it an interesting case for our study. The BPMN2.0 model (cf. Fig. 6) is based on an SAP Signavio workflow of KYC[3], enriched with textual descriptions created by business process experts and generative AI. KYC describes a process in the financial domain to ensure a potential customer's identification and due diligence procedures[4].

3.2 SOTA NLP LIR

The two embedding-based ranking methods are implemented as described in Sect. 2.2. Table 1 illustrates the input data used (cf. Sect. 2.1 for details about the groups). For the prompt, the business process descriptions were used, e.g. for use case 1, level 3: the 31 task descriptions were used individually as prompt to rank the 489 regulatory text passages (of use case 1) in their relevance for the analyzed task description.

Table 1. Input data by use case, process level, and regulatory text group

Process	Business data			Regulatory text				
	Lvl 1 process	Lvl 2 sub-pr	Lvl 3 tasks	10% Grp. A Compl. rel	Inform. rel	45% Grp. B	45% Grp. C	(100%) total
1: Insurance	1	7	31	21	28	220	220	489
2: Banking	1	7	19	24	7	140	140	311

3.3 GPT-4

The prompts are based on the design described in Sect. 2.2 and created in three iterations. After each iteration, the results are evaluated on a small test set of 12 regulatory texts. Depending on this analysis the prompt is adjusted and evaluated again. For iteration 2, the information about the regulatory text is extended with the automatically generated passages about the regulatory documents content and applicability. Additionally, the task description is enriched with the instruction to only match clear relations between regulatory text and business process. For iteration 3, the changes from iteration 2 are kept and the task description is extended by the information that recall is the most important

[3] https://www.signavio.com/de/workflow-beispiele/bankingknow-customer-kyc/.

[4] https://corpgov.law.harvard.edu/2016/02/07/fincen-know-your-customer-requirements/.

measurement for this task. The final prompts contain between 1500–2300 words, depending on the business case and requirements text. After three iterations, the resulting, general prompt is run on all 489 regulatory paragraphs for case study 1 and 311 regulatory paragraphs for case study 2 and evaluated against the gold standard. The input data used for the prompts is shown in Table 1.

4 Evaluation

4.1 Results

The quantitative evaluation of the study is based on accuracy, precision, and recall. Precision and recall are *"the two most frequent and basic measures for information retrieval effectiveness"* [17]. As our application is also similar to a text classification, we, like [13] in their generative AI study, add accuracy to our metrics. To calculate these metrics, the values True Positive (TP), False Positive (FP), True Negative (TN), False Negative (FN) are defined as follows: TP = relevant regulatory text is identified as relevant; FP = irrelevant regulatory text is identified as relevant; FN = relevant regulatory text is not identified as relevant; TN = irrelevant regulatory text is not identified as relevant.

Each process level (Fig. 1) is evaluated in order to assess the value of each method for each step and discuss possible combinations of approaches.

For the given business application, it is crucial to identify all relevant regulatory requirements. If too many are identified by the automated system, the following step of a human expert-revision can remove these false positives but if relevant texts are not retrieved they cannot be considered in further steps and thus would be left out of process compliance assessments. As [17] states: *"paralegals [...] are very concerned with trying to get as high recall as possible, and will tolerate fairly low precision results in order to get it."* Thus, we aim for a recall as close to one as possible, accepting low precision values as a trade-off.

Expert Analysis. Presenting our approach (cf. Sect. 2.2) to the experts responsible for the regulation implementation in the business, they confirm that the study results would aid their compliance assessment work.

For evaluating of the newly proposed methods, we compare their results with the gold standard retrieved by the expert analysis (cf. Sect. 3.1). Thus, we assume that the expert analysis did not mislabel anything and has therefore an accuracy, precision, and recall of 1.

SOTA NLP LIR. Table 2 shows the metrics achieved with the two in Sect. 2.2 introduced methods. To put the following numbers into perspective, the recall of the winning team in the (in Sect. 6 mentioned LIR competition) *COLIEE* 2023 task 1 was 0.41[5].

[5] https://sites.ualberta.ca/~rabelo/COLIEE2023/task1_results.html.

As the process relevance levels change from general to more specific, the recall values for both methods decrease. This might imply that as the relevance levels become more specific, it becomes more challenging for these methods to retrieve relevant information.

Table 2. SOTA NLP LIR results by process level and method

process level	method	use case 1			use case 2		
		Acc.	Prec.	Rec.	Acc.	Prec.	Rec.
level 1: process relevance	BM25+CE	0.76	0.19	0.43	0.74	0.11	0.23
	BI-E.+CE	0.75	0.18	0.41	0.74	0.13	0.29
level 2: sub-process relevance	BM25+CE	0.95	0.15	0.32	0.94	0.09	0.23
	BI-E.+CE	0.95	0.18	0.39	0.94	0.11	0.31
level 3: task/event relevance	BM25+CE	0.97	0.09	0.23	0.97	0.08	0.14
	BI-E.+CE	0.97	0.13	0.35	0.97	0.13	0.24

In general, the performance for use case 2 is worse than for use case 1. This could be an indicator that the process information provided for use case 2 yields less information. In both use cases and every process level but use case 1, level 1, the method BI-Encoder + Cross-Encoder outperforms the BM25 + Cross-Encoder method in terms of recall. Also, the results for use case 1, level 1 are very close (cf. Table 2). Therefore, BI-Encoder + Cross-Encoder is the recommended method for the SOTA NLP LIR aiming at business process relevance identification.

GPT-4. The output of the LLM is a binary judgment of the regulatory paragraphs relevance. For both use cases, GPT-4 achieves a very high recall on level 1, however for this multiple false positives have to be accepted, causing a lower precision. The results by process level (cf. Table 3) show that the more detailed the relevance Identification (cf. level 2 and 3), the lower the recall as the differentiation between the aspects of the level becomes less clear. For the analysis by type of regulatory text (cf. Table 4) an evaluation based on precision and recall would not be insightful, as only group A contains TP values. The accuracy for Group B is the lowest, as this is the most ambiguous: relevant to the business but irrelevant to the process.

Table 3. GPT-4 results by process level

	use case 1			use case 2		
	Acc.	Prec.	Rec.	Acc.	Prec.	Rec.
level 1: process relevance	0.81	0.34	1.00	0.90	0.51	0.97
level 2: sub-process relevance	0.89	0.71	0.81	0.76	0.58	0.62
level 3: task/event relevance	0.75	0.64	0.69	0.82	0.93	0.77

Table 4. Accuracy by type of regulatory text origin for business process level 1

	use case 1	use case 2
Group A (business relevant, process relevant)	1.00	0.97
Group B (business relevant, process irrelevant)	0.65	0.79
Group C (business irrelevant, process irrelevant)	0.92	1.00

4.2 Comparison

Compared to the gold standard created by expert analysis, none of the other three methods achieves recall results over all three process levels and for both use cases that would allow the method to replace the expert analysis but this was already expected and anticipated. For both methods, the results for use case 1 are overall better than use case 2, and there is a clear tendency that the more detailed the level, the less reliable the relevance judgment.

As described in Sect. 2.2 a generative AI like GPT-4 is able to consider vast amounts of context information as input for the relevance judgment. This seems to be advantageous in complex business settings as displayed by the two use cases of this study.

The SOTA NLP LIR method was designed similarly to best-performing methods for Legal Information Retrieval tasks. However, those tasks do not have to consider a complex business context for their retrieval. Being based on a simple prompt of the process level description, the implemented SOTA NLP LIR approach lacks the inclusion of the wider business context, which seems to cause its poorer performance compared to the other approaches.

Consequently, ranking the methods based on their recall results, GPT-4 performs best with recall values between 0.62–1.0, while SOTA NLP LIR reaches recall values between 0.24–0.41, depending on the use case and process level (1–3).

However, as introduced in Sect. 2.2, there are other aspects to be considered for a holistic analysis which will be discussed in Sect. 5.

5 Discussion

5.1 Implications

This section discusses the implications of the comparative study regarding applicability, automation, transparency, and reproducibility as introduced in Fig. 2. As the analyzed methods display extreme cases along these aspects, combinations and intermediate stages between the investigated methods are conceivable to achieve greater *applicability* for various use cases. If either method is combined with the expert analysis, e.g., for an automated pre-selection for the experts, it could decrease the manual workload for reference identification immensely.

Table 5. Selected process scenarios with recommended method (combination)

process usage	**process impact**	**reg. input and process and reg. dynamics**	**recommended method (combination)**
low-high	high	low	expert analysis
high	high	high	SOTA NLP LIR + exp. analysis
high	low	high	GPT-4 + exp. analysis

Therefore, we evaluate the results from all four methods in comparison and with regard to possible applicability as human-expert aid. Table 5 recommends method combinations for selected process scenarios aiming to balance the need for adaptability, regulatory compliance, and expert oversight in each scenario. The process scenarios are described by the following three characteristics:

- **process usage:** How often is process executed? Indicates the amount of data available (e.g. for fine-tuning) and the complexity of the processes (more executions tend to lead to more variations).
- **process impact:** What impact (monetary, reputation) would a violation in this process have? Indicates the need for a high recall value.
- **reg. input and process and reg. dynamics:** How many regulatory documents need to be considered and how often do they and the associated process change? Indicates the need for automation.

As shown in Table 5, each method has a value as human-expert aid for a certain application scenario. The **expert analysis** is the best choice in case of a process with high impact demanding a solution with high transparency and reproducibility, while at the same time, low dynamics and regulatory input allow for a purely manual solution. **SOTA NLP LIR + expert analysis** is recommended, if the impact of the process is high and, at the same time, the dynamics and regulatory input are also too high to be handled purely manually. **GPT-4 + expert analysis** is the method of choice if the process is frequently used, with a highly dynamic and vast regulatory input while at the same time the process impact is low, allowing for a less transparent automation support.

The generative AI shows reliable results on the process (level 1) and can thus be a real aid to experts for relevance judgments. Also, further improvements to both automation approaches could be achieved by fine-tuning the models to a given business and process setting.

5.2 Limitations

In the following, we discuss selected limitations of the presented study.

Subjectivity and Level of Detail in Business Process Models: A process model provides an abstraction of reality which depends on the modeler, cf. [22] for a more detailed analysis of this challenge. Yet, the regulatory relevance judgement depends on the abstraction levels of text and process model.

Comparability Between Process and Regulatory Information: The challenge of ontological alignment between business and regulatory texts has been observed in previous studies [28, 29]. Examples include differences in active vs. passive style, length of sentences, or being written from different perspectives.

Identifying the borders between business and processes relevance: False positive regulatory texts are mainly originating from group B, which are relevant for the business in general, but not for the specific process at hand.

Missing Reference Considerations: Legal texts often include references to other sections within the same document and across documents. As our study is designed to evaluate the relevance of each text paragraph, independently of each other, information exploiting references is currently not accessed.

Data Privacy Issues for Closed-Access Documents: GPT-4 poses challenges in handling secret (e.g. business internal requirements) or payable (e.g. ISO-Norm) documents.

Cost: No closer consideration of the cost for the methods in relation to the time saved for experts solving the task without aid was performed.

6 Related Work

Business process compliance is a complex problem, due to the *"scale and diversity of compliance requirements and additionally the fact that these requirements may frequently change"* [28]. The majority of existing business process compliance approaches are based on formalized constraints [9,11,28]. [6,30,34] extract machine-readable compliance requirements from legal natural language text, [20] compares process descriptions with model-based process descriptions, [1] is concerned with the interpretation improvement of textual process descriptions, and [33] assess compliance of process models with regulatory documents. [23] defines requirements for business process compliance monitoring, but does not cover how the compliance requirements (i.e. relevant laws, regulations) are identified. [10,28] describe that business processes and business requirements are designed and handled independently of each other, i.e., all of the above mentioned approaches take relevance as an implicit assumption.

While **legal information retrieval (LIR)** identifies relevant legal data based on other legal data (e.g. information on a current legal case), this study identifies relevant legal data based on business process data. Most LIR approaches are based on similarity computation between a search term and the text corpus [17], with the search term consisting of a few keywords, rather than a long, complex prompt as in our case. Current research [16,24,31] in the field of regulatory document ranking and retrieval show, that a *"combination of lexical and semantic retrieval models leads to the best results"* [2]. Major research in this area originates form the Competition on Legal Information Extraction/Entailment (COLIEE) [15]. The tasks about relevant text retrieval are related to our problem. However, the fact, that there are different tasks for case and statute law already indicates how different the approaches are even when retrieving relevant law texts based on other relevant law texts of the same type. Another recent work is concerned with classifying regulatory documents as business-relevant or business-irrelevant [5]. Our approach focuses on the identification of the relevant processes within the business and even more detailed, which process parts are affected. In summary, no LIR analysis exists to identify relevant requirements for a specific business process or even more specific task of a business process as presented in this work.

Our application area of **generative AI** falls into the field of text classification, as we want to classify a regulatory text as relevant or irrelevant, given a process scenario and context information. Recent work identifies *"ChatGPT has tremendous potential in text classification tasks"* [21] and *"great potential of ChatGPT as a data annotation tool"* [13]. As of January 2024, the latest GPT (Generative Pre-trained Transformer) version is GPT-4. OpenAI introduces their model with the promise of *"human-level performance on various professional and academic benchmarks"* [25], including applicability to the legal domain through e.g. *"passing a simulated bar exam"* [25]. Therefore, as applicability to classification tasks and the legal domain has been shown, but no combination of legal data classification, depending on complex process and context data has been performed yet, we choose GPT-4 as one method for this study.

7 Conclusion

This paper studies how legal and domain experts can be supported by generative AI and embedding-based ranking when assessing the relevance of regulatory documents for business processes. This is a challenging assessment due to the volume, variety, complexity and change of both regulatory documents and business processes. In order to tackle these challenges, the study provides recommendations of method combinations for selected process scenarios with different characteristics, i.e., usage, impact, dynamics, and regulatory input, allowing for and enabling different levels of automation and transparency/reproducibility. Generative AI can be a great human-aid in relevancy identification, especially on process level and also give indications and consideration ideas for the relevance judgment of the more detailed aspects of a process. However, due to its limitations in terms of transparency and reproducibility, we only recommend its usage in certain scenarios. Under certain circumstances, e.g., if the focus is more on the transparency of the support method embedding-based ranking can also be valuable assets to the relevance judgment for business processes.

The presented study can be extended in several ways. With respect to preprocessing, the presented approach does not include the search and scraping for relevant regulatory documents for a given business process, e.g., on the web or in closed-access ISO-Norms. This extension would allow for a holistic automated pre-selection of all regulatory requirements for a process that then needs to be evaluated for relevancy. As generative AI methods seem very sensitive to the prompt design and phrasing [14], multiple prompt runs with, e.g., para-phrases could be considered in the future. Finally, the inclusion of a confidence score in the responses could further aid the usability of the results during post-processing.

References

1. van der Aa, H., Leopold, H., Reijers, H.A.: Checking process compliance against natural language specifications using behavioral spaces. Inf. Syst. **78**, 83–95 (2018)
2. Althammer, S., Askari, A., Verberne, S., Hanbury, A.: Dossier@coliee 2021: Leveraging dense retrieval and summarization-based re-ranking for case law retrieval (2021)
3. Askari, A., Abolghasemi, A., Pasi, G., Kraaij, W., Verberne, S.: Injecting the BM25 score as text improves Bert-based re-rankers. In: ECIR Conference (2023)
4. Busch, K., Rochlitzer, A., Sola, D., Leopold, H.: Just tell me: prompt engineering in business process management. In: BPMDS Conference, pp. 3–11 (2023)
5. Dimlioglu, T., et al.: Automatic document classification via transformers for regulations compliance management in large utility companies. Neural Comput. Appl. **35**, 17167–17185 (2023)
6. Dragoni, M., Villata, S., Rizzi, W., Governatori, G.: Combining natural language processing approaches for rule extraction from legal documents. In: AICOL Workshops (2018)
7. Goebel, R., Kano, Y., Kim, M., Rabelo, J., Satoh, K., Yoshioka, M.: Summary of the competition on legal information, extraction/entailment (COLIEE) 2023. In: ICAIL Conference, pp. 472–480 (2023)

8. Gordon, D.G., Breaux, T.D.: The role of legal expertise in interpretation of legal requirements and definitions. In: IEEE Requirements Engineering Conference, pp. 273–282 (2014)
9. Governatori, G., Hoffmann, J., Sadiq, S.W., Weber, I.: Detecting regulatory compliance for business process models through semantic annotations. In: BPM Workshop, pp. 5–17 (2008)
10. Governatori, G., Sadiq, S.W.: The journey to business process compliance. In: Handbook of Research on Business Process Modeling, pp. 426–454 (2009)
11. Hashmi, M., Governatori, G., Wynn, M.T.: Normative requirements for regulatory compliance: an abstract formal framework. Inf. Syst. Front. **18**, 429–455 (2016)
12. Hliaoutakis, A., Varelas, G., Voutsakis, E., Petrakis, E.G.M., Milios, E.E.: Information retrieval by semantic similarity. Int. J. Semant. Web Inf. Syst. **2**, 55–73 (2006)
13. Huang, F., Kwak, H., An, J.: Is ChatGPT better than human annotators? potential and limitations of ChatGPT in explaining implicit hate speech. In: ACM Web Conference, pp. 294–297 (2023)
14. Jiang, Z., Xu, F.F., Araki, J., Neubig, G.: How can we know what language models know. Trans. Assoc. Comput. Linguist. **8**, 423–438 (2020)
15. Kim, M., Rabelo, J., Goebel, R., Yoshioka, M., Kano, Y., Satoh, K.: COLIEE 2022 summary: methods for legal document retrieval and entailment. In: JSAI-isAI Workshops, pp. 51–67 (2022)
16. Kim, M., Rabelo, J., Okeke, K., Goebel, R.: Legal information retrieval and entailment based on bm25, transformer and semantic thesaurus methods. Rev. Socionetwork Strateg. **16**, 157–174 (2022)
17. Klampanos, I.A.: Manning christopher, prabhakar raghavan, hinrich schütze: introduction to information retrieval. Inf. Retr. **12**, 609–612 (2009)
18. Kojima, T., Gu, S.S., Reid, M., Matsuo, Y., Iwasawa, Y.: Large language models are zero-shot reasoners. In: NeurIPS (2022)
19. Kuzi, S., Zhang, M., Li, C., Bendersky, M., Najork, M.: Leveraging semantic and lexical matching to improve the recall of document retrieval systems: a hybrid approach. CoRR (2020)
20. Leopold, H., van der Aa, H., Pittke, F., Raffel, M., Mendling, J., Reijers, H.A.: Searching textual and model-based process descriptions based on a unified data format. Softw. Syst. Model. **18**, 1179–1194 (2019)
21. Liu, Y., et al.: Summary of ChatGPT/GPT-4 research and perspective towards the future of large language models. CoRR (2023)
22. Looy, A.V., Backer, M.D., Poels, G., Snoeck, M.: Choosing the right business process maturity model. Inf. Manag. **50**, 466–488 (2013)
23. Ly, L.T., Maggi, F.M., Montali, M., Rinderle-Ma, S., van der Aalst, W.M.P.: Compliance monitoring in business processes: functionalities, application, and tool-support. Inf. Syst. **54**, 209–234 (2015)
24. Nigam, S.K., Goel, N., Bhattacharya, A.: nigam@coliee-22: legal case retrieval and entailment using cascading of lexical and semantic-based models. In: JSAI-isAI Workshops, pp. 96–108 (2022)
25. OpenAI: GPT-4 technical report. CoRR (2023)
26. Reimers, N., Gurevych, I.: Sentence-BERT: sentence embeddings using Siamese BERT-networks. In: EMNLP Conference (2019)
27. Robertson, S.E., Zaragoza, H.: The probabilistic relevance framework: BM25 and beyond. Found. Trends Inf. Retr. **3**, 333–389 (2009)
28. Sadiq, S.W., Governatori, G., Namiri, K.: Modeling control objectives for business process compliance. In: BPM Conference, pp. 149–164 (2007)

29. Sai, C., Winter, K., Fernanda, E., Rinderle-Ma, S.: Detecting deviations between external and internal regulatory requirements for improved process compliance assessment. In: Advanced Information Systems Engineering, pp. 1–16 (2023)
30. Sapkota, K., Aldea, A., Younas, M., Duce, D.A., Bañares-Alcántara, R.: Extracting meaningful entities from regulatory text: towards automating regulatory compliance. In: Workshop on Requirements Engineering and Law, pp. 29–32 (2012)
31. Schumann, G., Meyer, K., Gómez, J.M.: Query-based retrieval of German regulatory documents for internal auditing purposes. In: Data Science and Information Technology Conference, pp. 1–10 (2022)
32. Trautmann, D., Petrova, A., Schilder, F.: Legal prompt engineering for multilingual legal judgement prediction. CoRR (2022)
33. Winter, K., van der Aa, H., Rinderle-Ma, S., Weidlich, M.: Assessing the compliance of business process models with regulatory documents. In: Conceptual Modeling Conference, pp. 189–203 (2020). https://doi.org/10.1007/978-3-030-62522-1_14
34. Winter, K., Rinderle-Ma, S.: Deriving and combining mixed graphs from regulatory documents based on constraint relations. In: Advanced Information Systems Engineering, pp. 430–445 (2019)
35. Zhao, L., et al.: Natural language processing for requirements engineering: a systematic mapping study. ACM Comput. Surv. **54**, 1–41 (2021)

Conversational Systems for AI-Augmented Business Process Management

Angelo Casciani[1(✉)], Mario L. Bernardi[2], Marta Cimitile[3], and Andrea Marrella[1]

[1] Sapienza University of Rome, Rome, Italy
{casciani,marrella}@diag.uniroma1.it
[2] University of Sannio, Benevento, Italy
bernardi@unisannio.it
[3] Unitelma Sapienza, Rome, Italy
marta.cimitile@unitelma.it

Abstract. AI-augmented Business Process Management Systems (ABPMSs) are an emerging class of process-aware information systems empowered by AI technology for autonomously unfolding and adapting the execution flow of business processes (BPs). A central characteristic of an ABPMS is the ability to be *conversationally actionable*, i.e., to proactively interact with human users about BP-related actions, goals, and intentions. While today's trend is to support BP automation using reactive conversational agents, an ABPMS is required to create dynamic conversations that not only respond to user queries but even initiate conversations with users to inform them of the BP progression and make recommendations to improve BP performance. In this paper, we explore the extent to which state-of-the-art conversational systems (CSs) can be used to develop such proactive conversation features, and we discuss the research challenges and opportunities within this area.

Keywords: AI-augmented Business Process Management · Conversational Systems · Large Language Models · Process Mining

1 Introduction

In the era of Industry 4.0 (I4.0), the increased availability of event data tracing the execution of Business Processes (BPs), combined with advances in Artificial Intelligence (AI), is laying the ground for a new breed of AI-augmented BPM Systems (ABPMSs), capable of autonomously unfolding and adapting the BP execution flows. A recent research manifesto [19] describes the vision of ABPMSs and delineates the lifecycle of an ABPMS, which expands that of a classical BPMS in two directions. On the one hand, the traditional lifecycle phases (i.e., modeling, analysis, execution, monitoring, etc.) are continuously iterated, and empowered with AI capabilities. On the other hand, the lifecycle includes additional tasks that can only be realised with AI support, namely those of adaptation, explanation, and continuous improvement.

J. Araújo et al. (Eds.): RCIS 2024, LNBIP 513, pp. 183–200, 2024.
https://doi.org/10.1007/978-3-031-59465-6_12

In this transition, one particularly relevant aspect is that BP modelling is lifted to the more general notion of framing, which entail establishing multiple constraints encompassing procedural rules, best practices, and norms that must be considered during BP execution. Within the provided frame, an ABPMS is expected to be: (*i*) *autonomous* to act independently and proactively; (*ii*) *conversationally actionable* to seamlessly interact and cooperate with human users whenever the restrictions imposed by the frame cannot be met; (*iii*) *adaptive* to react to changes in its environment; (*iv*) (*self-*)*improving* to ensure the optimal achievement of its goals; (*v*) *explainable* to provide trust and, hence, foster collaboration with human users.

Among the most significant characteristics of an ABPMS is its ability to be *conversationally actionable*, i.e., being able to seamlessly interact with humans to not only respond to user queries and perform actions on their behalf but also initiate conversations with users to inform them of the BP progression, alert them of relevant BP changes, and make recommendations for interventions for improving the BP concerning relevant performance targets [19]. Indeed, integrating ABPMSs into a human workforce alters the role of human employees and dynamics, fueling a lack of trust, a notorious barrier to the adoption of automated technologies in Information Systems (ISs) [43]. In light of the considerations above, a possible solution to the lack of human trust in these ABPMSs can be the adoption of *Conversational Systems* (CSs).

CSs enable machines to engage with users in human-like dialogues to offer spoken, text-based, or multimodal conversational interactions with humans [51]. Thus, CSs can act as a natural language interface for the ABPMS towards the human and, consequently, boost the explainability of these ISs, since when users understand the reasoning behind the system's actions, they feel in control of the system and are more likely to trust it. In [16], the authors posit that ABPMSs can intensely benefit from the emergence of CSs, as they have the potential to empower the four main data-driven BPM approaches, namely: *Descriptive Process Analytics*, *Predictive Process Analytics*, *Prescriptive Process Optimization*, and *Augmented Process Execution*.

The main contribution of this paper is a survey for the analysis of the techniques developed in the field of CSs applied to BPM and the investigation of the related research problems and opportunities in the area. To achieve these objectives, we employed a rigorous search protocol across prominent digital libraries for each of the four identified topics, aiming to discover the state-of-the-art conversational techniques in BPM and to outline the research challenges to make an ABPMS conversationally actionable.

The rest of the paper is structured as follows. Section 2 introduces background knowledge about CSs and their taxonomies. Section 3 describes the adopted search protocol. Sections 4, 5, 6, and 7 explore CSs applied in each BPM area as identified in [16], discussing research challenges for improving ABPMSs conversational features. Section 8 reports the related work in the field. Finally, Sect. 9 concludes the paper by summarizing its key findings and discussing threats to the study's validity.

2 Background on Conversational Systems

This research area sits at the intersection of Natural Language Processing (NLP), Machine Learning (ML), and Information Retrieval, taking advantage of the techniques developed in these fields to make sense of user queries, provide context-aware responses, and engage, oftentimes, in multi-turns conversations [51].

Given these capabilities, CSs have historically attracted both scientific and industrial interests thanks to their potential for enhancing user interactions in various application domains, from customer support and healthcare to education and enterprises. The roots of the field can be traced back to the early attempts of the 1960s, with chatbots embedding predefined scripts to direct responses [70]. However, only in modern times, the vast availability of data about human conversations freely accessible on the Internet and the late breakthroughs in ML, and more specifically in Deep Learning, have enabled CSs to achieve goals initially deemed unattainable, leading to widespread popularity and generating considerable hype even among individuals without a technical background [33].

Several works have contributed significantly to the understanding and advancement of conversational agents. Indeed, they delve into the historical development and taxonomies, underlying technologies, and practical implementations of CSs. The first categorization that can be drawn for these systems is according to the problem they aim to address [29]:

- *Question answering* CSs respond to the user query in a direct and precise way, exploiting a huge amount of data coming from heterogeneous sources (i.e., Web documents) or local knowledge bases (i.e., business datasets).
- *Task completion* CSs execute a specific task requested by the user, spanning from setting reminders to scheduling meetings. In this case, the task is usually defined beforehand and the system is tailored to it.
- *Social chat* or *open dialogue* CSs engage in fluid and contextually appropriate conversations with users for the sake of entertainment or companionship, resembling human interaction similarly to the Turing test.

Another traditional categorization is based on the nature of the supported conversations, namely *single-turn* or *multi-turns* [71]. The former is relatively straightforward, focusing uniquely on the user query to generate an answer. In contrast, the latter considers the *context*, incorporating utterances from previous turns in the conversation to achieve a major degree of user engagement.

From a technological standpoint, many paradigms have been implemented to realize CSs [51]. In *rule-based* CSs, the dialogue is handled with a modularized architecture, and the conversation flow is defined in advance by designers who meticulously craft dialogue rules forming a set of if-then statements. These rules define the system's understanding and response to user inputs, yet limit its expressiveness. Over the years, *statistical data-driven* approaches have become predominant, learning conversational strategies from data but maintaining the overall modular framework. In particular, it is worth highlighting the possibility of addressing the dialogue formulation as a sequential decision-making process, modeling it as a Markov Decision Process, and employing Reinforcement Learning to find an optimal solution. Recently, researchers in CSs have shifted their focus towards the creation of *end-to-end neural* conversational agents, marking a significant change from traditional modular techniques that involve distinct components for understanding user input and generating responses. Instead, these systems directly map input utterances to output responses by leveraging Deep Neural Networks.

Among end-to-end neural approaches, we identify diverse methodologies [28].

- *Retrieval-based* techniques generate an answer to the user query by selecting the answer from a large set of candidate responses. Upon receiving a user utterance, the model encodes it together with the conversational context in a dense representation. Subsequently, it iterates over the whole set of candidate responses, assigning a score to each one depending on its appropriateness through a function that matches the context and the possible response. Eventually, the model produces in output the candidate utterance having the highest score.
- *Generation-based* techniques adopt an opposite approach, synthesizing the answer sequentially, word by word. Preeminent solutions implementing this paradigm primarily rely on the *encoder-decoder* architecture. The encoder translates the context into a hidden state, representing contextual information as a vector. Subsequently, the decoder selects a new word and adjusts consequently the hidden state at each time step in an auto-aggressive fashion. Even if they lag in performance compared to their counterparts, also non-auto-aggressive methods were explored to allow parallel token generation, considering each word conditionally independent in per-step distribution. The aforementioned encoder-decoder architecture can be implemented by means, respectively, of Long Short-Term Memory (LSTM) or Gated recurrent unit (GRU) for recurrent neural networks [62], or a stack of self-attention layers and cross-attention layers for transformer networks [67].
- *Hybrid* techniques overcome the limitations of the above methodologies, combining their strength. Indeed, retrieval-based approaches can provide high-quality responses but offer a limited hypothesis space of candidates while, on the contrary, generation-based methods can produce novel answers but with no guarantees about their quality. For this reason, hybrid techniques first retrieve instances of similar conversations from their dataset and, afterward, exploit them to support the generation of the response in many ways (e.g., Retrieval-Augmented Generation [45]).

Within generation-based approaches, the prominence and versatility of *Language Models* (LMs) have reached unprecedented heights, demanding a dedicated investigation. From the first statistical models, LMs have evolved into transformer models (e.g., BERT) pre-trained over massive textual datasets, demonstrating robust effectiveness in addressing NLP tasks via text generation. Notably, by increasing the size of these models (i.e., the number of parameters) over a specific threshold and moving to *Large Language Models* (LLMs) such as GPT-4 and Llama 2, researchers observed not only an important enhancement in performance but also the manifestation of peculiar capabilities that are not demonstrated by smaller LMs (in-context learning, instruction following, step-by-step reasoning) [75].

3 Search Protocol

To conduct this survey, a reproducible search protocol was employed to ensure a comprehensive exploration of the literature produced in the field of CSs applied to BPM, borrowed by the scientific methodology exposed by Kitchenham [40].

Initially, we formulated the research questions to define the scope of the search and produced a list of search strings. Afterward, we executed the search strings across diverse data sources. Ultimately, we applied inclusion criteria to select the studies acquired through the search. To the end of exploring the aforementioned research area, we identified the following research questions tailored to each BPM family identified in Sect. 1.

- **RQ1**: *Which conversational techniques are adopted in the BPM field?*
- **RQ2**: *What research challenges need to be addressed in creating actionable conversations for ABPMSs?*

In a nutshell, *RQ1* aims at discovering the most relevant conversational techniques implemented at the moment of writing for each BPM approach and *RQ2* explores future research challenges (RCs) and possible solutions.

Next, we formulated the search strings tailored to each data-driven BPM area as identified in [16]. Through an iterative trial-and-error process, we recognized the need to extend the scope of the search by incorporating broader terms to refer to the fields under examination. The search strings resulting from these considerations were defined as: *(Q1) AND (Q2)*, where *(Q1)* represents the fixed part that remains consistent across all search strings and *(Q2)* represents the variable part that changes based on the specific area of the search. Notably, *(Q1)* corresponds to: *("conversational" OR "dialogue" OR "chatbot" OR "natural language" OR "LLM" OR "language model" OR "ChatGPT")*. It is worth justifying the decision to consider "*ChatGPT*" as opposed to its competitors. We opted for its inclusion due to its widespread diffusion, often misinterpreted as a synonym for general LLM by non-technical users. Conversely, *(Q2)* is:

- *("descriptive process analytics" OR "process discovery" OR "conformance checking" OR "performance mining" OR "variant analysis" OR "process mining" OR "process modeling")* for Descriptive Process Analytics;
- *("predictive process analytics" OR "what-if analysis" OR "digital twin" OR "predictive process monitoring" OR "process analysis")* for Predictive Process Analytics;
- *("prescriptive process optimization" OR "process optimization" OR "prescriptive process monitoring" OR "process redesign")* for Prescriptive Process Optimization;

- *("process execution" OR "robotic process automation" OR "process automation" OR "process implementation")* for Augmented Process Execution.

First, each of the four search strings was employed in querying Google Scholar to retrieve studies where the search strings appeared in the *title*, *keywords*, or *abstract* of the paper. Subsequently, we double-checked and complemented the studies discovered in this primary search using widely recognized academic databases such as Scopus, ACM Digital Library, and IEEE Xplore. The search was completed in January 2024.

To maintain the focus on the most pertinent studies, to be considered a study must satisfy all the following inclusion criteria.

- **IN1**: *The study encompasses a technique for CSs in BPM.*
- **IN2**: *The study is peer-reviewed.*
- **IN3**: *The study is electronically available.*
- **IN4**: *The study is written in English.*
- **IN5**: *In case of multiple publications discussing the same technique, only the most comprehensive study is included.*

Table 1. Statistics of the search for each BPM area.

Phase	Descriptive Process Analytics	Predictive Process Analytics	Prescriptive Process Optimization	Augmented Process Execution	Total
Google Scholar	32	73	14	91	210
Scopus	356	236	119	198	909
ACM Digital Library	20	8	1	12	41
IEEE Xplore	25	19	3	34	81
All Publications	433	336	137	335	1241
Duplicates	60	61	7	63	191
All (-duplicates)	373	275	130	272	1050
Excluded	346	266	127	257	996
Final	27	9	3	15	54

Hence, the applied exclusion criterion was to eliminate studies that violated at least one of the aforementioned inclusion criteria. The results of our search protocol were documented in four distinct spreadsheets, each corresponding to a specific category. Comprehensive statistical data are presented in Table 1, showing that 54 studies were finally selected to conduct the survey analysis.

In the following sections, we present the outcomes of applying our search protocol for each BPM area considered, addressing the above research questions by introducing the conversational techniques developed for BPM, and identifying the research challenges (RCs), outlined in the concluding Table 2, along with potential solutions. The outcomes of the papers' extraction and selection for the survey are available at: https://zenodo.org/doi/10.5281/zenodo.10827054. Figure 1 illustrates the distribution of selected publications per year, categorized by BPM area.

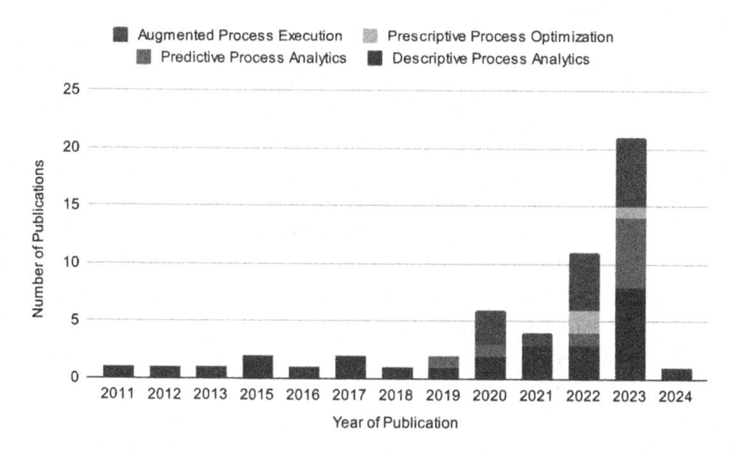

Fig. 1. Year distribution of the selected publications, grouped by BPM area.

4 Descriptive Process Analytics

The first data-driven BPM area we tackle in our analysis is *Descriptive Process Analytics* (DPA). This family of methodologies deals with the *as-is* description of BPs, i.e., their current state, supporting domain experts in identifying problems and potential improvements [16]. Its scope ranges from BP modeling to BP monitoring, including advanced process mining (PM) applications for analyzing the performance and conformance of BPs based on event logs produced during their execution [20].

Literature Analysis and Research Challenges. During interactions with business stakeholders, the ability to extract BP models from natural language descriptions proves to be highly time-efficient and effective. Over the years, researchers have developed many NLP techniques to this end, such as [27] for BPMN models. In [65], semantic unification is embedded to deal with partial and potentially contradictory information, while [14] generates BP models from descriptions in controlled natural language, enabling the BP discovery through user interactions. Other NLP methodologies to extract BP models from unstructured organizational documentation are exposed in [23,63]. In [56], the authors introduce an encoder-decoder translator to represent BP models in a middle representation and decode them into natural language descriptions, to mitigate information loss and semantic errors. With [26], a platform architecture enables the export of BP models into natural language and vice versa through a Web service interface, whereas, [37] presents a modeling environment facilitating the rapid generation of visual BPMN models from constrained natural language input during interviews and design workshops. [66] proposes a Machine Translation-inspired approach to simplify the BP modeling phase by generating BPMN diagrams from textual descriptions in natural language. [21] presents a method that combines BPMN with NLP to generate BPMN diagrams from natural language BP descriptions, employing Probabilistic Latent Semantic Analysis to cope with ambiguities in natural language. [54] introduces an automated approach utilizing NLP and Prolog to extract BP models in UML from user stories. [55] extends a data-driven pipeline for the automated generation of BP models from natural language text. Approaches to allow the natural language description of BP were also investigated for Declare, notably in [48] and in [1], whereby constraints can be expressed verbally and are converted via speech recognition into the closest set of Declare constraints. Prospective improvements in these methodologies may address the implicit ambiguity of natural language, which can result in unpredictable and varied representations of the same BP (**RC1**) [2]. A potential solution to this challenge involves refining the algorithms for semantic analysis and context-aware processing, ensuring a more consistent and standardized interpretation of natural language inputs.

The reverse approach was also explored, wherein BP models are expressed using natural language to enhance human comprehension. Relevant studies on this topic can be identified in [44] for prescriptive BP models and in [4] for declarative ones. Furthermore, [49] introduces an approach to automatically generate a natural language representation of BPMN models leveraging business rules in SBVR as an intermediate representation, and [24] utilizes PM, fuzzy linguistic protoforms, and natural language generation (NLG) to generate specialized textual explanations of BPs automatically. The challenges primarily revolve around comprehending model labels and explaining parallel behaviors in natural language without compromising the reader's grasp of the underlying BP semantics (**RC2**) [2]. To overcome the former, NLP techniques incorporating contextual analysis and domain-specific knowledge can be applied for precise word categorization. Meanwhile, addressing the latter challenge may involve adopting a structured and standardized template to articulate parallel behaviors, thereby enhancing clarity and maintaining human comprehensibility.

Moving to methodologies based on chatbots, it is worth mentioning [60], whereby chatbots are employed to enable adaptive learning of BPs in a multi-actor environment, and [25], which presents a conversational agent designed for tasks such as consistency, conformance, and model checking for declarative BP models. In tackling the challenge of bidirectional interaction between natural language descriptions and BP models (**RC3**), a promising direction can be embodied by multimodal LLMs. These models can comprehend BP models and respond to user queries in a grounded manner, offering visual aids such as charts for improved human interpretation. Moreover, their generative abilities enable them to transform inputs in natural language into the corresponding BP model, relying on advanced NLP techniques for semantic interpretation. Furthermore, the exploration of explainable AI (XAI) methodologies can enhance the transparency of the chatbot's decision-making process, fostering user trust and usability in dynamic process environments.

Another important trend is the development of conversational interfaces to enhance the understanding of PM findings and make them accessible to non-technical users. Notably, [8] addresses this challenge by developing a natural language querying interface in combination with Everflow and proposing a taxonomy of PM questions. [72] suggests a methodology to facilitate data extraction from PM using a natural language interface, eliminating the need for programming in a PM query language. [42] enhances the querying experience for domain analysts with limited technological expertise by introducing a natural language interface that utilizes graph-based storage techniques, i.e., labeled property graphs, and executes queries through the Cypher language. In [31], a solution is presented to automatically discover BPs models from textual documentation using a neural network with the Ordered Neurons LSTM architecture and a process-level language model objective. [22] suggests enhancing BP discovery with causal BP discovery and XAI for improving the interpretability of BP execution outcomes through LLMs. [57] reports a method facilitating advanced PM by automatically extracting BP information, including resources and business objects, from event data through semantic role labeling, employing an attribute classification technique. [58] introduces an approach for defining measurable Process Performance Indicators by combining NLP techniques and tailored matching strategies to integrate textual descriptions with event logs. The challenges in this domain involve the limited generalization of rule-based semantic parsing, necessitating new rules for novel questions, prompting the need for hybrid approaches mixing AI and rule-based techniques for NLP in PM, extending systems capabilities to support complex queries, multimodal conversational interfaces, and specialized evaluation frameworks (**RC4**). Moreover, in the analysis of event logs, the integration of LLMs may enable the automated execution of entity-extraction tasks and the computation of semantic similarity.

5 Predictive Process Analytics

We follow in our discussion with *Predictive Process Analytics* (PPA). This area is aimed at building predictive models to estimate the future state of the BP, enabling the prediction of its performance [16]. Thus, we can view this sub-field of BPM as framed within the activities of the *Process Analysis* step of the usual BPM lifecycle [20]. In particular, there are two preeminent techniques for this BPM approach. The first is *what-if digital process twins* that, constructing a simulation model representing the BP, tries to forecast the impact of changes to the BP concerning relevant KPIs. The second is *predictive process monitoring*, which leverages ML algorithms to learn a predictive model from historical data and uses it to generate predictions both at the case and BP levels.

Literature Analysis and Research Challenges. In [9], the authors introduce a chatbot for BP simulation that allows to conversationally specify what-if scenarios, simplifying the procedure for users without technical knowledge and enabling the comparison of the BP performance under these scenarios against a standard reference. In [46], the authors present a conversational system utilizing GPT-4 to automate the creation of digital twins in the data centers domain. Furthermore, [50] explores the application of ChatGPT in conjunction with Digital Twins in the construction industry, and [64] introduces neuro-symbolic reasoning for interacting with 3D digital twins using natural language in aircraft maintenance. Research opportunities in conversational what-if analysis concern the support for more extensive customization of the digital twin (**RC5**), e.g., allowing domain-specific modifications and tailoring the KPIs on that particular BP. Additionally, LLMs could play a crucial role by integrating with simulation engines, facilitating multimodal interaction, and allowing users to conversationally customize views over the digital twins.

Within the realm of predictive process monitoring, [12] introduces a text-aware approach combining ML and NLP to monitor knowledge-intensive BPs and incorporating structured features and unstructured textual information over the control flow, whereas [13] presents a text-aware technique to predict the next activity and timestamp in BP instances considering semantic information by including contextualized word embeddings. Moreover, [34] proposes the use of Attention-Based LSTM with Multi-Task Learning to improve business behavior prediction accuracy from historical event logs. In contrast to LSTM which relies solely on the last hidden state for predictions, [38] leverages the notion of attention but considers all hidden states to predict future behavior accurately. In [69], the authors investigate the integration of textual data into predictive process monitoring techniques to improve accuracy, while also prioritizing explainability, which boosts transparency and interpretability for black-box ML models at the cost of increased computation time. Potential research directions in predictive process monitoring could focus on enhancing explainability (**RC6**). Specifically, future endeavors may involve integrating explainability analysis with subsequent causality analysis to distinguish whether a variable is correlated or causally related to the outcome. Furthermore, investigating the connection between explainability analysis results and practical interventions could be explored [69].

6 Prescriptive Process Optimization

Prescriptive Process Optimization (PPO) is mainly concerned with the optimization of the process, especially through the translation of the findings from Predictive Process Analytics into actual actions to undertake for enhancing the execution of the process [16]. Drawing the parallelism with the traditional BPM lifecycle, this body of methodologies falls primarily into the *Process Redesign* phase, for the production of a *to-be* version of the process model [20]. In this case, these techniques refer to the training of a predictive model utilizing historical data, later employed for making predictions regarding the underlying process. Such predictions serve as input for a recommender system, which, based on this information, generates recommendations for subsequent courses of action. These recommendations may undergo automatic execution, be presented to domain experts, or be integrated into the prescriptive system to refine the model for subsequent recommendation generation. PPO methodologies can be divided into two broad classes. *Automated process optimization* is directed at proposing alterations to the BPs in order to achieve a balance between competing KPIs, such as reducing costs while simultaneously maximizing the quality. On the other hand, *prescriptive process monitoring* proposes recommendations about actions to perform for the process optimization with respect to the selected KPIs in real-time or near-real-time and, oftentimes, at the case level.

Given that PPO is in the early stages of development, a limited number of studies have explored its intersection with CSs.

Literature Analysis and Research Challenges. In [6], ChatGPT is employed for BP optimization within the domain of Additive Manufacturing (AM). The integration of ChatGPT into the manufacturing workflow is reported to result in improved efficiency, cost reduction, and increased accessibility in the field. In [52], the authors present a NLP approach for BP Redesign, aimed at the extraction of the redesign suggestions from end-user feedback in natural language and tested in a real-world use case with an extensive experiment program. Research opportunities include developing NLP-based approaches to automatically extract change proposals for BP Redesign, performing sentiment analysis on the suggestions, prioritizing them, and employing clustering to group suggestions based on similarity and frequency. In this domain, LLMs can be embedded to generate recommendations for BP improvement, working in conjunction with the predictive layer to assess and validate these suggestions (**RC7**).

In [73], crowd-wisdom and goal-driven methods from prescriptive process monitoring are applied to AI-powered BPs, blending classical BPM, goal-driven chatbots, and conversational recommendation systems, introducing a synthesized dataset derived from a real use case. Future research challenges in CSs for prescriptive process monitoring involve considering Reinforcement Learning to leverage implicit and explicit user feedback, handling complex utterances using AI planning for dynamic orchestration of automated tasks, and employing DL and NLP methodologies to map user utterances to activities, followed by the translation of the results back into natural language recommendations (**RC8**).

7 Augmented Process Execution

Augmented Process Execution (APE) brings a paradigm shift from the reactive execution that lies on the human, aided by system suggestions, to the inverse execution model, where the system proactively carries out the BP execution, supported by human operators. [16]. Now we are in the *Process Execution* stage of the BPM lifecycle, where the *to-be* BP is executed within the IS of the organization [20]. Furthermore, this is where the notion of ABPMS introduced in Sect. 1 comes in, with the domain experts that intervene in the execution only when the system needs it to disambiguate its behavior in a specific situation [19]. Within this sub-field, we can identify two categories of systems, depending on their interaction with the human operator. An *autonomic process execution system* operates within its predefined frame and resorts to human intervention and decision-making whenever it encounters an uncertain scenario. Conversely, an *autonomous process execution system* not only exercises full control over the BP within the frame, but it can even operate modifications on it to achieve specific business goals. In this context, the human assumes the supervisor role, intervening solely to avoid undesired consequences. Notably, Robotic Process Automation (RPA) can be considered a specific instance of AI-augmented BPM, aiming to realize more complex automation than a traditional BPMS. RPA operates on applications' user interface (UI) by creating software robots that automate mouse and keyboard interactions. This enables the automated execution of repetitive tasks on the UI, mitigating human errors stemming from mental lapses induced by boredom or exhaustion.

Literature Analysis and Research Challenges. In [15], the authors introduce a tool leveraging declarative design and AI planning to optimize complex BPs by composing with conversational agents and services. [47] implements a conversational agent using Rasa and Camunda Engine, designed to integrate with BPMSs and to simplify BP execution. [30] relies on LLMs to tackle various BPM tasks, and to assess the suitability of BP tasks for RPA. The paper [10] proposes leveraging GPT technology to generate new BP models, enhance decision-making in data-centric BPs, and improve overall BPM efficiency through task automation, insights provision, and operational enhancements. [7] introduces a no-code conversational interface facilitating collaboration between human users and bots in knowledge-intensive BPs, by enabling bots to identify user intents, orchestrate automation tasks, and include insights from conversations mining for performance monitoring and service quality improvement. [53] introduces an approach for verifying resource compliance requirements in BPs, leveraging GPT-4 for NLP and a customized compliance verification component. The study in [61] investigates the transformative role of AI and chatbots in procurement processes, and develops a chatbot to improve efficiency, reduce costs, and enhance supplier relations. In [32], a user-friendly natural language interface for querying runtime event data in BPMSs is introduced, enabling real-time insights without the need for backend knowledge and including a bootstrapping pipeline for the automatic instantiation of the natural language interface. In the domain of conversational agents for system-driven management, challenges arise in achieving seamless interaction, requiring advanced interfaces capable of understanding nuanced user requests. Ensuring efficient correction and optimization based on user-specified natural language instructions poses a significant challenge, emphasizing the need for automated solutions. Additionally, addressing challenges related to intent classification accuracy and building user trust through explicit approval requests for automation are critical aspects of future research (**RC9**). Furthermore, considerable variation in LLM responses suggests the need for further research into their behavior and reactions to diverse inputs for consistent and reliable performance in APE.

Shifting the focus on how CSs can empower RPA, a solution combining RPA and chatbots to automate iterative BPs is presented in [35]. [59] presents a conversational digital assistant framework for interactive automation that addresses the accessibility issues of RPA for business users through natural language interaction and a multi-agent orchestration model. [17] develops a multi-channel chatbot integrated with RPA, automating the end-to-end BP of product exploration, purchase, and transaction, for consistent operation across channels with minimal human intervention. [18] integrates chatbots with RPA in manufacturing to efficiently process and present data, offering a solution for rapid access to information in manufacturing sites through an architecture capable of managing complex queries and processes. Another relevant problem in RPA is the identification of suitable routines to automate, which is a time and cost-intensive manual task and is tackled in [3] by exploring its automation through NLP techniques. In [36], the authors address the challenge of making APIs, which provide access to RPA bots, accessible to non-technical users by introducing a data augmentation approach using LLMs for intent recognition. [74] presents a technique leveraging LLMs to enhance RPA capabilities through automatic workflow generation, ensuring reliable reasoning and maintaining data integrity. Given this plethora of works, we can indicate possible future improvement opportunities. Notably, a unified conversational interface could simplify the integration of various RPA automation solutions, fostering exploration of sophisticated orchestration models and autonomous agent composition through natural language (**RC10**). Moreover, employing NLP for identifying automation candidates in BPs highlights potential future improvements. Indeed, NLP techniques play a crucial role in supporting the design and execution of automation routines and advancing the automation of non-trivial tasks through cognitive automation (**RC11**).

8 Related Work

While the field is still undergoing significant development, numerous surveys have already delved into the applications and challenges of conversational techniques applied in the context of BPM. One of the first examples we can find in the literature is [2], a position paper that explores the potential of NLP in enhancing the advantages of BPM activities across various organizational levels. The authors report the principal research directions for a successful implementation of NLP in automating specific tasks that, otherwise, would require compelling effort to be performed. The paper also describes possible concrete applications, considering both the process perspective and its improvement through NLP.

The exploration of NLP in BPM is further addressed in two notable studies, namely, [5] and [11]. The former conducts a systematic literature review (SLR) to investigate the utilization of NLP techniques in extracting BPs and ensuring BP quality from unstructured text throughout the BPM lifecycle. The latter complements this by performing a qualitative analysis of state-of-the-art tools for BP extraction from unstructured documents, in the direction of uncovering existing limitations and challenges within the field.

Transitioning the focus from NLP to CSs, their realm has undergone significant transformation recently, particularly with the emergence of LLMs, which introduced new research perspectives in the area of BPM. In particular, the paper [39] introduces the notion of Large Process Model (LPM) that combines the correlation power of LLMs with the analytical precision of knowledge-based systems and automated reasoning approaches. The authors envision the integration of LLM applications at different stages of BPM and posit the feasibility of

Table 2. Research challenges for CSs in BPM.

BPM Area	Selected Papers	Identifier	Research Challenge
Descriptive Process Analytics	[27]; [65]; [14]; [63]; [23]; [56]; [26]; [37]; [66]; [21]; [54]; [55]; [48]; [1]; [44]; [4]; [49]; [24]; [60]; [25]; [8]; [72]; [42]; [31];[22]; [57]; [58]	RC1 RC2 RC3 RC4	*Unambiguous BP Discovery* *BP model semantics explanations* *From natural language to BP and vice versa* *Conversational interfaces for PM*
Predictive Process Analytics	[9]; [46]; [50]; [64]; [12]; [13]; [34]; [38]; [69]	RC5 RC6	*Conversational what-if analysis* *Explainable predictive process monitoring*
Prescriptive Process Optimization	[6]; [52]; [73]	RC7 RC8	*LLM-driven process redesign* *Multi-disciplinary integration for prescriptive process monitoring*
Augmented Process Execution	[15]; [47]; [30]; [10]; [7]; [53]; [61]; [32]; [35]; [59]; [17]; [18]; [3]; [36]; [74]	RC9 RC10 RC11	*Trustworthy conversational corrections* *Conversational RPA* *Cognitive automation*

implementing an LPM, while also underscoring inherent limitations and research challenges that must be addressed for its realization.

In [68], the authors focus on addressing the opportunities LLMs present in BPM by identifying six research directions in this respect. It is also worth mentioning the work in [41], where the authors perform a SLR and evaluate existing chatbots to assist conversational BP modeling leveraging a real-world test set. Based on this study, usage recommendations and further development in the identified area are consequently derived.

Our study distinguishes itself from previous literature in the same domain by focusing on challenges directly associated with the application of conversational techniques in BPM, offering a comprehensive and structured overview to foster future research to realize conversationally actionable ABPMSs.

9 Threats to Validity and Concluding Remarks

This survey has explored conversational techniques specifically designed for BPM, aiming to provide a comprehensive overview of this cutting-edge research domain. We opted for a survey instead of a SLR due to the relative novelty of the topic. Indeed, a survey approach seemed more appropriate to capture the current state of the field, as it may not have reached a level of maturity that meets the more stringent criteria associated with a complete SLR. To ensure the validity of our search protocol, we focused on addressing issues, such as incompleteness, by utilizing multidisciplinary search engines. Furthermore, we aimed to minimize selection bias by applying an accurate review of the studies and providing clear motivations for exclusions when necessary. It is also crucial to acknowledge the potential subjectivity inherent in the planning and implementation of the search protocol. To mitigate this threat, we clearly defined the survey's objectives, scope, and inclusion criteria, meticulously documenting each phase of the study to enhance transparency and reproducibility. Additionally, we adopted a validated methodology [40] for study selection and data extraction to mitigate interpretation biases. Our rigorous analysis pursued two primary objectives: (*i*) elucidating the principal conversational techniques applied across various BPM domains, and (*ii*) examining the extent to which these techniques can enable actionable conversations for ABPMSs. An overview of the identified RCs is outlined in Table 2.

Acknowledgments. This paper has been supported by the Sapienza projects DISPIPE and FOND-AIBPM, the PRIN 2022 project MOTOWN, and the PNRR MUR project PE0000013-FAIR. The work of Angelo Casciani has been carried out in the range of the Italian National Doctorate on AI run by Sapienza.

References

1. van der Aa, H., Balder, K.J., Maggi, F.M., Nolte, A.: Say it in your own words: defining declarative process models using speech recognition. In: BPM Forum (2020)
2. Van der Aa, H., Carmona Vargas, J., et al.: Challenges and opportunities of applying natural language processing in business process management. In: COLING (2018)
3. van der Aa, H., Leopold, H.: Automatically identifying process automation candidates using natural language processing. In: Koschmider, A., Schulte, S. (eds.) Blockchain and Robotic Process Automation. Springer, Cham (2022). https://doi.org/10.1007/978-3-030-81409-0_7
4. Ackermann, L., Schönig, S., Zeising, M., Jablonski, S.: Natural language generation for declarative process models. In: CAiSE Workshops (2015)
5. de Almeida Bordignon, A.C., Thom, L.H., Silva, T.S., et al.: Natural language processing in business process identification and modeling: a systematic literature review. In: Brazilian Symposium on Information Systems (2018)
6. Badini, S., Regondi, S., et al.: Assessing the capabilities of ChatGPT to improve additive manufacturing troubleshooting. Adv. Ind. Eng. Polymer Res. **6**(3), 278–287 (2023)
7. Bandlamudi, J., Mukherjee, K., Agarwal, P., et al.: Towards hybrid automation by bootstrapping conversational interfaces for IT operation tasks. In: AAAI (2023)
8. Barbieri, L., Madeira, E., Stroeh, K., van der Aalst, W.: A natural language querying interface for process mining. J. Intell. Inf. Sys. **61**(1), 113–142 (2023)
9. Barón-Espitia, D., Dumas, M., González-Rojas, O.: Coral: conversational what-if process analysis. In: ICPM (2022)
10. Beheshti, A., Yang, J., Sheng, Q.Z., et al.: ProcessGPT: transforming business process management with generative artificial intelligence. In: IEEE ICWS (2023)
11. Bellan, P., Dragoni, M., Ghidini, C.: A qualitative analysis of the state of the art in process extraction from text. DP@ AI* IA (2020)
12. Brennig, K., Benkert, K., Löhr, B., Müller, O.: Text-aware predictive process monitoring of knowledge-intensive processes: does control flow matter? In: BPM (2023)
13. Cabrera, L., Weinzierl, S., Zilker, S., Matzner, M.: Text-aware predictive process monitoring with contextualized word embeddings. In: BPM Workshops (2022)

14. Caporale, T.: A tool for natural language oriented business process modeling. In: 8th Central-European Workshop on Services and their Composition (2016)
15. Chakraborti, T., Agarwal, S., Khazaeni, Y., et al.: D3BA: a tool for optimizing business processes using non-deterministic planning. In: BPM Workshops (2020)
16. Chapela-Campa, D., Dumas, M.: From process mining to augmented process execution. Softw. Syst. Model. **22**, 1977–1986 (2023)
17. Dan, G., Claudiu, D., Alexandra, F., et al.: Multi-channel chatbot and robotic process automation. In: IEEE International Conference on Automation, Quality and Testing, Robotics (2022)
18. Do, S., Jeong, J.: Design and implementation of RPA based ChatMES system architecture for smart manufacturing. WSEAS Trans. Comput. Res. **10**, 88–92 (2022)
19. Dumas, M., Fournier, F., Limonad, L., Marrella, A., et al.: AI-augmented business process management systems: a research manifesto. ACM Trans. Man. Inf. Sys. **14**(1), 1–19 (2023)
20. Dumas, M., La Rosa, M., Mendling, J., Reijers, H.A., et al.: Fundamentals of Business Process Management, vol. 1. Springer, Heidelberg (2013). https://doi.org/10.1007/978-3-662-56509-4
21. Elmanaseer, S., Alkhatib, A.A., Albustanji, R.N.: A proposed technique for business process modeling diagram using natural language processing. In: ICIT (2023)
22. Fahland, D., Fournier, F., Limonad, L., et al.: Why are my Pizzas late? In: IJCAI (2023)
23. Ferreira, R.C.B., Thom, L.H., Fantinato, M.: A semi-automatic approach to identify business process elements in natural language texts. In: ICEIS (2017)
24. Fontenla-Seco, Y., Lama, M., Bugarín, A.: Process-to-text: a framework for the quantitative description of processes in natural language. In: TAILOR (2020)
25. Fontenla-Seco, Y., Winkler, S., Gianola, A., et al.: The droid you're looking for: C-4PM, a conversational agent for declarative process mining. In: BPM Forum (2023)
26. Freytag, T., Kanzler, B., Leger, N., Semling, D.: NLP as a service: an API to convert between process models and natural language text. In: BPM (PhD/Demos) (2021)
27. Friedrich, F., Mendling, J., Puhlmann, F.: Process model generation from natural language text. In: CAiSE (2011)
28. Fu, T., Gao, S., et al.: Learning towards conversational AI: a survey. AI Open **3**, 14–28 (2022)
29. Gao, J., Galley, M., Li, L.: Neural approaches to conversational AI. In: SIGIR (2018)
30. Grohs, M., Abb, L., Elsayed, N., Rehse, J.R.: Large language models can accomplish business process management tasks. In: BPM Workshops (2024)
31. Han, X., Hu, L., Mei, L., et al.: A-BPS: automatic business process discovery service using ordered neurons LSTM. In: IEEE ICWS (2020)
32. Han, X., Hu, L., Sen, J., et al.: Bootstrapping natural language querying on process automation data. In: IEEE SCC (2020)
33. Haque, M.U., Dharmadasa, I., et al.: "I think this is the most disruptive technology": exploring sentiments of ChatGPT early adopters using Twitter data. arXiv (2022)
34. Hnin, T., Oo, K.K.: Attention based LSTM with multi tasks learning for predictive process monitoring. In: International Workshop on Computer Science and Engineering (WCSE) (2019)

35. Hung, P.D., Trang, D.T., Khai, T.: Integrating Chatbot and RPA into enterprise applications based on open, flexible and extensible platforms. In: CDVE (2021)
36. Huo, S., Mukherjee, K., Bandlamudi, J., et al.: Accelerating the support of conversational interfaces for RPAs through APIs. In: BPM Forum (2023)
37. Ivanchikj, A., Serbout, S., Pautasso, C.: Live process modeling with the BPMN Sketch Miner. Softw. Syst. Model. **21**(5), 1877–1906 (2022)
38. Jalayer, A., Kahani, M., Beheshti, A., et al.: Attention mechanism in predictive business process monitoring. In: IEEE 24th EDOC (2020)
39. Kampik, T., Warmuth, C., Rebmann, A., et al.: Large process models: business process management in the age of generative AI. arXiv (2023)
40. Kitchenham, B.: Procedures for performing systematic reviews. **33**(2004). Keele University, Keele, UK (2004)
41. Klievtsova, N., Benzin, J.V., Kampik, T., et al.: Conversational process modelling: state of the art, applications, and implications in practice. In: BPM Forum (2023)
42. Kobeissi, M., Assy, N., Gaaloul, W., et al.: Natural language querying of process execution data. Inform. Syst. **116**, 102227 (2023)
43. Lee, J.D., See, K.A.: Trust in automation: designing for appropriate reliance. Human Factors **46**(1), 50–80 (2004)
44. Leopold, H., Mendling, J., Polyvyanyy, A.: Generating natural language texts from business process models. In: CAiSE (2012)
45. Lewis, P., Perez, E., Piktus, A., et al.: Retrieval-augmented generation for knowledge-intensive NLP tasks. In: Advances in Neural Information Processing Systems, vol. 33 (2020)
46. Li, M., Wang, R., Zhou, X., et al.: ChatTwin: toward automated digital twin generation for data center via large language models. In: ACM BuildSys (2023)
47. Lins, L.F., Melo, G., Oliveira, T., et al.: PACAs: process-aware conversational agents. In: BPM Workshops (2021)
48. López, H.A., Debois, S., Hildebrandt, T.T., Marquard, M.: The process highlighter: from texts to declarative processes and back. In: BPM (2018)
49. Malik, S., Bajwa, I.S.: Back to origin: Transformation of business process models to business rules. In: BPM Workshops (2012)
50. Mateev, M.: Predictive analytics based on Digital Twins, Generative AI, and ChatGPT. In: World Multi-Conference on Systemics, Cybernetics and Informatics, WMSCI (2023)
51. McTear, M.: Conversational AI: Dialogue Systems, Conversational Agents, and Chatbots. Springer, Cham (2022). https://doi.org/10.1007/978-3-031-02176-3
52. Mustansir, A., Shahzad, K., Malik, M.K.: Towards automatic business process redesign: an NLP based approach to extract redesign suggestions. Autom. Softw. Eng. **29**(1) (2022). https://doi.org/10.1007/s10515-021-00316-8
53. Mustroph, H., Barrientos, M., Winter, K., Rinderle-Ma, S.: Verifying resource compliance requirements from natural language text over event logs. In: BPM (2023)
54. Nasiri, S., Adadi, A., Lahmer, M.: Automatic generation of business process models from user stories. Int. J. Elect. Comp. Eng. **13**(1), 809 (2023)
55. Neuberger, J., Ackermann, L., Jablonski, S.: Beyond rule-based named entity recognition and relation extraction for process model generation from natural language text. In: CoopIS (2023)
56. Qian, C., Wen, L., Kumar, A.: BEPT: a behavior-based process translator for interpreting and understanding process models. In: CIKM (2019)
57. Rebmann, A., van der Aa, H.: Extracting semantic process information from the natural language in event logs. In: CAiSE (2021)

58. Resinas, M., del Río-Ortega, A., van der Aa, H.: From text to performance measurement: automatically computing process performance using textual descriptions and event logs. In: BPM (2023)
59. Rizk, Y., Isahagian, V., Boag, S., et al.: A conversational digital assistant for intelligent process automation. In: BPM Forum (2020)
60. Rooein, D., Bianchini, D., Leotta, F., et al.: aCHAT-WF: generating conversational agents for teaching business process models. Softw. Syst. Modeling **21**(3), 891–914 (2022). https://doi.org/10.1007/s10270-021-00925-7
61. Sai, B., Thanigaivelu, S., Shivaani, N., et al.: Integration of chatbots in the procurement stage of a supply chain. In: CSITSS (2022)
62. Sherstinsky, A.: Fundamentals of recurrent neural network (RNN) and long short-term memory (LSTM) network. Physica D Nonlinear Phenom. **404**, 132306 (2020)
63. Sintoris, K., Vergidis, K.: Extracting business process models using natural language processing (NLP) techniques. In: IEEE 19th Conference on Business Informatics (2017)
64. Siyaev, A., Valiev, D., Jo, G.S.: Interaction with industrial Digital Twin using neuro-symbolic reasoning. Sensors **23**(3), 1729 (2023)
65. Sokolov, K., Timofeev, D., Samochadin, A.: Process extraction from texts using semantic unification. In: IC3K (2015)
66. Sonbol, R., Rebdawi, G., Ghneim, N.: A Machine Translation Like Approach to Generate Business Process Model from Textual Description. SN CS **4**(3), 291 (2023)
67. Vaswani, A., Shazeer, N., Parmar, N., et al.: Attention is all you need. In: Advances in Neural Information Processing Systems, vol. 30 (2017)
68. Vidgof, M., Bachhofner, S., Mendling, J.: Large language models for business process management: opportunities and challenges. In: BPM Forum (2023)
69. Warmuth, C., Leopold, H.: On the potential of textual data for explainable predictive process monitoring. In: ICPM Workshops (2022)
70. Weizenbaum, J.: ELIZA-a computer program for the study of natural language communication between man and machine. Commun. ACM **9**(1), 36–45 (1966)
71. Yan, R.: "Chitty-Chitty-Chat Bot": Deep Learning for Conversational AI. In: IJCAI (2018)
72. Yeo, H., Khorasani, E., Sheinin, V., et al.: Natural language interface for process mining queries in healthcare. In: IEEE Big Data (2022)
73. Zeltyn, S., Shlomov, S., Yaeli, A., Oved, A.: Prescriptive process monitoring in intelligent process automation with chatbot orchestration. In: PMAI (2022)
74. Zeng, Z., Watson, W., Cho, N., et al.: FlowMind: automatic workflow generation with LLMs. In: ACM International Conference on AI in Finance (2023)
75. Zhao, W.X., Zhou, K., Li, J., et al.: A survey of large language models. arXiv (2023)

Data and Process Science

TimeFlows: Visualizing Process Chronologies from Vast Collections of Heterogeneous Information Objects

Max Lonysa Muller[1,2](✉), Erik Saaman[1], Jan Martijn E. M. van der Werf[2], Charles Jeurgens[3], and Hajo A. Reijers[2]

[1] Nationaal Archief, Den Haag, The Netherlands
{max.muller,erik.saaman}@nationaalarchief.nl
[2] Universiteit Utrecht, Utrecht, The Netherlands
{j.m.e.m.vanderwerf,h.a.reijers}@uu.nl
[3] Universiteit van Amsterdam, Amsterdam, The Netherlands
k.j.p.f.m.jeurgens@uva.nl

Abstract. In many fact-finding investigations, notably parliamentary inquiries, process chronologies are created to reconstruct how a controversial policy or decision came into existence. Current approaches, like timelines, lack the expressiveness to represent the variety of relations in which historic events may link to the overall chronology. This obfuscates the nature of the interdependence among the events, and the texts from which they are distilled. Based on explorative interviews with expert analysts, we propose an extended, rich set of relationships. We describe how these can be visualized as *TimeFlows*. We provide an example of such a visualization by illustrating the Childcare Benefits Scandal – an affair that deeply affected Dutch politics in recent years. This work extends the scope of existing process discovery research into the direction of unveiling non-repetitive processes from unstructured information objects.

Keywords: Timeline · Process Chronology · TimeFlow · Parliamentary Inquiry

1 Introduction

In various situations, it is worthwhile to determine retrospectively how a process has unfolded over time. The established discipline of *process mining* provides algorithms for discovering *work processes* from historic data. Such processes are cyclic in nature and can be discovered on the basis of events generated by IT systems. However, there are also processes worth reconstructing that are non-repetitive and lack a log of structured events.

Consider a *parliamentary inquiry*, which is an important use case for the work presented in the paper. A parliamentary inquiry is initiated to gain knowledge on a particular topic when a matter of national importance is at stake. The objective is to reconstruct the process that led to severe consequences, either for the government itself or for the citizens involved. The process in question, once

J. Araújo et al. (Eds.): RCIS 2024, LNBIP 513, pp. 203–219, 2024.
https://doi.org/10.1007/978-3-031-59465-6_13

reconstructed, is typically represented in the form of a *process chronology*, e.g. a timeline, a textual summary, or a fact sheet of events. The purpose of a process chronology is to capture what relevant events happened at different moments in time and how these events relate to one another.

The generation of a process chronology, either in the context of a parliamentary inquiry or any other sort of fact-finding investigation, may need to take place on the basis of a wide range of information objects. While such objects may be structured, it is often the case that relevant events are included in emails, text messages, policy briefs, reports, and other unstructured information objects. The analysis of such events is time-consuming as it is mostly done by hand and relies heavily on domain knowledge. A major challenge for these analysts is to present the process of interest in a concise and intuitive way. The objective of this paper is threefold: (1) to conceptualize the events and texts of a process chronology in a meaningful way, (2) to provide an intuitive visualization thereof, and (3) to set an agenda for process chronology research paving the way for automation. We view this work as a first step of a broader ambition to study requirements, design, and the automated generation of interactive process chronologies.

Our proposal links to earlier work, which deals with the generation of *timelines* on the basis of a set of textual documents as input. The Timeline Summarization (TLS) line of research aims at automatically identifying key dates of major events from a corpus of texts along with short descriptions of what happened on these dates [17]. Although applicable in settings where the datasets are small and homogeneous, a drawback of these linear timelines is that they quickly become difficult to interpret, e.g. when the amount of dates grows too large or the textual topics are heterogeneous [21].

More recently, in an attempt to break away from the simple, linear structure provided by TLS, *timelines with graph-based representations* (TGRs) were proposed. These structures provide a more intricate structure for the representation of related events (e.g. Figure 1). Pivotal work in this area was carried out in the early 2010 s by Shahaff and others [15]. A recent survey on the topic is provided in [8]. A drawback of TGRs, however, is that they center around one type of relation that ties the various events together. In reality, a wide range of relationships, e.g. causality, topicality, time order, etc., may be required to show how events contribute to the overall chronology of the process. In addition, the TGRs fail to relate to the original information objects that they are based on.

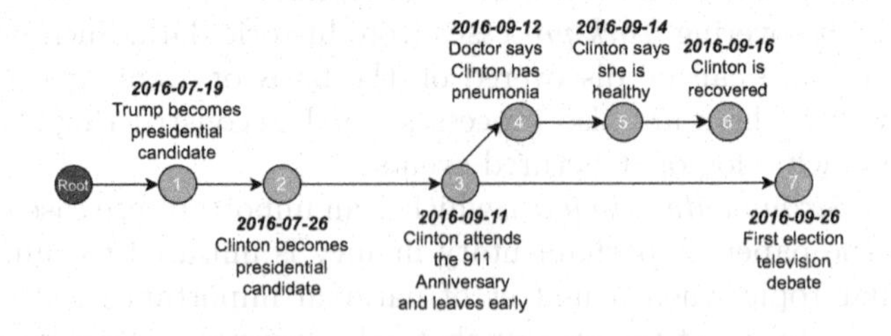

Fig. 1. This US elections TGR [10] links events through the *Subject Relation*.

In this paper, we propose *TimeFlows* as a way to visually represent process chronologies. They extend TGRs to represent collections of related events. Our proposal improves upon the state of the art in that it (1) provides flexibility for the inclusion of a wide range of relationships between events and (2) includes references to the constitutive information objects for the presented events. TimeFlows are applicable within a wide range of applications, including historical and forensic research. Their objective is to come up with a concise yet intuitive reconstruction of a process based on a heterogeneous and potentially large set of information objects.

This paper is structured as follows. In Sect. 2, we delve deeper into related work. Section 3 outlines our research method. It describes our approach to identify the types of relationships that are important for a process chronology through a set of structured interviews with expert analysts. Then, in Sect. 4, we (1) present the results obtained by exploratively coding the interviews, (2) motivate our choice of relationships to be included, and (3) provide a Process Delivery Diagram, which depicts how manual constructions of process chronologies takes place. In Sect. 5, we present a model for process chronologies and describe how TimeFlows can be used to visualize these. We illustrate our proposal by presenting a TimeFlow for the Dutch Childcare Benefits Scandal[1]. We conclude the article with an agenda for further research (Sect. 6).

2 Literature Review

Different models for TGRs have been proposed [1,9], all of which aim to represent what we have coined *process chronologies*. Several algorithms have been devised to automatically generate such graphs, as highlighted in a recent overview [8]. It includes the approaches described in [15,20].

The authors of [9] anchor their definition of a *narrative map* (a synonym of a temporal graph) by means of a weighted, directed, acyclic graph. Edges between events are made solely on the basis of topical or thematic coherence. The degree of coherence determines the weight assigned the edges.

In [1], the authors define yet another synonym for a temporal graph, namely a *Story Graph*. It is considerably more elaborate and complicated than the aforementioned definition. The authors define notions such as *Related Events* (by means of a user-defined coherence threshold), *Topic Theme*, and *Timeline Path*. The authors of [10] graphically depict four different types of structures to characterize a story. These consist of (a) a flat structure, (b) a timeline structure, (c) a graph structure, and (d) a tree structure. In all of these are graphs the nodes indicate events, while the edges denote a topical coherence among these events. As noted in [8], some frameworks allow for forms of interactivity. For instance, in [15], the authors enable the functionality of zoomability of the events, to inspect them at finer or coarser levels of granularity. Moreover, the authors of [16] propose the introduction of a *personalized coverage function* that can be employed

[1] See: https://en.wikipedia.org/wiki/Dutch_childcare_benefits_scandal.

to display temporal graphs that fit the users' needs and preferences, and suit their own queries.

What these approaches have in common, is that they all assume there is only one type of edge - a topical or thematic connection between events. Yet recent research has shown that analysts make different types of connections between events [7]. For instance, they also construct connections on the basis of causality, entities, or specific domain knowledge. In the present paper, we seek to find out which types of relations are most relevant to analysts, and propose a visualization technique process chronologies that captures the variety of relevant relations among events, called *TimeFlows*. Moreover, we investigate how to integrate the documents or texts – more broadly: information objects – on which these events are based within our conceptualization as well.

3 Research Method

We aim at grounding our proposal in data obtained through explorative, semi-structured interviews. These interviews are conducted with people having experience with manually constructing process chronologies in their professional work in their own, specific contexts. By analyzing the answers they have provided us, we seek to determine what kind of challenges analysts face when manually constructing process chronologies, in particular with respect to the relations they wish to establish between relevant events.

We collect data by recording interviews with four interviewee groups. These interviews are transcribed and exploratively coded. Based on the results, we construct a Process Delivery Diagram (PDD) in Sect. 3 and present our proposal for TimeFlows in Sect. 4.

3.1 Interviewee Groups

We selected interviewee groups on the basis of their professional experience with creating process chronologies. All participants have been actively involved in the creation of process chronologies based on unstructured information objects; three of them had an affinity with parliamentary inquiries. We conducted four interviews for our data collection:

(I1) Four researchers of a Dutch Ministry that have experience in creating chronologies for Parliament.
(I2) An interview with a junior council advisor who assisted a minister before and during the hearings for a parliamentary inquiries.
(I3) An investigative journalist who works for a well-known Dutch publication platform.
(I4) Two national government officials that have experience as a coordinator of several parliamentary inquiry research staff.

These interviewee groups are referred to as (I1) through (I4) throughout this paper. When we quote someone from a particular group, it is implied that their views correspond to the consensus opinion of all individuals comprising that particular group.

3.2 Semi-structured Interviews

The interviews were conducted in Dutch and focused on the approach of the different interviewee groups to create process chronologies. In particular, we asked about the different types of *relationships* the interviewees made between and among information objects and events, in order to be able to create meaningful chronologies.

The questions consisted of the following five sections. We list them below, and describe what kind of questions correspond to each section:

1. *Professional Information.* Interviewees were asked about job title, typical workday activities, and years of experience.
2. *Activities and Process.* These questions pertained to what activities their occupation consisted of. We also asked them specifically whether their job entailed making chronologies for parliamentary inquiries or other fact-finding investigations. If that was the case, we probed them about these activities, and asked how they conducted them.
3. *Textual Analysis.* In this section of the interview, we asked participants to describe the different steps they go through to discover and analyse texts they need to read in order to be able to create chronologies or policy document reconstructions.
4. *Relationships between Texts.* Questions within this section pertained to the relationships professionals make between texts while conducting research.
5. *Comparison between Texts and Events.* By asking these questions, we aimed at gaining an understanding of how the interviewees interpreted the level of granularity at which events correspond to texts.
6. *Operations.* This section would delve into the groups' tactics to single out particular passages in texts – like summarizing, highlighting, or splitting.

Overall, these questions were aimed at comprehending how the analysts carry out their day-to-day activities. We were specifically interested in what tactics and techniques the interviewees employ to create process chronologies, and what kinds of relations bore the most relevance to their work.

The following relationships were discussed during the interview. A substantial portion of them were based on the work in [7]. Other relationships were also incorporated in the interviews. The ones that form additions to the work of the authors of [7] are marked with an asterisk (*). In this context, A and B refer to distinct events.

- Temporal (A took place before B).
- Similarity (A is similar to B).
- Entity (A is about the same entity – a person, organization, or place – as B)
- Citation (A refers to B by means of a hyperlink to another document)
- Correspondence* (person P mentions A in a message to person Q, whom subsequently refers to B in a returned or forwarded message)
- Subject (A falls under the same theme or topic as B)
- Causal (A leads to B)

- Social* (A and B are discussed among persons $P_1, \ldots, P_n$)
- Speculative (A is connected to B)
- Domain knowledge (A is related to B through external domain knowledge X, which is not found or represented in the texts themselves)
- Categorical/Ontological* (A is filed under the same category or folder as B, for instance in a DMS or a file structure)

The interviewees were asked to (i) state which relationships they employed during their efforts to manufacture chronologies, and (ii) name the top three of most important relations (in no particular order). Sometimes, they would mention just two important relationships; on other occasions they named four of them.

4 Constructing Process Chronologies

In this section, we present the results of exploratively coding the interviews with the four interviewee groups. Based on the interviews, we constructed a Process Delivery Diagram (PDD) that describes how professionals currently construct chronologies. A PDD combines a process model and a conceptual model to indicate which concepts are created in which activities [19]. Activities in a PDD can have unordered subactivities to express that there is no predefined sequence in which these need to be carried out. In Sect. 4.1, we describe how process chronologies are currently constructed, in Sect. 4.2, we discuss what relations are used, both implicitly and explicitly, to extract concepts – such as events and entities – from the collected information objects.

4.1 Situationalized Process Chronologies

During the interviews, we asked participants their current method of constructing chronologies. The answers of the participants have been coded to discover the different activities and sub activities. This resulted in the PDD shown in Fig. 2. Each activity is supported by interviewee groups, indicated in parentheses.

The first activity is *Object Collection* (I1), which starts either by developing a research question for an information inquiry or by an external inquiry (I3). For example, the parliament defines an inquiry, and then propagates the request to the different departments (I1, I4). Typically, TimeFlows are requested to derive possible interpretations, for example to clarify the subjective images entities (including actors) can have in specific matters (I4). Based on the research question, an initial collection of *Information Objects* is retrieved from a plethora of sources, including Document Management Systems (I1) and open sources (I3). An *Information Object* can range from the agenda and minutes of meetings (I2), memos and email traffic (both within and between organisations) (I1), but can also include public sources such as news papers or social media (I3). The *Information Object* includes metadata about the object itself, such as creation data, version information and author information. The main challenge in this activity is that many information objects are unstructured, or poorly archived (I4). For

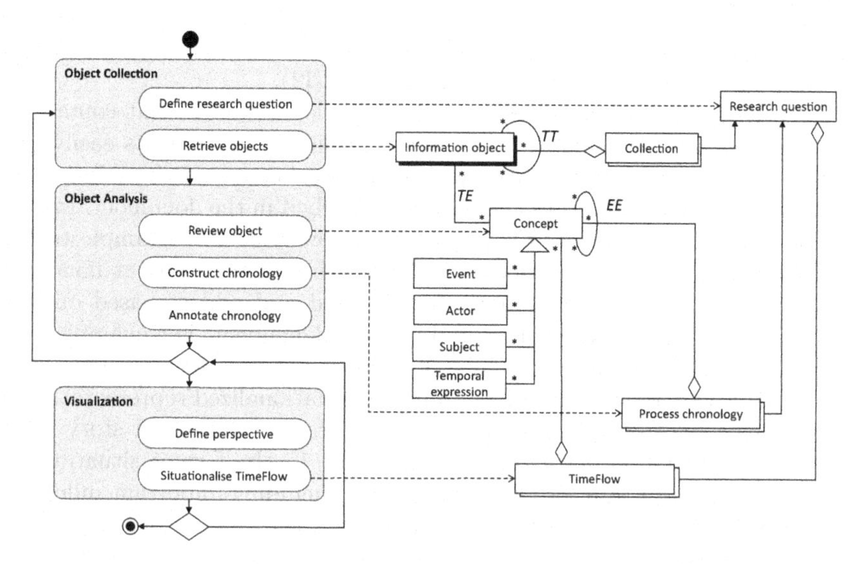

Fig. 2. Process Delivery Diagram of constructing TimeFlows based on the participants. The relations *TT*, *TE* and *EE* refer to the relations specified in Table 1. Together, these relations form a taxonomy.

example, topics can be discussed in meetings, but to discover the records carrying this information can be painstaking (I1). Another problem is versioning, for example the detection and selection of near-duplicate information objects (I4).

The next activity is *Object Analysis*. In this activity, all retrieved *Information Objects* are reviewed on relevance. This review is mostly a manual task ranging from scanning the object and its metadata to a careful read and interpretation (I1, I4). An initial scan is used to determine the relevance of each object (I1). Relevant objects are then analysed to extract the most important *concepts*, such as *events*, *entities*, *subjects*, and *temporal expressions* (I1, I3).

The extracted concepts are used to construct a *chronology*: a large model containing all events extracted from the relevant objects. Each event in the chronology is annotated with the relevant concepts, containing everything that could be relevant to it (I1), ranging from a single sentence of a document to complete documents (I4). In rare cases, sub-chronologies are created to describe how specific documents were established (I2).

The *Object Analysis* activity is typically a manual task, and thus very laborious and time consuming (I1, I4). As one of the participants stated: "it is like looking for a needle in an exponentially growing haystack" (I4). Additionally, objects as policy documents require a thorough analysis to fully grasp the implications of the document (I3). As such, the object analysis is subjective to the researchers who develop the chronology (I2).

The activities are very iterative, switching between the activities *Object Collection* and *Object Analysis*. Based on the chronology, new objects are retrieved to fill in gaps, to clarify events, or to provide an interpretation (I1). For example, if in the agenda of a meeting specific documents are mentioned, these are retrieved and added to the collection for analysis (I2).

Currently, a *Chronology* is described in a single document that contains all events and the concepts related to it. Such chronology documents easily grow to several hundreds of pages (I4). As a consequence, it is infeasible to fully understand the events, and thus the process described in the document. Instead, a specific perspective is created to construct an overview, for example to brief someone (I1, I2), or to gain a better understanding of the topic at hand (I4). For this activity, each event is manually screened and added, based on a set of criteria that is often left implicit and subject-dependent. For example, if an event could not have been known by a stakeholder at the time, then the event is left out (I2). Different perspectives result in situationalized representations of the chronology. For example, a different representation is used for story telling (I2) than for understanding a certain topic (I4). In the former situation, the representation mainly describes what happened, including important milestones and involved entities, whereas the latter may be a timeline to identify gaps, to refine specific concepts (I4). As one of the participants noted: a chronology is never static (I4).

4.2 Identified Relations Used in Document Analysis

As became apparent from the interviews, *Object Analysis* is the most laborious and time-consuming activity. All information objects need to be reviewed on relevance. Moreover, important concepts, such as events and entities, ought to be extracted. This is a manual task, as their interpretation is far from trivial. It requires in-depth knowledge, which is often tacit (I3, I4). As the participants indicated (I1, I3, I4), the concept extraction and chronology creation is a subjective task, where the researcher uses their tacit knowledge to decide the relevance of objects and the concepts it contains or discusses.

Equally important are the relations that explain the link between information objects and concepts. We distinguish three levels of relations:

1. *Between Information Objects*, i.e., relationships between different information objects (represented by level *TT* in Fig. 2).
2. *Between Information Objects and Concepts*, i.e., relationships that connect information objects to concepts (represented by level *TE* in Fig. 2).
3. *Between Concepts*, i.e., relationships between different concepts, such as events and entities (represented by level *EE* in Fig. 2).

As shown in Sect. 2, different types of relations can be identified, such as information objects that describe a certain event, or entities that participated in a meeting described by the information object. Therefore, we also asked the participants which relationships they use during the chronology construction. We extracted eight relations that are commonly used by the participants to obtain concepts from information objects and relate them.

Each of the participants created a top 3 of relation types. The results are shown in Fig. 3: *Temporal*, *Subject*, and *Entity* relationships were scored highest. The *Causal* relationship and *Correspondence* relations were mentioned less frequently. We choose to include all five of them within our framework.

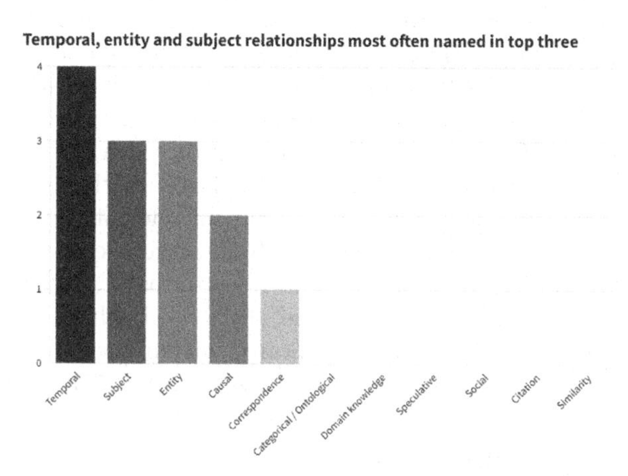

Fig. 3. Number of interviewee groups that mention the respective relationships between events as one of the top three most important ones

Our coding revealed three additional relations that were described by the participants, but not mentioned explicitly. For example, all participants indicated they extracted the concepts from information objects. In essence, this means that each information object is linked to several concepts that are extracted from it. We refer to this relation as the *Consists Of* relation. This is the first of the three additional relations.

Table 1. Relations identified by the participants and their use over the different levels.

		TT	EE	TE
Group A	Temporal	x	x	x
	Subject	x	x	x
	Entity	x	x	x
	Causal	x	x	x
	Correspondence	x	x	x
Group B	Succession	x		
	References To	x		
Group C	Consists Of			x

Two other relations were also identified while coding the interviews. In one of the interviews (I1), the participants explained that they considered email conversations, and added information based on the sender/receiver relation: "If we look at the most important message [of a chronology], we also look at policy documents, but certainly at email conversations." His colleague added: "Yes, so we look at who has responded to whom, and who has responded to that." Similarly, in another interview (I4), the participants indicated that they used the metadata of policy documents to derive the involved entities, and the dates on which these entities were involved. This analysis resulted in two additional relations: *Succession* and *Reference To*. The former indicates which objects directly follow each other, e.g. if an email is a direct reply to another email or if it is a forwarded message. The latter represents document referrals, such as documents discussed in meetings, or email messages with an attachment.

We categorize all eight relations within three groups: A, B, and C, according to the types of relationships they afford (see Table 1). In this upcoming section, we explain how these relations are incorporated within our conceptualization of process chronologies, and subsequently visualized within *TimeFlows*.

5 Visualizing TimeFlows

In this section, we conceptualize process chronologies to construct TimeFlows from collections of information objects. As several participants in the interviews mentioned, visualising process chronologies has many different purposes, ranging from understanding the role of entities in the process (I2) to exploratory studies to identify gaps (I4). The field of Visual Analytics "combines automated analysis techniques with interactive visualizations for an effective understanding, reasoning and decision making on the basis of very large and complex data sets" [6]. Thus, visualizing TimeFlows can be considered as an application of Visual Analytics, as depicted in Fig. 4. From the collection of information objects, a process chronology is extracted. This chronology is then visualized as a TimeFlow to gain new knowledge on the topic at hand. In the remainder of this section, we present our proposal to visualize process chronologies as TimeFlows, and illustrate its capabilities by presenting a TimeFlow for the Dutch Childcare Benefits Scandal.

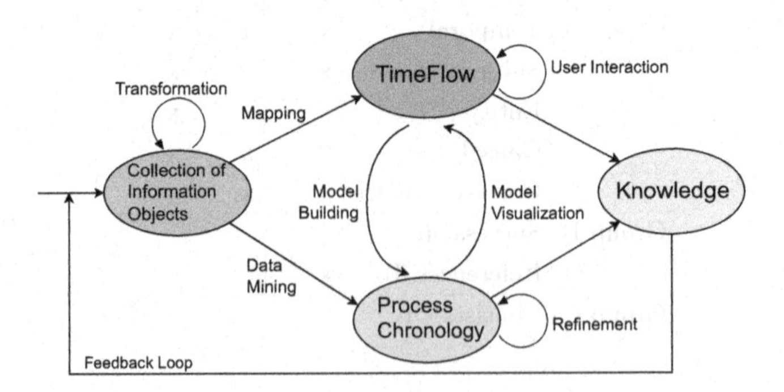

Fig. 4. The Visual Analytics feedback loop [6] applied to process chronologies.

5.1 TimeFlows to Visualize Process Chronologies

As presented in the previous section, a process chronology is a list of all annotated events extracted from a collection of information objects. Every event is annotated with the concepts it is related to, including entities and other events. Moreover, the events are enriched with the information objects that constitute them. A TimeFlow is a visual representation of a process chronology, to study a specific perspective. The following general design principles guide the visualization of TimeFlows:

1. We delineate *information objects* on the one hand, and *events* on the other. In their simplest form, events can be seen as "things that happen". Events leave traces in the form of information objects that can be viewed as manifestations of past events [4]. Events are depicted using white rounded rectangles, while information objects are shown in light grey rectangles with sharp corners.
2. Relations are shown by means of arrows. The arrows stemming from the relations in group A are distinguished from one another through the use of both icons and colors. The relations in group B and C are all shown in black, but have different stroke styles (solid, dashed, or dotted).
3. The progression of time is indicated by means of arrows from left to right between events, originating from relations in group A. Choosing to depict the flow of time from left to right stems from old conventions ingrained within Western culture [11]. Moreover, events are numbered according to their temporal order, which also makes it easy to refer to them.
4. The relationships depicted by the arrows in Group A are explained through the use of highlighted words in the information objects and events. They have the same color as the arrows of the respective relations, and indicate which words are of relevance to the relation.

The relations described in point 2 above are chosen and motivated in Sect. 4.2. Below, we describe the design considerations for each individual relationship.

- **Temporal-Semantic Relationship.** → We relate events and information objects with one another that bear the same *temporal expressions.* These are words that indicate a particular moment in time [18]. For instance, if one object contains the phrase "On the first of September, we had a meeting," and another object contains the fragment "Things went well during the meeting on 01/09, we'll meet again soon," then these objects are linked through the Temporal-Semantic Relationship by means of a light orange arrow with a clock icon. Other examples of temporal expressions are "tomorrow", "next week", "Halloween", and "The Summer holidays". (Levels: TT, EE, and TE.)
- **Subject Relationship.** → Events and information objects are linked with one another with a purple arrow and a book icon if the texts they comprise are related by subject. (Levels: TT, EE, and TE.)

- **Entity Relationship.** An Entity Relationship is made when the same concept, e.g. a name of a person, an organization, or a place, is mentioned in both of the events or objects. Those objects that belong to a particular event and mention the same concept, are assigned an Entity relationship by means of an orange arrow and a stick figure icon. (Levels: TT, EE, and TE.)
- **Causal Relationship.** This relationship is visualized with a turquoise arrow and an icon of Newton's Cradle on top. It depicts which events are causally related to another, and is also drawn between information objects and events when words that indicate a causal relationship are present. (Levels: TT, EE, and TE.)
- **Correspondence Relationship.** This relationship is depicted with a dark blue arrow and an icon of a letter on top. It is drawn when the correspondent of one object is also recognized as the correspondent of another object. (Levels: TT, EE, and TE.)
- **Succession Relationship.** The *Succession Relationship* is a black, filled arrow with its head pointing towards to an information object that is a response or forward of the original text. (Level: TT)
- **References To Relationship.** The *References To Relationship* is depicted by means of a shortly-spaced dotted black arrow. It ends at the information object that is referenced, and begins at the object that does the referencing. It is employed when, for instance, an e-mail contains an attachment. (Level: TT)
- **Consists Of Relationship.** The *Consists Of Relationship* is shown by means of a widely-spaced dotted black arrow, originating from the event and pointing towards the different information object the event consists of. (Level: TE)

As intended by the Visual Analytics framework, it is the user that constructs TimeFlows themselves, based on the information objects they have collected and their own perspective on the information objects they wish to study.

5.2 An Illustrative Example

We illustrate the proposal of TimeFlows to represent process chronologies by depicting a sequence of events and accompanying documents pertaining to what is known in the Netherlands as the "Childcare Benefits Scandal". This political affair has its origins in the year 2004 (when the Childcare Act came into effect), and concerned false allegations of fraud made by the Tax Authorities. It is estimated that between the years 2005 and 2019, this organization wrongfully accused more than 25.000 parents of making fraudulent childcare benefits claims[2], driving families into severe financial hardship. After a parliamentary inquiry was conducted in 2021, the third cabinet Rutte resigned over the affair.

[2] See https://www.trouw.nl/politiek/wie-wist-wat-in-de-toeslagenaffaire-de-kluwen-van-hoofdrolspelers-ontward~b721c834/.

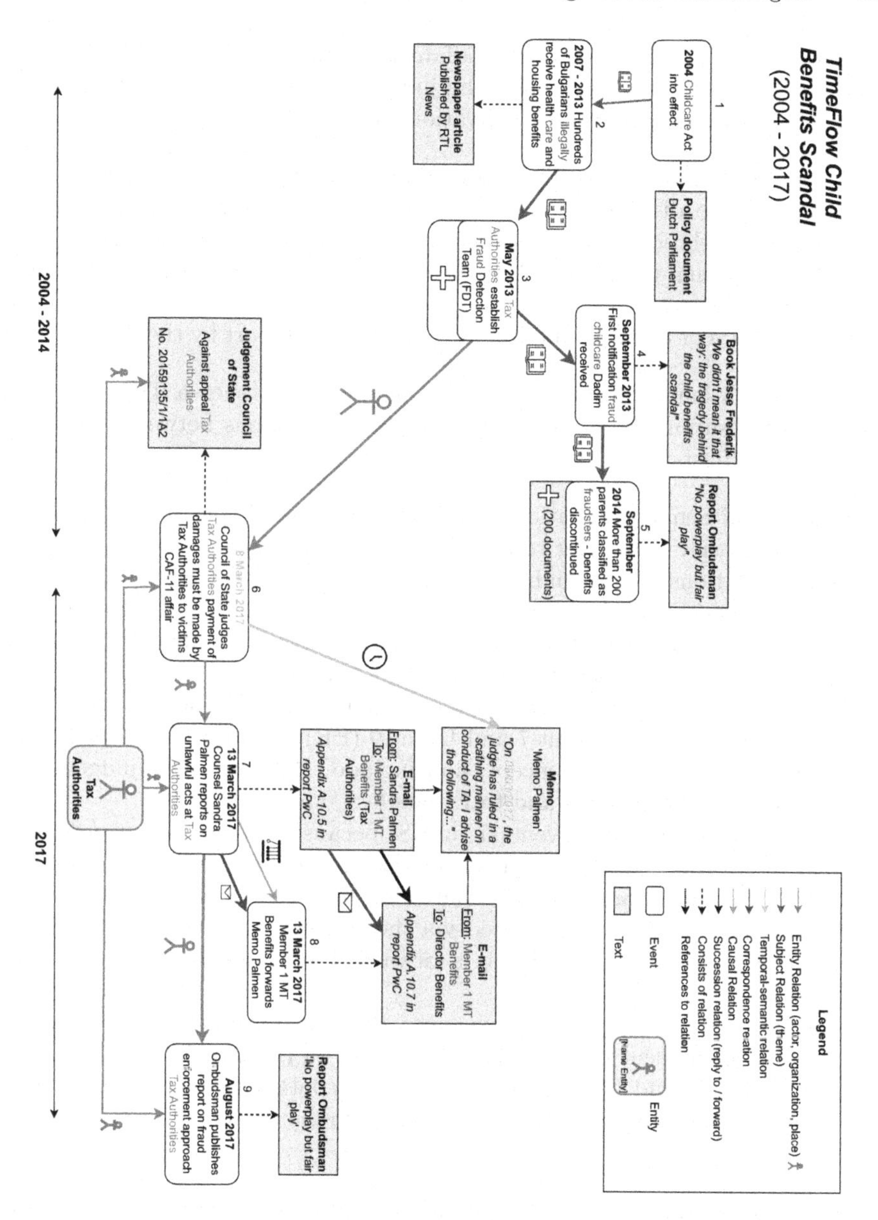

Fig. 5. A TimeFlow of the Childcare Benefits Scandal

This TimeFlow visualization is grounded in true events and real texts. A key document was the "Memo Palmen" [12,13]. The other texts we used are, in order of appearance: the Dutch Childcare Act of 2004 [3], a news article on the

'Bulgarian Fraud'[3], the book on the affair as a whole by Jesse Frederik [5], the report by the Ombudsman [2], and the judgement of the Council of State [14]. We make the following general observations of this TimeFlow:

- It consists of nine events. These are labelled by their temporal order.
- All events are distilled of one or more texts, shown with their respective Consists Of relations. In total, nine texts are explicitly shown. The existence of 200 additional texts is implied through the depiction of the + sign shown underneath Event 5.
- Note that Event 5 and Event 9 both consist of the report by the Ombudsman.

Moreover, additional information can be highlighted in such a TimeFlow. The current visualisation shows different examples of relations between events and information objects:

- A Temporal-Semantic relation is indicated between Event 6 and the Memo Palmen, as they both refer to the date 08-03-2017 (TE).
- Subject relations are drawn between Event 1 and Event 2, Event 2 and Event 3, Event 3 and Event 4, and between Event 4 and Event 5 (EE). For instance, the Subject relation between the first two events is justified because they contain the related words "Childcare" and "care".
- Entity relations are depicted between Event 3 and Event 6, Event 6 and Event 7, and between Event 7 and Event 9 (EE). We also single out the Tax Authorities as an entity concept, and include relations to the judgement of the Council of State (TE) and to Events 6, 7, and 9 (EE).
- A Causal relation links Events 7 and 8, because the act of forwarding the email - performed by Member 1 of the MT - is a direct causal consequence of Event 7, in which Sarah Palmen shares her findings (EE).
- Events 7 and 8 are also connected via the Correspondence relation, because the latter consists of an email that is forwarded from an email Event 7 consists of (EE).
- A Succession relation is shown between the Sarah Palmen's email, and the forwarded email sent by Member 1 of the MT of the Tax Authorities (TT).
- Both Palmen's and MT Member 1's emails contain Palmen's report as an attachment - thus both of these emails point to the report with the References To relation (TT).

As the events in this example show, multiple relation types are needed to understand and study what happened. As shown in Sect. 2, these relations are not shown in current visualisation techniques, such as the TGR in Fig. 1. Moreover, the constitutive documents clarify on what basis the events have been generated – something that is lacking in previous conceptualizations of timelines with graph-based representations (TGRs).

[3] See for example https://www.rtlnieuws.nl/nieuws/nederland/artikel/1173981/uitspraak-bulgarenfraude-hoe-zat-het-ook-alweer.

6 Further Research: Challenges

In this paper, we argue that the visual analytics framework (see Fig. 4) can be applied to study process chronologies using TimeFlows. However, to realize this, many challenges still need to be addressed. These challenges originate from the problems described by the interviewees.

Data Mining. The future software ought to be able to automatically mine important concepts of heterogeneous document sets. Relevant challenges in this context include *Named Entity Recognition and Linking* (NER and NEL), *Temporal Expression Recognition, Normalization, and Linking* (TERNL), *Topic Modelling* (TM), and *Event Extraction* (EvEx).

Mapping. The relations between the properties of the input data (information objects) and the visualizations (TimeFlows) are to be studied. A challenge lies in mapping more heterogeneous data sources to proper visualizations, including tabular data in Excel sheets, meeting minutes, and voice memos.

Model Visualization. In order that the TimeFlows are generated in a visually intuitive manner, research must be carried out on peoples' preferred shapes and colours, the positioning of the events, information objections, and relations, and the amount of relations shown so as to balance clarity on the one hand, yet avoid information overload on the other.

User Interaction. Here, the challenge lies in understanding how users can be supported in developing effective TimeFlows, e.g., the interaction with TimeFlows to enable users to obtain the information objects that bear relevance to each event, to move concepts and entities on the screen, to 'to focus on' or 'abstract away' specific events and investigate them at different levels of granularity.

Automated Model Building. Future TimeFlows software could allow users to study different perspectives on sequences of events and their information objects. They may single out a perspective or combine different perspectives to study a research question from different angles. To support this activity, generative techniques need to be developed that propose multiple perspectives on the collection of information objects. These perspectives can help in the understanding of the process and the events manifested within the information objects.

7 Conclusion

In this paper, we introduce *process chronologies* as non-repetitive processes, which lack an event log. Based on interviews with experts, we have conceptualized different relations between the concepts and unstructured information objects that underlie them. To convey the events in a non-repetitive process, we propose *TimeFlows* to visualize the relation between information objects and the concepts manifested in these objects. TimeFlows allow users to study non-repetitive processes from different perspectives. Although our analysis was conducted on explorative interviews with domain experts, we believe that an holistic approach that visual analytics with TimeFlows offers great potential aid the understanding of such processes. However, to fully support the understanding and study of non-repetitive processes, many challenges need to be overcome. We therefore propose a research agenda to guide future work in this direction, which is aimed at automating the time-consuming manual work for process chronology creation. We view the present work as a broader ambition to develop interactive software that allows analysts to inspect large amounts of heterogeneous data sources to retrospectively study how non-repetitive processes unfold over time.

References

1. Ansah, J., Liu, L., Kang, W., Kwashie, S., Li, J., Li, J.: A graph is worth a thousand words: telling event stories using timeline summarization graphs. In: The World Wide Web Conference, pp. 2565–2571 (2019)
2. van den Berg, W., Alhadjri, M., Mulder, M.: Geen powerplay, maar fair play. De Nationale Ombudsman (2017)
3. Dutch Government: Wet kinderopvang. Wettenbank (2004)
4. Ferejohn, M.T.: Formal Causes: Definition, Explanation, and Primacy in Socratic and Aristotelian Thought. OUP Oxford (2013)
5. Frederik, J.: Zo hadden we het niet bedoeld: De tragedie achter de toeslagenaffaire. De Correspondent Uitgevers (2021)
6. Keim, D., Andrienko, G., Fekete, J.D., Görg, C., Kohlhammer, J., Melancon, G.: Visual analytics: definition, process, and challenges. Inf. Vis. LNCS **4950**, 154–175 (2008). https://doi.org/10.1007/978-3-540-70956-5_7
7. Keith Norambuena, B.F., Mitra, T., North, C.: Design guidelines for narrative maps in sensemaking tasks. Inf. Vis. **21**(3), 220–245 (2022)
8. Keith Norambuena, B.F., Mitra, T., North, C.: A survey on event-based news narrative extraction. ACM Comput. Surv. **55**(14s), 1–39 (2023)
9. Keith Norambuena, B.F., Mitra, T.: Narrative maps: an algorithmic approach to represent and extract information narratives. Proc. ACM Hum.-Comput. Interact. **4**(CSCW3) (2021)
10. Liu, B., Han, F.X., N.D., Kong, L., Lai, K., Xu, Y.: Story forest: extracting events and telling stories from breaking news. ACM Trans. Knowl. Disc. Data **14**(3), 1–28 (2020)
11. Núñez, R., Cooperrider, K., Doan, D., Wassmann, J.: Contours of time: topographic construals of past, present, and future in the yupno valley of papua new guinea. Cognition **124**(1), 25–35 (2012)
12. PricewaterhouseCoopers: appendices bij pwc rapport reconstructie en tijdlijn van het 'memo-palmen'. Report PwC, pp. 1–301 (2021)
13. PricewaterhouseCoopers: Reconstructie en tijdlijn van het 'memo-palmen'. Report PwC, pp. 1–66 (2021)
14. Raad van State (Council of State): Uitspraak 201509135/1/a2. https://www.raadvanstate.nl/, pp. 1–8 (2017)
15. Shahaf, D., Guestrin, C.: Connecting the dots between news articles. In: IJCAI 2011, pp. 2734–2739. IJCAI/AAAI (2011)
16. Shahaf, D., Guestrin, C., Horvitz, E.: Trains of thought: generating information maps. In: WWW 2012, pp. 899–908. ACM (2012)

17. Steen, J., Markert, K.: Abstractive timeline summarization. In: Proceedings of the 2nd Workshop on New Frontiers in Summarization, pp. 21–31. Association for Computational Linguistics, Hong Kong, China (2019)
18. UzZaman, N., Allen, J.F.: Extracting events and temporal expressions from text. 2010 IEEE ICSC, pp. 1–8 (2010)
19. van de Weerd, I., Brinkkemper, S., Souer, J., Versendaal, J.: A situational implementation method for web-based content management system-applications: method engineering and validation in practice. Softw. Process. Improv. Pract. **11**(5), 521–538 (2006)
20. Xu, S., Wang, S., Zhang, Y.: Summarizing complex events: a cross-modal solution of storylines extraction and reconstruction. In: EMNLP, pp. 1281–1291. Association for Computational Linguistics, Seattle, Washington, USA (2013)
21. Yu, Y., Jatowt, A., Doucet, A., Sugiyama, K., Yoshikawa, M.: Multi-TimeLine summarization (MTLS): improving timeline summarization by generating multiple summaries. In: IJCNLP, pp. 377–387. Association for Computational Linguistics, Online (2021)

Imposing Rules in Process Discovery: An Inductive Mining Approach

Ali Norouzifar[1(✉)], Marcus Dees[2], and Wil van der Aalst[1]

[1] RWTH University, Aachen, Germany
{ali.norouzifar,wvdaalst}@pads.rwth-aachen.de

[2] UWV Employee Insurance Agency, Amsterdam, The Netherlands
Marcus.Dees@uwv.nl

Abstract. Process discovery aims to discover descriptive process models from event logs. These discovered process models depict the actual execution of a process and serve as a foundational element for conformance checking, performance analyses, and many other applications. While most of the current process discovery algorithms primarily rely on a single event log for model discovery, additional sources of information, such as process documentation and domain experts' knowledge, remain untapped. This valuable information is often overlooked in traditional process discovery approaches. In this paper, we propose a discovery technique incorporating such knowledge in a novel inductive mining approach. This method takes a set of user-defined or discovered rules as input and utilizes them to discover enhanced process models. Our proposed framework has been implemented and tested using several publicly available real-life event logs. Furthermore, to showcase the framework's effectiveness in a practical setting, we conducted a case study in collaboration with UWV, the Dutch employee insurance agency.

Keywords: process mining · process discovery · domain knowledge

1 Introduction

Process discovery seeks to identify process models that provide the most accurate representation of a given process. The quality of discovered process models is measured using evaluation metrics while ensuring comprehensibility for human understanding and alignment with domain experts' knowledge. Many state-of-the-art discovery approaches rely solely on event logs as their primary source of information, often neglecting additional valuable resources such as the knowledge of process experts and documentation detailing the process [1]. These overlooked resources can significantly enhance the discovery process [14].

This research was supported by the research training group "Dataninja" (Trustworthy AI for Seamless Problem Solving: Next Generation Intelligence Joins Robust Data Analysis) funded by the German federal state of North Rhine-Westphalia.

J. Araújo et al. (Eds.): RCIS 2024, LNBIP 513, pp. 220–236, 2024.
https://doi.org/10.1007/978-3-031-59465-6_14

Process experts often possess a common understanding of how the process functions and additional resources like process diagrams may be available alongside event logs. This information can often be expressed in human language, e.g., activities a and b cannot occur together, or activity a cannot occur after activity b. Automated methods can also be used to discover such relations between activities from an event log [4,11]. We propose a novel framework to impose such information in process discovery. Our framework leverages Inductive Mining (IM) techniques, with a distinctive feature that allows it to incorporate a set of rules as an additional input. We assume the rules are given by an oracle, e.g., user-defined or discovered rules using automated methods. The framework is designed independently from the source that provides the rules.

IM techniques discover block-structured process models that are both comprehensible for humans and offer guarantees, such as soundness [10]. The information flow in one recursion of IM approaches is illustrated in Fig. 1a. Overlooking other information sources, shown as process knowledge, results in some information loss. In each recursion of the IM techniques, a Directly Follows Graph (DFG) is derived from the event log that serves as the basis for determining the process structure. The conversion of an event log to a DFG can lead to information loss. Some variants include filtering mechanisms to eliminate infrequent behaviors (DFG') which contribute to more information loss. To avoid potential blocks in IM techniques when no cut is detected, heuristics are used to prevent the discovery process from becoming impeded, albeit at the cost of potentially generating over-generalized models.

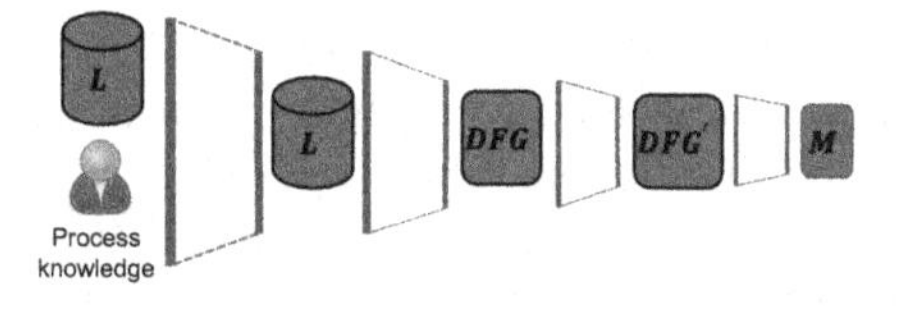

(a) **Information flow in traditional IM frameworks.**

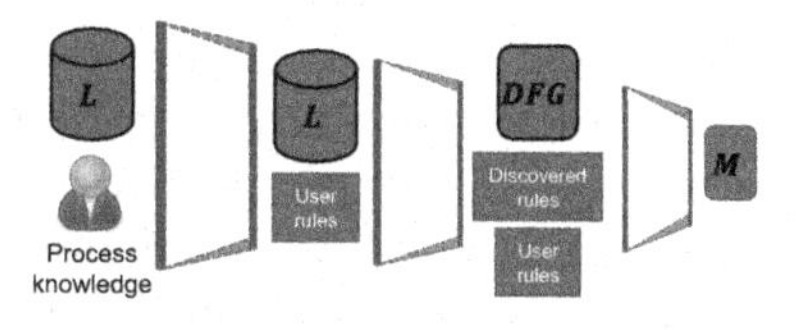

(b) **Information flow in our proposed framework.**

Fig. 1. Comparing the information flow in IM frameworks with our framework.

In this paper, we present an inductive mining framework designed to leverage encoded process knowledge expressed in the *Declare* language. This additional information is utilized to enhance the discovery of improved process models, as depicted in Fig. 1b. The information we can preserve in each step is illustrated in green color. The process knowledge can be encoded as user-defined rules, providing an extra source of information alongside the event log. Automated declarative process discovery methods can be employed to reveal process structures, compensating for the information loss in the extracted DFG. Importantly, our approach does not rely on fall-throughs. In cases where no perfect cut exists, it returns the most promising one based on specific cost functions similar to [12].

2 Related Work

The incorporation of process knowledge or additional information sources in process discovery has been explored in various formats [14]. This information may originate from domain experts' knowledge, business documents, or be derived through automated algorithms. Utilizing such information before the actual discovery process to preprocess event logs aligns with common steps in data analysis frameworks. However, the involvement of these information sources in process discovery remains limited within the literature.

In [8], the authors proposed an automatic approach for discovering artificially created events, revealing aspects not observed in the event log. This information guides the discovery algorithm towards generating more robust process models. Another approach, presented in [13], involves using prior knowledge to learn a control flow model in the form of information control nets. In [16] a method is introduced that leverages user knowledge in the form of relations between activities to construct a directly follows graph. Unlike [13] and [16] our focus is on Petri net models. Another strategy involves post-discovery process model repair based on predefined preferences [7]. Additionally, interactive process discovery and online process discovery are explored as related works [15].

Declare language [4,11] and compliance rule graph language [9] exemplify rule modeling techniques that offer high interpretability. In [6], the use of declarative rules provided by users or discovered by declarative mining algorithms is considered to enhance the quality of discovered process models. The rules are not used directly in the discovery, instead a discovered model is used, several modifications are applied and the best model that adheres to the rules is selected. In our proposed method, we directly use the rules in process discovery.

Automatic process discovery is a crucial research area, yet fundamental questions persist despite the plenty of proposed algorithms. Two types of inductive mining algorithms exist in the literature: those that output a unique cut in each recursion without quality evaluation [10], and those that select the best cut over a set of candidates based on quality measures [5,12] and [2]. We extend the idea proposed in [12] and make it capable of using rules in discovery recursions.

3 Motivating Examples

To motivate the research question addressed in this paper, we offer examples highlighting the necessity of our investigation. Figures 2a and 2b showcase a part of Petri net models discovered from publicly available real-life event logs, i.e., BPIC 2017 and BPIC 2018, using IMf algorithm with 0.2 as the infrequency filtering parameter. Additionally, Fig. 2c exemplifies a process model discovered using the same settings from an event log provided by UWV agency consisting of cases related to a claim handling process which is investigated in detail in the evaluation section.

In Fig. 2a, despite the sequential relation between final states identified by IMf, the event log analysis reveals that only one of the final states can occur which means either an application is accepted (*A_Submitted*), canceled

(*A_Cancelled*), or denied by the client (*A_Denied*). In Fig. 2b, the model over-generalizes the observed behavior and allows for some ordering of the activities which does not make sense both based on the event log analysis and a common sense we have based on the activity names. After *begin editing*, *calculate* should occur, followed by *finish editing* which makes a case ready to make a decision (activity *decide*). Activity *revoke decision* may then occur after making a decision. A case can have multiple repetitions of the explained procedure. The shown model allows for many behaviors that deviate from this procedure, e.g., *decide* before *calculate* and *finish editing*, or *revoke decision* before *decide*.

In the UWV event log, based on the domain knowledge a case must start with *Receive Claim* and *Start Claim*, however, in Fig. 2c, the process model allows for many activities before receiving a claim that does not make sense. *Block Claim 1*, *Block Claim 2*, and *Block Claim 3* have specific meanings in this process. The ordering of these activities does not make sense according to our investigations with process experts at UWV, e.g., *Block Claim 1* can only occur after starting a claim when some information is missing and should be followed by *Correct Claim* and *Unblock Claim 1* to make it ready to get accepted (*Accept Claim*). Figures 6a, 6b, and 9 show the models discovered by our proposed framework for BPIC 2017, BPIC 2018, and UWV event logs respectively.

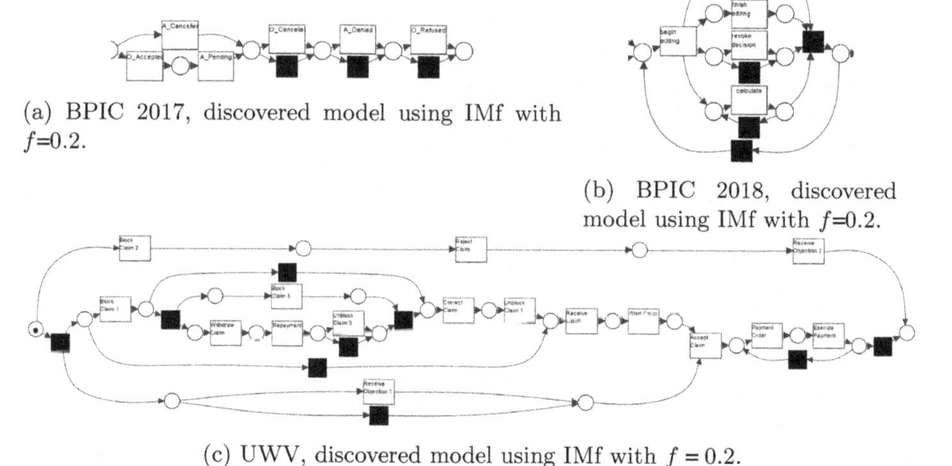

(a) BPIC 2017, discovered model using IMf with f=0.2.

(b) BPIC 2018, discovered model using IMf with f=0.2.

(c) UWV, discovered model using IMf with $f = 0.2$.

Fig. 2. Motivating examples, using IMf to discover process models for BPIC 2017, BPIC 2018, UWV event log.

4 Preliminaries

Considering $\mathcal{A}$ as the universe of activities, $s \in \mathcal{A}^*$ denotes a sequence of activities where $s(i)$ indicates the i-th element in this sequence. $\mathcal{P}(\Sigma)$ denotes the power set over set $\Sigma \subseteq \mathcal{A}$. We introduce an event log formally as a multiset of traces, i.e., a sequence of activities.

Definition 1 (Event log). *Let $\mathcal{A}$ be the universe of activities. A trace $\sigma = \langle a_1, a_2, \ldots, a_n \rangle \in \mathcal{A}^*$ is a finite sequence of activities. Each occurrence of an activity in a trace is an event. An event log $L \in \mathcal{B}(\mathcal{A}^*)$ is a multiset of traces. $\mathcal{L}$ is the universe of event logs.*

Since we are building our framework based on the inductive mining technique, process tree notation is used as the representation. Process trees can be converted to Petri nets or BPMN models as more popular process notations.

Definition 2 (Process tree). *Let $\bigoplus = \{\rightarrow, \times, \wedge, \circlearrowleft\}$ be the set of process tree operators and let $\tau \notin \mathcal{A}$ be the so-called silent transition, then*

- *activity $a \in \mathcal{A}$ is a process tree,*
- *the silent activity τ is a process tree,*
- *let $M_1, \cdots, M_n$ with $n > 0$ be process trees and let $\oplus \in \bigoplus$ be a process tree operator, then $\oplus(M_1, \cdots, M_n)$ is a process tree.*

$\mathcal{M}_\Sigma$ is the set of all possible process trees generated over a set of activities $\Sigma \subseteq \mathcal{A}$.

Each process tree operator $\oplus \in \{\rightarrow, \times, \wedge, \circlearrowleft\}$ has a semantic which generates a special type of behavior. The function $\phi : \mathcal{M}_\Sigma \rightarrow \mathcal{P}(\Sigma^*)$ extracts the set of traces allowed by a process tree which we refer to as the language of this process tree. If a process tree consists of an operator as the root node and single activities as children, the language of it is as follows:

- $\rightarrow$ denotes the sequential composition of children, e.g., $\phi(\rightarrow(a, b)) = \{\langle a, b \rangle\}$.
- $\times$ represents the exclusive choice between children, e.g., $\phi(\times(a, b)) = \{\langle a \rangle, \langle b \rangle\}$
- $\wedge$ denotes the concurrent composition of children, e.g., $\phi(\wedge(a, b)) = \{\langle a, b \rangle, \langle b, a \rangle\}$
- $\circlearrowleft$ represents the loop execution in which the first child is the body of the loop and the other children are redo children, e.g., $\phi(\circlearrowleft(a, b)) = \{\langle a \rangle, \langle a, b, a \rangle, \ldots\}$.

$M = \times(\rightarrow(a, b), \wedge(\times(c, \tau), d))$ is a more complex example such that $\phi(M) = \{\langle a, b \rangle, \langle d \rangle, \langle c, d \rangle, \langle d, c \rangle\}$.

Consider $\mathcal{G}(L)$ as a function that extracts the directly follows graph (Σ, E) from event log $L \in \mathcal{L}$ such that $\Sigma = \{a \in \sigma | \sigma \in L\}$ is the set of activities and $E = \{(a_1, a_2) | \; \exists \, \sigma \in L \wedge 1 \leq i < |\sigma| : \sigma(i) = a_1 \wedge \sigma(i+1) = a_2\}$ is the set of edges.

5 Inductive Miner with Rules (IMr)

In this paper, we adapt and extend the IMbi framework proposed in [12] to allow for rules being used in process discovery. The new framework is referred to as IMr in this paper. In Fig. 3, the main idea of this paper is illustrated. The IMbi framework, designed for two event logs, a desirable event log and an undesirable event log, is adapted in our approach. For the sake of simplicity, the undesirable event log is excluded[1], however, the approach is adaptable to scenarios where the undesirable event log is included. The algorithm finds binary cuts in each recursion like IMbi.

[1] The parameter *ratio* controls the relevance of the undesirable event log; setting the parameter $ratio = 0$ disregards L^-, focusing solely on the desirable event log L^+.

Definition 3 (Binary Cut). *Let $L \in \mathcal{L}$ be an event log. $\mathcal{G}(L)=(\Sigma, E)$ is the corresponding DFG. A binary cut $(\oplus, \Sigma_1, \Sigma_2)$ divides Σ into two partitions, such that $\Sigma_1 \cup \Sigma_2 = \Sigma$, $\Sigma_1 \cap \Sigma_2 = \varnothing$, and $\oplus \in \{\rightarrow, \times, \wedge, \circlearrowleft\}$ is a cut type operator.*

Algorithm 1 shows how IMr works. In each recursion, $explore(\mathcal{G}(L), R)$ explores the DFG extracted from event log L and returns a set of candidate cuts. We explain in this paper how the set of rules R can be used to prune the set of candidate cuts. We use the *ov_cost* function as defined in [12] to compare the cuts. The cost value for each candidate cut is determined by counting the number of deviating edges and estimating the number of missing edges required to modify $\mathcal{G}(L)$ to align with the candidate cut. Parameter $sup \in [0,1]$ specifies to what extent missing behaviors should be penalized. Among the set of candidate cuts, the cut with the minimum cost is selected. The algorithm continues with splitting the event log based on the selected cut (function $SPLIT$) and proceeding to the next recursion.

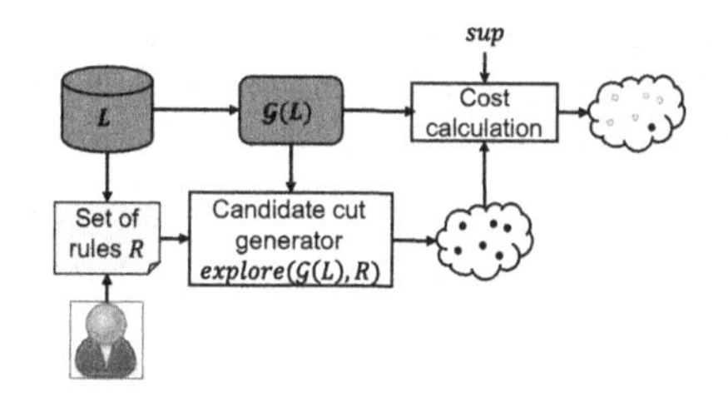

Fig. 3. One iteration of IMr, the framework proposed in this paper, identifies candidate cuts adhering to specified rules and selects the cut with minimum cost, incorporating cost functions from [12].

Algorithm 1 IMr algorithm

function $IMr(L, sup, R)$
  ▷ *$L \in \mathcal{L}$ is an event log, $sup \in [0,1]$ is a process discovery parameter and R is the set of rules.* ◁
  $base = checkBaseCase(L, sup)$
  if $checkBaseCase$ successful **then**
    return $base$
  $C = explore(\mathcal{G}(L), R)$
  $(\oplus, \Sigma_1, \Sigma_2) = \arg\min_{c \in C}\{ov\_cost_{\mathcal{G}(L)}(c, sup)\}$
  $L_1, L_2 = SPLIT(L, (\oplus, \Sigma_1, \Sigma_2))$
  return $\oplus(IMr(L_1, sup, R), IMr(L_2, sup, R))$

5.1 The Set of Rules

The main difference between IMbi and IMr is the use of the set of rules R in finding the set of candidate cuts. Each rule $r \in R$ is a constraint that limits the behavior. Process models may allow for traces that violate or satisfy a rule. We need a formal language to implement the idea, therefore, without loss of generality, we use declarative constraints in this paper. However, the concept is general, and any rule with clear semantics can be employed, provided that *there exists a clear mapping between rule satisfaction and the allowance of certain cut types.*

A declarative constraint is an instantiation of a template that involves one or more activities [4,11]. Templates are abstract parameterized patterns. In this paper, a subset of declarative templates is used including:

- *at-most*(a): a occurs at most once.
- *existence*(a): a occurs at least once.
- *response*(a, b): If a occurs, then b occurs after a.

- *precedence*(a, b): b occurs only if preceded by a.
- *co-existence*(a, b): a and b occur together.
- *not-co-existence*(a, b): a and b never occur together.
- *not-succession*(a, b): b cannot occur after a.
- *responded-existence*(a, b): If a occurs in the trace, then b occurs as well.

Including other declarative templates requires more sophisticated design choices and considerations, therefore, we only focus on a subset of them. Process experts can encode their knowledge and understanding in the form of declarative rules and use them in process discovery as we explain in this paper. In addition to user-defined rules, automated declarative process discovery algorithms such as Declare Miner [11] and MINERful [4] can be used to discover declarative constraints from event logs. If trace σ violates constraint r, we show it as $\sigma \nvDash r$. For example, consider $\sigma = \langle a, b, a, c\rangle$, and $r = response(a, b)$. $\sigma \nvDash r$ because the second a in trace σ is not followed by a b.

Definition 4 (Constraint violation). *Let* $L \in \mathcal{L}$ *be an event log and* $\mathcal{G}(L) = (\Sigma, E)$ *be the extracted DFG, and* R *be the set of rules.* $c \nvDash r$ *denotes that cut* $c = (\oplus, \Sigma_1, \Sigma_2) \in explore(\mathcal{G}(L), R)$ *violates the constraint* $r \in R$*, meaning that for all process trees* $M \in \mathcal{M}_c$ *where* $\mathcal{M}_c = \{\oplus(M_1, M_2) | M_1 \in \mathcal{M}_{\Sigma_1} \wedge M_2 \in \mathcal{M}_{\Sigma_2}\}$*, there is a trace* $\sigma \in \phi(M)$ *such that* $\sigma \nvDash r$.

For example, $c = (\rightarrow, \{b\}, \{a\})$ violates rule *response*(a, b), since all the models $M \in \mathcal{M}_c$ allow for trace $\langle b, a\rangle$ which violates *response*(a, b).

5.2 Candidate Cuts Pruning

Consider R as the set of all rules that are given by the user or are discovered using declarative process discovery algorithms. We remove a cut c from the set of candidate cuts *explore*$(\mathcal{G}(L), R)$ if there is a rule $r \in R$ such that $c \nvDash r$. In Table 1, it is shown with red color for each single activity constraint which rules should be rejected. Similarly, in Table 2, for each two activities constraint, it is shown which rules should be rejected. Next, we explain in more detail how the cut-pruning algorithm works. Please note that although declarative rules may capture certain long-term dependencies, our discovery algorithm's representational bias might fail to adequately represent them. Consequently, the pruned candidate set may not contain any rules. In such cases, the algorithm identifies the set of candidate cuts without considering the provided rule set. Further elaboration on this phenomenon can be found in Sect. 7.

at-most(a) a occurs at most once.

- $c = (\circlearrowleft, \Sigma_1, \Sigma_2) \nvDash$ *at-most*(a) if $a \in \Sigma_1$ ($a \in \Sigma_2$), because for any process tree $M \in \mathcal{M}_c$, there is a trace $\sigma \in \phi(M)$ which has multiple occurrences of activity a because the body (the redo part) of loop process trees can occur several times.

existence(a) a occurs at least once.

- $c = (\times, \Sigma_1, \Sigma_2) \nvDash$ *existence*(a) if $a \in \Sigma_1$ ($a \in \Sigma_2$), because for any process tree $M \in \mathcal{M}_c$, there is a trace $\sigma \in \phi(M)$ which only has activities in Σ_2 (Σ_1).

Table 1. The cuts that should be rejected are shown with red color for constraints with one activity.

	$(\rightarrow, \Sigma_1, \Sigma_2)$		$(\times, \Sigma_1, \Sigma_2)$		$(\wedge, \Sigma_1, \Sigma_2)$		$(\circlearrowleft, \Sigma_1, \Sigma_2)$	
	$a \in \Sigma_1$	$a \in \Sigma_2$	$a \in \Sigma_1$	$a \in \Sigma_2$	$a \in \Sigma_1$	$a \in \Sigma_2$	$a \in \Sigma_1$	$a \in \Sigma_2$
at-most(a)								
existence(a)								

Table 2. The cuts that should be rejected are shown with red color for constraints with two activities.

	$(\rightarrow, \Sigma_1, \Sigma_2)$				$(\times, \Sigma_1, \Sigma_2)$			
	$a \in \Sigma_1$ $b \in \Sigma_1$	$a \in \Sigma_2$ $b \in \Sigma_2$	$a \in \Sigma_1$ $b \in \Sigma_2$	$a \in \Sigma_2$ $b \in \Sigma_1$	$a \in \Sigma_1$ $b \in \Sigma_1$	$a \in \Sigma_2$ $b \in \Sigma_2$	$a \in \Sigma_1$ $b \in \Sigma_2$	$a \in \Sigma_2$ $b \in \Sigma_1$
response(a,b)								
precedence(a,b)								
co-existence(a,b)								
not-co-existence(a,b)								
not-succession(a,b)								
responded-existence(a,b)								
	$(\wedge, \Sigma_1, \Sigma_2)$				$(\circlearrowleft, \Sigma_1, \Sigma_2)$			
	$a \in \Sigma_1$ $b \in \Sigma_1$	$a \in \Sigma_2$ $b \in \Sigma_2$	$a \in \Sigma_1$ $b \in \Sigma_2$	$a \in \Sigma_2$ $b \in \Sigma_1$	$a \in \Sigma_1$ $b \in \Sigma_1$	$a \in \Sigma_2$ $b \in \Sigma_2$	$a \in \Sigma_1$ $b \in \Sigma_2$	$a \in \Sigma_2$ $b \in \Sigma_1$
response(a,b)								
precedence(a,b)								
co-existence(a,b)								
not-co-existence(a,b)								
not-succession(a,b)								
responded-existence(a,b)								

- c=$(\circlearrowleft, \Sigma_1, \Sigma_2)$$\not\models$*existence*$(a)$ if $a \subset \Sigma_2$, because for any process tree $M \in \mathcal{M}_c$, there is a trace $\sigma \in \phi(M)$ which does not trigger the redo part of the process tree and consists of only activities in Σ_1.

response(a,b) If a occurs in the trace, then b occurs as well.

- c=$(\times, \Sigma_1, \Sigma_2)$$\not\models$*response*$(a, b)$ if $a \in \Sigma_1$ and $b \in \Sigma_2$ (or $a \in \Sigma_2$ and $b \in \Sigma_1$), because for any process tree $M \in \mathcal{M}_c$, there is a trace $\sigma \in \phi(M)$ which has an occurrence of activity a but b cannot occur in this trace.
- c=$(\wedge, \Sigma_1, \Sigma_2)$$\not\models$*response*$(a, b)$ if $a \in \Sigma_1$ and $b \in \Sigma_2$ (or $a \in \Sigma_2$ and $b \in \Sigma_1$), because for any process tree $M \in \mathcal{M}_c$, there is a trace $\sigma \in \phi(M)$ in which activity b occurs before a and b does not occur again.
- c=$(\circlearrowleft, \Sigma_1, \Sigma_2)$$\not\models$*response*$(a, b)$ if $a \in \Sigma_1$ and $b \in \Sigma_2$, because for any process tree $M \in \mathcal{M}_c$, there is a trace $\sigma \in \phi(M)$ in which a occurs but b does not occur because the redo part is optional.

- $c{=}(\rightarrow, \Sigma_1, \Sigma_2) \nVdash response(a,b)$ if $b \in \Sigma_1$ and $a \in \Sigma_2$, because for any process tree $M \in \mathcal{M}_c$, there is a trace $\sigma \in \phi(M)$ in which activity b occurs first and then a occurs and b does not occur again.

precedence(a,b) b occurs only if preceded by a.

- $c{=}(\times, \Sigma_1, \Sigma_2) \nVdash precedence(a,b)$ if $a \in \Sigma_1$ and $b \in \Sigma_2$ (or $a \in \Sigma_2$ and $b \in \Sigma_1$), because for any process tree $M \in \mathcal{M}_c$, there is a trace $\sigma \in \phi(M)$ which has an occurrence of activity b but a cannot occur in this trace.
- $c{=}(\wedge, \Sigma_1, \Sigma_2) \nVdash precedence(a,b)$ if $a \in \Sigma_1$ and $b \in \Sigma_2$ (or $a \in \Sigma_2$ and $b \in \Sigma_1$), because for any process tree $M \in \mathcal{M}_c$, there is a trace $\sigma \in \phi(M)$ in which activity b occurs before a.
- $c{=}(\circlearrowleft, \Sigma_1, \Sigma_2) \nVdash precedence(a,b)$ if $b \in \Sigma_1$ and $a \in \Sigma_2$, because for any process tree $M \in \mathcal{M}_c$, there is a trace $\sigma \in \phi(M)$ in which b occurs but a does not occur because the redo part is optional.
- $c{=}(\rightarrow, \Sigma_1, \Sigma_2) \nVdash precedence(a,b)$ if $b \in \Sigma_1$ and $a \in \Sigma_2$, because for any process tree $M \in \mathcal{M}_c$, there is a trace $\sigma \in \phi(M)$ in which activity b occurs first and then a occurs.

co-existence(a,b) a and b occur together.

- $c{=}(\times, \Sigma_1, \Sigma_2) \nVdash co\text{-}existence(a,b)$ if $a \in \Sigma_1$ and $b \in \Sigma_2$ (or $a \in \Sigma_2$ and $b \in \Sigma_1$), because for any process tree $M \in \mathcal{M}_c$, there is a trace $\sigma \in \phi(M)$ which has an occurrence of only activity a or only activity b.
- $c{=}(\circlearrowleft, \Sigma_1, \Sigma_2) \nVdash co\text{-}existence(a,b)$ if $a \in \Sigma_1$ and $b \in \Sigma_2$ ($a \in \Sigma_2$ and $b \in \Sigma_1$), because for any process tree $M \in \mathcal{M}_c$, there is a trace $\sigma \in \phi(M)$ in which only activity a (b) occurs because the redo part is optional.

not-co-existence(a,b) a and b cannot occur together.

- $c{=}(\wedge, \Sigma_1, \Sigma_2) \nVdash not\text{-}co\text{-}existence(a,b)$ if $a \in \Sigma_1$ and $b \in \Sigma_2$ (or $a \in \Sigma_2$ and $b \in \Sigma_1$), because for any process tree $M \in \mathcal{M}_c$, there is a trace $\sigma \in \phi(M)$ in which both activities a and b occur.
- $c{=}(\circlearrowleft, \Sigma_1, \Sigma_2) \nVdash not\text{-}co\text{-}existence(a,b)$ if $a \in \Sigma_1$ and $b \in \Sigma_2$ (or $a \in \Sigma_2$ and $b \in \Sigma_1$), because for any process tree $M \in \mathcal{M}_c$, there is a trace $\sigma \in \phi(M)$ in which a and b both occur because the redo part is optional and may occur. Also, if $a \in \Sigma_1$ and $b \in \Sigma_1$ (or $a \in \Sigma_2$ and $b \in \Sigma_2$), since the do part or redo part of the loop tree can occur multiple times and the repetitions are independent, then activities a and b may occur in different repeats of the loop process trees.
- $c{=}(\rightarrow, \Sigma_1, \Sigma_2) \nVdash not\text{-}co\text{-}existence(a,b)$ if $a \in \Sigma_1$ and $b \in \Sigma_2$ (or $a \in \Sigma_2$ and $b \in \Sigma_1$), because for any process tree $M \in \mathcal{M}_c$, there is a trace $\sigma \in \phi(M)$ in which both activities a and b occur.

not-succession(a,b) b cannot occur after a.

- $c{=}(\wedge, \Sigma_1, \Sigma_2) \nVdash not\text{-}succession(a,b)$ if $a \in \Sigma_1$ and $b \in \Sigma_2$ (or $a \in \Sigma_2$ and $b \in \Sigma_1$), because for any process tree $M \in \mathcal{M}_c$, there is a trace $\sigma \in \phi(M)$ in which activity a occurs first and then b occurs after it.

Table 3. Event logs used in experiments.

	#activities	#events	#traces	#trace variants
BPIC 12	17	92,093	13,087	576
BPIC 17	18	433,444	31,509	2,630
BPIC 18	15	928,091	43,809	1,435
Hospital	18	451,359	100,000	1,020
Sepsis	16	15,214	1,050	846
UWV	16	1,309,719	144,046	484

- c=($\circlearrowleft$, Σ_1, Σ_2)$\not\models$*not-succession*(a, b) if $\{a, b\} \subseteq \Sigma_1 \cup \Sigma_2$, because for any process tree $M \in \mathcal{M}_c$, there is a trace $\sigma \in \phi(M)$ in which activity b occurs after activity a. Since the do part and redo part of the loop tree can occur multiple times and the repetitions are independent, then after the occurrence of activity a, activity b can eventually occur.
- c=($\rightarrow$, Σ_1, Σ_2)$\not\models$*not-succession*(a, b) if $a \in \Sigma_1$ and $b \in \Sigma_2$, because for any process tree $M \in \mathcal{M}_c$, there is a trace $\sigma \in \phi(M)$ in which activity a occurs first and then b occurs.

responded-existence(a,b) If a occurs in the trace, then b occurs as well.

- c=($\times$, Σ_1, Σ_2)$\not\models$*responded-existence*(a, b) if $a \in \Sigma_1$ and $b \in \Sigma_2$ (or $a \in \Sigma_2$ and $b \in \Sigma_1$), because for any process tree $M \in \mathcal{M}_c$, there is a trace $\sigma \in \phi(M)$ which has an occurrence of only activity a and b cannot occur in this trace.
- c=($\circlearrowleft$, Σ_1, Σ_2)$\not\models$*responded-existence*(a, b) if $a \in \Sigma_1$ and $b \in \Sigma_2$, because for any process tree $M \in \mathcal{M}_c$, there is a trace $\sigma \in \phi(M)$ in which only activity a occurs but b does not occur since the redo part is optional.

6 Evaluation

The proposed framework is implemented and is publicly available[2]. We used several real-life event logs to evaluate the framework including BPIC 2012 (application and offer sub-processes), BPIC 2017 (application and offer sub-processes), BPIC 2018 (application sub-process), Hospital Billing, Sepsis, and UWV event logs. All the event logs except UWV are publicly available[3]. Key statistics for these event logs are summarized in Table 3. The evaluation section is structured into two parts. Initially, we present results derived from publicly available event logs. Subsequently, we offer insights from a real-life case study conducted in collaboration with UWV, the Dutch employee insurance agency. Domain experts at UWV actively contributed to extracting a normative model for a claim-handling process and validating the obtained results.

[2] https://github.com/aliNorouzifar/IMr.
[3] https://data.4tu.nl/.

6.1 Real-Life Event Logs

Our framework is designed independently from the source which provides the rules. We use Declare Miner [11] with the subset of declarative templates introduced in this paper and select the rules with $confidence = 1$, i.e., the constraints for which there is no trace in the event log that deviates. However, one could employ various heuristics to account for noisy behavior or other considerations. Our initial comparison involves assessing models discovered by IMf [10], a state-of-the-art algorithm, against those produced by the IMr algorithm utilizing discovered declarative constraints. Experiments are conducted with the infrequency parameter f ranging from 0 to 1 in intervals of 0.1. In the IMr algorithm, we vary the *sup* parameter within the range of 0 to 1 at intervals of 0.1.

The visual representation of the discovered models' quality is depicted in Fig. 4a, employing well-known evaluation metrics. Specifically, the x-axis represents alignment fitness [3], the y-axis represents precision [3], and the contours on the plot illustrate the F1-score derived from these two values. Various shapes differentiate between event logs, i.e., ○: BPIC17, ◇: BPIC18, □: BPIC12, △: Hospital, and ☆: Sepsis, shapes with blue color represent IMf models and shapes with red color represent IMr models while color intensity corresponds to the parameters of the discovery algorithms, i.e., f in IMf and *sup* in IMr. Notably, the figures indicate that, in general, IMr models outperform IMf models across the three evaluation metrics illustrated in the plot.

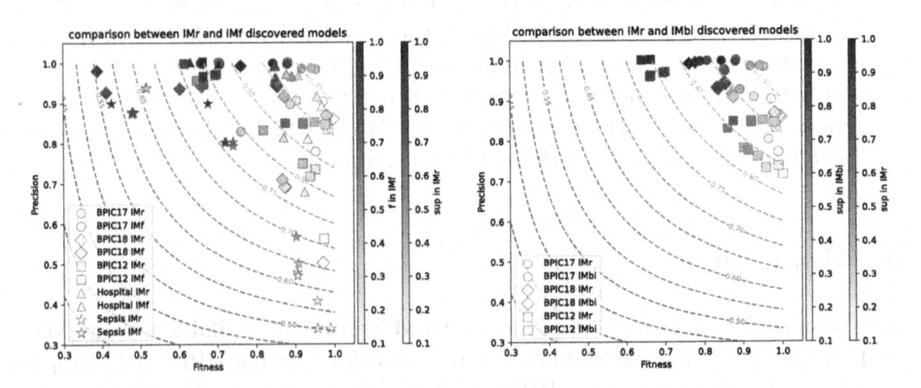

(a) Comparison between process models discovered using IMr and IMf.

(b) Comparison between process models discovered using IMr and IMbi.

Fig. 4. Comparison between process models discovered using IMr, IMf, and IMbi. As can be seen, IMr (red shapes) models perform better than IMbi (green shapes) and IMf (blue shapes) models. (Color figure online)

In Fig. 4b, the models discovered with IMr are shown with red color and the models discovered with IMbi are shown with green color. For the Hospital Billing and Sepsis event logs, it was infeasible to discover process models using the IMbi algorithm considering a maximum run time of one hour. The discovered model using BPIC 12 and BPIC 17 shows that IMr models score better. For BPIC 2018, although the discovered models are very similar, the run time of the IMr algorithm is about four times shorter.

Experiments exceeding the one-hour maximum run-time were terminated. While the IMf algorithm is notably fast, delivering a model within seconds, the computational costs for IMbi and IMr are considerably higher. In Table 4, some statistics are presented to compare the run time of IMbi with IMr[4]. Notably, a comparison between the number of candidate cuts in the initial recursion of both IMbi and IMr demonstrates the effect of utilizing rules in efficiently pruning the search space while preserving model quality, as indicated in Fig. 4b. The candidate cut search is independent of the *sup* parameter. The table additionally includes information on the average run-time duration for each event log that shows IMr is considerably faster because of the reduced number of candidate cuts.

Table 4. Run time statistics IMr and IMbi

	BPIC 12	BPIC 17	BPIC 18	Hospital	Sepsis
$\|C\|$ in the first iteration of IMbi	3601	4659	8103	106771	153502
$\|C\|$ in the first iteration of IMr	12	47	19	8	329
average run time of IMbi	20 s	176 s	801 s	>1 h	>1 h
average run time of IMr	11 s	55 s	201 s	152 s	9 s

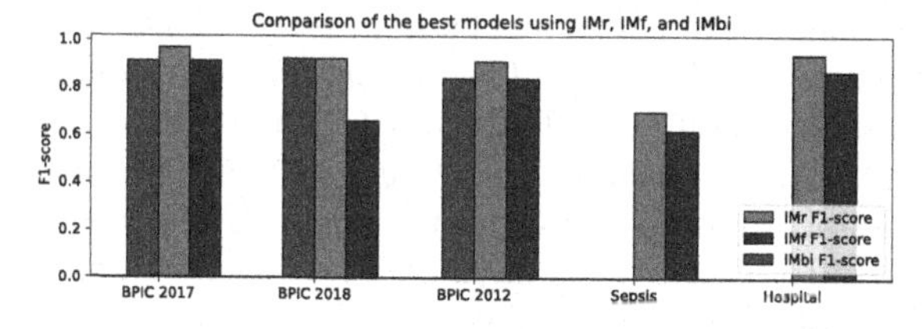

Fig. 5. Comparison between best models discovered by IMr, IMf, and IMbi (for Sepsis and Hospital event logs it was not feasible to discover a model in an hour using IMbi).

In Fig. 5, we chose the best models for each event log using different parameters in IMf, IMr, and IMbi experiments. The criterion for selecting the best model involves choosing the one with the highest F1-score among those with a minimum alignment fitness of 0.9. The bar chart in this figure compares alignment fitness, precision, and F1-score for the selected models.

In Fig. 2, we provided examples to motivate our goal. To emphasize the impact of our proposed IMr framework, a part of the discovered models is presented in Fig. 6. In Fig. 6a, it is evident that only one of the activities

[4] The time required for rule extraction is not included.

A_Cancelled, *A_Denied*, or *A_Pending* can occur. In Fig. 6b, the sequential order of transitions *begin editing*, *calculate*, *finish editing*, and *decide* is more coherent, aligning seamlessly with our understanding of the process.

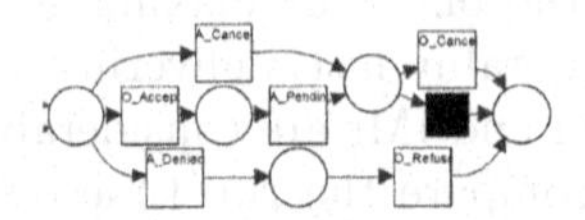

(a) A part of the discovered model for BPIC 2017 event log using IMr with $sup = 0.2$.

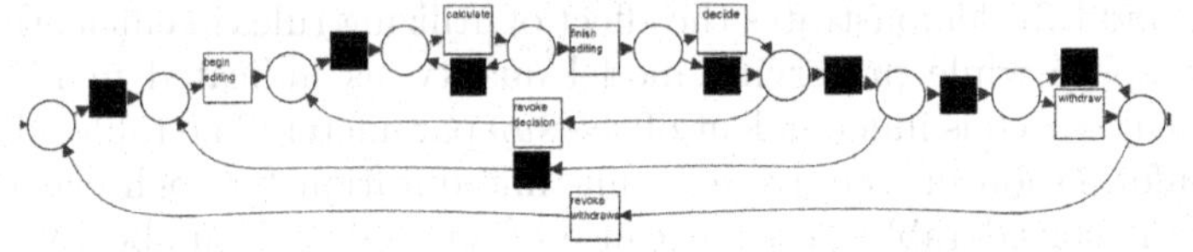

(b) A part of the discovered model for BPIC 2018 event log using IMr with $sup = 0.1$.

Fig. 6. Discovered models using IMr for BPIC 2017 and BPIC 2018

6.2 Case Study UWV

This case study is a collaborative effort with the UWV agency, involving the analysis of an event log pertaining to one of their claim-handling processes, encompassing data for 144,046 clients. Figure 7 illustrates the normative model derived from a comprehensive examination of the event log, supplemented by insights from process experts who actively contributed their domain knowledge throughout this case study. After a claim is received the process is started, some cases are blocked (*Block type 1*) and consequently, some corrections are performed and the case is unblocked afterward. Some other cases after starting a claim have another type of blocking (*Block type 2*) which is followed by rejecting the claim and possibly an objection from the client. If the case is accepted, then the client is entitled to receive some payments (between one to three payments). Some clients file an objection after receiving the payments which continues with withdrawing the claim and repaying the received money to UWV. Optionally a third type of blocking might also occur (*Block type 3*) to prevent any pending payments to the client from being made.

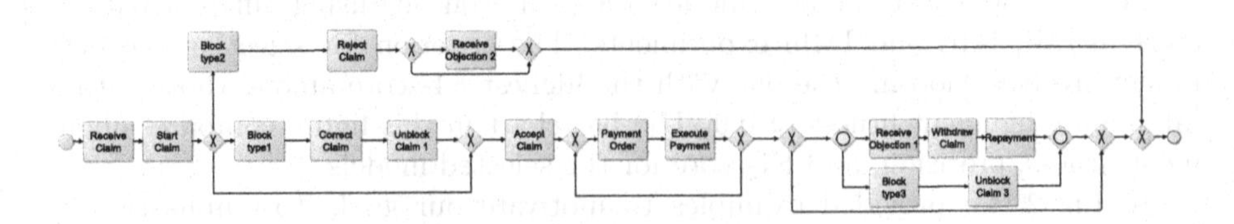

Fig. 7. UWV normative model.

IMf and IMr with different parameter settings are utilized to discover process models from the UWV event log. The set of rules R extracted from the event log using Declare Miner [11] with $confidence$=1 is used in IMr. Some examples of discovered declarative constraints are *precedence*(*Block type* 1, *Unblock Claim* 1), *response*(*Block type* 2, *Reject Claim*), *not-succession*(*Execute Payment*, *Accept Payment*), and *not-co-existence*(*Block type* 1, *Block type* 2). These rules align with the normative model and can be used to guide IMr to discover better models.

A comparative analysis of these models is presented in Fig. 8. The circle shape ○ in the figure represents the original event log without any filtering applied. Although IMr models exhibit superior scores, the difference does not seem significant. This is primarily attributed to the most frequent trace variant, constituting 86% of the data, significantly influencing the alignment fitness value. Both IMr and IMf discovered models replay this trace variant. However, upon filtering out this prevalent trace variant, the contrast becomes clearer, evident in the figure with cross-shaped points 𝕏. The IMf models have lower fitness values which shows their difficulties in modeling these trace variants. In contrast, IMr models show a robust representation of the process, as the removal of this frequent trace variant has a marginal impact on the overall model quality.

It was infeasible to discover process models in one hour using IMbi, therefore the comparison between IMbi models and IMr models is excluded from the experiments. The average run-time of the IMr algorithm in different experiments is 348 s. The best model discovered from the complete event log with IMf using $f = 0.5$ has an alignment fitness of 96.7, a precision of 93.9, and an F1-score of 95.3. The best model discovered with IMr using $sup = 0.5$ has an alignment fitness of 99.6, a precision of 93.8, and an F1-score of 96.6. This is illustrated in Fig. 9. This model represents the process much better than Fig. 2c, especially if we compare it with the normative model illustrated in Fig. 7.

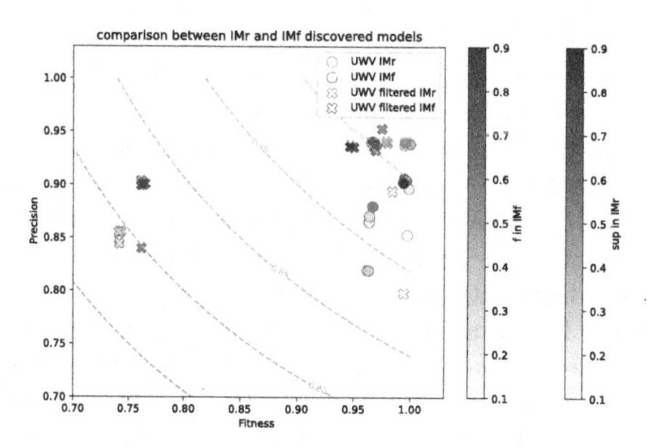

Fig. 8. Comparison between process models discovered for UWV event log using IMr and IMf.

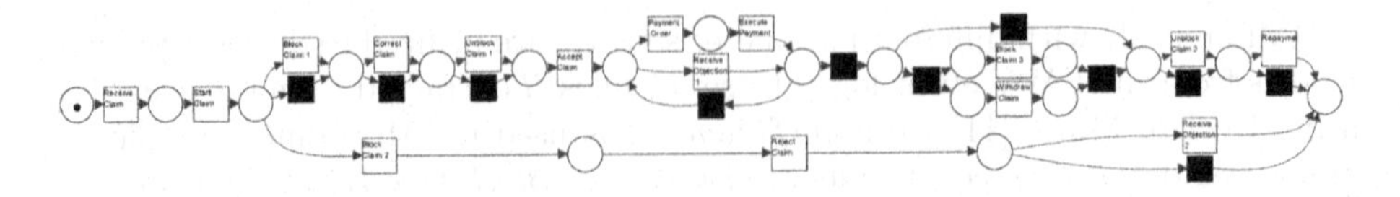

Fig. 9. Discovered model for UWV event log using IMr with $sup = 0.5$.

7 Open Challenges

Due to the representational bias of inductive mining algorithms, the IMr algorithm does not take into account long-term dependencies in the discovery procedure. However, declarative rules may represent long-term dependencies. Consider the event log $L_1 = [\langle a, c, d\rangle, \langle b, c, e\rangle]$ and set of rules $R_1 = \{\textit{not-co-existence}(a, e), \textit{not-co-existence}(b, d)\}$. All traces in L_1 satisfy all the rules in R_1. In the first recursion of the IMbi algorithm, the set of possible cuts without considering the rules in R_1 is $C = \{\rightarrow(\{a\}, \{b, c, d, e\}), \rightarrow(\{b\}, \{a, c, d, e\}), \rightarrow(\{a, b\}, \{c, d, e\}), \rightarrow(\{a, b, c\}, \{d, e\}), \rightarrow(\{a, b, c, d\}, \{e\}), \rightarrow(\{a, b, c, e\}, \{d\})\}$. If we use IMr with the set of rules R_1, because of the long-term dependencies between a and e, and between b and d, all the cuts are rejected. Applying any of these cuts results in traces that do not satisfy at least one rule. In such cases, IMr continues with ignoring the set of rules in that specific recursion.

Consider the event log $L_2 = [\langle c, a, c, b, c\rangle]$, and set of rules $R_2 = \{\textit{response}(a, b)\}$. The dependency between a and b is long-term and observed in different runs of a loop. Considering $sup = 0.2$, the first recursion of IMr finds $\circlearrowleft(\{c\}, \{a, b\})$ which splits the event log as $L_3 = [\langle c\rangle^3]$ and $L_4 = [\langle a\rangle, \langle b\rangle]$. The only possible candidate cut in $IMr(L_4, 0.2, R_2)$ is $\times(\{a\}, \{b\})$ which is rejected based on R_2. The dependency is between different runs of the process. The discovered model using IMbi without considering the rules is $\circlearrowleft(\{c\}, \times(\{a\}, \{b\}))$ that can generate traces $\langle c, a, c\rangle$, and $\langle c, b, c, a, c\rangle$ which deviates $\textit{response}(a, b)$.

Our framework always guarantees a sound model. The declarative rules in IMr are used to guide the algorithm to understand the order of activities. Therefore, it cannot guarantee that the final model satisfies all the input rules. For example, consider the event log $L_5 = [\langle a, c, b\rangle^{50}, \langle d, c\rangle^{50}]$ and $R_3 = \{\textit{precedence}(a, b)\}$. IMr with $sup = 0.2$ and R_3 discovers $\rightarrow(\times(a, d), \rightarrow(c, \times(b, \tau)))$. This model allows for trace $\langle d, c, b\rangle$ which violates $\textit{precedence}(a, b)$. In case the strict version of the algorithm is used which outputs no model if all cuts are rejected in a recursion, we provide the following guarantees: $\textit{at-most}(a)$: in all traces, a can occur at most one time, $\textit{existence}(a)$: in all traces, a can occur, $\textit{response}(a, b)$: each time a occurs, b can occur after it, $\textit{precedence}(a, b)$: each time b occurs, it is possible that a occurred before it, $\textit{co-existence}(a, b)$: a and b can occur together, $\textit{not-co-existence}(a, b)$: a and b never occur together, $\textit{not-succession}(a, b)$:

b cannot occur after a, *responded-existence*(a, b): if a occurs in the trace, then b can occur as well.

8 Conclusion

The proposed framework is based on the inductive mining idea and makes a contribution to the field by introducing a novel variant of inductive mining. This variant incorporates the use of user-defined or discovered rules during the process discovery recursions. Through extensive evaluation, our results demonstrate that the discovered models surpass current approaches, yielding more accurate process models that align closely with process knowledge available prior to process discovery. This framework holds the potential for extension into an interactive process discovery framework. In such a setting, domain experts can interactively examine discovered models, utilizing a set of rules to guide the discovery algorithm toward refining the process model. This iterative process enhances the adaptability of the framework and ensures alignment with evolving domain expertise. Moreover, the discussed approach offers the possibility of handling scenarios involving multiple event logs. The framework can be extended to discover a process model that supports a desirable event log while simultaneously avoiding an undesirable event log.

References

1. Beerepoot, I., et al.: The biggest business process management problems to solve before we die. Comput. Ind. **146**, 103837 (2023)
2. Brons, D., Scheepens, R., Fahland, D.: Striking a new balance in accuracy and simplicity with the probabilistic inductive miner. In: 3rd International Conference on Process Mining, ICPM 2021, pp. 32–39. IEEE (2021)
3. Carmona, J., van Dongen, B.F., Solti, A., Weidlich, M.: Conformance Checking - Relating Processes and Models. Springer, Heidelberg (2018). https://doi.org/10.1007/978-3-319-99414-7
4. Ciccio, C.D., Mecella, M.: On the discovery of declarative control flows for artful processes. ACM Trans. Manag. Inf. Syst. **5**(4), 24:1–24:37 (2015)
5. van Detten, J.N., Schumacher, P., Leemans, S.J.J.: An approximate inductive miner. In: 5th International Conference on Process Mining, ICPM 2023, pp. 129–136. IEEE (2023)
6. Dixit, P.M., Buijs, J.C.A.M., van der Aalst, W.M.P., Hompes, B., Buurman, H.: Enhancing process mining results using domain knowledge. In: Proceedings of the 5th International Symposium on Data-driven Process Discovery and Analysis (SIMPDA 2015). CEUR Workshop Proceedings, vol. 1527, pp. 79–94. CEUR-WS.org (2015)
7. Fahland, D., van der Aalst, W.M.P.: Repairing process models to reflect reality. In: Barros, A., Gal, A., Kindler, E. (eds.) BPM 2012. LNCS, vol. 7481, pp. 229–245. Springer, Heidelberg (2012). https://doi.org/10.1007/978-3-642-32885-5_19
8. Goedertier, S., Martens, D., Vanthienen, J., Baesens, B.: Robust process discovery with artificial negative events. J. Mach. Learn. Res. **10**, 1305–1340 (2009)

9. Knuplesch, D., Reichert, M., Ly, L.T., Kumar, A., Rinderle-Ma, S.: Visual modeling of business process compliance rules with the support of multiple perspectives. In: Ng, W., Storey, V.C., Trujillo, J.C. (eds.) ER 2013. LNCS, vol. 8217, pp. 106–120. Springer, Heidelberg (2013). https://doi.org/10.1007/978-3-642-41924-9_10
10. Leemans, S.J.J., Fahland, D., van der Aalst, W.M.P.: Discovering block-structured process models from event logs containing infrequent behaviour. In: Lohmann, N., Song, M., Wohed, P. (eds.) BPM 2013, vol. 171, pp. 66–78. Springer, Heidelberg (2013). https://doi.org/10.1007/978-3-319-06257-0_6
11. Maggi, F.M., Bose, R.P.J.C., van der Aalst, W.M.P.: Efficient discovery of understandable declarative process models from event logs. In: Ralyte, J., Franch, X., Brinkkemper, S., Wrycza, S. (eds.) CAiSE 2012. LNCS, vol. 7328, pp. 270–285. Springer, Heidelberg (2012). https://doi.org/10.1007/978-3-642-31095-9_18
12. Norouzifar, A., van der Aalst, W.M.P.: Discovering process models that support desired behavior and avoid undesired behavior. In: Proceedings of the 38th ACM/SIGAPP Symposium on Applied Computing, SAC 2023, pp. 365–368. ACM (2023)
13. Rembert, A.J., Omokpo, A., Mazzoleni, P., Goodwin, R.: Process discovery using prior knowledge. In: Basu, S., Pautasso, C., Zhang, L., Fu, X. (eds.) ICSOC 2013. LNCS, vol. 8274, pp. 328–342. Springer, Heidelberg (2013). https://doi.org/10.1007/978-3-642-45005-1_23
14. Schuster, D., van Zelst, S.J., van der Aalst, W.M.P.: Utilizing domain knowledge in data-driven process discovery: a literature review. Comput. Ind. **137**, 103612 (2022)
15. Schuster, D., van Zelst, S.J., van der Aalst, W.M.P.: Cortado: a dedicated process mining tool for interactive process discovery. SoftwareX **22**, 101373 (2023)
16. Yahya, B.N., Bae, H., Sul, S.O., Wu, J.Z.: Process discovery by synthesizing activity proximity and user's domain knowledge. In: Song, M., Wynn, M.T., Liu, J. (eds.) AP-BPM 2013, vol. 159, pp. 92–105. Springer, Cham (2013). https://doi.org/10.1007/978-3-319-02922-1_7

An Approach for Discovering Data-Driven Object Lifecycle Processes

Marius Breitmayer(✉), Lisa Arnold, David Goth, and Manfred Reichert

Institute of Databases and Information Systems, Ulm University, Ulm, Germany
{marius.breitmayer,lisa.arnold,david.goth,manfred.reichert}@uni-ulm.de

Abstract. The discovery of process models from event logs has been a well-understood topic regarding activity-centric processes. For alternative paradigms (e.g., data- or object-centric processes as implemented in many information systems), however, this model discovery still poses several challenges. One of these challenges concerns the discovery of object behavior expressed in terms of object lifecycle processes. In particular, this discovery requires the consideration of different granularity levels (i.e., object states and object attributes). This paper presents an approach for discovering object lifecycle processes. The approach divides the discovery of object lifecycle processes into subproblems by preprocessing event logs to enable the use of well-known discovery algorithms. Overall, object-centric process mining gives insights into data-driven and object-centric processes as implemented in many information systems.

Keywords: Process Discovery · Object Lifecycle Processes · Data-centric Processes · Object-centric Processes

1 Introduction

Process Mining bridges the gap between Data and Process Science by extracting process-related information from the data recorded by information systems in event logs [1]. The latter document how a particular process was executed. Existing approaches for process discovery use event logs to reconstruct process models, but focus on the control flow of activity-centric processes (i.e., the sequence in which activities are executed). However, the actual specification of an activity (e.g., the actual data required) is often considered as a black-box. Consequently, alternative paradigms like object-centric processes emerged, which represent a process in terms of multiple interacting objects. Usually, in these approaches, the individual behavior of objects is data-driven and represented in terms of object lifecycle processes [24].

The object behavior modeled by an object lifecycle process covers the way in which objects transition between states (i.e., the state level), and object attributes are acquired (i.e., the step level) during process execution. Discovering object lifecycle processes from event logs enables a better understanding of how object-centric processes are actually executed in information systems. However, to adequately represent object behavior, lifecycle process discovery must account for the

J. Araújo et al. (Eds.): RCIS 2024, LNBIP 513, pp. 237–254, 2024.
https://doi.org/10.1007/978-3-031-59465-6_15

different granularity levels of object lifecycle processes. To deal with this challenge, the following main research question has been identified:

RQ: How can different granularity levels (i.e., state and step level) be considered when discovering data-driven object lifecycle processes?

The approach presented in this paper supports the discovery of object lifecycle processes by transforming event logs so that well-known process discovery algorithms can be applied. It considers both the state and the step level of object lifecycle processes by preprocessing an event log, discovering the two levels separately, and then connecting the discovered models into object lifecycle processes.

The paper is structured as follows: Sect. 2 introduces object lifecycle processes as used in the PHILharmonicFlows framework for object-centric process management. Section 3 describes the approach, whereas Sect. 4 deals with model discovery at the different granularity levels. Section 5 evaluates the approach using different event logs. In Sect. 6, we relate the approach to existing approaches. Section 7 summarizes the paper and provides an outlook.

2 Fundamentals

PHILharmonicFlows enriches the concept of data-driven processes with the concept of *objects*. For each real-world business object, a corresponding object exists in a PHILharmonicFlows process model. In a human resource management scenario, for example, objects such as *Application*, *Job Offer*, *Review*, and *Interview* may be involved in a process. An object, in turn, comprises (data) *attributes* and an object lifecycle process, describing its behavior in terms of a state-based process model (see Fig. 1). Object *Application* (see Fig. 1), for example, includes attributes *Job Posting*, *Files*, *Experience*, *Reason*, and *Date*. Moreover, Fig. 1 depicts the corresponding lifecycle process. It describes the behavior in terms of *object states* (e.g. *Edit*, *Submit*, *Evaluate*, *Invite*, and *Decline*) and *object state transitions*. Moreover, a state may comprise several *steps* (e.g., *Job Posting*, and *Files* in state *Edit*), each referring to exactly one object attribute. From an operational perspective, the steps of an object lifecycle process specify which attributes need to be set before completing a state and transitioning to the next one. Accordingly, the acquisition of data in PHILharmonicFlows is based on both the states and steps of the corresponding object lifecycle processes. As example consider the lifecycle process of object *Application* (see Fig. 1):

Edit constitutes the initial state of the *Application* object as it has no incoming transitions. After an applicant has provided data for steps *Job Posting* and *Files*, the *Application* may transition to state *Submit*. In state *Submit*, in turn, the application is sent to the responsible personnel officer. Based on attribute *Experience* (e.g., provided via a CV provided in attribute *Files*) the personnel officer decides on the application in state *Evaluate*. If the experience level is above the minimal requirements, the applicant is invited and a *Date* is set (i.e., the *Application* transitions to state *Invite*). Otherwise, the application is declined with a *Reason* (i.e., the *Application* transitions to state *Declined*).

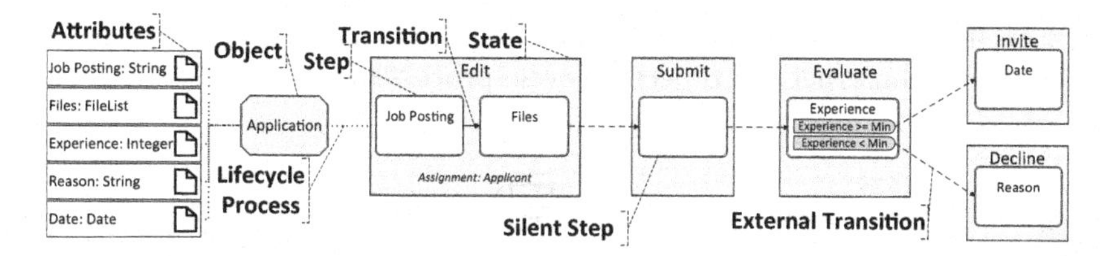

Fig. 1. (Simplified) Object Lifecycle Process of Object *Application*

The example of object *Application* illustrates the importance of data during the execution of object lifecycle processes [7,22].

In general, a business process comprises multiple interacting objects (e.g., objects *Application*, *Job Offer*, *Review*, and *Interview*), including their object lifecycle processes. In PHILharmonicFlows, a data model captures such objects, their semantic relations, and cardinality constraints [14]. At runtime, objects may be instantiated multiple times, each representing an individual *object instance* [7]. The lifecycle processes of different object instances are executed concurrently, and their execution is synchronized by a coordination process.

3 Proposed Approach

Object lifecycle process discovery leverages execution data (i.e., event logs) of an object-centric process to reconstruct the object lifecycle process model [1]. A single object lifecycle process is derived considering the dependencies between events recorded in an event log. For activity-centric processes, where activities are considered as black-boxes, algorithms for discovering process models from event logs exist [1,3,5,15,25]. Regarding object-centric processes, for which object lifecycle process models represent the behavior of the respective object, process discovery is of great relevance as well. Due to the object-centric and data-driven execution of object lifecycle processes as well as the execution flexibility enabled by frameworks like PHILharmonicFlows, the discovery of object lifecycle processes poses additional challenges compared to traditional process discovery. In particular, it is unclear whether and how existing process discovery algorithms are suited for discovering object lifecycle processes.

On the one hand, event logs constitute the main component of any process discovery endeavor, and, therefore, must satisfy certain requirements. On the other, object lifecycle processes are multi-dimensional, i.e., they represent object behavior in terms of states, state transitions, and steps. The approach presented in this paper enables the application of traditional process discovery algorithms to object lifecycle processes by splitting the discovery problem (and corresponding event logs) into the state level (see Sect. 4.1: how does an object transition between states?) and the step level (see Sect. 4.2: which data is provided how within a particular object state?) according to the granularity of the object lifecycle process. Splitting the discovery problem enables the application of existing process discovery algorithms to object lifecycle processes (see

Sect. 4.3). The discovered process models for both granularity levels are then combined to reconstruct the object lifecycle process model (see Sect. 4.4).

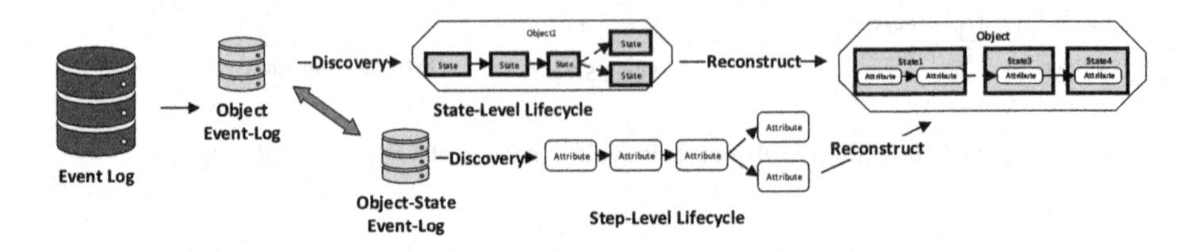

Fig. 2. Proposed Approach

4 Object Lifecycle Process Discovery

Most existing process discovery algorithms are tailored towards activity-centric process models and, therefore, require an event log whose entries have at least a case identifier, an activity label, and a timestamp [1]. We enable the application of existing process discovery algorithms to the multi-dimensionality of object lifecycle processes by preprocessing corresponding event logs. In other words, we separately discover the behavior of objects on state and step level, and then combine the discovered models in order to reconstruct the entire object lifecycle process. We illustrate this approach along the object lifecycle process depicted in Fig. 1. Moreover, an excerpt of an event log is depicted in Table 1. For example, this event log comprises information on the execution of the lifecycle process instances of object *Application* (i.e., case identifiers in column *ObjectInstanceId*). To be more precise, the event log includes data on the object states, corresponding steps (see column Attribute), actions, and a *Timestamp* per event.

Table 1. Excerpt Event Log for Object *Application*

ObjectInstanceID	Object	State	Attribute	AttributeValue	Action	Timestamp
A1	Application	Edit	Job Posting	J1	SetValue	24.08.2011 20:01
A1	Application	Edit	Files	CV.zip	SetValue	29.08.2011 01:44
A1	Application	Submit			ChangeState	22.09.2011 15:15
A1	Application	Evaluate			ChangeState	16.10.2011 05:13
...	...	...	...	...	...	...
A25	Application	Evaluate			ChangeState	11.12.2011 22:45
A25	Application	Decline	Reason	A reason	SetValue	26.12.2011 22:34

4.1 Discovering Object Behavior on State Level

Process discovery on the state level deals with the way an object transitions between states. To ensure that the approach is applicable to event logs in which state transitions are not explicitly recorded, we focus on a minimal set of required information, i.e., the object instance, its state, and a corresponding timestamp.

On the state level, the event log is first filtered for those events recorded for the object for which the lifecycle process shall be discovered (e.g., object *Application*). Then the state level discovery of individual object lifecycle processes is enabled by ordering the event log per object instance (i.e., case) and timestamp. As a result, all instances are recorded sequentially in the event log, and the event log is grouped per instance. Then, for each group state transitions are identified if two subsequent events correspond to different states (i.e., the object instance has changed its state between these two events) and events not associated with a state transition (i.e., multiple events recorded in sequence for a certain state) are removed. Note that this removes undesired loops (i.e., due to multiple events recorded in sequence in the same state) in the discovered model. In detail, for each object instance we duplicate column *State* and shift the copy down by one row, enabling checking whether the documented state differs between two events using column *PreviousValue* (see Table 2). If the values in columns *State* and *PreviousValue* differ, a state transition took place. If they do not differ, the events may be removed from the preprocessed event log as they do not correspond to a state transition. The first entry of column *PreviousValue* for each object instance recorded in the event log remains empty allowing for the identification of the first state. Algorithm 1 illustrates this approach in pseudo code. The resulting event logs for the state level of each object instance are then merged to obtain a state level event log for a single object lifecycle process. Algorithm 2 describes all preprocessing steps for the state level in pseudo code.

Algorithm 1. Clean Trace

```
Require: log, mode
  if mode = statelevel then                                ▷ State Level Discovery
      log(PreviousValue) = log.Shift(state)                ▷ Copy State Column & Shift
      log(StateChange) = (log(state)! = log(PreviousValue))   ▷ State Changed?
      log.Filter(StateChange) = True                       ▷ Filter relevant Event
      return log
  else                                                     ▷ Step Level Discovery
      log(PreviousValue) = log.Shift(step)                 ▷ Copy Step Column & Shift
      log(StepChange) = (log(step)! = log(PreviousValue))     ▷ Step Changed?
      log.Filter(StepChange) = True                        ▷ Filter relevant Event
      return log
  end if
```

Algorithm 2. State Level Preprocessing

```
Require: processlog, object
  processlog.Filter(Object) = object
  processlog.SortAscending(ObjectInstanceID, Timestamp)
  case_groups = processlog.GroupBy(ObjectInstanceID)
  CleanTrace(case_groups, statelevel)                      ▷ Alg. 1
  clean_log = Merge(case_groups)
  return clean_log
```

Table 2 describes the event log after applying Algorithm 2. Note that the additional columns *Previous Value* and *ValueChange?* only demonstrate the identification of state transitions during state level preprocessing. Consequently, they are not considered during process discovery.

Note that the obtained event log representing the object behavior on the state level satisfies the requirements of event logs in general [1]. In particular, existing process discovery algorithms can be applied to discover the state level behavior of objects leveraging the entries in column *State* (see Table 2) as activities. Figure 3 depicts the state level model of object *Application* discovered with the Heuristic Miner [25].

Table 2. Event Log for Object *Application* preprocessed on State Level

ObjectInstanceID	Object	State	Attribute	AttributeValue	Action	Timestamp	PreviousValue	ValueChange?
A1	Application	Edit	JobPosting	J1	SetValue	24.08.2011 20:01		TRUE
A1	Application	Submit			ChangeState	22.09.2011 15:15	Edit	TRUE
A1	Application	Evaluate			ChangeState	16.10.2011 05:13	Submit	TRUE
A1	Application	Invite	Date	01.12.2001	ChangeState	28.11.2011 10:57	Evaluate	TRUE
A10	Application	Edit	JobPosting	J3	SetValue	06.09.2011 15:10		TRUE
...	...	...	...	...	...	...	...	...
A9	Application	Decline	Reason	A reason	ChangeState	29.12.2011 22:17	Evaluate	TRUE

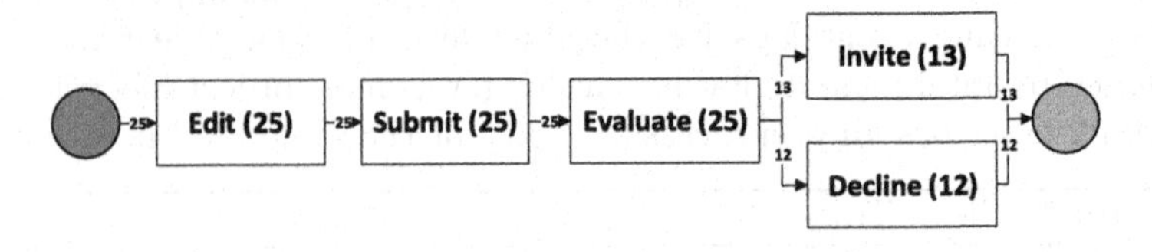

Fig. 3. Object *Application* - State Level (Heuristic Net)

4.2 Discovering Object Behavior on Step Level

As opposed to the approach for discovering object lifecycle processes on the state level, this section focuses on the step level (i.e., which object attributes are written in which order when processing a specific state).

We discover object behavior on the step level (i.e., the attributes required in a specific state and the sequence in which these attributes are actually written) for each individual state. In general, a state may only be completed once all required attributes (i.e., steps) are written, but the sequence in which attributes are written is not enforced. The ordering of steps within a state is, however, utilized to auto-generate forms (including their logic) at runtime [22].

Discovery on the step level is based on preprocessing the event log by only considering those events related to writing attributes within a particular state (e.g., state *Edit* of object *Application*). In other words, we focus on a particular state of the object lifecycle process by filtering the event log with respect to a specific object and a particular state of that object (e.g., state *Edit* of object *Application*). The event log is ordered by *ObjectInstanceID* and *Timestamp*.

On the step level, events need not correspond to initially writing an attribute value as attribute updates may be recorded as well. Consequently, we remove events not associated with writing attribute values (i.e., column *Attribute* = blank) from the event log. The remaining events are then grouped per object instance. The resulting event log then only comprises events associated with writing attribute values. To further ensure that the discovered process model does not contain undesired loops (e.g., due to multiple events recorded for writing the same object attribute), Algorithm 1 is applied to the step level. Algorithm 1 removes sequential events associated with the same attribute (e.g., an attribute is written and updated immediately). Finally, the preprocessed step level event logs per object instance and state are merged creating the step level log for a particular object state. Algorithm 3 describes the preprocessing on step level in pseudo code. The most important difference to the state level concerns the possibility to only consider the most frequent sequence in which attributes (i.e., steps) are provided (i.e., the happy path) [3]. This allows identifying the attributes required to complete a state and transition to the next one. If the multiple execution sequences of steps are of interest (e.g., for identifying variants or if-then-else constructs in the object lifecycle process), the happy path option may be deactivated. Table 3 describes the event log of object *Application* in state *Edit* after applying the step level preprocessing with the happy path configuration. Figure 4 depicts the state level model of object *Application* discovered with the Heuristic Miner [25].

Algorithm 3. Step Level Preprocessing

Require: processlog, object, state, happypath
  $processlog.Filter(Object, State) = object, state$
  $processlog.SortAscending(ObjectInstanceID, Timestamp)$
  $FilterBlankAttributes(processlog)$
  $case\_groups = processlog.GroupBy(ObjectInstanceID)$
  $CleanTrace(case\_groups, steplevel)$ ▷ Alg. 1
  $clean\_log = Merge(case\_groups)$
  if $happypath$ = True **then**
    $FilterTopVariant(clean\_log)$
  end if
  return $clean\_log$

Table 3. Event Log State *Edit* of Object *Application* preprocessed on Step Level

ObjectInstanceID	Object	State	Attribute	AttributeValue	Action	Timestamp	PreviousValue	ValueChange?
A1	Application	Edit	JobPosting	J1	SetValue	24.08.2011 20:01		TRUE
A1	Application	Edit	Files	CV.zip	SetValue	29.08.2011 01:44	JobPosting	TRUE
A10	Application	Edit	JobPosting	J3	SetValue	06.09.2011 15:10		TRUE
A10	Application	Edit	Files	App.pdf	SetValue	25.09.2011 15:22	JobPosting	TRUE
...	...	...	...	...	...	...	...	...
A9	Application	Edit	JobPosting	J3	SetValue	06.09.2011 15:10		TRUE
A9	Application	Edit	Files	CV.pdf	SetValue	06.10.2011 18:38	JobPosting	TRUE

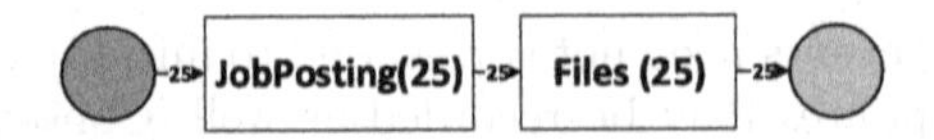

Fig. 4. Object *Application* State *Edit* - Step Level (Heuristic Net)

4.3 Algorithm Selection

In general, the event log preprocessing methods presented in Sects. 4.1 and 4.2 allow applying existing process discovery algorithms to object-centric event logs as well. As always, from the variety of existing discovery algorithms, certain algorithms might be more suitable for the discovery of object lifecycle processes than others. In particular, existing algorithms are often designed to tackle certain problems (e.g., noise or incompleteness) and, consequently, have strengths and weaknesses [1,3,5,15,25].

To determine which process discovery algorithm suit best to object lifecycle processes, we applied the following algorithms to a large real-world object-centric event log from an e-learning application. We preprocessed this log according to the approach presented in Sects. 4.1 and 4.2:

a) α-Algorithm [5]
b) Heuristic Miner[1] [25]
c) Inductive Miner [15]
d) Directly-Follows Graph (DFG) [3]

The event log has been obtained from the real-world e-learning platform *PHoodle*, an object-centric information system implemented with PHILharmonicFlows [7]. Over the course of 4 months, *PHoodle* replaced the existing e-learning platform *Moodle* for more than 100 students from Management Science, providing an object-centric process support to users including event log recording. As major advantage of employing this scenario to determine the most suitable process discovery algorithm, the object lifecycle processes used in *PHoodle* are publicly available, providing an appropriate baseline for testing[2]. Figures 5 (a) - (d) depict the discovered *state level* process models obtained with a) α-algorithm, b) Heuristic Miner, c) Inductive Miner, and d) a directly-follows graph (DFG) for object *Exercise* of the *PHoodle* scenario.

[1] Using the following thresholds: Dependency (0.5), AND (0.65), Loop 2 (0.5).
[2] https://cloudstore.uni-ulm.de/s/HL7dBkT2TpXt6b6.

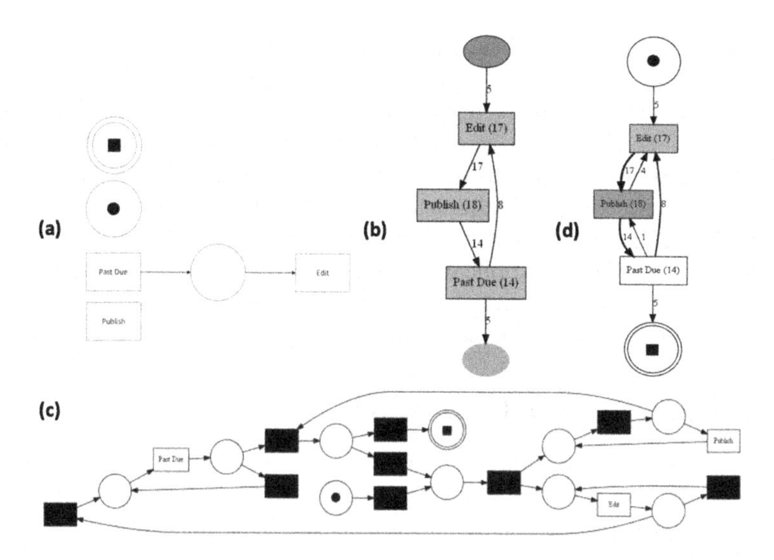

Fig. 5. Discovered State Level models for object *Exercise*, using a) α-Algorithm, b) Heuristic Miner, c) Inductive Miner, d) Directly Follows Graph

Remember that the execution of object lifecycle processes might be varying due to its built-in runtime flexibility [7]. Consequently, the discovered model should abstract from infrequent behavior recorded in the event log. Considering the discovered process models (see Fig. 5), the α-algorithm is unable to discover a sound process. The Heuristic Miner (see (b) in Fig. 2) can discover an executable process model that captures the actual state level behavior of the object lifecycle process and abstracts from infrequent executions. The model discovered using the Inductive Miner (see (c) in Fig. 2) is executable. However, it incorporates silent transitions to include infrequent behavior. In the real-world scenario considered, this led to models allowing for behavior not specified in the baseline model even if rather simple object lifecycle processes were to be discovered. The discovered DFG (see (d) in Fig. 2) also corresponds to an executable process model but does not allow for the customization of thresholds crucial for the flexible execution of object lifecycle processes.

Additionally, the comprehensibility of discovered process models has been evaluated using 13 domain experts [11]. The 13 domain experts have been asked to evaluate to which degree they are able to recognize each process model on a 5-Point Likert scale (see Table 4).

Table 4. Domain Expert Process Recognition Legacy Software System (N=13) [11].

	Inductive (Tree)	Inductive (BPMN)	DFG	Heuristic (thold = 0.75)	Heuristic (thold = 0.9)	Heuristic (thold = 0.95)
Mean (SD)	3.08 (1.07)	3.08 (0.73)	2.31 (1.2)	4.15 (0.77)	4.46 (0.63)	3.38 (1.27)

In a nutshell, we compared the process models discovered with different algorithms in the context of a real-world scenario with the baseline models. The comparison revealed that most algorithms yield suitable models if the flexible execution of object lifecycle processes is not used for an object. However, in case the flexibility of object lifecycle processes has been used (see Fig. 5), the Heuristic Miner [25] discovers the most reliable object lifecycle processes. The most likely reason for this is its ability to abstract infrequency by incorporating frequencies. Considering the real-world application and the expert study (see Table 4), the Heuristic Miner is most suitable. The results of all process discovery algorithms considered, together with the event logs used, can be found online.[3]

4.4 Combining State and Step Level

After discovering one model on the state level as well as one model for each state of an object (i.e., the step level), the discovered models are merged to reconstruct the object lifecycle process. Each state discovered in the state level model is associated with the respective model discovered on the step level and added to the process model. Figure 6 depicts the merged model of object *Application*. Note that we combined the discovered models to better illustrate the discovery and combination of state and step level models.

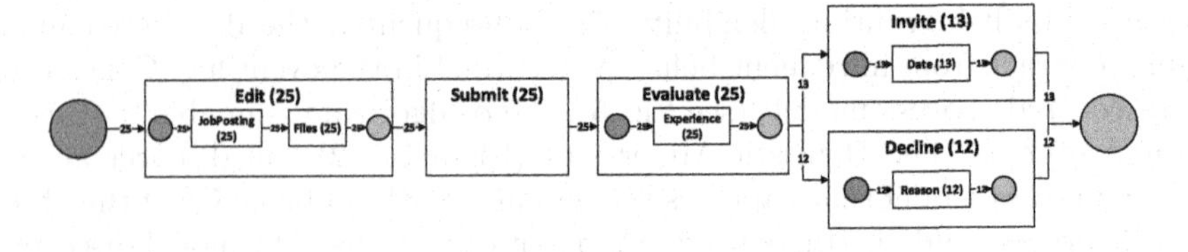

Fig. 6. Object *Application* - Combined Object Lifecycle Process (Heuristic Net)

5 Evaluation

To evaluate the presented approach, we applied it to event logs from three different scenarios:

1. Human Resource Management
2. E-Learning (Phoodle)
3. SAP Procure-to-Pay

Scenario 1 corresponds to a (simplified) process from the Human Resources domain, which we obtained from a joint project with an ERP vendor. Scenario 2 corresponds to real-world event log recorded from an e-learning system implemented with PHILharmonicFlows, whereas the event log of Scenario 3 was obtained from a simulation of a procure-to-pay process (in SAP) [20]. We illustrate the results using object lifecycle processes discovered with the Heuristic

[3] https://cloudstore.uni-ulm.de/s/HL7dBkT2TpXt6b6.

Miner [25]. The event logs considered for the evaluation, preprocessing results, the implementation using Python as well as the process models discovered with alternative algorithms (see Sect. 4.3) are provided.[4]

5.1 Scenario 1: Human Resource Management

The object-centric process from Scenario 1 comprises objects *Company*, *Job Posting*, and *Application* as well as their corresponding lifecycle processes. Note that object *Application* has been used as the running example in this paper. The corresponding event log recorded 279 events of 32 object instances.

After applying the *state level* and *step level* preprocessing (see Sects. 4.1 and 4.2) to all objects, the Heuristic Miner discovers the object lifecycle processes depicted in Figs. 6 and 7.

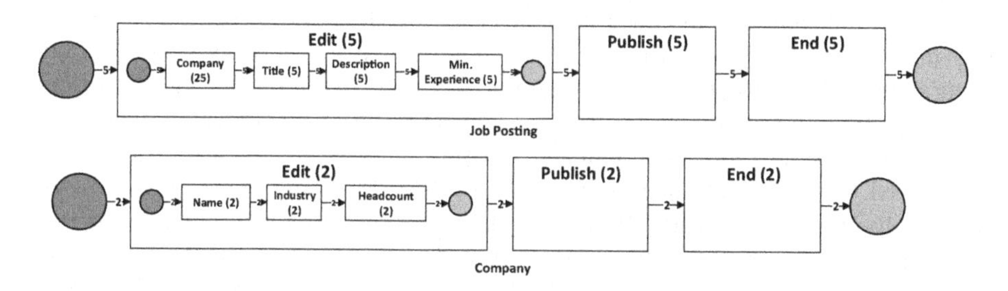

Fig. 7. Discovered Object Lifecycle Processes for Objects *Job Posting* & *Company*

Table 5 compares the number of states, steps, and transitions of the discovered model with the object lifecycle processes. In Scenario 1, our approach enabled the discovery of process models with the same number of states, steps, and transitions (i.e., the discovered models match the ones used in the corresponding information system).

Table 5. Heuristic Miner Results - Scenario 1 (Human Resource Management)

					Heuristic Miner			
Object	States	State - Transitions	Steps	Step - Transitions	States	State - Transitions	Steps	Step - Transitions
Company	3	2	4	3	3	2	4	3
Job Posting	3	2	3	2	3	2	3	2
Application	5	4	5	1	5	4	5	1

5.2 Scenario 2: E-Learning (Phoodle)

In Scenario 2, we applied the approach to a real-world event log recorded by an e-learning application implemented with PHILharmonicFlows [7]. Over the

[4] https://cloudstore.uni-ulm.de/s/HL7dBkT2TpXt6b6.

course of one semester, *Phoodle* orchestrated a *Lecture* (including, for example, *Exercises, Submissions, Tutorials, Downloads*) for more than 100 students from Management Science. During this study, an event log was recorded that comprises roughly 40.000 events from 848 object instances of 9 different objects. Figure 8 depicts the object lifecycle process discovered for object *Exercise.*

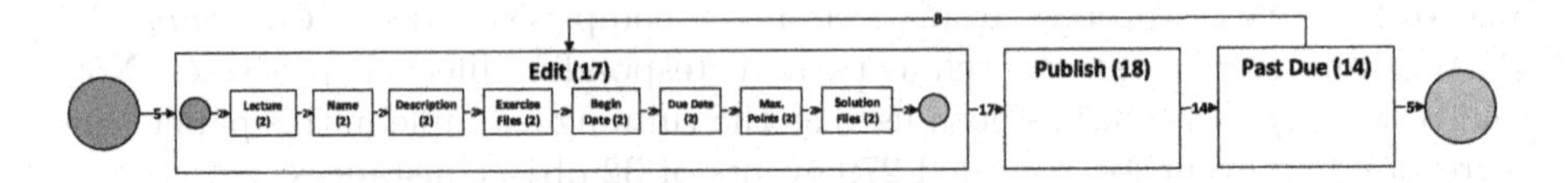

Fig. 8. Discovered Object Lifecycle Processes for Object *Exercise*

Comparing the lifecycle process models discovered for object *Exercise* with the actual lifecycle process used in the study, two deviations could be identified on the state level. First, state *End* has not been discovered as corresponding events (i.e., events in state *End*) were not recorded in the event log. This deviation can be explained with the real-world setting of Scenario 2 in which corresponding states have not been reached. Second, backwards transitions have been discovered that have not been part of the real-world model. This deviation was potentially caused by ad-hoc changes of instances enabling this behavior at runtime [7]. Table 6 compares the number of states, steps, and transitions between the model discovered with our approach and the actual object lifecycle process model used during the study. An in-depth investigation regarding these deviations revealed two main reasons for them: First, most deviations result from behavior not recorded in the event log (e.g., transitions not being used, or states not being reached). To be more precise, the object lifecycle processes used in the study allowed for behavior never used during the study. Regarding object *Exercise*, for example, no object instance reaches state *End.* Second, deviations might result from the application of the Heuristic Miner. When recording insufficient events that reflect a particular behavior (e.g., a backwards transition), the Heuristic Miner abstracts this behavior. Adapting corresponding thresholds during discovery, however, tackles this challenge.

Table 6. Heuristic Miner Results - Scenario 2 (E-Learning)

					Heuristic Miner			
Object	States	State - Transitions	Steps	Step - Transitions	States	State - Transitions	Steps	Step - Transitions
Lecture	3	4	3	2	2	1	3	2
Employee	2	1	2	1	2	1	3	0
Person	2	1	2	1	2	1	2	0
Exercise	4	6	7	6	3	3	8	7
Tutor	2	1	2	1	2	1	2	1
Download	2	2	3	2	2	0	3	2
Tutorial	2	2	2	1	2	1	3	1
Attendance	4	5	4	1	3	3	5	2
Submission	4	5	4	2	4	3	5	1

5.3 Scenario 3: SAP Procure-to-Pay

In this scenario, we applied the approach to a publicly available event log simulated for an SAP P2P process [20]. This event log follows the OCEL 2.0 specification [9] and comprises 10 event types, 7 object types, 14.671 events, and 9.543 objects [20] describing an end-to-end P2P process. This includes the initiation of a purchase requisition, the issuing of purchase orders, and the generation of receipts for goods and invoices as well as the payment of invoices. As opposed to the previous scenarios, the event log does not fully comply with the state level and step level representation of object behavior. The event log includes *object types* and corresponding *attributes*. However, no information regarding the states of object types is provided. Consequently, we may only apply the step level discovery to this event log. Moreover, due to the data-driven nature of object lifecycle processes, we focused on event log entries (i.e., the 33.475 entries) dealing with the initial writing or update of object attributes in the context of 6 object types. Note that no such data was available for object type *Material.*

Figure 9 depicts the (step level) results obtained with our approach for the remaining 6 objects. The approach enabled the discovery of object lifecycle processes for all six object types in the event log. The discovered models show all attributes required in the context of an object to conclude the process.

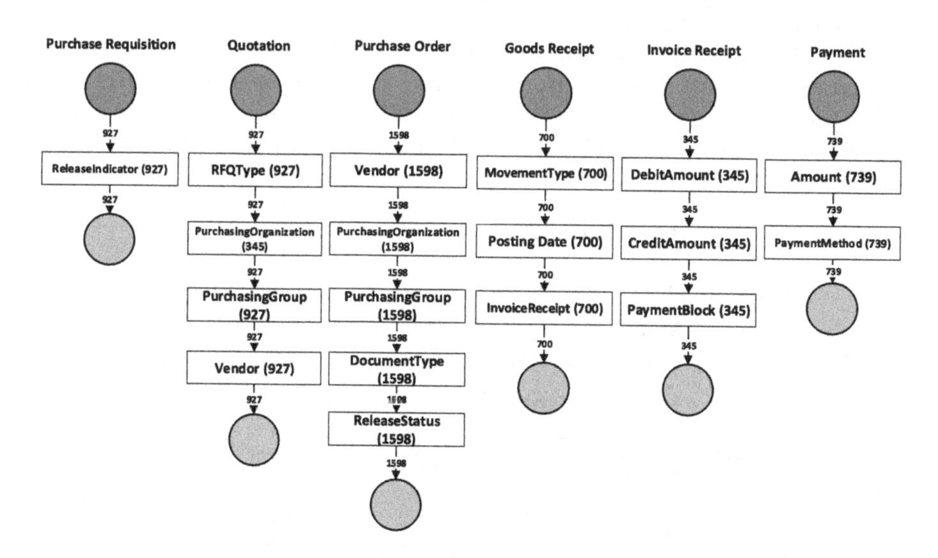

Fig. 9. Discovered Object Lifecycle Processes Procure-to-Pay (SAP, OCEL 2.0)

As *OCEL 2.0* does not capture object states, Table 7 only compares the discovered attributes with the ones outlined in the corresponding *object type* specification. Overall, the approach enabled us to accurately capture all specified attributes for an object as well as to reveal the order in which these attributes may be represented in an object lifecycle model. If object states are available, a more accurate representation of object lifecycle processes becomes possible.

Table 7. Heuristic Miner Results - Scenario 3 (SAP Procure-to-Pay)

		Heuristic Miner
Object Type	Attributes	Attributes
Purchase Requisition	1	1
Quotation	4	4
Purchase Order	5	5
Goods Receipt	3	3
Invoice Receipt	3	3
Payment	2	2

5.4 Footprint Comparison

To further evaluate the process models discovered for Scenarios 1 and 2, we compared their footprints with the actual object lifecycle process models and calculated the corresponding footprint-based conformance [1,12]. Note that this enables a more sophisticated (i.e., model-to-model) evaluation of the discovered models. For Scenario 3, no a-priori models are available and, therefore, no footprint comparison is possible.

Table 8 describes the conformance values for object *Application* from Scenario 1 as well as for object *Exercise* from Scenario 2. Note that certain states (e.g., states *Publish*, and *Past Due* as depicted in Fig. 8) of object *Exercise* do not comprise any steps and, therefore, no conformance values are calculated.

Table 8. Footprint Comparison Results

Scenario	Object	State/Step Level	Conformance
1	Application	State Level	1.0
1	Application	Step Level - Edit	1.0
1	Application	Step Level - Evaluate	1.0
1	Application	Step Level - Invite	1.0
1	Application	Step Level - Decline	1.0
2	Exercise	State Level	0.5
2	Exercise	Step Level - Edit	0.969

Scenario 1: For object *Application*, the conformance between the lifecycle process model and the discovered model equals to 1 for both the state and the step level. In other words, the model discovered with the presented approach is identical to the real-world object lifecycle processes models.

Scenario 2: The conformance values for object *Exercise* in Scenario 2 are below the ones of Scenario 1. On the state level, the conformance value can be explained as follows: One of the 4 states of object *Exercise* (i.e., state *End*) has not been recorded in the event log. Consequently, this state is not represented in the discovered model. Additionally, as certain transitions (i.e., backwards transitions) were only executed infrequently, they are not part of the discovered state level model. Using alternative thresholds for the Heuristic Miner, however, allows coping with this limitation.

On the step level, the conformance of 0.969 indicates that the discovered lifecycle process model is very close to the behavior of the a-priori model. The only difference between the two models occurs due to the additional attribute (i.e., step *Solution Files* in Fig. 8) recorded in the event log as well. Therefore, this step is part of the discovered object lifecycle process model. Furthermore, the sequence in which the steps are ordered also matches between the two models.

5.5 Threats to Validity

The evaluation revealed a limitation regarding the requirements towards the event logs. Events in the event log need to include information about *object*, *state*, *attribute*, and *timestamp* for the approach to discover complete object lifecycle processes. However, even with partial information recorded in the event log (see Scenario 3 of the evaluation), the approach discovers useful models.

Additionally, depending on the event log, alternative discovery algorithms may yield more suitable results. While the Heuristic Miner discovered the most suitable models in the scenarios we considered, this may not be generalized. However, as long as sound models are discovered, dividing the discovery of object lifecycle processes into several subproblems, solving them, and reconstructing an object lifecycle process is independent from the discovery algorithm applied.

6 Related Work

In general, object lifecycle processes are both object-centric and data-driven, i.e., their discovery relates to research from both areas.

Regarding object-centric processes, [2] motivates the use of object-centric event logs, which are better aligned with reality. An event log notation is introduced that allows relating activities to one or multiple objects. *Directly follows multigraphs* are used for discovering objects and their interactions.

[13] introduces the *OCEL Standard*, a generic and scalable structure for object-centric event logs, tackling challenges such as poor scalability of previously introduced approaches (e.g., *XOC* [16]). *OCEL 2.0* [9] enhances the previous OCEL Standard by enabling object changes (e.g., dynamic object attribute changes over time) and by capturing more information on object relationships.

OCEL-specific process mining approaches (e.g. discovery, conformance checking, log filtering, or log flattening), as well as a corresponding tool capable of applying these techniques are introduced in [8]. [4] constructs object-centric Petri nets as a variation of colored Petri nets, enabling process mining techniques. $OC\pi$ [6], for example, is capable of discovering object-centric Petri nets. [10] presents a literature review of challenges regarding the preprocessing as well as the discovery of object-centric processes. OCEL discovery approaches flatten the event log for each object as well. In contrast, our approach relies on object lifecycle processes as implemented in many object- and process-aware information systems instead of object-centric Petri nets. Object-centric Petri nets assume objects executing activities, whereas object lifecycle processes focus on the data-driven execution of objects in terms of states and steps.

Regarding the discovery of data-centric processes several approaches exist. [17] identifies artifacts from an event log and discovers lifecycles using existing discovery algorithms. As opposed to our approach, these lifecycles do not consider the state and step level of object lifecycle processes. [21] and [18] separate the discovery problems into subproblems enabling the application of existing algorithms. However, the discovered models correspond to artifacts representing them as Petri nets instead of object lifecycle process models. Finally, [19] presents the discovery of processes using the Neo4J graph database language *Cypher* to enable process discovery based on graph databases. However, the models do not conform with the state and step level of object lifecycle processes.

7 Summary and Outlook

This paper presented an approach for discovering object lifecycle processes. We introduced preprocessing methods enabling the discovery of object lifecycle processes based on well-established algorithms that were initially designed for activity-centric processes. In this context, the biggest challenge during discovery, also identified in the research question, is to cope with different granularity levels (i.e., the state and step level) of object lifecycle processes. Consequently, the presented approach divides the problem of discovering object lifecycle processes into two subproblems (i.e., the discovery tasks on the state and on the step level) that can be solved with traditional process mining algorithms. To enable this approach, we preprocess the event log and divide it into multiple, smaller event logs. Flattening the event log in this particular way enables transforming it into event logs suitable for existing process discovery algorithms.

We showed that different discovery algorithms may be applied to the preprocessed event logs, and we evaluated the obtained results. Most algorithms are able to discover models with the Heuristic Miner returning the most suitable models in the considered scenarios. We apply the discovery algorithm to discover one model for the state level and one for every state (i.e., the step level). Finally, we reconstruct the object lifecycle process from the discovered models by enriching the state level model with the corresponding step level behavior.

We evaluated the approach by applying it to 3 different scenarios including a scenario without deviations (i.e., Human Resources), a real-world implementation (i.e., E-learning), and a simulation of a procure-to-pay process. If the initial object lifecycle processes have been available, we compared their footprints with the ones from the discovered process models to evaluate the quality of the latter.

In future work, we plan to extend the presented approach. First, we aim to discover additional layers of object lifecycle processes. To be more precise, we plan to enable the discovery of the data model (i.e., objects and their semantic relations, including cardinality constraints) as well as the coordination process (i.e., the process and rules setting out the constraints for enacting dependent object lifecycle processes in a given application context) [23]. Note that the discovery of coordination processes will enable a better understanding of how objects become instantiated and how they are processed by information systems in general. Relationships between object instances might require an adjusted preprocessing approach to unravel all relationships between instances to obtain a coherent event log suitable for the discovery of coordination processes.

References

1. van der Aalst, W.M.P.: Process Mining: Data Science in Action, vol. 2. Springer, Heidelberg (2016)
2. van der Aalst, W.M.P.: Object-centric process mining: dealing with divergence and convergence in event data. In: Olveczky, P., Salaun, G. (eds.) SEFM. LNCS, vol. 11724, pp. 3–25. Springer, Heidelberg (2019). https://doi.org/10.1007/978-3-030-30446-1_1
3. van der Aalst, W.M.P.: A practitioner's guide to process mining: limitations of the directly-follows graph. Procedia Comput. Sci. **164**, 321–328 (2019)
4. van der Aalst, W.M.P., Berti, A.: Discovering object-centric petri nets. Fundamenta Informaticae **175**, 1–40 (2020)
5. van der Aalst, W.M.P., Weijters, T., Maruster, L.: Workflow mining: discovering process models from event logs. IEEE ToKDE **16**(9), 1128–1142 (2004)
6. Adams, J., van der Aalst, W.M.P.: Ocπ: object-centric process insights. In: Bernardinello, L., Petrucci, L. (eds.) Application and Theory of Petri Nets and Concurrency, pp. 139–150. Springer, Heidelberg (2022). https://doi.org/10.1007/978-3-031-06653-5_8
7. Andrews, K., Steinau, S., Reichert, M.: Enabling runtime flexibility in data-centric and data-driven process execution engines. Inf. Syst. **101**, 101447 (2021)
8. Berti, A., van der Aalst, W.M.P.: OC-PM: analyzing object-centric event logs and process models. Int. J. STTT **25**(1), 1–17 (2023)
9. Berti, A., et al.: Ocel (object-centric event log) 2.0 specification (2023). https://www.ocel-standard.org/2.0/ocel20_specification.pdf
10. Berti, A., Montalli, M., van der Aalst, W.M.P.: Advancements and challenges in object-centric process mining: a systematic literature review. arXiv e-prints (2023)
11. Breitmayer, M., Arnold, L., La Rocca, S., Reichert, M.: Deriving event logs from legacy software systems. In: Montali, M., Senderovich, A., Weidlich, M. (eds.) 4th International Conference on Process Mining. ICPM 2022 Workshops. Springer, Heidelberg (2022). https://doi.org/10.1007/978-3-031-27815-0_30
12. Carmona, J., van Dongen, B., Solti, A., Weidlich, M.: Conformance Checking: Relating Processes and Models. Springer, Heidelberg (2018). https://doi.org/10.1007/978-3-319-99414-7
13. Ghahfarokhi, A., Park, G., Berti, A., van der Aalst, W.M.P.: OCEL a standard for object-centric event logs. In: Bellatreche, L., et al. (eds.) New Trends in Database and Information Systems, pp. 169–175. Springer, Heidelberg (2021). https://doi.org/10.1007/978-3-030-85082-1_16

14. Künzle, V., Reichert, M.: PHILharmonicFlows: towards a framework for object-aware process management. JSME **23**(4), 205–244 (2011)
15. Leemans, S., Fahland, D., van der Aalst, W.M.P.: Discovering block-structured process models from event logs - a constructive approach. In: Colom, J.M., Desel, J. (eds.) Application and Theory of Petri Nets and Concurrency, pp. 311–329. Springer, Heidelberg (2013). https://doi.org/10.1007/978-3-642-38697-8_17
16. Li, G., de Carvalho, R., van der Aalst, W.M.P.: Automatic discovery of object-centric behavioral constraint models. In: Abramowicz, W. (eds.) BIS, pp. 43–58. Springer, Heidelberg (2017). https://doi.org/10.1007/978-3-319-59336-4_4
17. Lu, X., Nagelkerke, M., van de Wiel, D., Fahland, D.: Discovering interacting artifacts from ERP systems. IEEE ToSC **8**(6), 861–873 (2015)
18. Nooijen, E., van Dongen, B., Fahland, D.: Automatic discovery of data-centric and artifact-centric processes. In: La Rosa, M., Soffer, P. (eds.) BPM Workshops, pp. 316–327. Springer, Heidelberg (2013). https://doi.org/10.1007/978-3-642-36285-9_36
19. Nour Eldin, A., Assy, N., Kobeissi, M., Baudot, J., Gaaloul, W.: Enabling multi-process discovery on graph databases. In: CoopIS. pp. 112–130. Springer (2022)
20. Park, G., Unterberg, L.: Procure-To-Payment (P2P) object-centric event log in OCEL 2.0 Standard (2023). https://doi.org/10.5281/zenodo.8412920
21. Popova, V., Fahland, D., Dumas, M.: Artifact lifecycle discovery. CoRR arxiv:1303.2554 (2013)
22. Steinau, S., Andrews, K., Reichert, M.: Executing lifecycle processes in object-aware process management. In: Ceravolo, P., van Keulen, M., Stoffel, K. (eds.) Data-Driven Process Discovery and Analysis, pp. 25–44. Springer, Heidelberg (2017). https://doi.org/10.1007/978-3-030-11638-5_2
23. Steinau, S., Andrews, K., Reichert, M.: Modeling process interactions with coordination processes. In: Panetto, H., Debruyne, C., Proper, H., Ardagna, C., Roman, D., Meersman, R. (eds.) OTM, pp. 21–39. Springer, Heidelberg (2018). https://doi.org/10.1007/978-3-030-02610-3_2
24. Steinau, S., Marrella, A., Andrews, K., Leotta, F., Mecella, M., Reichert, M.: Dalec: a framework for the systematic evaluation of data-centric approaches to process management software. Softw. Syst. Model. **18**(4), 2679–2716 (2019)
25. Weijters, T., van der Aalst, W.M.P., de Medeiros, A.: Process Mining with the HeuristicsMiner Algorithm. Technische Universiteit Eindhoven (2006)

Security

US4USec: A User Story Model for Usable Security

Mohamad Gharib(✉)

University of Tartu, Tartu, Estonia
mohamad.gharib@ut.ee

Abstract. Constant integration of new technologies in our daily lives exposes us to various security threats. While numerous security solutions have been developed to protect us from these threats, they fail due to users' insufficient comprehension of how to employ them optimally. This challenge often stems from inadequate capture of Usable Security (USec) requirements, leading to these requirements being overlooked or not properly considered in the final solution, resulting in barely usable security solutions. A viable solution is to adeptly capturing USec requirements. Although techniques like User Stories (US) have gained popularity for focusing on users' needs, they encounter difficulties when dealing with non-functional requirements (NFR), like USec. This occurs due to the lack of well-defined US models explicitly tailored to address these particular requirements. This paper aims to tackle this issue by proposing US4USec, a US model tailored for USec. US4USec has been constructed based on best practices for the consideration and integration of NFR into US models that have been identified via a Systematic Literature Review (SLR). The coverage and completeness of US4USec have been demonstrated by applying it to a set of security US.

Keywords: Usable Security · User Story · Security-aware systems · Human-centric design · Requirements Engineering

1 Introduction

Novel technologies are constantly integrating into our daily lives. Despite their aim to simplify our lives, these technologies also expose us to an array of security threats [1]. To counter these threats, various security solutions, including numerous techniques and mechanisms, have been developed to safeguard users [2]. However, a significant body of research (e.g., [1–4]) has highlighted a prevalent issue: many of these solutions often fall short of their intended goals due to a lack of user understanding of their proper usage. Consequently, the effectiveness of any security solution relies heavily on end users being able to utilize it correctly without impeding their primary tasks [3]. Specifically, users not only need to use the security solution but also use it correctly [4].

Although the field of 'Usable Security (USec)' has been working on resolving this issue for more than two decades [3], existing solutions continue to face

J. Araújo et al. (Eds.): RCIS 2024, LNBIP 513, pp. 257–272, 2024.
https://doi.org/10.1007/978-3-031-59465-6_16

numerous challenges to achieve this objective [1]. Such challenges often arise from inadequate capture and analysis of USec requirements [5], struggling to effectively balance the interplay of security and usability, leading to these requirements being ill-defined and, in turn, overlooked or not properly considered in the final solution [4], resulting in barely usable security solutions. This problem could be solved if the users specify their USec requirements, which allows for designing more USec solutions that fit their needs and capabilities. However, such requirements might not be visible to the users [5], and the users need to be adequately empowered and assisted to specify them.

Unlike many conventional requirement elicitation approaches that tend to deliver system-focused requirements [6], User Stories (US) center on users' perspectives and needs [5,7], which makes them a possible solution for the aforementioned problem. However, US captures users' requirements at a high-abstraction level [7], lacks expressiveness, and mostly captures user-visible functional requirements [5]. Accordingly, US faces various challenges when dealing with NFR. The importance of considering NFR in US has been highlighted in many studies (e.g., [8,9]), as the ill-definition or neglection of NFR will lead to a poorly designed system that may eventually fail. In this context, numerous studies have aimed to consider and/or integrate NFR into US [8,9]. Although most of these studies present valuable concepts and methods for considering NFR aspects into US, only a few of them have put forth US templates or models that provide precise syntax and semantics for such consideration, and none of them offers a US model specialized for USec to the best of our knowledge.

This paper aims to tackle this issue by proposing US4USec, a US model tailored for USec. US4USec has been constructed based on best practices for the consideration and integration of NFR into US models that have been identified via an SLR. Specifically, the SLR was conducted to identify the aforementioned best practices, analyze their strengths and weaknesses, and identify to what extent they apply to the consideration/integration of USec aspects into US.

Our research methodology has been developed following a Design Science Research (DSR) approach [10], which identifies the problem that needs to be solved, motivates the development of the solution as a design artifact, and evaluates the application of the design artifact via appropriate means. DSR approach suggests the following steps: *1. Identification of the problem:* as discussed earlier there is a need for developing US4USec; *2. Artifact design:* we develop US4USec by identifying best practices for its construction via an SLR, deriving a set of requirements for US4USec, and assuring that such requirements have been realized in the developed US4USec. *3. Artifact evaluation:* we demonstrate the coverage and completeness of US4USec by applying it to a set of security US.

The rest of the paper is organized as follows; Sect. 2 presents the research method, and Sect. 3 describes the SLR. Section 4 discusses key requirements for constructing the US4USec model, and we construct the model based on these requirements in Sect. 5. In Sect. 6, we evaluate the coverage and completeness of US4USec, and discuss threats to our research validity in Sect. 7. Finally, we conclude the paper and discuss future work in Sect. 8.

2 Research Method

Our research method (depicted in Fig. 1) is composed of four main steps (**S**):

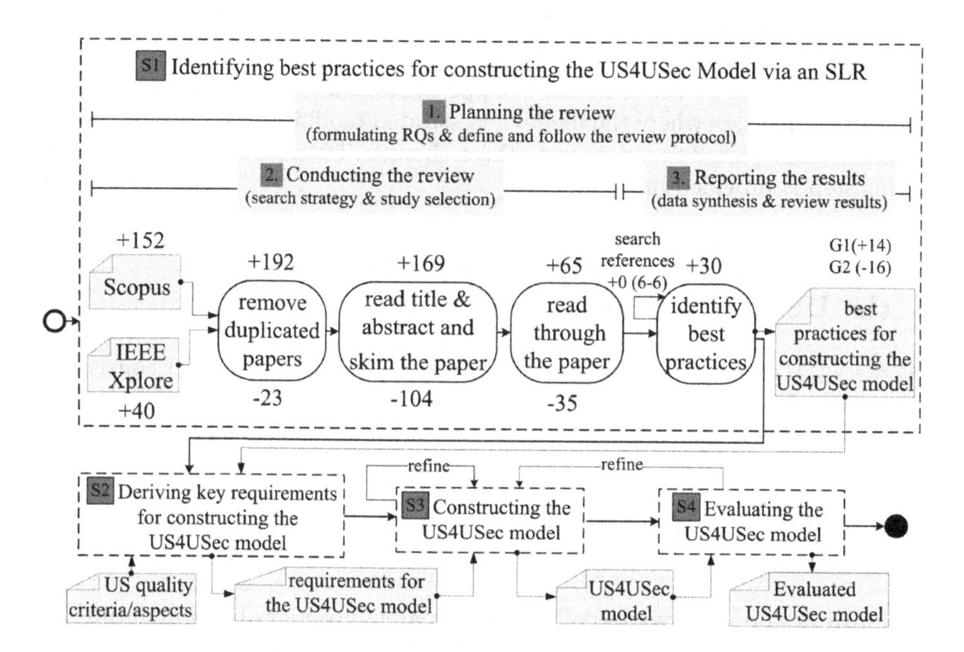

Fig. 1. The process and key activities of the research method

Step 1. Identifying best practices for constructing the US4USec model via a Systematic Literature Review (SLR): After an initial analysis of relevant literature, it was determined to conduct an SLR since relevant work is scarce and scattered widely in the literature and it is better to follow a well-accepted approach for identifying relevant studies. The main aim of the SLR is to identify existing studies that offer best practices for considering/integrating NFR into US, specifying the weaknesses and strengths of such practices, and to what extent they can be used to construct the US4USec model. Detailed information about this step is provided in Sect. 3.

Step 2. Deriving the requirements for constructing the US4USec model: takes the identified best practices and US quality criteria, then, analyzes them to specify the requirements that the US4USec model should satisfy. Detailed information about this step is provided in Sect. 4.

Step 3. Constructing the US4USec model: takes the identified requirements and uses them to construct the US4USec model. In particular, we used a traditional US template as starting point, and kept extending its syntax and semantics to realize the identified requirements. The construction process was iterative as realizing some requirements require unforeseen modifications in other requirements. However, we kept refining the model until all requirements are satisfied. Detailed information about this step is provided in Sect. 5.

Step 4. Evaluating the US4USed Model: takes the US4USec model that resulted from *step 3* as input, and empirically evaluated the coverage and completeness of US4USec model by applying it to a set of security US. Please note that in cases where the model necessitated modifications or refinements, *step 3* was iterated to incorporate these changes. Detailed information about this step is provided in Sect. 6.

3 Identifying Best Practices for Constructing the US4USec Model via an SLR

Following [11], the SLR process (depicted in the top part of Fig. 1) consists of three main phases: 1. *Planning the review*, in which, we formulate the research questions (RQs), define and follow the review protocol. 2. *Conducting the review* starts by defining the search terms and the literature sources, then, conducting the search, followed by the study selection activity, and 3. *Reporting the results of the review*, extracts detailed information from selected studies, and then using it to answer the RQs[1]. In what follows, we describe each of these phases.

1. **Planning the review,** we defined the research objectives by formulating the research questions, and how the review will be carried out by defining the review protocol. This includes two key activities:
   1.1. **Formulating the research questions (RQs)** is a very critical task since these RQs are used to guide the entire systematic review methodology [11]. The main aim of this review is to identify best practices that have been presented in the literature concerning the consideration/integration of NFR (especially security and usability) into US. Consequently, we need to select the most mature studies concerning the used methods for such consideration/integration of NFR into US, the strengths and weaknesses of these methods, and to what extent they apply to the consideration/integration of USec aspects into US. To this end, we formulate the following three RQs:
      RQ1 Which methods are used to consider/integrate NFR (especially security and usability) into traditional US use?
      RQ2 What are the strengths and weaknesses of the identified methods?
      RQ3 What are the best practices that can be adopted/adapted to be used for the consideration/integration of USec aspects into US?

[1] Detailed information about papers selection, summary of their contributions, and pros and cons of each used method of the final selected papers can be found at https://zenodo.org/records/10806824.

1.2. **Defining and following the review protocol,** a review protocol should specify the strategy that will be used to search for relevant studies; study inclusion and exclusion criteria, study selection criteria; and data extraction strategies [11]. In the rest of this section, we discuss how we perform each of these activities.

2. **Conducting the review.** This phase is composed of two main activities: 1- search strategy; and 2- study selection, where each of them is composed of several sub-activities.

   2.1. **Search strategy** aims to identify the most relevant studies to our RQs using an objective and repeatable strategy. The search strategy consists of three main activities:

   1. **Identify the search terms.** Following [11], we derived the main search terms from the research questions. In particular, we used the Boolean AND to link the major terms, and we used the Boolean OR to incorporate alternative synonyms of such terms. The resulting search terms are: `user AND (story OR stories) AND (requirement OR requirements) AND (security OR usability OR (usable AND security) OR non-functional)`.
   2. **Identify the literature resources.** Initially, five electronic database sources were selected for this research. These include Scopus, IEEE Xplore, Web of Science (WoS), Springer, and ACM Library. These databases index main scientific publications in the fields of computer science, software/requirements engineering, and information science. However, after searching the first two databases (Scopus and IEEE Xplore), it was determined not to consider the remaining three databases given the number of duplicated papers as well as the value of the new papers found for this research in the second database. This decision was confirmed when we searched the references of the selected papers and we were not able to add any new paper that can contribute to answering any of our RQs.
   3. **Conduct the search.** We have used the search terms to search the two selected databases, and all returned studies were considered.

   2.2. **Study selection.** Using the search terms to search the electronic database sources returned 152 and 40 papers from Scopus and IEEE Xplore respectively resulting in 192 papers. After removing 23 duplicated papers, we read the title, abstract, and skimmed through the rest of the remaining 169 papers applying the inclusion and exclusion criteria (shown in Table 1). Specifically, we excluded all the papers that were not published in English, not related to any of our RQs, and when we were able to identify multiple versions of the same paper, only the most complete one was included. We included papers that are relevant to our RQs and contain ideas that might be useful for the consideration/integration of NFR into US. The outcome of this selection stage was 65 papers, which were fully read and only papers that contained sufficient information to contribute to at least one of the RQs were included. This resulted in the

Table 1. Inclusion and exclusion criteria

	Inclusion criteria		Exclusion criteria
a.	Papers related to at least one of the RQs	a.	Papers that are not published in English
b.	Papers contain information concerning the consideration/integration of NFR into US	b.	If a paper has several versions only the most complete one is included

selection of 30 papers. The reference lists of these papers were carefully checked, and only six papers were identified. However, after reading these papers, none of them contained sufficient information to answer any RQ, which resulted in excluding them.

3. **Reporting the results.** The final phase of the systematic review involves synthesizing the data and reporting the results, as follows:
   - **3.1. Data synthesis** aims at combining findings of the selected studies in a way that allows answering the RQs. To facilitate this activity, each of the 30 selected papers has been analyzed and its contribution concerning the consideration/integration of NFR into US is summarized. Based on this analysis, the papers were classified into two groups: 1- **Group 1 (G1)** contains 14 papers that propose useful ideas and extensions for the consideration/integration of NFR into US, and 2- **Group 2 (G2)** contains 16 papers that do not provide significant contribution concerning the consideration/integration of NFR into US. Consequently, papers in **G1** were surveyed to identify key methods for the consideration/integration of NFR into US to answer **RQ1**. Then, each of the identified methods has been analyzed to identify its strengths and/or weaknesses to answer **RQ2**. Taking into consideration the strengths and weaknesses of these methods, their fit for the consideration/integration of USec aspects into US was discussed to answer **RQ3**.
   - **3.2. Review results and discussion:** The results are organized in Table 2 in terms of six sections. Each section starts with a key method for the consideration/integration of NFR into US to answer **RQ1**, followed by its strengths and weaknesses to answer **RQ2**. Finally, we discuss whether the method can be adopted/adapted to be used in US4USec to answer **RQ3**. In brief, the analysis has identified six key methods that facilitate, consider, or integrate NFR into US. After considering their strengths and weaknesses, four of them were fully considered and one was partially considered to be used for the construction of the US4USec.

Table 2. Review results - Answering RQs

1.	**Using traditional US to represent NFR:** several researchers have suggested the use of traditional US to represent various types of NFR (e.g., usability and security).
⊕	Can be used for general NFR related to the entire system.
⊖	If the NFR is tied to a specific functional requirement (FR), maintaining the connection between the FR and USec is challenging as each of them is represented as a distinct US.
×	Does not fit the needs of US4USec as it may require to explicitly capture the relationships between FR, security, and USec requirement(s).
2.	**Extending traditional US to integrate NFR aspects:** several researchers extend US with additional information including NFR [12], or extending US to consider usability requirements [8]. Others suggested integrating elements of User eXperience (UX) factors/aspects, User eXperience Design (UxD) artifacts [13], and even visual interface elements (mockups) [5] into US.
⊕	There are few methods that offer precise syntax and semantics to integrate NFR aspects into US.
⊕	If precise syntax and semantics is provided, the relationships between functional, security, and USec features/requirements are explicitly captured.
⊖	Most existing methods offers ad-hoc solutions for integrating NFR into US.
⊖	These method assumes that users can specify their NFR, which might not be an easy task as discussed earlier.
✓	US4USec should provide precise syntax and semantics for the integration of USec into US.
3.	**Adding or modifying the AC of a US to integrate/consider NFR AC:** several researchers extend the AC to integrate Nielsen heuristics [14] as a means to verify usability requirements [8], or used a specialized usability AC [9].
⊕	Can guide the acceptance tests to verify if a US is developed as expected.
⊕	It does not rely on users, since experts (e.g., product owner) handle the AC specification, while users and other stakeholders may improve these criteria.
⊖	It may necessitate replicating the same AC for multiple US.
✓	US4USec should include an AC section, which should be designed to accommodate the AC for the USec requirements.
4.	**Assisting/empowering users to express/specify their NFR:** several researchers propose the use of NFR elicitation/reasoning/validation taxonomies [15], a searchable repository of security requirements [16], and Heuristics [8,13] to facilitate the incorporation of NFR aspects into US.
⊕	Users may not be able to express/specify their NFR, and they need to be empowered and assisted to do that [8,13,15,16].
⊖	Assisting users may influence their behavior and their specified requirements.
✓	US4USec could benefit from the use of USec heuristics to facilitate the incorporation of USec aspects into US.

(continued)

Table 2. (*continued*)

5.	**Considering more than the user perspective and/or engaging others besides the user in furnishing the US and its corresponding AC:** several works suggested considering the developers' perspectives [5], involving an expert group to evaluate US considering quality aspects, and it is well-accepted that product owners are, usually, responsible for writing AC [16].
⊕	US should be mainly furnished by users with minimum assistance. This can be achieved by providing a concrete US template/model that assists but not direct users when specifying their requirements.
⊕	Stakeholders other than users, can contribute to the US, and any sections that are not expected to be furnished by the user, e.g., the AC section.
✓	US4USec should consider the perspectives of other stakeholders in the design of the US template/model, and it could involve them in furnishing some parts of the AC section.
6.	**The use of the Persona concept:** several works suggested the use of the Persona concept in US (e.g., [8,12,13]).
⊕	A persona can be created by defining user representations after learning and analyzing users' goals and behavior [12].
⊖	The role concept has been more commonly used in US and it can describe a stakeholder or persona [6].
✓	US4USec will adopt the role concept as it can cover the persona concept.

Method: ⊕ strength ⊖ weakness **US4Usec:** ✓: fit ✓: partial fit: ✗: does not fit .

4 Deriving Key Requirements for Constructing the US4USec Model

US4USec should offer a well-defined model with concrete syntax and semantics that enable capturing USec requirements. Consequently, we take the best practices identified in the previous section and couple them with numerous criteria for building quality US to derive the requirements that the US4USec model should satisfy. Concerning the criteria, we draw inspiration from the Quality User Story (QUS) framework [17]. Although QUS has been developed for assessing and enhancing the quality of a US, not the quality of a US template/model, it provides several criteria that can be adapted to assess the quality of a US template/model. These criteria are classified under three main themes (similar to the qualities proposed in [18]): 1. Syntactic quality focuses on the structure of a US without considering its meaning; 2. Semantic quality focuses on the relations and meaning of components (parts) of the US; and 3. Pragmatic quality focuses on choosing the most effective way to communicate a US as well as how it will be interpreted by the audience (stakeholders). Table 3 lists the requirements derived from the best practices, followed by the syntactic, semantic, and pragmatic requirements to be considered while developing the US4USec model.

Table 3. Requirements for constructing the US4USec model

Best practices - requirements	
R1.	US4USec should provide precise syntax and semantics for integrating USec into US.
R2.	US4USec should include an AC section designed to accommodate the AC for the USec features.
R3.	US4USec should facilitate the selection of relevant USec heuristics while integrating USec features into US.
R4.	US4USec should specify the responsibilities of different stakeholders for furnishing its content.
R5.	US4USec should adopt a concept to represent the active entity in US, which can be specialized/generalized into more specific/general concept.
Syntactic, semantic and pragmatic requirements	
R6.	US4USec expresses a requirement for at most one functional feature, one security feature, and one or more corresponding USec features - (**Atomic (Syntactic)**).
R7.	US4USec should explicitly capture the relationships among functional (if any), security, and its corresponding USec features, as well as their respective AC at the other hand - (**Explicit dependencies (Pragmatic)**).
R8.	US4USec only specifies the problem (features to be developed) and the criteria to consider them solved/satisfied - (**Oriented (Semantic)**).
R9.	US4USec should be flexible enough to allow producing complete US4USec regardless of the user expertise - (**Uniform (Pragmatic)**).

5 Constructing the US4USec Model

A traditional US has the following template [7]: As a `<type of user>`, I want `<goal>` so that `<benefit>`, and it is, usually, accompanied by acceptance criteria (AC), which offers a checklist that ensures the feature behaves as intended from an end-user perspective. Specifically, a US describes the desired outcome, while its corresponding AC specifies the criteria to achieve that outcome. AC can be used for FR and NFR as discussed earlier, where functional acceptance criteria ensure that the system performs the intended tasks, while non-functional acceptance criteria describe its acceptable performance concerning the NFR of concern (e.g., security, usability). AC can have various structures such as rules (resembling traditional requirements that need to be met) or scenarios. A well-adopted scenario template is the Given-When-Then (GWT), which has the following structure: Given that `<some context>`, when `<some action is carried out>`, then `<a set of observable outcomes should occur>`. Nevertheless, any well-structured template for AC can be adopted.

Starting from the traditional US template and considering the requirements identified in the previous section, the traditional template is extended to integrate related security and USec features (R1). Similarly, the AC section is extended to accommodate the AC for the security and USec AC (R2). To facilitate the integration of USec features into US4USec, a set of USec heuristics relevant to the security feature of concern should be offered (R3). Concerning R4, US4USec should be mainly furnished by the user, and the AC should be mainly furnished by experts. The model uses the role concept following traditional US (R5). The US4USec model expresses a requirement for at most one functional feature[2], one security feature, and one or more corresponding USec features (R6). The US4USec model captures an explicit relationship between the functional (if any) and its corresponding security feature, and it also captures an explicit relationship between the security feature and its corresponding USec feature(s). Additionally, the model captures explicit relationships between the aforementioned features and their corresponding AC (R7).

Concerning R8, the US4USec model differentiates between the problem (features to be developed) and the criteria to consider them solved. Finally, the US4USec model allows producing complete US4USec regardless of the user expertise. This is achieved by allowing an expert to fill in the relevant Usable Security Heuristics (USHs) if the user is not capable of doing that. Yet, the filled-in USHs have to be reviewed and approved by the user. Please note that the related section `.. [assuring that ⟨specific USHs⟩ concerning the ⟨security goal⟩ are satisfied]` is repeated in both the US and the security AC. However, only one of them is required to be filled, i.e., if the user did not fill it in the US section, an expert will fill it in the security AC section.

The US4USec model is depicted in Fig. 2 with a simplified representation of how the requirements were realized in its different components. In brief, the model is separated into two parts, one dedicated to specifying the features, and the other for specifying their AC. The functional feature and its relevant AC are optional since the security feature may not always be related to a functional feature. Consequently, the model provides [0|1]:1:1-n relationships between its functional, security, and USec features. Concerning the AC, besides the functional AC, the model provides 1:1-n relationships between the security feature and the AC for USec. Specifically, USec (the specific USHs) are listed with respect to the security feature, then, treated individually within AC for USec (USecAC). Please note that when the achievement of the functional feature and the security feature can be verified through the same set of desired observable results, their AC can be combined into a single AC.

The construction of the US4USec model was an iterative process, and we kept refining it until it was considered complete for capturing USec requirements, i.e., it provided all necessary and sufficient concepts for capturing the functional (if any), security, and USec features along with the interrelationships as well as their relationships with different AC. Besides using the best practices to derive the requirements for constructing the US4USec model, we also considered ideas from the selected studies to formulate the syntax and semantics of the model. Moreover, we had to refine the model several times while we were applying it to the security US set as we were able to identify several shortcomings in it.

[2] A security feature may not always depend on a functional feature. Consequently, the functional feature and its AC are optional.

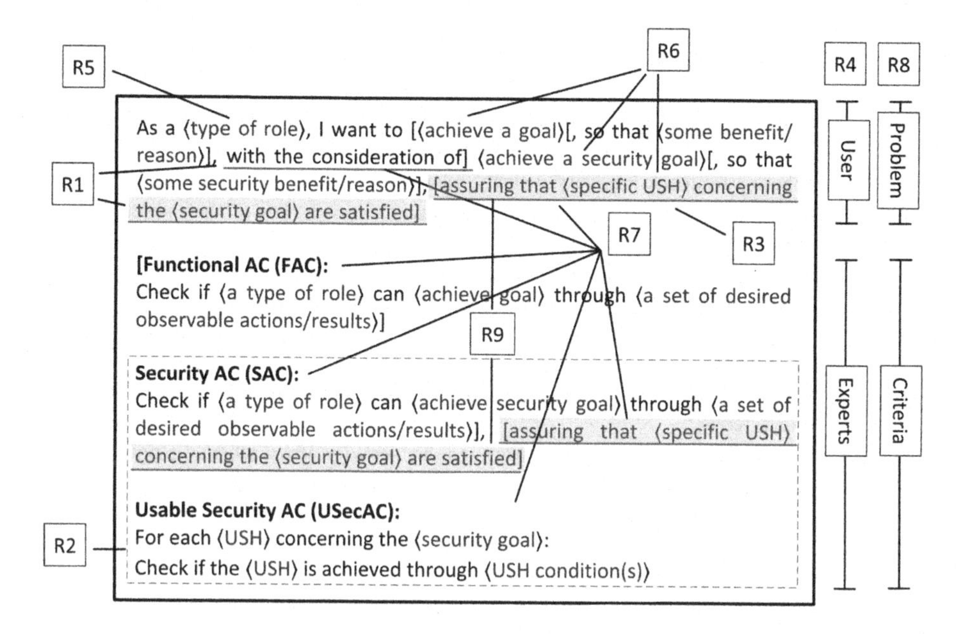

Fig. 2. The US4USec model

In what follows, we describe the list of USH along with their corresponding USH Criteria (USHC) to be provided by the US4USec model. Then, we demonstrate the applicability of the model by relying on a realistic example.

USec Heuristics. A heuristic can be defined as "enabling a person to discover or learn something for themselves", and heuristics have been used to represent features of the real world, for easier comprehension of the problem at hand [19]. Consequently, Usable Security Heuristics (USH) can be defined as a set of guidelines, standards, and best practices that facilitate designing secure systems while prioritizing user experience and usability [19,20].

Nielsen ground-breaking usability heuristics [14], one of the most used usability guidelines for user interface design, was used to promote a better user-to-system dialogue [13]. Nielsen's heuristics form the basis of most heuristics later developed in the area of USec [19]. For example, Yeratziotis et al. [20] conducted a literature review in the fields of usability, security, usable security, and privacy, and have identified 13 high-level heuristics for USec. Moreover, Mujinga et al. [19] propose a heuristic model that contributes to the design of USec in the context of online banking security. Their model considers 16 heuristics that have been developed based on Nielsen's heuristics for general information systems (e.g., [20]) taking into consideration several security design principles.

We have surveyed and analyzed the USec heuristics identified in both works [19,20], which mostly overlap. Based on our analysis, we have selected a list of USec heuristics (shown in Table 4) and their corresponding USH Criteria (USHC) that can be used to assess the aforementioned USH.

Table 4. USH and their corresponding USHC

USH1.	**Visibility:** the system should keep users informed about their security status.
USH1C1.	Is there a mechanism to assist/guide the behavior of the user?
USH1C2.	Is there a feedback for every security-related action?
USH2.	**Revocability:** the system should allow users to revoke any of their security actions.
USH2C1.	Can users easily reverse their security actions?
USH2C2.	Do security options make it obvious whether reverse is possible?
USH3.	**Clarity:** the system should inform users in advance about the consequences of any security actions.
USH3C1.	Does the system warn users if they are about to make a potentially serious security error?
USH3C2.	Does the system prevent users from making security errors whenever possible?
USH4.	**Expressiveness:** the system should guide users on security in a manner that still gives them freedom of expression.
USH4C1.	Does the system correctly anticipate and prompt for the users' probable next security-related activity?
USH4C2.	Is there a clear understanding of the systems security capabilities?
USH5.	**Learnability:** the system should ensure that security actions are easy to learn and remember.
USH5C1.	Are security operations easy to learn and use?
USH5C2.	Are there security defaults option?
USH6.	**Minimalist Design:** the system should offer users relevant information relating to their security actions.
USH6C1.	Is only the security information essential to decision-making displayed to the user?
USH6C2.	Are security labels brief, familiar, and descriptive?
USH7.	**Errors:** the system should provide users with detailed security error messages that they can understand and act upon.
USH7C1.	Do error messages inform the user about the severity of the error?
USH7C2.	Do error messages suggest the cause of the problem, and how it can be corrected?
USH8.	**Satisfaction:** the system should ensure that users have a good experience when using security and that they are in control.
USH8C1.	Do security-related prompts imply that the user is in control?
USH9.	**User Suitability:** the system should provide options for users with diverse levels of skill and experience in security.
USH9C1.	If the system supports both novice and expert users, are multiple levels of security error messages detail available?
USH9C2.	Can users customize security to meet their individual preferences?
USH10.	**User Language:** the system should use plain language that users can understand with regards to security.
USH10C1.	Are security actions named consistently across all prompts in the design?
USH10C2.	Is security information accurate, complete and understandable?
USH11.	**User Assistance:** the system should make security help apparent to users.
USH11C1.	Is there a visible security help?

To demonstrate the applicability of the US4USec model, we consider a novice user, who wants to change his password (functional feature) to maintain the safety of her account, and she wants to create a strong password (security feature) to make it hard to be guessed or cracked by others. Using the US4USec model, and considering relevant USH and their corresponding USHC, the US4USec can be furnished, as follows:

As a ⟨a novice system user⟩, I want to ⟨change my password⟩, so that ⟨I can maintain the safety of my account⟩, with the consideration of ⟨creating a strong password⟩, so that ⟨the password is hard to be guessed or cracked by others⟩, assuring that ⟨creating a strong password⟩ should satisfy ⟨USH1. (Visibility), USH4. (Expressiveness), and USH7. (Errors)⟩

Functional and Security AC (FSAC):
Check if ⟨a novice system user⟩ can ⟨create a new strong password⟩ through ⟨the existence of password change mechanism, the existence of rules to enforce the creation of strong password, and the automatic generation of a confirmation message/notification that the new password has been set⟩

Usable Security AC (USecAC):
Check if the ⟨USH1. (Visibility)⟩ is achieved through ⟨USH1C1. there is a mechanism to assist/guide the behavior of the user⟩ and ⟨USH1C2. there is a feedback for every security-related action⟩

Check if the ⟨USH4. (Expressiveness)⟩ is achieved through ⟨USH4C1. the system correctly anticipates and prompts for the user's probable next security-related activity⟩ and ⟨USH4C2. there is a clear understanding of the systems security capabilities⟩

Check if the ⟨USH7. (Errors)⟩ is achieved through ⟨USH7C1. error messages inform the user of the severity of the error⟩ and ⟨USH7C2. error messages suggest the cause of the problem, and how it can be corrected⟩

6 Evaluating the US4USec Model

Evaluation aims at ensuring that the developed artifact meets the purposes it was developed for, i.e., it fulfills its intended usage. In our case, the US4USec model can capture and analyze USec requirements. Following [21], we evaluated the US4USec model by applying it to a set of security US[3] to demonstrate the coverage of the US4USec model, i.e., verify whether each element of the model is required?, and the completeness of the model, i.e., verify whether we need to consider other elements for the US4USec model?

[3] The list of 35 security US is available at https://github.com/OWASP/user-security-stories/blob/master/user-security-stories.md.

Out of the 35 security US, only 12 have USec features, and we have slightly modified three US to introduce USec features. Consequently, we have 15 security US that have USec features, on which, we have applied our model. As previously mentioned, the US4USec model has been refined several times while we were applying it to the security US. Specifically, we refined the interrelationships among the functional (if any), security, and USec features as well as their relationships with different AC. The final version of the model has adopted [0|1]:1:1-n relationships between its functional, security, and USec features. Moreover, we linked the security feature/goal with the AC for USec (USecAC) with 1:1-n either in the main body of the US or via the Security AC (SAC). Additionally, we noticed that the functional feature is not always present in the security US, thus, we have made it and its relevant FAC optional in the model.

Applying the US4USec model to the security US[4] demonstrates that each of the elements of the model is required either for capturing a feature, its AC or the relationships among features or between them and their AC. Specifically, we did not have a single case, where an element of the model was not required to capture an aspect of a security US. Accordingly, we were not able to identify any unnecessary or superfluous element, i.e., each element of the model is required. Similarly, all considered security US were covered by using the elements of the model, i.e., they can capture all features as well as their interrelationships and the relationships between the features and different AC. Specifically, we did not have a single case, where an aspect of a security US was not captured by an element of the model. This means that the US4USec model can be considered complete for capturing all features, AC, and their interrelationships, i.e., we do not need to extend the US4USec mode with any new elements.

7 Threats to Validity

In this section, we present and discuss six threats to the validity of our work, where the first four are relevant to the conducted SLR, and the last two are relevant to the developed US4Usec artifact:

1. **Systematic error.** May occur while designing and conducting an SLR. To avoid this threat, the review protocol has been carefully designed based on well-adopted methods, and it has been strictly followed during the different phases of the review.
2. **Potential studies selection bias.** Refers to situations where positive research results are more likely to be reported than negative ones. Our review focused on finding the most relevant studies related to integrating NFR into US, and there are no positive or negative research results in such a case.
3. **Completeness.** It is almost impossible to guarantee completeness in SLRs, yet we did our best to have sufficiently complete results by carefully designing our review protocol and search strategy to cover as much as possible of the

[4] The results of applying the US4USec model to the set of security US can be found at https://zenodo.org/records/10806824.

related studies. Moreover, we performed a manual scan of references of all the selected studies to identify those studies that might be missed.

4. **Limited number of selected studies.** The number of selected studies (30, 14 of which have been used to derive best practices) can raise concerns about the SLR results. However, it is a relatively new area of research, and the contributions are rather limited.
5. **No extensive evaluation.** The model has been applied to only one set of security US, which might threaten the generalization of our findings. Probably, applying the model to a more comprehensive set might reveal undetected inadequacies in the model. However, we were not able to find any other publicly available US security set.
6. **Reliability threat.** Is concerned with to what extent the study is dependent on the researcher(s). The search terms and sources, inclusion and exclusion criteria, studies selection process, as well as the results of applying the model to the security US set are all available, and any researcher can repeat the study and should obtain comparable results.

8 Conclusion and Future Work

We aimed to tackle the problem of capturing and analyzing USec requirements by proposing US4USec, a US model tailored for USec requirements. US4USec aims to assist users while specifying their USec requirements by providing them with a model that offers the necessary and sufficient concepts for capturing such requirements. In this paper, we have evaluated the coverage and completeness of US4USec by applying it to a set of security US, which needs to be complemented in future empirical validation. Additionally, we are planning to contact the authors of the selected studies as well as researchers who made significant contributions to the design and quality assessment of US to get their feedback concerning the US4USec model. We also plan to refine and extend both the USH and their corresponding USHC with the help of experts in the domain since the current USH and USHC are a bit abstract and provide a "One-Size-Fits-All" solution. Finally, we will develop a web-based tool to facilitate US4USec use.

Acknowledgment. This study was performed within the framework of COST Action CA22104 (Behavioral Next Generation in Wireless Networks for Cyber Security), supported by COST (European Cooperation in Science and Technology; www.cost.eu).

References

1. Lennartsson, M., Kavrestad, J., Nohlberg, M.: Exploring the meaning of usable security - a literature review. Info. Comput. Secur. **29**(4), 647–663 (2021)
2. Jean Camp, L.: Mental models of privacy and security. IEEE Technol. Soc. Mag. **28**(3), 37–46 (2009)
3. Groen, E.C., et al.: Achieving Usable Security and Privacy Through Human-Centered Design. In: Gerber, N., Stöver, A., Marky, K. (eds.) Human Factors in Privacy Research, pp. 83–113. Springer, Cham (2023). https://doi.org/10.1007/978-3-031-28643-8_5

4. Gutfleisch, M., Klemmer, J.H., Busch, N., Acar, Y., Sasse, M.A., Fahl, S.: How does usable security (not) end up in software products? Results from a qualitative interview study. In: Proceedings of the IEEE Symposium on Security and Privacy, pp. 893–910 (2022)
5. Medeiros, J., Vasconcelos, A., Goulao, M., Silva, C., Araujo, J.: An approach based on design practices to specify requirements in agile projects. In: The ACM Symposium on Applied Computing, pp. 1114–1121 (2017)
6. Hudson, W.: User stories don't help users: introducing persona stories. Interactions **20**(6), 50–53 (2013)
7. Cohn, M.: User Stories Applied for Agile Software Development (2004)
8. Choma, J., Zaina, L.A.M., Beraldo, D.: UserX story: incorporating UX aspects into user stories elaboration. In: Kurosu, M. (ed.) HCI 2016. LNCS, vol. 9731, pp. 131–140. Springer, Cham (2016). https://doi.org/10.1007/978-3-319-39510-4_13
9. Moreno, A.M., Yagüe, A.: Agile user stories enriched with usability. In: Wohlin, C. (ed.) XP 2012. LNBIP, vol. 111, pp. 168–176. Springer, Heidelberg (2012). https://doi.org/10.1007/978-3-642-30350-0_12
10. Hevner, A.R., March, S.T., Park, J., Ram, S.: Design science in information systems research. MIS Q. **28**(1), 75–105 (2004)
11. Kitchenham, B., Brereton, P., Budgen, D., Turner, M., Bailey, J., Limkman, S.: Systematic literature reviews in software engineering - a systematic literature review. Inf. Softw. Technol. **51**(1), 7–15 (2009)
12. Marques, A.B., Costa, A.F., Santos, I., Maria Castro De Andrade, R.: Enriching user stories with usability features in a remote agile project: a case study. In: ACM International Conference Proceeding Series, pp. 1–10 (2022)
13. Lopes, L.A., Pinheiro, E.G., Da Silva, T.S., Zaina, L.A.M.: Using UxD artefacts to support the writing of user stories: findings of an empirical study with agile developers. In: ACM International Conference Proceeding Series, vol. Part F1477, pp. 1–4. Association for Computing Machinery (2018)
14. Nielsen, J.: 10 Usability Heuristics for User Interface. TR (1995)
15. Domah, D., Mitropoulos, F.J.: The NERV methodology: a lightweight process for addressing non-functional requirements in agile software development. In: IEEE SOUTHEASTCON, pp. 1–7 (2015)
16. Ionita, D., van der Velden, C., Ikkink, HJ.K., Neven, E., Daneva, M., Kuipers, M.: Towards risk-driven security requirements management in agile software development. In: Cappiello, C., Ruiz, M. (eds.) Information Systems Engineering in Responsible Information Systems, CAiSE 2019. LNBIP, vol. 350, pp. 133–144. Springer, Cham (2019). https://doi.org/10.1007/978-3-030-21297-1_12
17. Lucassen, G., Dalpiaz, F., Martijn, J., Van Der Werf, E.M., Brinkkemper, S.: Forging high-quality user stories: towards a discipline for agile requirements. In: Requirements Engineering Conference, pp. 126–135. IEEE (2015)
18. Lindland, O.I., Sindre, G., Solvberg, A.: Understanding quality in conceptual modeling. IEEE Softw. **11**(2), 42–49 (1994)
19. Mujinga, M., Eloff, M.M., Kroeze, J.H.: Towards a heuristic model for usable and secure online banking. In: Proceedings of the 24th Australasian Conference on Information Systems, pp. 1–12 (2013)
20. Yeratziotis, A., Pottas, D., van Greunen, D.: A usable security heuristic evaluation for the online health social networking paradigm. Int. J. Hum. Comput. Interact. **28**(10), 678–694 (2012)
21. Wautelet, Y., Heng, S., Kolp, M., Mirbel, I.: Unifying and extending user story models. In: Jarke, M., et al. (eds.) CAiSE 2014. LNCS, vol. 8484, pp. 211–225. Springer, Cham (2014). https://doi.org/10.1007/978-3-319-07881-6_15

Do Cialdini's Persuasion Principles Still Influence Trust and Risk-Taking When Social Engineering is Knowingly Possible?

Amina Mollazehi[1], Israa Abuelezz[1], Mahmoud Barhamgi[1], Khaled M. Khan[1], and Raian Ali[2](✉)

[1] College of Engineering, Qatar University, Doha, Qatar
[2] College of Science and Engineering, Hamad Bin Khalifa University, Doha, Qatar
raali2@hbku.edu.qa

Abstract. Despite recognizing the applicability of Cialdini's principles in social engineering context, studies on their effectiveness needed more tailored and validated tests, primary data collection, and multicultural samples. Cialdini's six persuasion principles include reciprocity, commitment, liking, scarcity, social proof, and authority. We designed and face validated 12 scenarios representing the presence and absence of each principle in a situation where an acquaintance prompts online group members to install an app for testing and improving it. Through an online survey with 314 UK and 328 Arab participants, we collected data on the impact of persuasion principles on risk taking, i.e., to accept installing and trying the app, and trust in the requester, who might be knowingly a social engineer. Results across both cultural frameworks indicate significant impacts, with Social Proof and Authority being the most influential, and Scarcity the least, yet still significant. Interestingly, the principles not only influenced the decision to take the risk but also affected trust in the potential social engineer. This holds true even in less intuitive scenarios, representing Scarcity and Commitment/Consistency principles. This applies to two distinctive cultural frameworks, Arab and British, increasing robustness. The research also investigates the relationship between security attitudes, measured through SA-6 scale, and susceptibility to these principles, in terms of trust and risk taking, revealing surprising results of positive correlations. These findings emphasize the need for cybersecurity strategies that include awareness of psychological manipulation alongside technical knowledge, catering to different cultural contexts.

Keywords: Social engineering · Cialdini's principles · Risk-taking · Persuasion · Cybersecurity · Arab · UK

1 Introduction

In the dynamic field of cybersecurity, understanding the influence of persuasion and manipulation on human decision-making is paramount, particularly in contexts where security risks are involved [1]. Social engineering represents a critical vector for information security breaches, exploiting human psychology rather than technological vulnerabilities [2, 3]. The tactics of manipulation and persuasion are essential tools for social

J. Araújo et al. (Eds.): RCIS 2024, LNBIP 513, pp. 273–288, 2024.
https://doi.org/10.1007/978-3-031-59465-6_17

engineers, aiming to influence individuals' behavior. Social engineering in the context of cybersecurity refers to the manipulative tactics used by cyber attackers to deceive individuals into compromising security measures [2]. The fusion of psychological persuasion techniques with cybersecurity challenges underscores the necessity for advanced protective measures against increasingly complex cyber threats [4]. The complexity of human factors in information security is further underscored by studies focusing on the psychological aspects of social engineering. Attackers exploit basic human tendencies like trust, fear of authority, and the desire to help and responses to these interventions vary based on individual perceptions of fear, trust, and commitment [4, 5]. Understanding psychological aspects is essential in shaping cybersecurity strategies.

Cialdini's six principles of persuasion [6] have been foundational in the study of how individuals are influenced. The principles include Social Proof, where decisions are influenced by a perception of what others are doing; Likeability, focusing on the impact of requests from individuals who are particularly likable or relatable; Authority, which assesses the weight given to suggestions from credible and knowledgeable sources; Commitment and Consistency, highlighting a tendency to align with our previous commitments and behaviors; Reciprocity, the inclination to return favors or respond positively when someone has previously been accommodating; and Scarcity, which examines how perceived rarity or limited availability of something can make it more attractive. These principles provide a framework for understanding the various psychological triggers that can significantly persuade our decisions and actions. Traditionally, the impact of these principles on individual susceptibility has been a subject of extensive research within contexts such as marketing, health, and behavioral change, where the desired behavior is typically positive. Zalake et al. [7] focused on how virtual humans' verbal persuasion strategies influence users' intentions to adopt health behaviors. The study examined college students' intentions to engage in mental health coping skills by employing Cialdini's principles. The study found that these strategies effectively influenced students' perceived behavioral control, subsequently impacting their intentions to perform the suggested health behavior. However, when these persuasion principles are applied within the cybersecurity domain, the dynamics become different as the same strategies that can guide positive behaviors may lead individuals toward actions that entail significant risk. Additionally, the impact of Cialdini's principles on trust is yet to be explored. This study seeks to uncover how persuasion techniques can impact individuals' trust and willingness to take risks within cybersecurity contexts, particularly when the potential for security threats is known to the target. Our investigation aims to deepen the understanding of these dynamics and their implications for developing robust cybersecurity defenses to strengthen resistance against the biases and prejudices that might be triggered by principles of persuasion. Such strategies showed success such as building resistance to persuasion in the contexts of gaming and gambling [8].

Security attitudes refer to the tendency to follow security recommendations to reduce vulnerability to social engineering attacks. However, the practical application of measures that indicate a higher security attitude, e.g., seeking to learn and consulting with experts, does not necessarily result in lower susceptibility to social engineering. According to Workman [5], individuals may not consistently act in their best interest with regard to security, especially when they are under pressure or when they face persuasive tactics.

In addition, Workman's findings suggest that the principle of authority can influence individuals to perform actions that go against their security training. Similarly, the principle of social proof can be manipulated to create a false sense of security or urgency, leading to risky behavior. Grassegger and Nedbal [9] investigate the role of information security awareness in employees' resistance to social engineering attacks. They argue that while technical measures are important, promoting security awareness is crucial for effective information security. Their findings indicate a weak correlation between leadership and security awareness and no significant association between information security policies, training measures, and security awareness. The Persuasion Knowledge Model [10] describes how even well-informed individuals can be vulnerable if they cannot identify persuasion attempts. Therefore, while security attitudes are important, their impact on reducing susceptibility in high-pressure situations requires a more nuanced examination.

Significant research has been conducted on the role of subjects' personality in their susceptibility to persuasion principles in various domains, including cybersecurity. Lawson et al. [11] explore how personality traits interact with persuasion tactics in email phishing. The research found that emails using the Liking principle were the most effective for phishing while Authority and Scarcity combined were more likely to arouse suspicion. The study confirms that high extroversion is a significant predictor of increased susceptibility to phishing attacks.

The exploration of Cialdini's principles of persuasion across various cultural, age, and gender demographics reveals nuanced interactions between these universal principles and specific contextual factors. A study by Spasova [12] investigates how young consumers respond to persuasion principles in advertising, particularly focusing on the impact of gender and age. The research finds that different genders and age groups exhibit various susceptibilities to persuasion principles, with women being more responsive to principles like reciprocity, commitment and consistency, likability, while men show more responsiveness to authority and commitment and consistency. Age also influences the effectiveness of these principles, with young adults showing a greater influence by social proof, and older adults by commitment and consistency. Cultural factors significantly influence decision-making and susceptibility to persuasion. Oyibo et al. [13] highlight the contrast between Nigerian and Canadian responses to persuasive strategies. Their findings demonstrate that while Nigerians exhibit higher susceptibility to authority and scarcity, Canadians are more responsive to reciprocity, liking, and consensus. The contrast suggests a deeper cultural inclination in Nigerian society towards respect for authority and urgency, whereas Canadian society shows a preference for mutual benefit and social validation. Hence, questioning the cross-cultural impact of Cialdini's persuasion techniques in the context of social engineering and cybersecurity is also valid. Therefore, in this paper, we include samples from both the UK and Arab populations.

In Arab cultures, the interplay of traditional values and modern influences creates a unique landscape for persuasion. Obeidat et al. [14] explore how cultural differences impact managerial practices in Arabian countries. Their study underscores the significant influence of Islamic values and traditional practices on Arabian culture. Parallel to this, the research into Arab Muslims' susceptibility to persuasion by Alnunu et al. [15] underscores the effectiveness of strategies that resonate with traditional cultural values such as reciprocity. The contrast in perceptions and approaches between Arab and British professionals, as examined by Rees-Caldwell et al. [16], offers a compelling argument about cultural relativity in professional settings. Their findings reveal distinct project management approaches between Arab and British professionals, with Arab managers prioritizing communication, whereas British counterparts focus more on scope and planning. This contrast in approaches suggests that persuasive communication techniques effective in one cultural context may not yield the same results in another. Furthermore, the intricate relationship between culture and decision-making processes, as reviewed by Yates and de Oliveira [17], highlights the impact of cultural norms and cognitive styles on individual and collective decision-making behaviors. The study emphasizes the need to consider cultural variations in individualism, collectivism, and other social norms when examining decision-making in cybersecurity contexts. It underscores the pivotal role that cultural background plays in shaping responses to persuasion techniques, suggesting that strategies effective in one culture may not necessarily translate to another due to differing cultural norms and values.

According to Henrich et al. [18] the majority of psychological and human factors research consist of individuals who are western, educated, industrialized, rich, and democratic (WEIRD). This focus risks overlooking the distinctive characteristics of collective cultures such as those in Arab societies, where communal norms and a preference for uncertainty avoidance play a crucial role in decision-making processes [6, 13].

While Cialdini's principles have been studied for their inherent appeal in influencing behavior, their effectiveness in influencing trust and decision-making in risk-related scenarios has not been explored. In addition, while studies on persuasion were conducted in various cultural frameworks, comparing them faces a methodological concern as they have been conducted using different measurement tools, samples, and timing. Hence, making comparisons with the same tools, timing, and sample characteristics would be a methodological strength, and this is a focus in this paper. To further explore the nuanced interplay between persuasion and cybersecurity, our study delves into the comparative analysis of cultural influences, focusing specifically on the UK and Arab populations, who exhibit different cultural traits, according to Hofsted [19] and also academic studies [20]. This comparative approach is essential for developing a more personalized and culturally-sensitive understanding of persuasion and its countermeasures, which will enhance cybersecurity behaviors [15, 16].

To bridge the identified research gaps, our exploratory study is designed to address two key questions, on Arabian and British samples:

RQ1: To what extent do Cialdini's principles of persuasion influence trust and risk-taking when potential social engineering is knowingly present?

RQ2: What is the correlation between security attitude and the levels of trust and risk-taking when potential social engineering, leveraging Cialdini's principles, is knowingly present?

2 Methodology

In our study, we designed 12 scenarios to explore the influence of Cialdini's principles, with six scenarios demonstrating their presence and six their absence. Each scenario was carefully designed to focus solely on one principle without incorporating additional persuasive elements. Participants were introduced to a situation where an online forum member, who is a software developer, requests help in installing and testing their new mobile app. The order of these scenarios was randomized for each participant to avoid order effects potentially impacting the responses, such as through fatigue or learning biases. The survey participants were asked to respond regarding their willingness to install this app and their level of trust in the requesting software engineer, in each of the 12 scenarios. Importantly, participants were informed that installing the app could pose security risks, ensuring they were aware of the potential implications of their decision. A selection criterion for participants was that they perceive at least minimal risk in installing and testing the app.

2.1 Recruitment and Data Collection

The survey for this study, developed on SurveyMonkey [21], involved participants from the Arab Gulf Cooperation Council (GCC) and the United Kingdom (UK), recruited with the assistance of TGM Research, a firm specialized in collecting data for research studies [22]. The selection of these two distinct cultural backgrounds in this study was based on their variations in morals and values, as evidenced by Hofstede's chart which highlights significant differences in Power Distance and Individualism between the UK and GCC countries [23]. Participants who completed the survey were compensated. Our goal was to achieve a balanced sample in terms of age and gender distribution, both within and across the Arab and UK populations. This process was also implemented throughout data collection.

To participate in the survey, individuals underwent a brief eligibility screening. This included criteria such as being older than 18 years of age and being either born or currently residing in the GCC countries (Saudi Arabia, Qatar, Bahrain, Kuwait, Oman, and the UAE) or the UK (England, Scotland, Wales, and Northern Ireland). Additionally, participants confirmed they identify as Arabs GCC or UK in terms of norms and culture. All survey participants were required to give informed consent and were informed of their right to withdraw from the survey at any stage. The survey included specific attention checks to ensure the integrity of the responses. Participants who did not pass attention checks were not included in the final analysis. The study received approval from the relevant Institutional Review Board (IRB) at Hamad Bin Khalifa University, Qatar. The final dataset comprised 314 participants from the UK and 328 from Arab GCC countries.

2.2 Measures

Demographic Measures. The participants provided demographic information, including their gender and age.

Cialdini's Principles Scenario. The second part presented participants with a scenario set within a social media group, where a member solicits assistance to install and evaluate a mobile app designed to enhance lifestyle choices. To ensure the scenarios were understood as intended and the perceived risks were clear, we conducted face validation with individuals familiar with Cialdini's principles from both Arab and UK backgrounds to test whether the map the principles to the scenarios and ensure they understood the scenarios, through 'think aloud' feedback, as we intended by their design. We then examined participants' responses to 12 scenarios to explore Cialdini's principles of persuasion, with each principle presented in both its presence and absence. After each scenario, participants were asked specific questions, allowing us to understand their reactions and decisions in relation to these persuasion principles.

We measured risk-taking by asking participants, *'In a similar scenario, how likely are you to install the app and give it a try?'*. Additionally, to assess their level of trust, we inquired, *'In a similar scenario, how much do you trust Oliver's [Majid's for the Arabs] transparency and intentions?'*. Both items were evaluated using a six-point Likert Scale. The scale for the first item spanned from "very unlikely" to 6 "very likely". For the second item, the scale ranged from 1 "Complete distrust" to 6 "Complete trust". Of note, participants were fully informed that installing the app entails security risks. The complete survey questions and documentation for this study and its dataset can be accessed on the following Open Science Framework (OSF) link: https://osf.io/vubty/

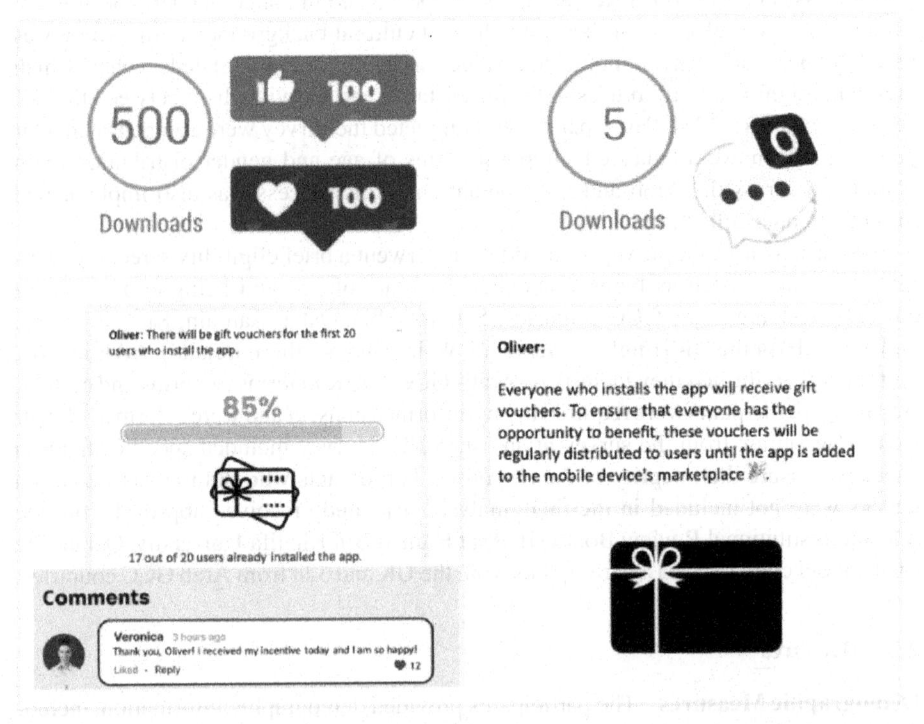

Fig. 1. The Presence (left) and Absence (right) of Social Proof (top) and Scarcity (bottom)

To assess understanding of the principles, visual representations were presented. Figure 1 illustrates four scenarios examining social proof and scarcity in Cialdini's principles. The top left shows high app engagement, suggesting social proof with many downloads and likes. In contrast, the top right shows fewer downloads, suggesting less social proof. The bottom left represents scarcity with a limited offer, promoting urgency, while the bottom right offers vouchers to all, eliminating scarcity.

Security Attitudes. The Security Attitudes (SA-6) scale used in our survey is a six-item measure designed to assess individual attitudes towards cybersecurity [24]. It was developed through a standardized process that included empirical research, ensuring it captures a range of responses with demonstrated reliability and validity. In the survey, participants were presented with a series of statements designed to measure their security attitudes and behaviors. Examples of these scale items include: '*I seek out opportunities to learn about security measures that are relevant to me*' and '*I am extremely motivated to take all the steps needed to keep my online data and accounts safe*'. Respondents rated their agreement on a five-point Likert scale from 1 "strongly disagree" to 5 "strongly agree". This scale correlates significantly with self-reported measures of security behavior intentions and actual secure behaviors, serving as an effective tool for quantifying and comparing people's attitudes toward using recommended security tools and practices. The original study reported a Cronbach's alpha of 0.84 for the SA-6 scale, demonstrating strong internal consistency. In our data, the Cronbach's alpha for the SA-6 scale was calculated to be 0.86 for the UK and 0.80 for the Arab, indicating a comparable level of internal consistency in measuring security attitudes in the original paper.

2.3 Data Analysis

The data pre-processing procedure involved removing duplicate entries and removed incomplete responses to maintain data integrity. Additionally, necessary variables were created, and responses to questions, such as risk taking (app installation) and level of trust were transformed into numerical format when needed, to facilitate subsequent analysis. Regarding the SA-6 scale, responses were aggregated into a single score by computing the mean of the six items. These tasks were performed using Microsoft Excel.

To investigate the influence of Cialdini's principles on trust and risk-taking rates across various cultural contexts, paired sample t-tests were conducted. This choice of statistical test is justified by the large sample sizes, which align with the Central Limit Theorem (CLT) [25]. This theorem suggests that when sample sizes are larger than 30, the means of the samples will closely resemble a normal distribution, making it appropriate to use parametric tests. The magnitude of these effects was quantified using Cohen's d values, providing a measure of effect size relevant for practical interpretation. However, since our data is from a 6-point Likert scale, we also performed the Wilcoxon signed-rank test, to ensure the reliability of our findings and the results were similar.

To mitigate the risk of Type I errors associated with multiple statistical comparisons, the Bonferroni correction was employed. This correction adjusts the p-value threshold by dividing the conventional alpha level (0.05) by the number of tests performed (six). Consequently, the revised significance threshold was established at 0.008.

Furthermore, Spearman's correlation coefficient was utilized to explore the associations between user security attitudes, trust, and risk-taking. This analysis was conducted for both UK and Arab participant samples. Data analysis was performed using JASP software, version 0.18.1 [26].

3 Results

3.1 Descriptive Statistics

The demographic characteristics of the participants are summarized in Table 1. Gender distribution varies between the two groups, with Arab participants consisting of a higher percentage of males (56.40%) compared to UK participants (39.49%), while the UK sample has a higher proportion of females (60.51%). The mean age for the Arab participants is 35.59 years with a range of 18–57 years, and for UK participants, the mean age is slightly higher at 37.71 years with a range of 18–59 years.

Table 1. Participant Demographics

Variables	*Participants (N = 642)*	
	Arab (N = 328)	*UK (N = 314)*
Gender (%)		
Male	187 (56.40)	124 (39.49)
Female	143 (43.60)	190 (60.51)
Age		
M (SD)	35.59 (10.08)	37.71 (12.16)
Range	18–57	18–59

3.2 Impact of Cialdini's Persuasion Principles on App Installation and Trust

Table 2 presents the paired-sample t-test results within the UK cultural context, assessing the influence of Cialdini's persuasion principles on app installation, which is indicative of risk-taking behavior. The analysis revealed that the presence of Social Proof significantly increased the likelihood of app installation ($t(313) = 23.91$, $p < .001$), with a substantial effect size (Cohen's $d = 1.35$). Similarly, the principles of Likeability ($t(313) = 15.03$, $p < .001$, Cohen's $d = 0.85$) and Authority ($t(313) = 21.42$, $p < .001$, Cohen's $d = 1.21$) were found to be significantly related to higher rates of app installation, indicating their strong influence on participants' decision-making.

The results further indicated that the Commitment principle significantly increased the likelihood of app installation with a moderate effect size ($t(313) = 8.19$, $p < .001$, Cohen's $d = 0.46$). Similarly, Reciprocity had a significant effect ($t(313) = 18.67$, $p < .001$, Cohen's $d = 1.05$). However, the Scarcity principle, while yielding a statistically significant result, produced the smallest effect size ($t(313) = 2.96$, $p = 0.002$, Cohen's $d = 0.17$) among the principles tested.

Within the same UK sample, the influence of the principles on trust was examined, as detailed in Table 3. Analysis revealed that the presence of social proof principle notably increased trust levels ($t(313) = 19.62$, $p < .001$, Cohen's $d = 1.11$). The principles

Table 2. Impact of Cialdini's Principles on Risk Taking in the UK Sample (N = 314)

Cialdini principles	Presence		Absence		t(313)	p	Cohen's d
	Mean	SD	Mean	SD			
Social Proof	3.92	1.29	2.15	1.15	23.91	<.001	1.35
Likeability	3.50	1.37	2.43	1.16	15.03	<.001	0.85
Authority	3.71	1.45	2.13	1.10	21.42	<.001	1.21
Commitment	3.33	1.30	2.85	1.23	8.19	<.001	0.46
Reciprocity	3.57	1.33	2.27	1.12	18.67	<.001	1.05
Scarcity	3.56	1.50	3.40	1.48	2.96	0.002	0.17

of Likeability ($t(313) = 10.83$, $p < .001$, Cohen's $d = 0.61$) and Authority ($t(313) = 20.49$, $p < .001$, Cohen's $d = 1.16$) were also found to have a positive effect on trust. Meanwhile, Commitment ($t(313) = 5.32$, $p < .001$, Cohen's $d = 0.30$) and Scarcity ($t(313) = 2.76$, $p = 0.003$, Cohen's $d = 0.16$) showed statistically significant result.

Table 3. Impact of Cialdini's Principles on Trust in the UK Sample (N = 314)

Cialdini principles	Presence		Absence		t(313)	p	Cohen's d
	Mean	SD	Mean	SD			
Social Proof	3.84	1.00	2.55	1.08	19.62	<.001	1.11
Likeability	3.41	1.08	2.65	1.04	10.83	<.001	0.61
Authority	3.80	1.16	2.39	1.07	20.49	<.001	1.16
Commitment	3.44	0.97	3.20	1.04	5.32	<.001	0.30
Reciprocity	3.67	1.04	2.49	1.03	19.79	<.001	1.12
Scarcity	3.46	1.15	3.32	1.11	2.76	0.003	0.16

In the Arab sample, as detailed in Table 4, a paired-sample t-test was conducted to evaluate the effect of Cialdini's persuasion principles on app installation. Significant impacts were observed across all principles tested. Social Proof ($t(327) = 23.60$, $p < .001$, Cohen's $d = 1.30$) and Authority ($t(327) = 19.84$, $p < .001$, Cohen's $d = 1.10$) stood out as having particularly strong influences. The principles of Likeability ($t(327) = 17.61$, $p < .001$, Cohen's $d = 0.96$) and Reciprocity ($t(327) = 18.52$, $p < .001$, Cohen's $d = 1.02$) also produced notable effects on the decision to install the app. The principles of Commitment ($t(327) = 7.94$, $p < .001$, Cohen's $d = 0.44$) and Scarcity ($t(327) = 2.52$, $p = .006$, Cohen's $d = 0.14$), while yielding significant results, had relatively smaller effect sizes.

Table 5 presents the results from an analysis within an Arab context, examining the impact of Cialdini's persuasion principles on trust. The findings indicated that presence of social proof significantly enhanced trust levels ($t(327) = 20.57$, $p < .001$, Cohen's

Table 4. Impact of Cialdini's Principles on Risk Taking in the Arab Sample (N = 328)

Cialdini principles	Presence		Absence		t(327)	p	Cohen's d
	Mean	SD	Mean	SD			
Social Proof	4.76	1.13	2.72	1.40	23.60	<.001	1.30
Likeability	4.45	1.21	3.20	1.28	17.61	<.001	0.96
Authority	4.52	1.30	2.77	1.38	19.84	<.001	1.10
Commitment	4.23	1.15	3.79	1.23	7.94	<.001	0.44
Reciprocity	4.31	1.22	2.79	1.36	18.52	<.001	1.02
Scarcity	4.30	1.37	4.15	1.33	2.52	0.006	0.14

d = 1.14). Similarly, Authority (t(327) = 21.87, $p < .001$, Cohen's d = 1.15) and Reciprocity (t(327) = 18.15, $p < .001$, Cohen's d = 1.00) also positively influenced trust. The principle of Likeability showed a moderate, yet significant, effect on trust (t(327) = 10.13, $p < .001$, Cohen's d = 0.56). Furthermore, Commitment (t(327) = 5.68, $p < .001$, Cohen's d = 0.31) and Scarcity (t(327) = 2.90, p = .002, Cohen's d = 0.16), while producing significant results, had relatively smaller effects on trust.

The results from the paired sample t-test were similar with those obtained from the Wilcoxon signed-rank test, thereby supporting the robustness of our findings. For the Wilcoxon test results, please refer to the supplementary materials at OSF link provided in Sect. 2.2.

Table 5. Impact of Cialdini's Principles on Trust in the Arab Sample (N = 328)

Cialdini principles	Presence		Absence		t(327)	p	Cohen's d
	Mean	SD	Mean	SD			
Social Proof	4.30	1.08	2.70	1.24	20.57	<.001	1.14
Likeability	3.85	1.16	3.11	1.14	10.13	<.001	0.56
Authority	4.26	1.18	2.70	1.21	20.87	<.001	1.15
Commitment	3.94	1.09	3.63	1.08	5.68	<.001	0.31
Reciprocity	4.06	1.12	2.72	1.22	18.15	<.001	1.00
Scarcity	3.86	1.17	3.70	1.17	2.90	0.002	0.16

3.3 Security Attitude Vs. Risk Taking and Trust Under Cialdini's Principles

Table 6 displays Spearman's correlation coefficients that detail the associations between trust levels, security attitudes, and the likelihood of app installation in relation to the presence of Cialdini's persuasion principles, as observed in UK and Arab population samples.

Table 6. Spearman's Correlations of Security Attitude (SA-6) with Trust and App Installation in each Cialdini's scenario in UK and Arab Samples

Cialdini's Principle	UK		Arab	
	Trust	Risk Taking	Trust	Risk Taking
Social Proof	0.10	0.10	0.11*	0.10
Likeability	0.14*	0.18**	0.10	0.15**
Authority	0.14*	0.15**	0.09	0.13*
Commitment Consistency	0.09	0.14*	0.01	0.03
Reciprocity	0.07	0.13*	0.10	0.10
Scarcity	0.18**	0.11*	0.07	0.11*

Significance levels: $^{*}p < .05$, $^{**}p < .01$.

In the Arab sample, Social Proof exhibited a significant but weak correlation with trust (r = 0.11), suggesting a slight influence on trust. The Likability principle in the UK context showed a significant positive correlation with both trust (r = 0.14) and app installation risk (r = 0.18), suggesting that positive security attitudes may lead to increased trust and likelihood of app installation. In the Arab sample, Likability showed a significant correlation with app installation risk (r = 0.15).

For Authority, the UK sample revealed significant positive correlations with both trust (r = 0.14) and app installation risk (r = 0.15), suggesting a tendency to trust and take risks with apps endorsed by authorities. In the Arab sample, Authority had a correlation with app installation risk (r = 0.13), highlighting its role in influencing risk-taking behavior.

Commitment Consistency in the UK sample correlated significantly with app installation risk (r = 0.14), suggesting its influence on risk-taking. Reciprocity had significant correlation with app installation risk (r = 0.13), indicating a potential influence on risk-taking behavior.

Lastly, Scarcity in the UK sample had significant correlations with trust (r = 0.18) and app installation risk (r = 0.11), indicating its influence on both trust and risk-taking behavior. In the Arab sample, Scarcity showed a significant correlation with app installation risk (r = 0.11), emphasizing its role in influencing risk-taking behavior.

4 Discussion

The study confirmed the impact of Cialdini's persuasion principles on decision-making regarding risk-taking and trust, even when the participants are aware of potential cybersecurity risks. This underscores the enduring effect of these principles in shaping individual's behavior in risk-taking scenarios [1]. Addressing our first research question, this finding explains how Cialdini's principles, notably the principles of social proof and authority, exert a significant influence on trust and risk-taking decisions, even in situations where security risks are knowingly present. However, we must consider the potential gap between participants' responses to hypothetical scenarios and their actual behavior in real-world situations, i.e. the intention-behavior gap [27]. The responses might reflect aspirational rather than genuine reactions, particularly under the unexpected pressure of actual phishing attempts [28]. Additionally, cultural norms might lead participants to underreport susceptibility to tactics like Scarcity due to the stigma associated with greed or materialism [29], suggesting a need for caution in interpreting these results. This study provides insights into how Cialdini's principles influence trust and risk-taking across different cultures, with a focus on the UK and Arab populations. The majority of existing research in this area has focused primarily on the psychological and behavioral aspects of potential victims of social engineering, with less emphasis on how specific principles adopted by attackers influence their decision-making and trust-building processes.

Our research addresses this gap by exploring how each principle influences individual behavior and comparing their effects in two different cultural contexts. Our second research question delves into how security awareness influences risk-taking in the face of known threats. The findings reveal that security knowledge doesn't necessarily lead to cautious actions against persuasive tactics, underscoring the complexity of human decisions in cybersecurity. This underscores the need for strategies that not only raise awareness but also enhance individuals' ability to critically evaluate and resist persuasive threats. This phenomenon has also been observed in other decision-making domains where harm is a factor, such as gambling [8]. In our study, we investigated how each of Cialdini's principles influenced the likelihood of risk-taking and the development of trust, and examined existing differences through statistical analysis. In particular, the principles of social proof and authority appeared to be significantly influential in both the UK and Arab contexts. This aligns with the expectations from the literature, which suggests that these principles have a universal appeal in influencing human behavior [4, 5]. Our results, however, provide a nuanced perspective on their application, as the effect size for authority was slightly more pronounced in the Arab context, potentially reflecting cultural differences in attitudes towards hierarchy and expertise. In contrast, scarcity, although statistically significant, showed the smallest effect size in both cultural contexts. This suggests that scarcity has a more modest influence, possibly surpassed by the stronger effects of authority and social proof, especially in cyber-security attacks scenarios.

According to our analysis, there is a positive relationship between security attitude, trust, and risk-taking behaviors in certain scenarios. This finding may seem unexpected at first glance as a strong security mindset is commonly linked with being cautious, rather than taking risks. However, this can be clarified by examining the underlying factors that influence security attitude which is often measured by an individual's tendency to the authorities and experts' advice. This tendency to value expert guidance, while beneficial, can also make individuals more susceptible to authority. Workman's research [5] highlights the influence of authority in shaping individual responses to security threats. Our findings reflect this view that even individuals with strong security attitudes may fall target to persuasive techniques that mimic authority. This positive correlation between security attitudes and susceptibility to authority highlights the complexity of human factors in cybersecurity. The claim made by Workman that social proof can result in risky behavior finds similarities in our study. We observed that social proof can significantly influence an individual's trust and decision-making when it comes to taking risks.

In their research, Grassegger and Nedbal [9] explore the importance of information security awareness in strengthening employee resistance to social engineering. Their findings reveal a weak correlation between leadership and security awareness, as well as a lack of significant association between policies, training measures, and actual security awareness. These findings are significant to our research as they suggest that traditional security measures and the promotion of security attitudes, while undoubtedly important, may not be sufficient. Our study suggests that maintaining security awareness without understanding persuasive tactics may associate with increased vulnerability. This correlation emphasizes the need for a more sophisticated approach to security education - one that promotes awareness and provides individuals with the skills to critically assess and respond to persuasive scenarios of social engineering attacks.

In the Arab dataset, the significant correlations for likeability and authority reflect the cultural nuances in decision-making processes discussed by Yates and de Oliveira [17]. Their study highlights how cultural norms and values, such as the collectivist nature, influence the decision-making process. This cultural tendency towards collectivism and respect for authority can lead individuals to place greater value on likeability and authority in their decision-making processes. This perspective highlights why certain principles are more effective in different cultural settings. This underlines the importance of understanding how cultural factors intertwine with security attitudes in shaping technology adoption behaviors, a concept also high-lighted in studies like those by Oyibo et al. [13], where cultural context significantly shapes the effectiveness of persuasion techniques. They found distinct patterns of susceptibility to persuasive strategies in Nigerian and Canadian contexts. Their study indicated a higher influence of authority and scarcity on Nigerians, while Canadians showed a greater response to reciprocity and consensus. These findings point to the need for culturally tailored cybersecurity strategies, reflecting the diverse ways different cultures respond to persuasion techniques. It also suggests expanding research beyond WEIRD contexts, as previously discussed.

The results of our study show the influential role of Cialdini's principles of trust and risk-taking in cybersecurity and underscore the need for innovative strategies to counter social engineering threats. This aligns with the literature which highlights the complexity of human factors in cybersecurity, such as the studies on psychological aspects of social engineering [4, 5]. Additionally, the importance of debiasing suggests that cybersecurity strategies need to train individuals to identify and resist manipulation tactics. This approach is similar to the Inoculation Theory [30], which advocates for 'vaccinating' people against persuasion through awareness and training. Furthermore, the diverse responses to persuasion in different cultural contexts [13, 14] underscore the need for culturally sensitive cybersecurity measures. Our findings contribute to this do-main by highlighting the effectiveness of specific persuasion strategies in different cultures, thereby supporting the development of targeted and culturally-informed cybersecurity defenses.

While this study offers significant insights into persuasion principles within Arab and UK contexts, it is important to acknowledge its limitations and the measures taken to mitigate them. The focus on these specific cultural groups, while providing a comprehensive comparison, might not fully capture the diversity of global contexts. To counterbalance this, we selected these groups to represent a broad spectrum of cultural dynamics. Additionally, we recognize the potential for fatigue effects due to the survey's length. To reduce potential fatigue, we limited the questionnaire to just two questions following each of the 12 scenarios. To minimize habitual responses and learning effects, we randomized the order of the 12 scenarios, presenting them in a different sequence to each participant. The online survey was chosen because it has the capability to reach a wider audience, and it was created with reliable, face-validated scenarios. However, we understand that this method may not capture all the complexities of real-life interactions. Therefore, future research should consider expanding the cross-cultural perspective and using a range of methods, such as qualitative interviews, to gain a more thorough and nuanced understanding of how persuasion works in the field of cybersecurity.

This study adds to the body of research on human factors in cybersecurity, particularly in understanding how universal psychological principles can be exploited in social engineering. By bridging research gaps identified in previous studies, these findings emphasize the necessity of integrating psychological principles into cybersecurity defenses, especially considering the limited exploration of these principles in high-risk scenarios.

5 Conclusion

This study's investigation into Cialdini's persuasion principles within cybersecurity, specifically within the UK and Arab GCC cultural landscapes, reveals novel insights into how these psychological influence principles are nuanced by cultural contexts, affecting trust and risk-taking in cybersecurity scenarios. It identifies the need for cybersecurity strategies that are not only technically sound but also culturally attuned and grounded in the principles of social psychology. This study not only reaffirms the significant influence of Cialdini's principles on human behavior but also highlights the study's importance in advocating for an integration of psychological literacy into cybersecurity practices. It highlights a critical gap in conventional cybersecurity education, particularly in addressing psychological manipulation tactics, emphasizing the limited exploration of these principles in high-risk scenarios and across diverse cultural settings. The research points to the importance of understanding security attitudes and their impact on decision-making in cybersecurity. As cyber threats continue to evolve, the intersection of human psychology, cultural context, and cybersecurity risk becomes increasingly critical. This study's findings underscore the necessity of innovative strategies to counter social engineering threats, enriching the cybersecurity domain by bridging research gaps identified in previous studies. Future research should extend to other cultural contexts, exploring the relationship between personality, culture, and susceptibility to persuasion, and how security attitudes influence these dynamics in cybersecurity. Such investigations are crucial for developing targeted and culturally informed cybersecurity defenses, moving beyond conventional approaches to counter the sophisticated tactics employed in social engineering. This approach will foster more holistic strategies, effectively countering the nuanced threats posed by social engineering. By integrating psychological principles into cybersecurity education and practice, we can enhance the resilience of individuals and organizations against the evolving landscape of cyber threats.

Acknowledgement. This publication was supported by NPRP 14 Cluster grant # NPRP 14C-0916-210015 from the Qatar National Research Fund (a member of Qatar Foundation). The findings herein reflect the work and are solely the responsibility of the authors.

References

1. Mahmoud, S., Alez, R.A., EL-Refai, F.: Persuasion based recommendation system. J. Al-Azhar Univ. Eng. Sect. **12**(44), 894–899 (2017)
2. Lohani, S.: Social engineering: hacking into humans. Int. J. Adv. Stud. Sci. Res. **4**(1), 385–393 (2019)
3. Thornburgh, T.: Social engineering: the dark art. In: Proceedings of the 1st Annual Conference on Information Security Curriculum Development, pp. 133–135 (2004)
4. Peltier, T.R.: Social engineering: concepts and solutions. Inf. Secur. J. **15**(5), 13 (2006)
5. Workman, M.: A test of interventions for security threats from social engineering. Inf. Manag. Comput. Secur. **16**(5), 463–483 (2008)
6. Cialdini, R.B.: Influence: The psychology of persuasion. Collins, New York (2007)
7. Zalake, M., De Siqueira, A.G., Vaddiparti, K., Antonenko, P., Lok, B.: Towards understanding how virtual human's verbal persuasion strategies influence user intentions to perform health behavior. In: Proceedings of the 21st ACM International Conference on Intelligent Virtual Agents, pp. 216–223 (2021)
8. Cemiloglu, D., Gurgun, S., Arden-Close, E., Jiang, N., Ali, R.: Explainability as a psychological inoculation: building resistance to digital persuasion in online gambling through explainable interfaces. Int. J. Human–Comput. Interact. 1–19 (2023)
9. Grassegger, T., Nedbal, D.: The role of employees' information security awareness on the intention to resist social engineering. Procedia Comput. Sci. **181**, 59–66 (2021)
10. Friestad, M., Wright, P.: The persuasion knowledge model: how people cope with persuasion attempts. J. Consum. Res. **21**(1), 1–31 (1994)
11. Lawson, P., Pearson, C.J., Crowson, A., Mayhorn, C.B.: Email phishing and signal detection: How persuasion principles and personality influence response patterns and accuracy. Appl. Ergon. **86**, 103084 (2020)
12. Spasova, L.: Impact of gender and age on susceptibility to persuasion principles in advertisement. Econ. Sociol. **15**(3), 89–107 (2022)
13. Oyibo, K., Adaji, I., Orji, R., Olabenjo, B., Vassileva, J.: Susceptibility to persuasive strategies: a comparative analysis of Nigerians vs. Canadians. In: Proceedings of the 26th Conference on User Modeling, Adaptation and Personalization, pp. 229–238 (2018)
14. Obeidat, B.Y., Shannak, R.O., Masa'deh, R., Al-Jarrah, I.: Toward better understanding for Arabian culture: implications based on Hofstede's cultural model. Eur. J. Social Sci. **28**(4), 512–522 (2012)
15. Alnunu, M., Amin, A., Abu-Rayya, H.M.: The susceptibility to Persuasion strategies among Arab Muslims: the role of culture and acculturation. Front. Psychol. **12**, 574115 (2021)
16. Pinnington, A.H., Rees-Caldwell, K.: National Culture Differences in Project Management: Comparing British and Arab Project Managers Managers' Perceptions of Different Planning Areas (2013)
17. Yates, J.F., De Oliveira, S.: Culture and decision making. Organ. Behav. Hum. Decis. Process. **136**, 106–118 (2016)
18. Henrich, J., Heine, S.J., Norenzayan, A.: The weirdest people in the world? Behav. Brain Sci. **33**(2–3), 61–83 (2010)
19. Country comparison graphs country comparison graphs. (in en-GB). Geert Hofstede. https://www.hofstede-insights.com/country-comparison-tool. Accessed 16 Mar 2024

20. Harb, C.: The Arab region: cultures, values, and identities. In: Handbook of Arab American Psychology, pp. 3–18. Routledge (2015)
21. Waclawski, E.: How I use it: survey monkey. Occup. Med. **62**(6), 477 (2012)
22. Thien, P.: MOBILE Panel Sample and ONLINE Surveys TGM Research. TGM Research, https://tgmresearch.com/. Accessed 16 Mar 2024
23. Hofstede, G.: Culture's Consequences: International Differences in Work-Related Values. Sage, Thousands Oaks (1984)
24. Faklaris, C., Dabbish, L.A., Hong, J.I.: A {self-report} measure of {end-user} security attitudes ({{{{{SA-6}}}}}). In: Fifteenth Symposium on Usable Privacy and Security (SOUPS 2019), pp. 61–77 (2019)
25. Kwak, S.G., Kim, J.H.: Central limit theorem: the cornerstone of modern statistics. Korean J. Anesthesiol. **70**(2), 144 (2017)
26. JASP - A Fresh Way to Do Statistics. JASP - Free and User-Friendly Statistical Software. https://jasp-stats.org/. Accessed 16 Mar 2024
27. Jenkins, J.L., Durcikova, A., Nunamaker, J.: Mitigating the security intention-behavior gap: the moderating role of required effort on the intention-behavior relationship. Association for Information Systems (2021)
28. Gerdenitsch, C., Wurhofer, D., Tscheligi, M.: Working conditions and cybersecurity: time pressure, autonomy and threat appraisal shaping employees' security behavior. Cyberpsychol. J. Psychosocial Res. Cyberspace **17**(4) (2023)
29. Jonas, E., Sullivan, D., Greenberg, J.: Generosity, greed, norms, and death–differential effects of mortality salience on charitable behavior. J. Econ. Psychol. **35**, 47–57 (2013)
30. McGuire, W.J.: The effectiveness of supportive and refutational defenses in immunizing and restoring beliefs against persuasion. Sociometry **24**(2), 184–197 (1961)

Classifying Healthcare and Social Organizations in Cybersecurity Profiles

Steve Ahouanmenou[1(✉)], Amy Van Looy[1], Geert Poels[1,3], Petra Andries[2], and Thomas Standaert[2]

[1] Faculty of Economics and Business Administration, Department of Business Informatics and Operations Management, Ghent University, Ghent, Belgium
Steve.ahouanmenou@ugent.be

[2] Faculty of Economics and Business Administration, Department of Marketing, Innovation and Organization, Ghent University, Ghent, Belgium

[3] CVAMO Core Lab, Flanders Make @UGent, Ghent, Belgium

Abstract. While cybersecurity is of high relevance for all organizations, special care is needed in the healthcare and social realm when coping with sensitive patient data. This study contributes to this under-investigated yet relevant field by examining how cybersecurity measures have been implemented within healthcare and social organizations. We rely on a combination of clustering analysis, discriminant analysis, and Tukey HSD testing to analyze survey data on 265 organizations in Flanders, Belgium. The resulting five clusters unveil five distinct approaches or organizational profiles and three major differentiators. The data suggests that the extent to which training, regular software updates, and data backup are implemented best describes the underlying cybersecurity profiles. Our findings reveal that a significant majority of surveyed organizations are situated in the lower echelons of the cybersecurity implementation differentiators, while only a minority of organizations demonstrate commendable levels of implementation. By enriching cybersecurity insights within the healthcare and social domain, our findings and their implications could resonate deeply, urging researchers to expand their research to bolster cyber resilience in specific sectors.

Keywords: Cybersecurity · Healthcare · Social

1 Introduction

Cybersecurity incidents are on the rise in healthcare and social organizations, making cybersecurity a growing concern for senior executives, practitioners, and academics worldwide [1]. For instance, in March 2023, a university hospital in Brussels became one of the latest organizations targeted in the surge of cyberattacks against European hospitals [2]. Also, the vulnerability of healthcare organizations has drastically expanded due to the substantial number of connected medical devices and the proliferation of the Internet of Medical Things [3]. Hence, the need for protecting patient data from immediate cybersecurity risks has substantially increased with cloud adoption [4].

J. Araújo et al. (Eds.): RCIS 2024, LNBIP 513, pp. 289–304, 2024.
https://doi.org/10.1007/978-3-031-59465-6_18

Meanwhile, several studies have focused on understanding why and how cybercriminals attack, and what must be done to bolster a healthcare organization's defence [5]. For instance, Coronado and Wong [6] have looked at the cybersecurity risk management aspect by providing an effective strategy for healthcare organizations. Alternatively, Busdicker and Upendra [7] have provided a technology and medical device angle, while also formulating a useful set of good security practices that are specifically applied to healthcare technology. Other authors such as Tervoort et al. [8] have investigated the nature of the cyberattacks applied to the healthcare sector to apply customed solutions and embed security in the technology by design.

However, to our knowledge, no study which identified and characterized cybersecurity profiles of healthcare and social organizations based on their implementation of cybersecurity measures was found. Our objective is to identify key differentiators which could be leveraged in defining organizations' profiles. Nevertheless, although comprehensive cybersecurity frameworks and information security best practices are available, such as the NIST Framework [9], and ISO 27001 [10], healthcare and social organizations would benefit from more insight into profiles concerning cybersecurity, due to their unique management structure with limited business orientation, their collection of large sensitive information and the limited cybersecurity resources at their disposal. Moreover, the overabundance of data linked to a lack of comprehensive cybersecurity guidance [11] compromises daily the safeguard of sensitive information, such as patient data or social security information in a sector characterized by a crucial digital transformation journey [12].

To address this gap, we will focus on classifying healthcare and social organizations alongside the implementation of cybersecurity measures (RQ1), while also deriving differentiators to determine the cybersecurity profiles of those organizations (RQ2). These differentiators will help us identify tangible cybersecurity profiles representing different existing cybersecurity approaches. The research questions are as follows:

- RQ1. Which groups of healthcare and social organizations exist based on their actual implementation of cybersecurity measures?
- RQ2. What are the differences between these groups in terms of the implementation of cybersecurity measures?

This research relied on a quantitative research design to arise with various types of organizations based on their cybersecurity set of practices (RQ1) and the related differences (RQ2). The statistical findings will be derived based on a combination of cluster analysis and discriminant analysis (RQ1), followed by an analysis of variance (including Tukey Honest Significant Difference) to discover substantial differences among the statistical clusters (RQ2). Our scientific contribution resides in proposing a schema or framework that describes various cybersecurity profiles of healthcare and social organizations. From a practical viewpoint, managers can use this framework to see which cybersecurity profile their organization matches best for.

We proceed as follows. Section 2 provides the background for our analysis by defining the main terminology of the paper and a review of prior studies. The methodology is described in Sect. 3, and we present the results for organizations' profiles in Sect. 4. Our contributions are discussed in Sect. 5, and we conclude in Sect. 6.

2 Background

We first define the key terms, before elaborating on prior studies and cybersecurity measures.

2.1 Definitions

A cybersecurity measure refers to any action, process, or technology implemented to safeguard computer systems, networks, and data from unauthorized access, attacks, damage, or theft. In this paper, a cybersecurity measure refers to a mechanism, a procedure, an abstract metric or a somewhat subjective attribute, such as how well an organization's systems are secured against external threats, or how effective the organization's incident response team is [13]. Examples are the percentage of fully patched systems within an organizational framework, the temporal gap between patch release and system implementation, or the extent of system access potentially enabled by a vulnerability within the system [14].

A cybersecurity profile can be defined as a description of a set of organizations grouped based on similar characteristics or attributes [15]. In our study, the cybersecurity profiles are formed based on the implementation of key cybersecurity measures that we have beforehand identified. Hence, a profile is determined according to the implementation of cybersecurity measures within healthcare and social organizations.

The differentiators of the profiles are the selected cybersecurity measures that play a significant role in the classification of healthcare and social organizations. Thus, the differentiators constitute the main characteristics to describe cybersecurity profiles, according to the comparison of the variations of the level of implementation of these specific measures in organizations. In other words, differentiators are the dimensions or factors [16] for distinguishing healthcare and social organizations profiles.

2.2 Prior Studies

In addition to previous references on the literature review of information security and privacy in the healthcare sector [17], the body of knowledge comprises some studies related to the classification of cybersecurity measures, albeit not our sector under study. For instance, Aman and Al Shukaili [18] have identified essential cybersecurity factors that are necessary in the public sector to implement a robust cybersecurity strategy. Alternatively, Atoum et al. [19] have focused on defining a classification scheme for organizations, to allow them to select cybersecurity standards tailored to their needs. Next, papers exist that relate to the cybersecurity profiles but of individuals. For instance, Nieto and Rios [20] have proposed a methodology to assess the cybersecurity profiles of individuals based on their use of IOT devices, while Soumelidou and Tsohou [21] have provided controls for the creation of cybersecurity profiles of individuals. Also, Zamfirescu et al. [22] have suggested four cybersecurity profiles based on the digital activity of the individuals.

A cybersecurity vulnerability profile for healthcare organizations, not individuals, was presented in (Majkowski & Feldman, n.d.). However, this profile serves another goal, which is providing a framework for decreasing cybersecurity vulnerabilities. On a broader scale, the European Union Agency for Cybersecurity (ENISA) has provided several practical guidance. For instance, ENISA has made available a tool to assess the level of residual risk associated to a processing activity based on the implementation of related cybersecurity measures [24]. ENISA has also made public an assessment tool for Small and Medium Businesses to enhance their overall cybersecurity maturity level [25]. These tools have different objectives than profiling healthcare organizations based on which cybersecurity measures they have implemented, which is the goal of our study.

2.3 Cybersecurity Measures

We use the extent of implementation of cybersecurity measures to distinguish between healthcare and social organizations with different cybersecurity profiles. The selection of cybersecurity measures in Table 1 relates to the minimal security requirements of the US National Institute for Science and Technology, which has specified several security-related areas regarding protecting the confidentiality, integrity, and availability of federal information systems, as well as the information processed, stored, and transmitted by those systems [26].

3 Methodology

3.1 Data Collection

Our statistical analysis relied on primary survey data collected in 2023 in Flanders, namely the Dutch-speaking region of Belgium. The survey questionnaire's goal was to better understand the participants' cybersecurity posture. We contacted several organizations in the healthcare and social sectors. After three reminders and intensive telephone follow-up, we obtained 265 responses. Removing observations with missing values on those variables used in our analysis leads to a final sample of 178 organizations.

Table 2 shows how we inquired about the three security measures included in Table 1. All variables were binary (yes/no) in nature. Statistical analyses were conducted using SPSS (version 29).

To execute meaningful comparisons concerning the distance from the mean value, we first transformed the observations into standardized scores (z-scores) [27].

Table 1. Overview of cybersecurity measures used in the study.

1/Training	10/Audit
2/Password enforcement	11/Security testing
3/Regular software update	12/Procedure of risk identification
4/Use of biometric techniques	13/Procedure of risk detection
5/Use of encryption techniques	14/Procedure of asset protection
6/Data backup	15/Procedure of incident response
7/Access Management	16/Recovery plan
8/Use of VPN	17/Cybersecurity insurance
9/Log review	18/Cybersecurity expertise

3.2 Classification Approach for RQ1

For RQ1, we classified organizations along the 18 cybersecurity variables using cluster analysis and a discriminant analysis. Both statistical techniques are usually combined for exploratory and confirmatory classification, respectively. Cluster analysis comprises a range of methods for classifying multivariate data into subgroups [28], and so revealing the characteristics of any structure or patterns present. These techniques have proven useful in a wide range of areas, such as medicine, psychology, market research and bioinformatics [29].

Discriminant analysis seeks to determine the factors that determine group separation [30]. Hence, we first identified groups based on a cluster analysis and then validated our selection with a discriminant analysis.

3.3 ANOVA Approach for RQ2

For RQ2, we examined whether the clusters of RQ1 showed any statistically significant difference to observe the trends of variables against clusters. We inspected which clusters specifically contrasted from each other per cybersecurity measure. When an analysis of variance (ANOVA) gave a significant result, we obtained an indication that at least one group differed from the other groups. Yet, the test did not inform us of the pattern of differences between the means.

To further analyze this pattern of difference between means, the ANOVA test was followed by the Tukey Honest Significant Difference (HSD) test [31], mainly because this is the most used test involving the comparison of means [32].

Table 2. Operationalization of the variables related to cybersecurity measures.

	Survey Question: ***"In the past 12 months, did your organization take the following cybersecurity measures?"***	
	Variable	**Description**
Technical Measures	1/ Security testing	ICT security testing (performing penetration testing, security alarm system testing, reviewing security measures, testing backup systems)
	2/ Password enforcement	Strong password authentication (e.g., minimum length, special characters, regular modification, multi-factor authentication)
	3/ Regular software update	Regular software updates
	4/ Use of biometric techniques	User identification and authentication via biometric techniques (e.g., fingerprint, voice, or facial recognition)
	5/ Use of encryption techniques	Encryption techniques for data, documents, or emails
	6/ Data backup	Data backup to a separate location or in the cloud
	7/ Access Management	Enterprise network access management (for devices or users)
	8/ Use of VPN	VPN (Virtual Private Network) to ensure secure data transmission of data
	9/ Log review	Maintain log files to analyze cyber attacks
	10/ Audit	ICT security analysis (periodic audit of vulnerability in the face of cyber-attacks)
Administrative (Non - Technical Measures)	11/ Training	Training/activities to make employees aware of the importance of cybersecurity
	12/ Procedure of risk identification	Procedures to identify potential security risks within the enterprise (e.g., documenting sensitive data sources, critical enterprise processes)
	13/ Procedure of risk detection	Procedures to detect cyber-attacks (continuous monitoring of security risks, techniques, and protocols to detect security attacks)
	14/ Procedure of asset protection	Procedures to protect the enterprise from cyber-attacks (e.g., access management, identification management, backups, encryption, updates)
	15/ Procedure of incident response	Procedures to respond to cyber-attacks (incident analysis, threat elimination, crisis communication)
	16/ Recovery plan	Procedures to recover from cyber-attacks (restore backups, reinstall systems, change passwords, change firewall)
	17/ Cybersecurity insurance	Does your company have insurance against cyber-attacks?
	18/ Cybersecurity expertise	Your own staff (including the staff of the parent company or subsidiaries) performs ICT security related activities

4 Results

4.1 Classification of Healthcare and Social Organizations (RQ1)

Exploratory Classification by Cluster Analysis

We used trial and error to choose the algorithmic method and number of clusters that best fit our data.

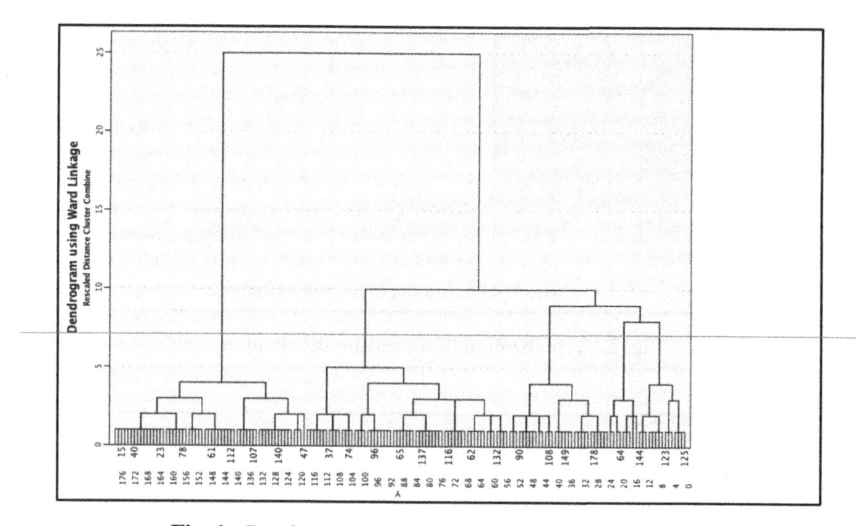

Fig. 1. Dendrogram of the hierarchal clustering method.

We first used z-scores to normalize the variable values before examining the clustering alternatives. We applied K-means, which uses the median or mean as a cluster center to represent each cluster. K-means is appropriate to find mutually exclusive clusters of spherical shape based on distance [33]. We also considered the hierarchical algorithm (Fig. 1), which begins with 'n' clusters and sequentially combines similar clusters until only one cluster is obtained [28]. Hierarchical solutions can be either disruptive or cumulative.

Nevertheless, both K-means and the hierarchical average between-groups linkage clustering methods suggested five clusters (or cybersecurity measures profiles). We eventually opted for the K-means solution given some hierarchical clustering limitations (e.g., poor scaling in both memory and computing time for large data sizes, and not always generating clear and obvious clusters) [34]. The five resulting clusters respectively covered 91, 65, 11, 9 and 2 cases (Fig. 2).

Furthermore, visualizing cluster centers is crucial in a clustering exercise for obtaining interpretable insights.

It aids in understanding each cluster's characteristics, validating results, and communicating findings effectively. Such a visual inspection also helps validate clustering quality and highlights the importance of features in defining clusters. Additionally, it assists in outlier detection, optimization assessment, and comparing different clustering algorithms. Figure 3 shows the distribution visualization of the healthcare and

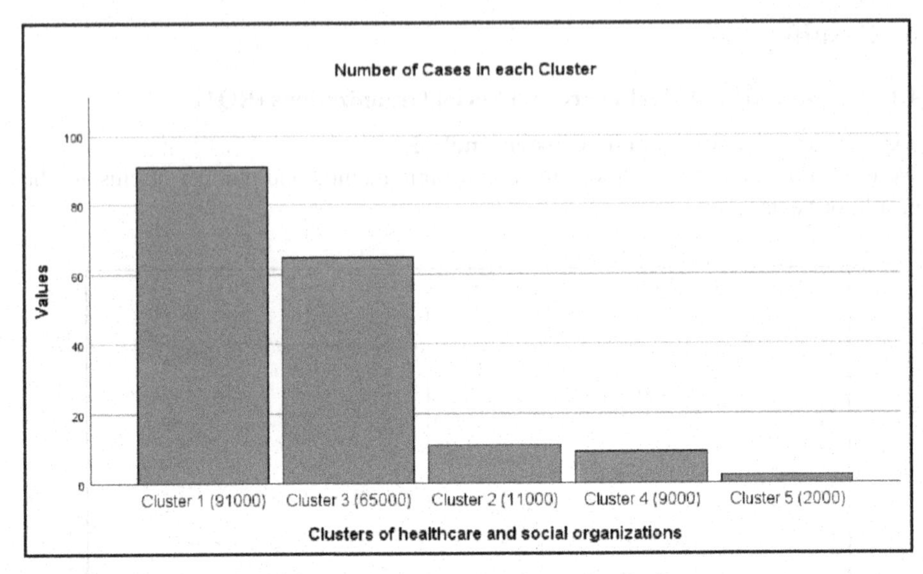

Fig. 2. Visualization of the cluster distribution.

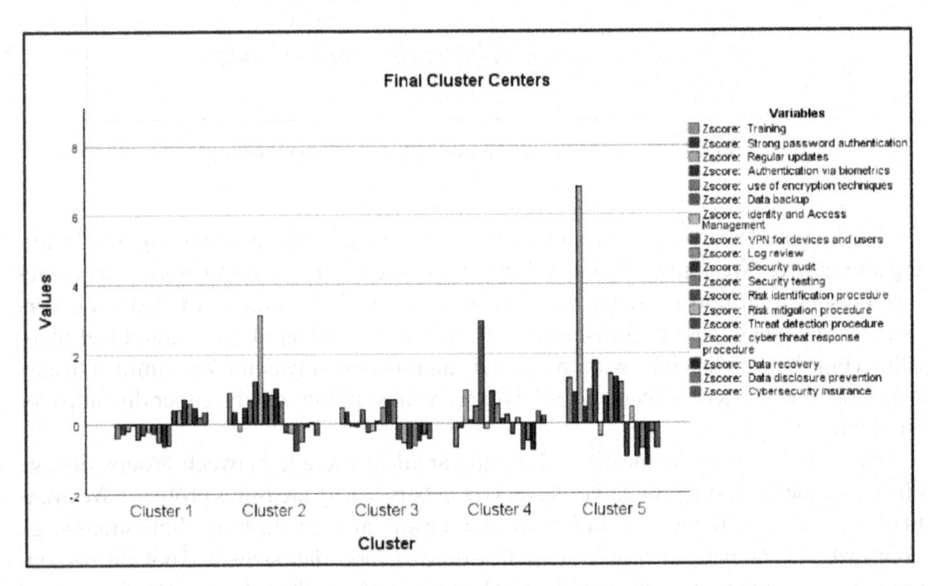

Fig. 3. Visualization of cluster centers.

social organizations per cluster center and cybersecurity measure. This enhanced the interpretability and communication of clustering results, contributing to better-informed decision-making in our data analysis.

Confirming Classification by Discriminant Analysis

As an extra check, a discriminant analysis was conducted to predict which cybersecurity measures best fit which cluster. If the anticipated group membership coincided with the group membership resulting from cluster analysis, then our classification was approved. The independent variables (i.e., discriminators or predictors) were the same 18 variables related to cybersecurity measures as in the K-means cluster analysis. The categorical membership variable (i.e., from the cluster analysis) served as the dependent variable in the analysis. The range was set from 1 to 5.

We used both the regular and stepwise discriminant methods, each calculating two discriminant functions. The discriminant functions turned out to be significant ($p < .001$) for both methods, explaining 98.9% and 92.7% of the total variability between the clusters in the regular method, whereas 92.1% and 91% were found in the stepwise method. We thus reflected on the relevant results of two applied methods (i.e., regular, and stepwise), but observed no significant differences between them and both were highly reliable. Consequently, this discriminant analysis provided extra evidence for our clustering solution by positioning all respondents with a score close to their centroid.

4.2 Differences Between the Cybersecurity Profiles (RQ2)

Assessing the Significance of Each Variable Related to Cybersecurity Measures
For each variable item related to the cybersecurity measures, we examined whether the clusters of RQ1 showed any statistically significant difference by testing the following analysis of variance (ANOVA) hypotheses:

- H0: no significant difference between the clusters.
- Ha: at least one significant difference between the clusters.

We rejected the null hypothesis if the P-value associated with the F-ratio was smaller than 0.05. All variable items were significant ($P < 0.005$), except for "authentication via biometrics" ($P = 0.537$) and "data disclosure prevention" ($P = 0.051$). This finding enabled us to scope our analysis on merely those variables that were statistically significant in our total group of organizations to observe the exact differences.

Interpretation of the Differences Uncovered via Feature Analysis
We uncovered the main characteristics of our classification by comparing the means and significance per variable option and cluster (Fig. 3) (Fig. 4).

In addition to the significance of variables via their representation in the healthcare and social organizations, we were interested in the differences between the clusters which were statistically significant.

The Tukey HSD test helped us identify which variables could be determinants to predict the results observed in each cluster and therefore narrowed our research of differentiators.

We labeled a variable as being a differentiator for our classification when the variable options followed a different pattern across the clusters. As a result, we observed three differentiators: training, regular software updates, and data backup. Table 3 visualizes the revealed patterns among the clusters by means of a color code (i.e., with red referring to lower values, orange referring to average values, and green referring to higher values for cybersecurity measures).

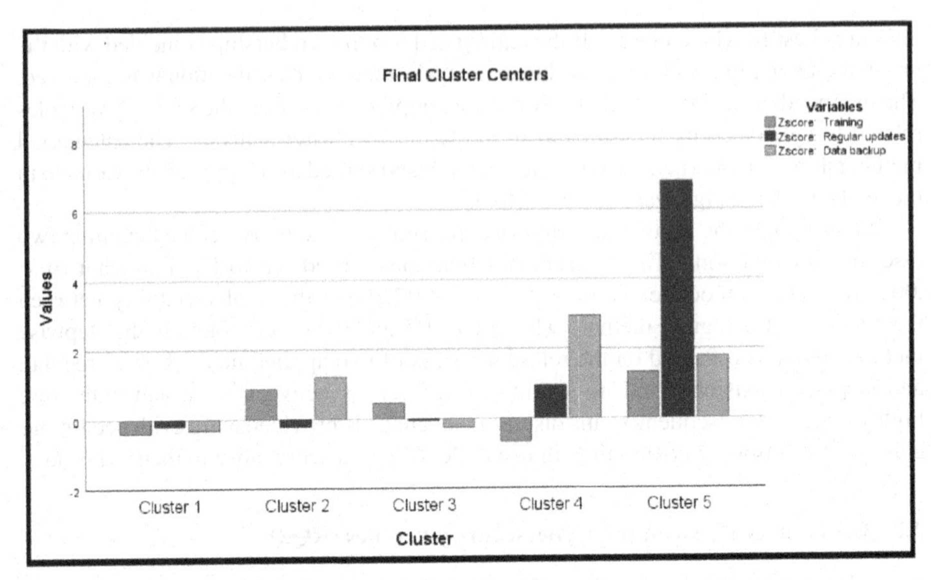

Fig. 4. Visualization of cybersecurity differentiators among clusters.

Table 3. Overview of profiles along differentiators by translating the feature analysis into a textual comparison.

Clusters	Cluster 1	Cluster 3	Cluster 2	Cluster 4	Cluster 5
Profiles	Profile A	Profile B	Profile C	Profile D	Profile E
Training	Lack of training	Moderate level of training	Moderate level of training	Severe lack of training	Superior level of training
Regular updates	Severe lack of regular updates	Severe lack of regular updates	Little or no regular updates	Moderate level of regular up-dates	High level of regular updates
Data backup	Lack of data backup	Lack of data backup	Moderate frequency of data backup	Frequent data backup	Moderate frequency of data backup

The resulting labelling of our classification is summarized in Table 4, showing five profiles of healthcare and social organizations that are interpreted in terms of their extent of implementation of the cybersecurity measures that were identified as the differentiators between the five clusters of organizations.

For comprehensiveness and conciseness, each cluster was assigned to a label linked to its description (i.e., starter, practitioner, promoter, performer, advocate). These labels are not to be seen as gradation levels [35].

Table 4. Cybersecurity tags for healthcare and social organizations.

Cybersecurity tags	Profile	Cluster	Description
Advocate	E	5	High level of regular software updates, superior culture of training and moderate frequency of data backup
Performer	D	4	High frequency of data backup, moderate level of regular software updates but a severe lack of training
Promoter	C	2	Moderate levels of training and frequency of data backup but a severe lack of regular software updates
Practitioner	B	3	Moderate level of training but severe lack of updates and data backup
Starter	A	1	Severe lack of training, regular software updates, and data backup measures

5 Discussion

5.1 Scientific Contributions

Based on our empirical data, we have used 18 cybersecurity measures as input for statistical classification, resulting in five clusters of social and healthcare organizations. Further analysis uncovered three major differentiators between those clusters. This then allowed characterizing the five clusters in terms of cybersecurity profiles for healthcare and social organizations.

Our findings fit into the broader cybersecurity literature. For instance, the Cyber-fundamentals Framework (CyFun) [36] also pays attention to the differentiators of our study, showing the effectiveness of our observations and the relevance of these measures to the healthcare and social sector.

In our analysis, the cybersecurity profiles tagged as “advocate” and “performer” merely encompass a small minority of organizations (namely 6% in our data). This could indicate that there is limited appetite for cybersecurity matters in this sector, meaning the absence of a clear cybersecurity governance structure within healthcare and social organizations. It could also mean that there is a lack of resources to implement cybersecurity measures.

The cybersecurity profiles tagged as “starter”, “practitioner” and “promoter” represent the majority of the surveyed organizations. These organizations spend less resources on training and awareness as well as on data protection measures, such as data backup and regular security updates. This could be a translation of the current level of the sector when it comes to these differentiators. Alternatively, our observation could point out the urgency when it comes to education and training of data protection measures in the healthcare and social realm.

Also, it is worthwhile to consider that two differentiators are related to technical measures and one linked to administrative (non-technical) measures. On the one hand, this could indicate how managers and practitioners prioritize the implementation of cybersecurity measures. Therefore, more campaigns could be done to illustrate the importance of information security policies and procedures. On the other hand, it could be an illustration of the current imperatives that the sector is facing due to their scope of work and the types of past attacks.

Finally, the cybersecurity profiles identified as "starter", "practitioner", "promoter", "advocate", "performer" do not reflect any form of hierarchy among them. Additional parameters will need to supplement our current analysis to allow us to translate these profiles into maturity levels.

5.2 Practical Implications

The empirical results of our research could be of direct use to practitioners and managers of healthcare and social organizations. First, they introduce conceptual clarifications of effective cybersecurity measures, which could be considered as key milestones when determining the organizational profile of these organizations. Secondly, we note that the formation of the cybersecurity profiles relies on the selection of the most distinctive variations between similar groups of organizations. Therefore, organizations external to our study may benefit from investigating the impact of the adoption of these differentiators on their current cybersecurity posture. Thirdly, the research could be useful for a cybersecurity strategy definition exercise of a hospital, healthcare entity or social organization.

Alternatively, our work can stimulate healthcare and social organizations to improve a limited set of key areas in cybersecurity (as per the differentiators) by self-assessing their profile, because the sector is often recognized as having a lack of resources to secure its network against cyberattacks [37, 38].

Finally, existing maturity assessment tools provided by ENISA and reflected in Sect. 2.2 are either focused on small and medium businesses or granular to the level of a data processing activity. That is why enlarging this research towards a cybersecurity maturity assessment for the healthcare sector could be useful for healthcare practitioners. We anticipate that this may allow them to save a tremendous amount of time and resource in avoiding tailoring the existing tools to the specific structure of the healthcare realm.

5.3 Research Limitation and Future Research Directions

Although the dataset that we utilized was of extreme value for our analysis, our research decisions have also been constrained by differences in cluster sizes. Also, we note that the dataset represents perceptions rather than objective data, with only one employee representing each institution. Next, we acknowledge that subjective and self-reported data do not necessarily represent actual case situations. This means that an employee's opinions can be biased concerning one's role in the organization, or regarding one's projected expectations. Additionally, the dataset is binary for the variables of cybersecurity measures. Therefore, we may face accuracy limitations in profiling organizations due to the use of z-score for consistency across the dataset of the survey. Finally, because the dataset only covers Flemish organizations, more research is needed to allow for a worldwide generalization.

Due to these limitations, future research can improve the findings. For instance, the benchmark of cybersecurity measures with existing cybersecurity best practices will ensure consistency across the respondents on their understanding of the cybersecurity measures. Extending cloud controls [39] to the measures analyze will provide additional insights on the differentiators and a comprehensive description of the profiles. Using worldwide datasets that contain data with a higher degree of detail would also grant tangible value to the study. Such extensions in terms of geography and variables can lead to a further validation of the differentiators. Furthermore, exploring social aspects which could emerge from the results and conducting qualitative research methods, such as an expert panel followed by a multiple case study design can validate our current approach.

6 Conclusion

In this research, we have analyzed the cybersecurity implementations within healthcare and social organizations based on statistical clustering and ANOVA-based testing to extract meaningful insights. We have delineated five distinct profiles for characterizing different cybersecurity approaches in healthcare and social organizations. Subsequently, our investigation has delved into discerning the main differentiators among these profiles, showing three cybersecurity differentiators that are related to training protocols, regular update practices, and data backup strategies. Our empirical evidence has also revealed that 95% of the sampled healthcare and social organizations exhibit the lowest levels of the identified differentiators, indicating that substantial progress can still be made to improve their cybersecurity posture. Hence, we point healthcare and social entity practitioners and managers to important cybersecurity reflections along with a reflective assessment of their current posture, while also considering sector-specific indicators. Future research endeavors could aim to transcend the outlined limitations, advocating for global datasets offering enhanced granularity, such as a wider array of organizational types and information security processes. Despite these limitations, our study serves as a catalyst, urging researchers to strengthen cybersecurity guidelines tailored to the distinct requirements of the healthcare and social sector.

Acknowledgments. This work has (partly) been made possible by the financial support of the Flemish government to the Center for R&D Monitoring (ECOOM). Any opinions expressed in this paper are the authors.

Disclosure of Interests. The authors have no competing interests to declare.

References

1. Giansanti, D.: Cybersecurity and the digital-health: The challenge of this millennium. Healthcare (Switzerland) **9**(1) (2021). https://doi.org/10.3390/HEALTHCARE9010062
2. Hospital in Brussels latest victim in spate of European healthcare cyberattacks. Accessed 11 Dec 2023. https://therecord.media/brussels-hospital-cyberattack-belgium-saint-pierre
3. Chenthara, S., Ahmed, K., Wang, H., Whittaker, F.: Security and privacy-preserving challenges of e-health solutions in cloud computing. IEEE Access **7**, 74361–74382 (2019). https://doi.org/10.1109/ACCESS.2019.2919982
4. Abrar, H., et al.: Risk analysis of cloud sourcing in healthcare and public health industry. IEEE Access **6**, 19140–19150 (2018). https://doi.org/10.1109/ACCESS.2018.2805919
5. McConomy, B.C., Leber, D.E.: Cybersecurity in healthcare. In:Clinical Informatics Study Guide, pp. 241–253 (2022). https://doi.org/10.1007/978-3-030-93765-2_17
6. Coronado, A.J., Wong, T.L.: Healthcare cybersecurity risk management: keys to an effective plan. Biomed. Instrum. Technol. **48**(HORIZONS SPRING), 26–30 (2014). https://doi.org/10.2345/0899-8205-48.S1.26
7. Busdicker, M., Upendra, P.: The role of healthcare technology management in facilitating medical device cybersecurity. Biomed. Instrum. Technol. **51**(Horizons), 19–25 (2017). https://doi.org/10.2345/0899-8205-51.S6.19
8. Tervoort, T., De Oliveira, M.T., Pieters, W., Van Gelder, P., Olabarriaga, S.D., Marquering, H.: Solutions for mitigating cybersecurity risks caused by legacy software in medical devices: a scoping review. IEEE Access **8**, 84352–84361 (2020). https://doi.org/10.1109/ACCESS.2020.2984376
9. Adopting the NIST Cybersecurity Framework in Healthcare. Accessed 28 May 2021. https://www.esecurityplanet.com/network-security/healthcare-industry-hit-most-frequently-by-cyber-attacks.html
10. ISO/IEC 27001:2022 - Information security, cybersecurity and privacy protection — Information security management systems — Requirements. Accessed 17 Mar 2024. https://www.iso.org/standard/27001
11. Dias, F.M., Martens, M.L., de P. Monken, S.F., da Silva, L.F., Santibanez-Gonzalez, E.D.R.: Risk management focusing on the best practices of data security systems for healthcare. Int. J. Innov. **9**(1), 45–78 (2021). https://doi.org/10.5585/IJI.V9I1.18246
12. Frumento, E.: Cybersecurity and the evolutions of healthcare: Challenges and threats behind its evolution. In: Andreoni, G., Perego, P., Frumento, E. (eds.) M_Health Current and Future Applications. EICC, pp. 35–69. Springer, Cham (2019). https://doi.org/10.1007/978-3-030-02182-5_4
13. Black, P.E., Scarfone, K., Souppaya, M.: Cyber security metrics and measures (2008)
14. Schatz, D., Bashroush, R., Wall, J.: Towards a more representative definition of cyber security. J. Dig. Forensics Secur. Law **12**(2), 8 (2017). https://doi.org/10.15394/jdfsl.2017.1476
15. Stouffer, K., Zimmerman, T., Tang, C., Lubell, J., Cichonski, J., Mccarthy, J.: NISTIR 8183 cybersecurity framework manufacturing profile (2019). https://doi.org/10.6028/NIST.IR.8183
16. Chang, S.E., Ho, C.B.: Organizational factors to the effectiveness of implementing information security management. Ind. Manag. Data Syst. **106**(3), 345–361 (2006). https://doi.org/10.1108/02635570610653498
17. Ahouanmenou, S., Van Looy, A., Poels, G.: Information security and privacy in hospitals: a literature mapping and review of research gaps. Inf. Health Soc. Care **48**(1), 30–46 (2023). https://doi.org/10.1080/17538157.2022.2049274
18. Aman, W., Al Shukaili, J.: A classification of essential factors for the development and implementation of cyber security strategy in public sector organizations. Int. J. Adv. Comput. Sci. Appl. **12**(8), 2021 (2021). https://doi.org/10.14569/IJACSA.2021.0120820

19. Atoum, I., Otoom, A.A., Otoom, A.: A classification scheme for cybersecurity models. Int. J. Secur. Appl. **11**(1), 109–120 (2017). https://doi.org/10.14257/ijsia.2017.11.1.10
20. Nieto, A., Rios, R.: Cybersecurity profiles based on human-centric IoT devices. Hum.-centric Comput. Inf. Sci. **9**(1), 1–23 (2019). https://doi.org/10.1186/S13673-019-0200-Y/FIGURES/10
21. Soumelidou, A., Tsohou, A.: Towards the creation of a profile of the information privacy aware user through a systematic literature review of information privacy awareness. Telemat. Inf. **61**, 101592 (2021). https://doi.org/10.1016/j.tele.2021.101592
22. Zamfirescu, R.G., Rughinis, C., Hosszu, A., Cristea, D.: Cyber-security profiles of European users: a survey. In: Proceedings - 2019 22nd International Conference on Control Systems and Computer Science, CSCS 2019, pp. 438–442 (2019). https://doi.org/10.1109/CSCS.2019.00080
23. Majkowski, G., Feldman, S.S.: Getting in Front of Cybersecurity Frameworks with a Cyber Vulnerability Profile: Assessing Risk from a Different Perspective. Accessed 18 Jan 2024. https://www.forbes.com/sites/thomasbrewster/2016/02/18/ransomware-hollywood-payment-locky-28
24. On-line tool for the security of personal data processing—ENISA. Accessed 17 Mar 2024. https://www.enisa.europa.eu/risk-level-tool/assessment
25. Cybersecurity Maturity Assessment for Small and Medium Enterprises—ENISA. Accessed 17 Mar 2024. https://www.enisa.europa.eu/cybersecurity-maturity-assessment-for-small-and-medium-enterprises#//
26. Gutierrez, C.M., Jeffrey, W.: FIPS PUB 200 Minimum Security Requirements for Federal Information and Information Systems (2006)
27. Colan, S.D.: The why and how of Z scores. J. Am. Soc. Echocardiogr. **26**(1), 38–40 (2013). https://doi.org/10.1016/j.echo.2012.11.005
28. Everitt, B. S., Landau, S., Leese, M., Stahl, D.: Cluster Analysis, 5th edn., pp. 1–330 (2011). https://doi.org/10.1002/9780470977811
29. Blashfield, R.K.: The growth of cluster analysis: Tryon, ward, and johnson. Multivar. Behav. Res. **15**(4), 439–458 (1980). https://doi.org/10.1207/S15327906MBR1504_4
30. Brown, M.T., Tinsley, H.E.A.: Discriminant analysis (leisure research). J. Leis. Res. **15**(4), 290–310 (1983). https://doi.org/10.1080/00222216.1983.11969564
31. Chmiel, D., Wallan, S., Haberland, M.: tukey_hsd: an accurate implementation of the tukey honestly significant difference test in python. J. Open Source Softw. **7**(75), 4383 (2022). https://doi.org/10.21105/joss.04383
32. Prasad Kumar Mahapatra, A., et al.: Multiple comparison test by Tukey's honestly significant difference (HSD): do the confident level control type I error. Int. J. Stat. Appl. Math. **6**(1), 59–65 (2021). https://doi.org/10.22271/maths.2021.v6.i1a.636
33. Wu, J.: Cluster Analysis and K-means Clustering: An Introduction, pp. 1–16 (2012). https://doi.org/10.1007/978-3-642-29807-3_1
34. Blashfield, R.K., Albenderfer, M.S.: The literature on cluster analysis. Multivar. Behav. Res. **13**(3), 271–295 (1978). https://doi.org/10.1207/S15327906MBR1303_2
35. CSA Cyber Trust mark Certification | TÜV SÜD PSB. Accessed 20 Dec 2023. https://www.tuvsud.com/en-sg/services/cyber-security/csa-cyber-trust-mark
36. CyFun Self-assessment Tool | CCB Safeonweb. Accessed 11 Dec 2023. https://atwork.safeonweb.be/tools-resources/cyberfundamentals-framework/cyfun-self-assessment-tool
37. Tully, J., Selzer, J., Phillips, J.P., O'Connor, P., Dameff, C.: Healthcare challenges in the era of cybersecurity. Health Secur. **18**(3), 228–231 (2020). https://doi.org/10.1089/HS.2019.0123

38. Shingari, N., Verma, S., Mago, B., Javeid, M.S.: A review of cybersecurity challenges and recommendations in the healthcare sector. In: 2023 International Conference on Business Analytics for Technology and Security (ICBATS), pp. 1–8. IEEE (2023). https://doi.org/10.1109/ICBATS57792.2023.10111096
39. CSA. Accessed 17 Mar 2024. https://cloudsecurityalliance.org/research/cloud-controls-matrix

Sustainability

A Reference Architecture for Digital Product Passports at Batch Level to Support Manufacturing Supply Chains

Malina Wiesner[1], João Moreira[1], Renata Guizzardi[1](✉), and Paul Scholz[2]

[1] University of Twente, Enschede, The Netherlands
{j.luizrebelomoreira,r.guizzardi}@utwente.nl
[2] Hilti Group, Thüringen, Austria
paul.scholz@hilti.com

Abstract. Despite the availability of metrics and measurement tools, the lack of formal models and standardization concerning product lifecycle information poses a challenge in assessing how sustainable the supply chain operations of a company are. Digital Product Passports (DPPs) emerge as a promising solution to track and ensure accurate product information is maintained throughout the whole product lifecycle. DPPs are digital representations of the accumulated information of a particular product from its inception to end-of-life. We investigated the topic from the perspective of a large European manufacturer, examining how different phases of the product lifecycle can be supported by static and dynamic data at different levels of granularity. As a result of our research, we propose a reference architecture that supports the development of DPPs with emphasis on product components at batch level granularity. The validation of the proposed architecture shows that the approach provides an opportunity for manufacturers to address sustainability issues in resource-intensive manufacturing supply chains while actively reusing legacy infrastructure.

Keywords: Digital Product Passport · Sustainable Manufacturing Supply Chain · Reference architecture · Interoperability

1 Introduction

Sustainability has become a major concern of the European industrial landscape, driven by consumer demands, regulation and reporting standards, and efficiency and cost. Most emissions are caused by operations in the supply chain [1], which are among the biggest levers for the European industries to become sustainable in the long run. To obtain the right level of sustainability, a company needs to be able to assess how sustainable it is. But existing research underlines the lack of formal models and standardisation especially around product life cycle information [2]. This hinders the measurement of sustainability, despite the available

J. Araújo et al. (Eds.): RCIS 2024, LNBIP 513, pp. 307–323, 2024.
https://doi.org/10.1007/978-3-031-59465-6_19

metrics and measurement tools. The issue is exacerbated by the frequent presence of inadequate information infrastructure, which fails to support every stage of the product life cycle. As a result, information exchange at the interfaces between domains (i.e. between product design, engineering and manufacturing functions) and across company borders (i.e. concerning all supply chain partners in a network) fails to overcome data silos, and the information to properly assess sustainability is often missing or incomplete [3].

Industry 4.0/5.0 technologies are enablers of sustainable manufacturing, particularly addressing information transparency in the context of supply chains. The Digital Product Passport (DPP) [4] emerges as a possible solution to guarantee that the correct information about products and materials can be specified and maintained. DPP is promoted and regulated by the European Commission to boost transparency and encourage circularity by sharing product details throughout its lifecycle. Beginning in 2026, the initial lineup of products, starting with renewable batteries, will be required to meet DPP information criteria set out by supporting regulations. In short, a DPP consists in a digital representation of the necessary data about a particular product. Product lifecycle and circularity assessments as well as the operationalisation of circular business models may be supported by their implementation [5].

This research targets the question on how to develop the digital capabilities required for DPPs in the general domain of manufacturing supply chain. In this context, the study explores the opportunities and implications from the perspective of a renowned European manufacturer, while focusing on generalizing our insights to other manufacturers. In particular, we investigate how the scientific aspects of the different product lifecycle phases can be supported by static and dynamic data, and different levels of granularity: from raw materials to batches and components, to the final product. We made a thorough analysis of the common existing processes and information systems to understand the processes that may be affected by the DPPs, and the typical information systems that can serve as information sources for them at the manufacturer. Following this architectural analysis, we perform a gap analysis, considering the current state-of-affairs at the manufacturer and the DPP information requirements. As a result, we propose a reference architecture for developing DPP systems, and we develop a proof-of-concept of such architecture to support the validation through a panel of experts on real use cases.

This paper is structured as follows. Section 2 motivates the research on Sustainability in Manufacturing Supply Chains and explains how sustainability may be supported by Digital Product Passports. Section 3 describes the performed analysis and introduces our Component-based Product Passport approach. Section 4 presents the validation of the approach through a proof-of-concept and expert opinion. Section 5 concludes this paper.

2 Towards Sustainable Manufacturing Supply Chains

Supply Chain Management (SCM) can be defined as the planning and management of all activities involved in sourcing and procurement, conversion, and all logistics management activities with the purpose of synchronising supply and demand through information and coordination mechanisms [6]. Supply chains have significant environmental impact, with 8 key sectors responsible for over 50% of annual greenhouse gas emissions worldwide [1]. The Greenhouse Gas (GHG) Protocol acknowledges challenges in accurately assessing CO_2 emissions across the entire value chain and recommends using varied calculation methods based on data availability. The overall suggestion is to tailor calculation methods based on data quality, emphasizing that sustainability achievements are directly tied to information quality [3].

Green material purchasing (alongside product design and manufacturing practices) is crucial for sustainable manufacturing supply chains. This involves raw materials, parts, and all supplies contributing to the final product. Life Cycle Assessment may be used to assess the environmental and social footprint of products and associated manufacturing processes. To perform Life Cycle Assessment, it is necessary to collect data on the environmental effects of a product or service during its entire life cycle, from the extraction of raw materials to its disposal. This includes information about the use of resources and raw materials extraction, energy consumption, CO_2 emissions, and waste production as well as product, component and material composition. Focusing on maintaining product information, DPPs may provide an effective solution supporting Life Cycle Assessment (Fig. 1).

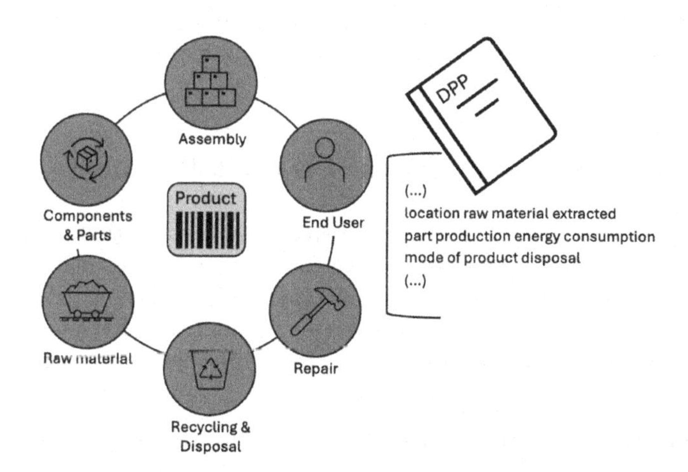

Fig. 1. Digital Product Passports for Supply Chain Lifecycle Assessment

The main idea of a DPP is to enable tracking all associated information of a product, from the conception to the end of life. In other words, DPPs enable transparency and traceability of products and materials through the product value chain, by creating the digital representation and necessary data connections. Thus, product lifecycle and circularity assessments, as well as the operationalisation of circular business models may be supported by DPP implementation [5].

In general, DPPs are applied in 4 product lifecycle phases: product design, manufacturing, operation/maintenance, and recycling. In manufacturing, Product Lifecycle Management (PLM) plays a relevant role for DPP on sustainability data [7], relying on data that are created and modified through the whole product lifecycle.

Information requirements for the DPP can be defined through comparing related research on the topic [8–11], and using industry feedback gathered from the project CIRPASS in consideration of related regulatory initiatives. CIRPASS plays a pivotal role in laying the groundwork for DPPs aligned with relevant standards such as the Ecodesign for Sustainable Product Regulation (ESPR) [12]. The ESPR offers guidance on emissions calculation and gives a first indication of what might be expected from a DPP dataset [9]. The level of granularity for the DPP is specified by the ESPR to be either on product model (e.g. product model A), batch (e.g. product A, from plant X in year 2023) or item (e.g. product A, Serial number 12345678910). Research on product information classification emphasizes that a significant portion of product data is static [11].

The ESPR defines high-level DPP information requirements [9]: Durability, Reliability, Reusability, Upgradability, Repairability, Maintenance, Hazardous Substance Identification, Efficiency, Recycled Content, and Environmental Impact, including recycling and waste generation. Comparing research and the results of CIRPASS, these information requirements extend into detailed categories such as Product Identification, Manufacturer Information, Specifications, Material Composition, Design, Usage, Repair History, and Environmental Certifications [12,13]. These high-level requirements may be refined to arrive at requirements that are more specific to the manufacturer and to the analysis context.

3 Component-Based Digital Product Passport

This section covers the analysis of the current (As-Is) situation of an European manufacturer, presents a gap analysis to note what is currently missing, and introduces a reference (target) architecture that addresses the DPP requirements by proposing data collection and integration capabilities that enable manufacturers to harness the product information from their legacy systems.

3.1 Architectural Analysis of Manufacturing Supply Chain

The manufacturer in our research is an emblematic European-based enterprise that manufactures construction products through discrete manufacturing and assembly, by employing a high-mix and low-volume production strategy. This choice allowed us to understand the challenges associated with producing a diverse product range in smaller quantities, which poses unique considerations and demands, especially in the context of the DPP use case when compared to high-volume production with limited product variations. We focused on two key processes at the manufacturing plant: a) material flow and b) change management.

Material Flow covers the different steps material goes through until it becomes part of the final product. Deliveries can comprise raw materials or components that need to go through one or more manufacturing steps at the plant first, for instance heat treatment or different steel cutting procedures, before being assembled into the final product. Moreover, deliveries can also comprise of components and spare parts that directly go into the product, for instance screws and electronic components (such as batteries). Most components and spare parts delivered to the plant pass through incoming quality control.

Components move through the plant into smaller batches, starting from batches on palettes to smaller handling units. Subsequently, materials are either batch-processed through different machining stages and then repacked into smaller handling units for transportation to assembly lines; or materials move directly into warehousing and repacking. Once at the assembly line, components are assembled into the final product, which is then stored in the warehouse before being shipped to distribution centres.

Most of the processes with regards to material flow occur within the warehouse, while the remaining steps are the responsibility of in-house logistics, production, or assembly. There are three key supporting systems related to the flow of materials: Enterprise Warehouse Management (EWM), Enterprise Resource Planning (ERP), and Manufacturing Execution System (MES). Data is mostly centralised in the ERP, with information related to both inbound quality control and production orders. The EWM manages warehouse and in-house logistics processes such as supply to production and assembly lines, tracking batch locations and transactions. The MES supports production and assembly processes, sharing information with the ERP.

In scenarios where a batch of components is identified as faulty (this can happen at different stages of the material flow, including faults registered by a supplier and internal production issues), it becomes essential to trace their final destination. Therefore, a capability is required to establish ***connections between batches*** from suppliers or internal production, individual components, and the final product. If a batch of components is responsible for issues in final products, it is crucial to not only identify the problematic batch but also the specific set of products in which these components were used. Establishing links between products, components and batches allows for the seamless integration of information about raw materials from suppliers to the final product, ensuring that product information is not solely reliant on static master data such as the bill of materials. Batch level granularity can be achieved mainly through data integration of existing systems. However, achieving item-level granularity for product KPIs entails a significant investment in capabilities, both internally and within the supply chain, including advanced track and trace systems and the placement of further IoT components on shop floor level. Moreover, for manufacturing companies, like the one under study, with relatively low production volumes per product model, the return on investment may not justify the effort required.

Change management involves making alterations to a product model, potentially leading to a new version or revision level. This is crucial for the DPP as it shapes the product model and the foundational information of the actual product. Key stakeholders include internal supply chain participants, like engineering, quality, procurement, health, safety and environmental departments. The main processes are initiated by plant quality and procurement, driven for example by quality issues or supply shortages that can influence product design. This process highlights how new product versions emerge and showcases the complexity of the Product Lifecycle Management (PLM), given the number of stakeholders involved. The manufacturer's high-mix, low-volume strategy results in relatively specific average data per product model due to its low volume. Additionally, the manufacturer's product models have extended life cycles, making changes infrequent.

The ERP, the Computer Aided Design (CAD) system and project management software support this process. Depending on the change required, the CAD system is used to create new drawing of a product of component, while the ERP system is used to register the change in revision. For procurement, the manufacturer uses a cloud-based procurement platform, which allows the connection and collaboration with suppliers and enables streamlining of procurement processes. It is important to note that the usage of the procurement platform varies among the manufacturer's suppliers at present, based on their size and digital capabilities. Specifically for sustainability reporting, a dedicated questionnaire is sent to suppliers to gather additionally necessary information.

3.2 DPP Information Model and Requirements

We propose the high-level information model illustrated in Sect. 3.2 with the respective source systems to address the DPP information requirements in line with typical information structures at manufacturing plants today.

Figure 2 shows the three core systems MES, ERP and EWM, which are the single source of truth for the information, with: a) MES holding information coming directly from production and assembly lines (such as the task completed); b) ERP holding information about the production orders, the orders and suppliers as well as information about the final product; and c) the EWM system is the source for warehouse information, including information about the handling units.

Going in more detail in Fig. 2, the *production order* is a central entity, linking most information of the product. The *serial number* (see within ERP/Product) is the unique ID for a manufactured product, and links information about the *history of the serial number*, which includes any *movements* or status changes registered in the respective IT system. The *Warehouse task* info (see within EWM) links to both the *final product* and the *production order*. Warehouse tasks can be internal movements of products or components, but can also represent the movement of final products to the outside of the plant. *Warehouse task* also relates to an information element called *handling unit*, which represents smaller holding units to transport components and products within the plant,

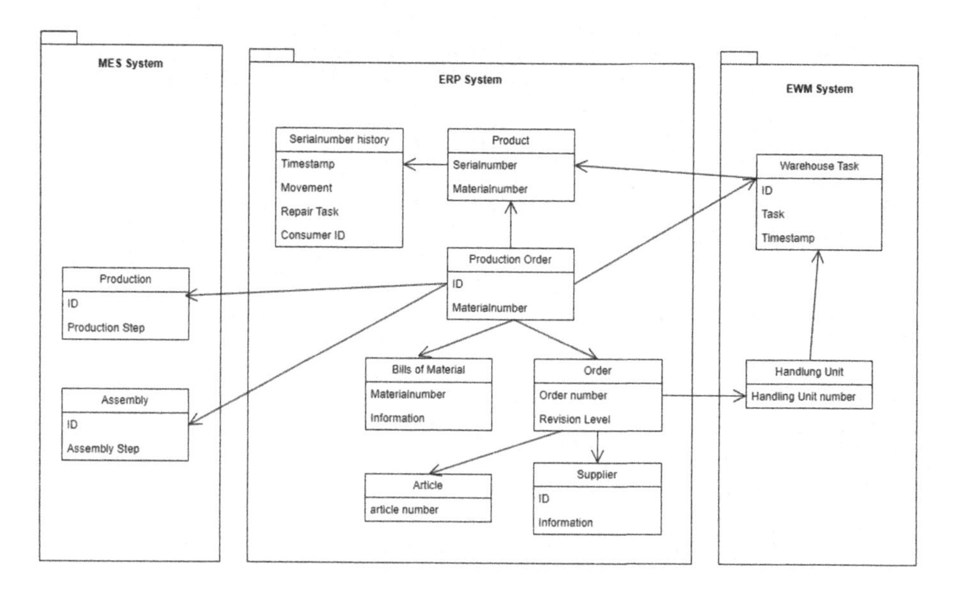

Fig. 2. DPP information model and associated data sources

used by supply production and assembly lines. The *handling unit* is also linked to the *order number* for raw material and components delivered by *suppliers* to the plant. The *order number* in turn relates to information about the *order*, including *supplier* information available in the respective systems. Finally, the *production order* links to information from *production* and *assembly lines* (see within MES for the last two entities). The *material number* is linked to the bill of materials within the respective systems, which lists all materials and parts associated with one product model.

The DPP information model of Fig. 2 adopts a batch-level granularity for data points, such as raw material origin. This makes it possible to consider different product model granularity, besides providing information about the origin of components and materials, which may change periodically based on supplier choices and sourcing. Considering potential future reporting obligations, such as dedicated supply chain regulations, there is a likelihood that this information will require a finer level of granularity. In other words, the batch-level granularity proposed, ensures that individual products can be linked to specific information related to the orders associated with their respective components and materials.

The high level information requirements covered in Sect. 2 were refined into more detailed requirements. Figure 3 shows these information requirements mapped to the phases of the product lifecycle, and also classifying data into static against dynamic data, and direct against aggregated data. Three different combinations are identified: a) Static data collected directly from the source, with the source either the manufacturer's internal system or a system at one of the other participants in the value chain; b) Static data which is aggregated from multiple sources, such as the product environmental footprint, which is partly

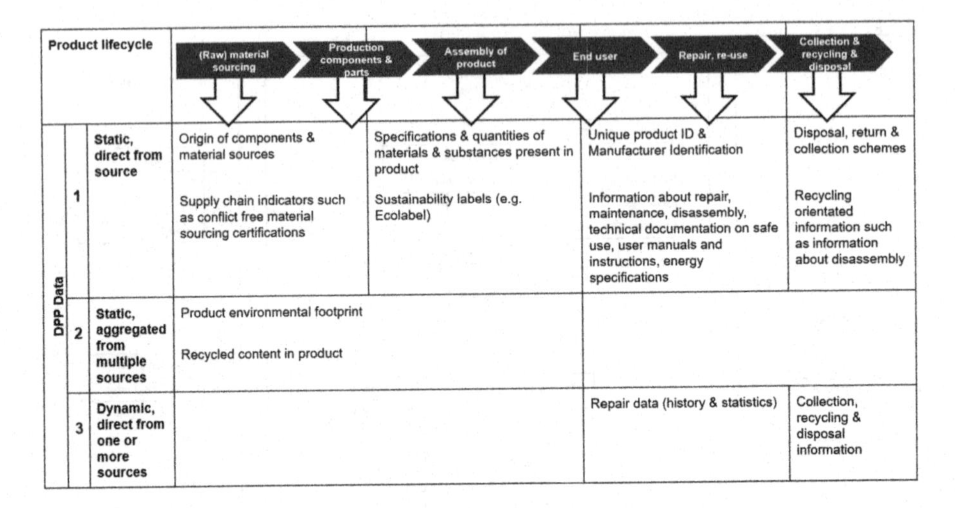

Product lifecycle
(Raw) material sourcing
Production components & parts
Assembly of product
End user
Repair, re-use
Collection & recycling & disposal
DPP Data
1
Static, direct from source
Origin of components & material sources
Supply chain indicators such as conflict free material sourcing certifications
Specifications & quantities of materials & substances present in product
Sustainability labels (e.g. Ecolabel)
Unique product ID & Manufacturer Identification
Information about repair, maintenance, disassembly, technical documentation on safe use, user manuals and instructions, energy specifications
Disposal, return & collection schemes
Recycling orientated information such as information about disassembly
2
Static, aggregated from multiple sources
Product environmental footprint
Recycled content in product
3
Dynamic, direct from one or more sources
Repair data (history & statistics)
Collection, recycling & disposal information

Fig. 3. DPP information requirements aligned to product lifecycle

based on emissions generated in the supply chain prior to the manufacturer in the value chain, and partly based on emissions generated during the manufacturer's internal production and assembly processes; and c) Dynamic data, coming from one or more sources, such as repair data during the use of the product. Service centres external to the OEM may also be allowed to perform certain types of repair services, which means that third parties may also generate dynamic event data about the product in the future.

Static and dynamic data can be distinguished by how regularly they change. Static data usually does not change or changes only seldom, while dynamic data is expected to change at a relevant frequency [14]. Data related to the product's initial lifecycle stages remains relatively stable, and therefore, is static. Data pertaining to usage, repairs, and processes involving product collection, recycling, and disposal are expected to exhibit dynamic behaviour. For static data, a granularity level at the product model or batch level should be adequate due to the limited expected changes over time, resulting in minimal variations among batches. However, adopting a finer granularity at the product item level is possible for dynamic data because movements in and out of repair centers are already tracked based on serial numbers by the manufacturer.

The chosen granularity level imposes specific IT infrastructure requirements: (a) For static data originating from a single source within the manufacturer, suitable interfaces are needed to feed this data into the DPP back-end system; (b) Static data that is aggregated from multiple sources requires a data connection and a data processing layer within the DPP back-end to perform the necessary processing and aggregations; (c) Dynamic data in the later stages of the product lifecycle demands bidirectional data connectivity to allow stakeholders to both read and write data to the DPP system.

We finalize this subsection by presenting an Enterprise Architecture Model[1], resulting from our gap analysis.

3.3 Gap Analysis for DPP System Development

Based on the DPP information model (see Sect. 3.2), the most significant issue regards establishing an internal data collection infrastructure to support DPP information collection. However, the As-Is analysis shows that the existing digital capabilities within the manufacturer present challenges in implementing this infrastructure. Within this context, we identified 3 main gaps to be considered (Fig. 4).

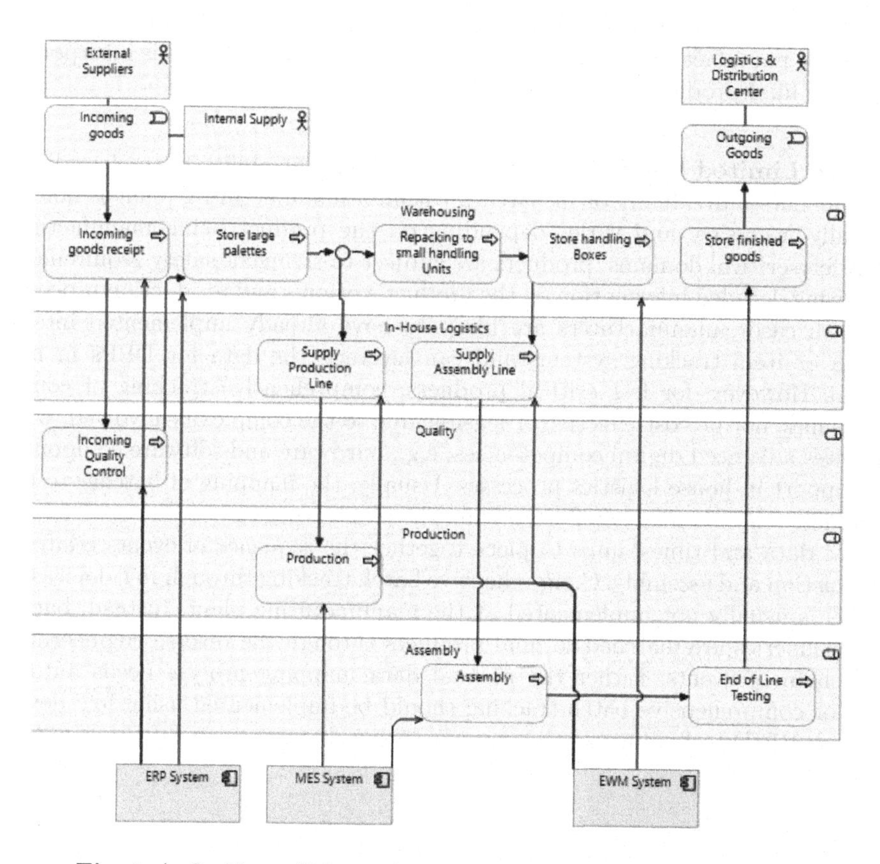

Fig. 4. As-Is: Flow of Materials process and supporting IT systems

[1] We use ArchiMate 3.0.1 (https://pubs.opengroup.org/architecture/archimate301-doc/) to model our As-Is EA and our proposed Reference Architecture.

Gap 1: Different Data Granularity Across the Product Lifecycle. The DPP system requires information to be available at product level, whether it is aggregated from batch, model, or item level of that product. Typically, data from supply chain operations is provided to the manufacturer by suppliers in aggregated form, such as X tons of steel delivered with Y kilos of CO_2 emissions generated in the process. In contrast, the manufacturer possesses data at a more granular level, primarily at the production batch level. Therefore, required information for a DPP is also available at the batch level. However, once the product is fully assembled and in use, data is generated at the individual product item level, including usage and repair information on the individual serial number. Information from operations earlier in the product's lifecycle, including the supply chain, production, and assembly processes must be aggregated on product-level. For instance, if a supplier provides data on the amount of steel used, the manufacturer needs to aggregate how much of that steel is incorporated into the final product.

Gap 2: Limited Batch Tracking Capability. The ability to track and trace components individually or in batches within a manufacturing plant is not universally necessary and varies depending on the product being manufactured. In safety-critical domains, products are subject to stringent safety requirements, requiring detailed information on the product's origins and production processes. In such cases, manufacturers are likely to have already implemented internal batch or item tracking systems and can leverage the data for DPPs in more detail. However, for less critical products, comprehensive tracking of components may not be cost-effective or feasible due to the complexity involved, which requires advanced digital competencies, e.g., hardware and software components to support in-house logistics processes. Usually, the mapping of batches to final products relies on manual efforts within the plant, where personnel use ERP and EWM data and timestamps to piece together the sequence of events related to production and assembly. Comprehensive batch tracking through IoT devices like RFID is usually not implemented at the manufacturing plant. Instead, batches and deliveries are matched to final products through the reverse engineering of the chain of events. Either the manual data mapping process needs automation or comprehensive batch tracking should be implemented using IoT devices such as RFID technology at the manufacturing plant to meet DPP information requirements.

Gap 3: Missing Product Data. Missing data can result in several requirements depending on the data source. If the data comes from external partners in the value chain, suitable interfaces for data exchange need to be implemented. Opportunities such as Data Spaces and Blockchain could help with supplier collaboration in the mid to long term future. If the data input is not done properly inside the manufacturer, this likely results in requirements for process optimisation at the time of data creation.

3.4 DPP Reference Architecture

In this section, we propose a reference (target) architecture for the DPP system in manufacturing supply chain, aligning it with existing manufacturing infrastructure and the required digital capabilities. The architecture is illustrated in Fig. 5 through an ArchiMate viewpoint, and is based on [15], which defines the scope and expected outcomes within the broader product lifecycle context. The lifecycle phases are represented as business functions related to the data acquisition capability, and are served by the discussed applications, which interact with data integration services for data exchange. This architecture aims to address the information requirements discussed in Sect. 3.2, and the gaps 1 and 2 presented in the last section. Also, flexibility in operationalising the DPP system is considered, so that third-party service providers can be included for additional benefits of supply chain transparency through material traceability. The decision regarding whether to directly adopt a DPP service from established providers is not within the scope of this research, given the still ongoing evolution of DPP requirements. Instead, the focus is on internal digital capabilities necessary for data collection and integration, which remain essential regardless of whether a third-party or manufacturer system is eventually chosen. It primarily focuses on the manufacturer's internal digital capabilities related to data extraction from production and assembly processes and supporting systems.

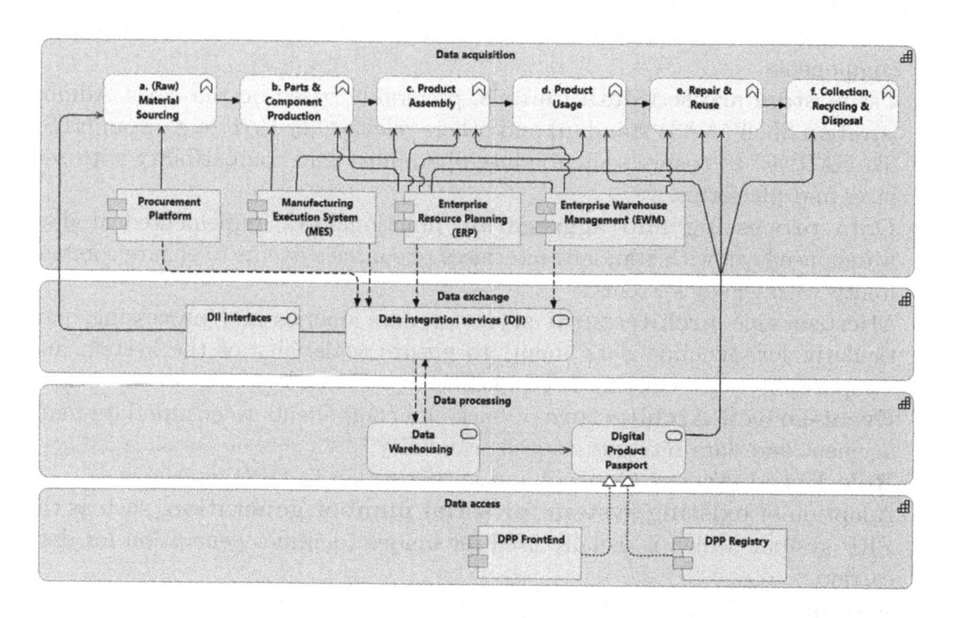

Fig. 5. DPP Reference Architecture for Manufacturing Supply Chain

This architecture addresses the digital capabilities required for the DPP in manufacturing supply chain, giving emphasis to: (1) gathering and mapping information from the supply chain and (2) integrating data from third parties involved in later stages of the product lifecycle (e.g., external repair centres) into the DPP system. The architecture considers the proposed battery passport [14], and consists of four capability layers: data collection, data exchange, data processing, and data access. The data collection layer gathers information from various sources in the product's lifecycle, especially from early stages suppliers and stakeholders. Data exchange occurs through direct interfaces or the DPP system. The data processing layer encompasses the main DPP back-end, responsible for mapping, processing, and aggregating data from diverse sources. The data access layer includes a DPP repository, necessary for compliance with European Commission policies. The focus is on the Manufacturer's DPP, comprising the back-end system and its connection to the front-end. External interactions with the system can happen through the API or the front-end. Stakeholders in the earlier stages of the product lifecycle provide data through these interfaces, resembling the current scenario where suppliers transmit data to the manufacturer, often through systems procurement platforms. The architecture adopts these design principles:

- **Standardised APIs**, preferably REST-based with JSON-LD payload for interfaces between DPP front-end and back-end. Interfaces between the different back-end components and adjacent systems can be based on current data warehousing and data integration capabilities as well as vendor specific aspects that may influence the choice of service provider for integration components.
- Use of **standardised data models**, preferably based on the Asset Administration Shell (AAS) standard and others relevant for IIoT (e.g., from ETSI, like SAREF) to ensure semantic interoperability and compatibility with services and platforms.
- **Data processing and aggregation** functionalities implemented at data warehouse level with standard interfaces to source systems to ensure compatibility with legacy systems.
- **Microservice architectural style** for data queries and processing, particularly for dynamic data input, to ensure scalability of the system and compatibility with existing IoT pipelines.
- **Event-Driven Architecture** of back-end components to ensure data management and data integrity as well scalability
- **Role-Based Access Control** and authorisation to DPP interfaces.
- Adoption of **existing system for serial number generation**, such as the ERP system, which is usually used for unique identifier generation for data carrier.
- **Automated link between product and batches** either via automated mapping of ERP and EWM] data on data warehouse level, or via life batch tracking system using IoT as next level solution.

At the heart of the proposed architecture lies the standardised representation of products through a standardised data model. The architecture follows the AAS meta model as baseline for the realisation of the data model [16]. The AAS meta model is considered the standardized digital representation of an asset, aimed at providing interoperability of Industrie 4.0/5.0 components. The architecture recommends data warehousing for legacy system data integration, preferably with structure aligned to standardized models. Service components act as intermediaries connecting the data warehouse, source systems, and the DPP service. Microservices offer flexibility and scalability. Events aid data integrity, serving as notification and data logging for version control of DPP data. Microservices align with DT implementation, ensuring modularity and scalability. The choice of data warehousing for data integration does not contradict a possible decentralised approach to the DPP ecosystem. While data warehousing centralises data storage and management, it does so in a way that complements the broader decentralised architecture. Data warehousing offers a centralised repository for structured data (and can consume data lakes), ensuring data consistency, security, and efficient querying. However, it does not imply centralisation of control or processing. Decentralisation is realized through the microservices and event-driven design. These service components operate independently, handling specific tasks and processing data asynchronously. In addition, the reference architecture fits into the manufacturer's legacy infrastructure and emphasises the importance of interfaces between data sources, the data warehouse, and service components.

4 Validation

4.1 Proof of Concept

The primary goal of the Proof of Concept (PoC) is to showcase the proposed architecture's ability to integrate data. All PoC artefacts described here are available online[2]. The PoC features design core components implemented as Python programs, while the data sources and the data warehouse are implemented through separate SQL-based databases.

The PoC demonstrates how a DPP can be implemented based on manufacturers' legacy infrastructure, i.e., based on the ERP, EWM and MES systems used in the company. Given the proposed granularity levels represented in the reference architecture, the PoC reflects a DPP on product item level (e.g., product with serial number 12345) and subsequent information aggregated on batch level (e.g., batch 1234 delivered by supply X) or individual component level (e.g., component with ID 1234 from supply X), if available. The DPP also requires information to be aggregated on product level based on information on product model level (e.g. material composition, the product contains 1 KG steel from 6 components based on bill of material specifications).

To implement the DPP service as part of the PoC, two python programs are provided. The first establishes the front-end and processes the user input as well as the data that is returned from the second program, which demonstrates the service component responsible for fetching data from the data warehouse

[2] https://github.com/malina-w/PoC_DPP_public.

component defined in the target architecture. The second program queries the data from the sql database defined by the simplified product representation. This demonstrates the benefits of standardised data representation. Figure 6 shows a screenshot of the PoC frontend, detailing the material breakdown of a particular product.

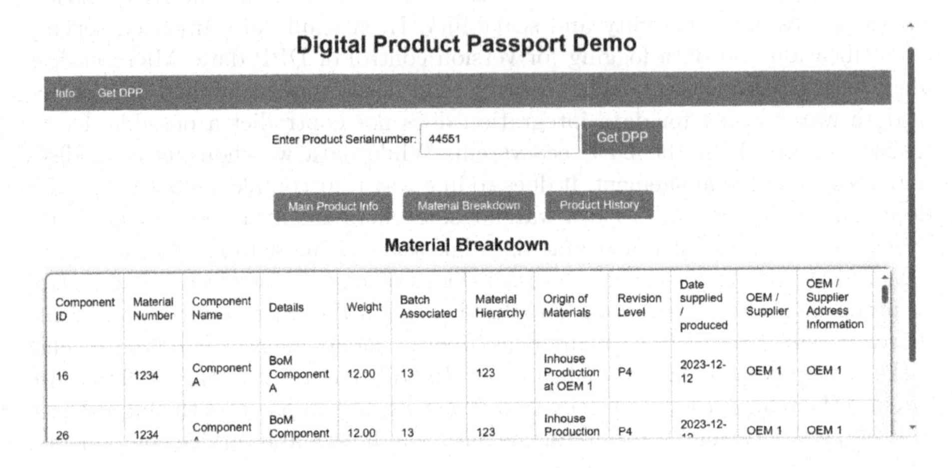

Digital Product Passport Demo
Info
Get DPP
Enter Product Serialnumber: 44551
Get DPP
Main Product Info
Material Breakdown
Product History
Material Breakdown
Component ID
Material Number
Component Name
Details
Weight
Batch Associated
Material Hierarchy
Origin of Materials
Revision Level
Date supplied / produced
OEM / Supplier
OEM / Supplier Address Information
16
1234
Component A
BoM Component A
12.00
13
123
Inhouse Production at OEM 1
P4
2023-12-12
OEM 1
OEM 1
26
1234
Component
BoM Component
12.00
13
123
Inhouse Production
P4
2023-12-
OEM 1
OEM 1

Fig. 6. Screenshot of the DPP system with a product event history

The DPP PoC is compatible with the DPP Information Model and with the Target Architecture. As such, the DPP data model is based on the AAS meta model [16]. To demonstrate the benefits of standardized representation of products, this metamodel is used as basis for the SQL database. The Asset element of the AAS can capture information about the product itself, while submodels define the structure and behavior of the asset. Submodels in the AAS can be utilized to specify the different aspects of the product, such as its physical characteristics, functional capabilities, and lifecycle information.

4.2 Panel of Experts Validation

We conducted a preliminary validation of the proposed architecture and PoC, using a panel of experts composed of 7 people, selected based on their roles and backgrounds at the manufacturer in question, and thus offering diverse perspectives on similar requirements. Example of areas of responsibility of the experts are sustainability, data integration, value creation, quality control, and logistics. The validation comprised of the following 3 steps: a) presentation of the proposed target architecture for the manufacturer's DPP; b) presentation of the implemented proof of concept; and c) brief interviews and survey with the experts to get their input on how the target architecture and proof of concepts support their requirements and goals. The survey was designed with answers using a typical likert scale with five possible answers, namely: not at all, fairly, moderately, well, and very well.

In general, the survey results show that:

1. The capability of the architecture to integrate, collect and process data was perceived well, with four out of seven experts rating it as 'very well'.
2. The capability of the proposed IT architecture to merge and aggregate data from different sources (mainly legacy systems) was 'moderately well scored', indicating that the demonstrated interfaces need further specifications to operationalise the target architecture for the manufacturers specific systems.
3. The capability of the proposed architecture to support simple and flexible information access scored 'very well' by the majority of experts. This indicates that the standardised data model and data warehousing components are perceived as useful by the experts.
4. The capability of the proposed solution to present information with the right level of granularity was rated 'well' by five out of seven experts, indicating that the proposed level of granularity for the manufacturer matches the expectations of the experts for the DPP with some limitations, and raising the need for additional attention for data points of the DPP that have not been further implemented into the PoC.
5. The potential of the target architecture to generate future insights (for instance, root cause analysis for quality issues) scored 'very well', indicating that stakeholders see the proposed architecture as suitable to generate benefits beyond the DPP use case.

We collected positive opinions and suggestions for improvement from the experts. For example, on the positive side, the machine connectivity expert expressed the chosen microservices design and event-based data management as appropriate because of the compatibility with existing IoT components. Both the data warehouse and interfacing components were pointed out as useful for standardising reporting initiatives across departments, helping to overcome obstacles concerning of individual and manual data flows, often aggregating similar types of information. On the other hand, the expert also commented on the intrinsic complexities of products and components, noting that in the context of DPPs, a product may well be a component from the manufacturers perspective. Thus, the classification of an asset being a product or a component may be relative. Additionally, more detailed information such as time periods and quantity ranges are necessary in addition to information such as serial number range to indicate which products and batches are impacted by a quality incident and should also be considered in the DPP.

5 Conclusion

The Digital Product Passport (DPP) initiative endorsed by the European Commission aims to ensure accurate product information throughout its lifecycle, promoting transparency and circularity, and thus leading to more sustainable supply chains. This paper focuses on how to develop the digital capabilities necessary for DPPs in such context. We investigated how information can be acquired from the different product lifecycle phases to address the DPP information requirements. We proposed a model mapping each phase to 3 levels about the nature of the DPP data, namely: (1) static, direct from the source; (2) static, aggregated from multiple sources, and (3) dynamic, direct from one or more sources. In addition, we performed an architectural analysis of common processes and information systems in manufacturing supply chains, and we identified the main information sources for the DPP. These results culminated in a proposed reference architecture and a proof-of-concept on a DPP system development for manufacturing supply chain, which was validated through expert panels. This DPP reference architecture is a component-based approach that addresses common gaps like data granularity and batch tracking issues. This approach enables manufacturers to tackle sustainability concerns related to DPP in resource-heavy supply chains on the basis of common legacy infrastructure.

The main limitations and future research were identified as follows. Firstly, following the presented preliminary validation, more concrete results are expected once the product is effectively implemented at the manufacturer. Secondly, the interaction between the DPP and environmental management systems should be investigated more closely. Thirdly, more research is needed on the system's interaction with external parties, leaving opportunities for data ecosystems such as International Data Spaces. In this context, questions emerge about the business case for comprehensive supply chain traceability, particularly in understanding manufacturers' motivations to participate in such initiatives. Finally, further investigation is required on how to (re)use operational ontologies, particularly standardized ones, to address the DPP information requirements.

References

1. World Economic Forum and Boston Consulting Group: Net-zero challenge: the supply chain opportunity (2021). https://www.weforum.org/reports/net-zero-challenge-the-supply-chain-opportunity
2. Guinée, J.: Life cycle sustainability assessment: what is it and what are its challenges? In: Clift, R., Druckman, A. (eds.) Taking Stock of Industrial Ecology, pp. 45–68. Springer, Cham (2016). https://doi.org/10.1007/978-3-319-20571-7_3
3. Rachuri, S., Sriram, R.D., Sarkar, P.: Metrics, standards and industry best practices for sustainable manufacturing systems. In: 2009 IEEE International Conference on Automation Science and Engineering, pp. 472–477. IEEE (2009)
4. Jansen, M., Gerstenberger, B., Bitter-Krahe, J., et al.: Current approaches to the digital product passport for a circular economy: an overview of projects and initiatives. Working paper, Wuppertal Institut für Klima, Umwelt, Energie, Wuppertal (2022). https://nbn-resolving.org/urn:nbn:de:bsz:wup4-opus-80426
5. CIRPASS: Benchmark of existing DPP-oriented reference architectures (2023). https://cirpassproject.eu/project-results/. Accessed 23 Jul 2023
6. Rausch-Phan, M.T., Siegfried, P.: Sustainable Supply Chain Management. BGG, Springer, Cham (2022). https://doi.org/10.1007/978-3-030-92156-9
7. Rosich, M.B., Le Duigou, J., Bosch-Mauchand, M.: Implementing sustainable supply chain in PLM. In: Emmanouilidis, C., Taisch, M., Kiritsis, D. (eds.) APMS 2012. IAICT, vol. 398, pp. 168–175. Springer, Heidelberg (2013). https://doi.org/10.1007/978-3-642-40361-3_22

8. Jansen, M., Meisen, T., Plociennik, C., et al.: Stop guessing in the dark: identified requirements for digital product passport systems. Systems **11**(3) (2023). ISSN 2079-8954. https://doi.org/10.3390/systems11030123. https://www.mdpi.com/2079-8954/11/3/123
9. Götz, T., Berg, H., Jansen, M., et al.: Digital product passport: the ticket to achieving a climate neutral and circular European economy? Technical report, University of Cambridge Institute for Sustainability Leadership, Cambridge (2022). http://nbn-resolving.de/urn:nbn:de:bsz:wup4-opus-80497
10. Berger, K., Baumgartner, R.J., Weinzerl, M., et al.: Data requirements and availabilities for a digital battery passport - a value chain actor perspective. Cleaner Prod. Lett. **4**, 100032 (2023). ISSN 2666-7916. https://doi.org/10.1016/j.clpl.2023.100032. https://www.sciencedirect.com/science/article/pii/S2666791623000052
11. Stratmann, L., Hoeborn, G., Pahl, C., et al.: Classification of product data for a digital product passport in the manufacturing industry. In: Herberger, D., Hübner, M., Stich, V. (eds.) Proceedings of the Conference on Production Systems and Logistics, CPSL 2023, pp. 448–458 (2023). https://doi.org/10.15488/13463. https://www.repo.uni-hannover.de/handle/123456789/13573
12. CIRPASS: D2.1 mapping of legal and voluntary requirements and screening of emerging DPP-related pilots (2023). https://cirpassproject.eu/project-results/. Accessed 23 Aug 2023
13. CIRPASS: Deriving an initial set of information requirements to serve as a basis for future discussions (2023). https://cirpassproject.eu/project-results/. Accessed 23 Aug 2023
14. acatech. Battery passport: content guidance (2023). https://www.acatech.de/publikation/battery-passport-content-guidance/. Accessed 20 Aug 2023
15. Battery Pass Consortium: Battery passport content guidance (2023). https://thebatterypass.eu/assets/images/content-guidance/pdf/2023_Battery_Passport_Content_Guidance.pdf. Accessed 24 Aug 2023
16. Plattform Industrie 4.0: Details of the asset administration shell - Part 1: the exchange of information between partners in the value chain of Industrie 4.0. Technical report (2023). Accessed 13 Aug 2023

The Effects of Class Balance on the Training Energy Consumption of Logistic Regression Models

María Gutiérrez(✉), Coral Calero, Félix García, and Mª Ángeles Moraga

University of Castilla-La Mancha, 13003 Ciudad Real, Spain
{maria.ggutierrez,coral.calero,felix.garcia,mariaangeles.moraga}@uclm.es

Abstract. The presence of Artificial Intelligence and specifically Machine Learning (ML) has increased in all manner of software applications, and it already plays a major role in a variety of systems pertaining to Information Science such as public transport, disease diagnosis support and other medical problems. This increase in use has raised concerns about possible environmental impacts, since ML models require to be trained in datacentres that can impose a high ecological toll. With the aim of uncovering new ways of reducing the energy consumption of ML models, in this study we will explore the energetic impact of class balance for binary classification tasks by comparing a set of logistic regression models (LRMs) trained on a synthetic balanced dataset against another set trained on a synthetic, unbalanced dataset. We focus on the total energy and time required to complete the task, and discover that the order in energy efficiency of the models remained consistent regardless of class balance, but those trained on the unbalanced dataset required between 1.42 and 1.5 times more energy to complete the tasks, despite requiring only around 1 s more of runtime. We finish by analysing the results and proposing using synthetic datasets to estimate the energy cost of different hyperparameter options for LRMs.

Keywords: Green AI · Machine Learning · Sustainable Software

1 Introduction

Information Science (IS) has made use of Artificial Intelligence techniques for a long time now, and ever since the introduction of the first Machine Learning (ML) models we have discovered that they are a versatile tool due to their abilities to quickly process great amounts of data and extract relevant patterns. Therefore, ML has been applied to many fields: in transport, it features in problems such as timetable planning [1] and automatic identification of delays in trains [25]; in medicine it is being used as support in diagnosing all manner of diseases [8,26,27] and in all manner of pandemic-related problems such as contact tracing [23] and pandemic spread simulation [5,14].

J. Araújo et al. (Eds.): RCIS 2024, LNBIP 513, pp. 324–337, 2024.
https://doi.org/10.1007/978-3-031-59465-6_20

The current technological environment has reached a point of data availability and ML performance that has turn ML into the obvious solution for a wide selection of problems, and in the wake of these successes, the ecological impacts are becoming evident. ML models require to be trained before they're able to do useful work, and the datacentres that carry out this training can become significant stressors on their surrounding environments due to the high amounts of energy and computational resources that this process demands. In an attempt to measure the extent of these impacts, many researchers have turned their attention to the CO_2 emissions of large language models, with some studies estimating that the emissions caused by a single pre-training cycle of BERT_{BASE} [6] can amount up to those of an intercontinental flight [28], while the Large Language Model (LLM) BLOOM was estimated to emit 24.69 tonnes of CO_2eq during the total 1,08 million GPU hours required for its training [17]. On top of producing huge quantities of CO_2, datacentres consume equally huge amounts of water: the complete training of GPT-3 has been estimated to require 700,000 liters of freshwater [16], and both Google and Microsoft have reported recent, sharp increases in their water use (23% increase for Google along 2023, and 34% for Microsoft in 2022 [10, 21]).

In March 2022, Microsoft published an update on its progress towards meeting its 2030 sustainability commitments, and the company admitted that its total CO_2 emissions had increased by about 23%, an increase that it explains by its "expanded global datacentre footprint to meet the increased demand for Microsoft's cloud business" [22]. Their sustainability report for 2023 is not published yet, but considering that the 20% total emissions increase is consistent every year (as they note in that same update), and that this year has been peppered with new Microsoft AI-related services such as Microsoft Copilot and an expansion to its Cloud Partner Program [7, 20], it doesn't seem likely that this trend is going down. And while Microsoft is only one company, OpenAI and Google are also following this trend in order to meet the increasing demand for applications that integrate AI services such as ChatGPT, which in turn will increase the ecological damage caused by ML development.

As we can see, the effects of developing energy-intensive ML systems go far beyond a company's energy bill, so it's not surprising that many researchers are pouring their efforts into sustainable ML systems. To further our collective understanding of ML's energetic behaviour, in this paper we will study the energy consumption required for training a logistic regression model. Using one of our previous studies on logistic regression models [12] as a starting point, we will explore the relationships between time, training data and model energy consumption by varying training data characteristics such as class balance, and measuring said training on a local desktop computer. And since our research is oriented by the need to provide clear and actionable directives for developers, we will use our results to provide suggestions and directives for energy/performance trade-offs that can help developers to make energy-conscious decisions, with the aim of framing our entire research inside the methodologies of the ML life-cycle.

After this section, we briefly present some related works in the area of software energy efficiency, specially in AI-related software. Next there is Sect. 4, explaining the study that was used as a base for this replica, give the reader a basic understanding of the principles and decisions necessary to understand the present study, including a detailed explanation of the tasks and the datasets used as data sources. In Sect. 5 we expose and analyse our results, and then we finish with our conclusions and plans for future works.

2 Related Works

Over the last years, research has shown that AI systems can have a great ecological impact that should not be overlooked. As we have already mentioned, a state-of-the-art LLM such as BLOOM can emit tonnes of CO_2 to the environment [17], on account of the huge amounts of GPU hours that it requires for its training. These estimations are product of the different methodologies that have been developed to help us to get a grip on the magnitude of the problem, such as frameworks for estimating carbon emissions [13,15], which take advantage of public data about the energy consumption of a local electrical grid and cross-reference it with the known locations of datacentres, making them valuable tools to estimate the consumption of cloud-based ML services. For models that run entirely on local machines there are methodologies for energy estimation that use performance and monitoring tools such as RAPL and ARM Streamline to build complex models that can estimate the energy consumption of different ML models, such as decision trees [9] and neural networks [3,24]. The newest research makes a case for factoring the water needed to cool datacentres into the impacts of AI, and a study has already proposed methodology for freshwater consumption that gives an estimation of 700,000 liters of clean water used for GPT-3's training at Microsoft's datacentres [16].

This kind of research is focused on exploring the best ways for characterising and accounting for the environmental cost of AI, but there are also studies that integrate these tools into the current practises of ML development in order to formulate best practises for energy efficiency, like the studies that explore the trade-off between energy consumption and model performance [2] and those that emphasise the importance of using empirical, data-centric techniques for energy efficiency [29]. Other studies denounce the lack of sustainability awareness and standardised methodologies that seems to still have a hold on the field, like this study [4] that mined Hugging Face's repository of ML models to understand the state of carbon-footprint reporting among the models hosted in the repository, or this one [13] that proposes energy-efficiency leaderboards to encourage the systematic reporting carbon emissions among ML developers.

3 Background from Previous Study

We published a study [12] in which we performed a set of tests to explore what happens to the energy consumption of training a logistic regression model

(LRM) when one changes the optimisation function for said training. We used Scikit-Learn and its LogisticRegression model for the implementation, and prepared three binary classification tasks that were identical except for the optimisation function used during the model training: stochastic average gradient descent (SAG), limited-memory BFGS (LBFGS) and Newtonian conjugate gradient method (Newton-CG). We used the FEETINGS [19] framework to measure the energy that each model required to complete the task, and compared the results to learn about the relationship between the choice of optimisation function, the model's performance and energy consumption. We found out some interesting facts about this relationship:

- The most energy-efficient model was the one using LBFGS, while the least energy-efficient was the model using Newton-CG.
- The fastest model used LBFGS, while the slowest used Newton-CG.
- Ordered from worst to best performance, the models were SAG, LBFGS, Newton-CG.
- The model using LBFGS consumed 34.81% of the energy that the model using Newton-CG consumed.
- The model using LBFGS achieved an F1-score of 99.914% for class 0, while the model using Newton-CG scored 99.929% (difference of 0.016% points, in favor for Newton-CG). Since we're aware of the unbalanced dataset affecting the performance, we also included the F1-scores for class 1: 51.485% for LBGFS and 63.636% for Newton-CG (difference of 12.151% points, in favor for Newton-CG).

Using this data, we established **two general guidelines for using LRMs in energy-mindful ML systems**:

- SAG should be avoided from both the point of view of both energy and performance
- The choice between LBFGS and Newton-CG should be made with careful consideration regarding whether the advantage that Newton-CG has over LBFGS's performance is really necessary it for the use case.

However, as it can be gleaned from the performance data that we've just exposed, the dataset that we trained the models on was heavily unbalanced. This was a dataset meant for detecting credit card fraud with less than 1% of all instances belonging to class 1 (FRAUD), which we originally decided to overlook because we needed a non-image-based dataset that was as big as we could find (this dataset was 150.83 MB). Since LRMs are rarely used to process images nowadays and we wanted a use case representative of how LRMs are used in real-world applications, we ended up selecting this dataset. Aware that the class unbalance would negatively affect its performance, we chose to showcase the F1-scores for both classes along with the global F1-score and accuracy in an effort of being transparent about the flaws in the study and stated our assumption that the energy consumption wouldn't be affected by the unbalance.

In this new study, we test whether class balance truly has an effect on the energy consumption of a binary classification task. We will reproduce the tasks from the original study but using new datasets, so that we have one set of models trained on a synthetic, balanced dataset and compare them against a second round of tasks trained on a matching, synthetic unbalanced dataset. We will also compare our new results those from the previous study, which will allow us to propose guidelines on how to exploit the influence of dataset properties to reduce the energy consumption of training a LRM that's fit for use.

4 Description of the Laboratory Study

The experimental tasks consisted in training a LRM on a dataset for binary classification, using a different optimisation function for the training (SAG, LBFGS or Newton-CG) in each task. Other than that, all three tasks are the same: same dataset, same model hyperparameters same and 20–80 train-test split evaluation. We followed an energy measurement framework during the planning and execution of the tasks, which guided us in the process of obtaining rigorous and systematic recordings of the time and energy that each LRM required for its training, as well as performance metrics such as accuracy and F1-scores. This framework is called FEETINGS [19], which was developed specifically for measuring energy consumption of software artifacts. An important feature of this framework required us to repeat each task 20 times, and consider the average energy consumption across all 20 repetitions as the authoritative figure, which is a practise that minimises the influence of noise (such as the energy consumption variations caused by background processes and hardware operation) in the results. To carry out the measurements without risking human error, we automatised the execution of all tasks with a batch script. For now and going on, we will be presenting information using terminology from FEETINGS, so let's take a moment to explain them: the classification tasks that comprise the study were executed on a desktop computer without special capabilities; this computer will be referred to as the DUT (Device Under Testing). We used a hardware device to measure the energy consumption of the DUT during the execution of the tasks, sectioned into its main components: the Solid State Drive (SSD), the graphic card and the processor. The entire setup, including the measuring device, is the same used in [12], which can be consulted for further details on the framework, the setup and the original case study.

The objective of our work is to study the relationship between class balance, energy consumption and performance in LRMs, so the tasks are the same as in the original study [12]. We reused the code from the original tasks, switching out the old dataset for two synthetic datasets that we generated specifically for this study, and executed the programs using the same computer, OS, measurement methodology, measurement device and programming environment as the original study. This guarantees that the only variable that's been altered between this study and the old one is the dataset, and will provide us with empirical data to test our hypothesis that class balance doesn't affect the energy consumed during the training of a LRM.

Originally, we planned to execute these tasks with a balanced dataset and compare the results with those from our original study, but we were unable to find a dataset that was balanced, composed of natural data and that was similar enough to the unbalanced dataset from the original study in terms of size and attributes. This means that attempting to compare their energy consumption would be meaningless, since there would be too many variables that could be influencing said consumption. Instead of that, we decided to use a data generator to create two synthetic datasets, one of which we manipulated to have purposefully unbalanced classes. This way we can ensure a fair and meaningful comparison between the energy consumption of the tasks, since the only difference in the datasets they use is the class balance. These datasets are publicly available in [11].

We used WEKA's random Radial Basis Function generator [30] to create synthetic data and populate two datasets that will be the data source for the binary classification tasks in this study. Since this tool generates balanced datasets by default, creating an unbalanced one required some creativity: we generated two balanced datasets and mixed their instances into a third, picking the instances at random but only allowing a small percentage of those picks to belong to class c0. Using the data generation tool and this unbalancing method we ended up with two binary classification datasets, both of them with 105.126 instances and 185 attributes (around 150 MB each). They are identical in everything except class balance: the balanced dataset has 52,13% instances belonging to class c0, while the unbalanced dataset only has 4,04% of its instances belonging to class c0. It also should be noted that, while these two datasets are identical to each other except in class balance, they are not mirrors of the unbalanced, real dataset we used in [12]. While the three datasets are about 150 MB and are composed exclusively of numerical data, they don't have the same shape: the synthetic datasets have 105.952 instances and 185 attributes, while the real dataset used in our previous study has 284.807 instances and 30 attributes.

Putting all of this data together gives us a base of knowledge to reach accurate conclusions and build future energy-conscious guidelines that are based in empirical data.

5 Results

In this section we present and analyse the results from our study. Most of this data will be in the shape of tables: a column for the energy consumed by the "SSD", another for "Graphic card", another for the "Processor" and finally one for the "DUT" (consumption of the entire system as a whole). The table also features a column for "Time", which shows the total execution time for each task. Time is measured in seconds (s), and all energy consumption is measured in watts per second (W * s).

5.1 Comparing the Synthetic Balanced and Unbalanced Datasets

In this section we expose the results of the classification tasks. Tables 1 and 2 contain a recap of the results of running the tasks with the balanced and unbalanced datasets respectively, and the data is presented with the terminology explained in Sect. 5. Looking at Tables 1 and 2, we can quickly spot a few interesting observations, starting with how the data shows that LRMs trained with unbalanced datasets tend to consume a little more energy than the ones trained on balanced datasets: LBFGS consumes 1.26 times more energy during the task using the unbalanced dataset, and Newton-CG consumes 1.19 times more energy for the unbalanced task in comparison with the balanced task. However, SAG consumes 1.02 times more energy in the balanced task rather than the unbalanced one; it's the only exception, and it is so small that we have found similar and even greater differences between the 20 measurements of SAG that we took as part of the experiment. We think that it is fair to consider SAG's energy consumption as being unaffected by class balance.

Table 1. Mean energy consumption for each task: Balanced dataset

Solver	Time (s)	Energy consumption (W * s)			
		SDD	GraphCard	Processor	DUT
LBFGS	4,28	10,26	191,01	307,58	848,42
SAG	11,51	40,41	62,34	402,14	1117,69
Newton-CG	5,33	12,56	180,57	468,43	1273,68

Table 2. Mean energy consumption for each task: Unbalanced dataset

Solver	Time (s)	Energy consumption (W * s)			
		SDD	GraphCard	Processor	DUT
LBFGS	5,14	19,18	30,91	393,36	1068,08
SAG	11,07	43,72	70,88	416,28	1042,01
Newton-CG	6,39	23,80	40,31	565,02	1524,17

Understanding the computational complexity of the solvers can help us explain the energy efficiency of the models. Newton-CG is an optimisation function that calculates second derivatives, while LBFGS and SAG only calculate first derivatives, so it's unsurprising that the heavier computational demands of Newton-CG will be reflected in the energy consumed during the tasks. This is in fact the case in our results, with the tables showing that, even though the LRMs using Newton-CG were only around 1 s slower than those using LBFGS, they consumed 1,5 and 1,42 times more energy during the execution of the tasks (balanced and unbalanced tasks, respectively). In fact, Newton-CG is so computationally intensive that it was the most energy-consuming solver by a fair margin in all tasks, despite being around half as fast as SAG.

If we compare LBFGS and SAG, we're faced with a more uncertain scenario. LBFGS is an optimisation function that approximates a Hessian matrix using limited memory and bounded values, which makes it a very fast and efficient optimisation option that outperforms SAG in both time and energy consumption. Regardless of class balance, SAG requires more than double the time to finish a task than LBFGS, and in the task using the balanced dataset SAG consumed 1,3 times more energy than LBFGS. However, if we pay attention to the results for the tasks using the unbalanced dataset in Table 2, we see that in these tasks, the model using SAG actually consumed less that the one using LBFGS, but only by 2%, which is small enough for us to say that SAG and LBFGS consume the same energy in the unbalanced dataset. This is even more obvious if we take into account that, in the tasks with the balanced dataset, LBGFS consumes 24.09% less energy than SAG, which is a much more significant difference: LBFGS is more efficient than SAG in general, but since it is also more computationally intensive, if a task is big enough it can achieve similar energy demands to SAG.

5.2 Comparing the Synthetic Datasets with the Original Dataset

Let's now compare the Tables 1 and 2 to the results from our previous study [12], in which we tested the effects that different solvers have on the energy consumption of LRMs. In that study we trained the LRMs with an unbalanced dataset composed of real data, collected by a bank for the purpose of detecting fraudulent credit card transactions -let's remember that in the current study we're using synthetic datasets. We reproduce the results of [12] in Table 3 for the convenience of the reader.

Table 3. Mean energy consumption for each task in [12].

Solver	Time (s)	Energy consumption (W * s)			
		SDD	GraphCard	Processor	DUT
LBFGS	6,33	373,35	10,52	27,44	961,36
SAG	9,51	561,33	16,21	65,47	1798,36
Newton-CG	13,72	801,23	22,09	87,35	2761,71

First, it's worth noting that the relationships between the models in regards to their comparative energy consumption have remained constant for both studies. That is, the model with the lowest energy consumption is LBFGS in both the previous study using real data and in this one, which uses synthetic datasets.

The same thing happens with Newton-CG, which was the most energy-expensive model regardless of using real or synthetic data. This can easily be seen in the left graph in Fig. 1, in which we can also graphically see how similar is SAG's energy consumption for the unbalanced and balanced tasks.

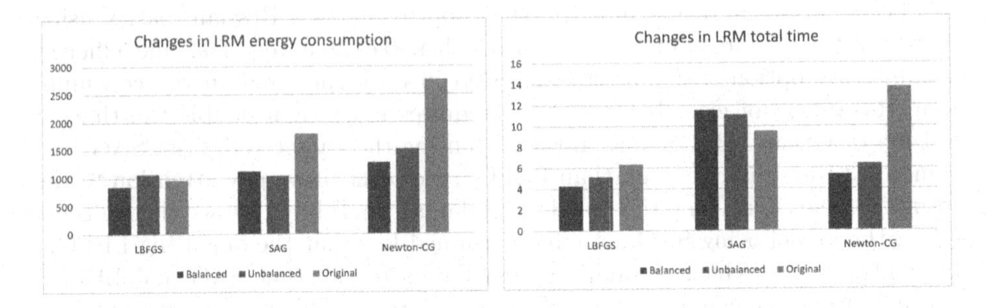

Fig. 1. Energy consumption and runtime for all LRMs in the study.

The changes in runtime are also interesting. Looking at the right graph in Fig. 1, we can easily appreciate that the models using synthetic datasets have a different relationship between runtime and energy from those using a real dataset: the LRMs that use the real dataset see their energy consumption increase as the runtime increases, but we don't see this happen in the models trained on synthetic data. In that set of LRMs, the slowest model (SAG) never is the most energy-consuming (Newton-CG), and the fastest model (LBFGS) can have an energy consumption similar to SAG. What could be influencing this behaviour? We believe that the most likely explanation lies in the convergence of the models. When using the real dataset, the LRMs almost always failed to converge during training, while the opposite was true for the LRMs training on the synthetic datasets (regardless of class balance). That means that the training runs that used the real dataset continued on and on until they reached the limit of 500 iterations per run, while the runs that used synthetic data could stop as soon as convergence was reached. This, of course, drives up the runtime and energy consumption of the LRMs in the previous study. Considering that the only differences between these datasets are the nature of the data (real data opposed to synthetic data) and the proportion of attributes and instances (as we explain in Sect. 4), we think that this is a topic that deserves further investigation.

6 Threats to Validity

In this study there are two main threats to validity: our ability to obtain accurate energy measurements from which to draw conclusions, and the correct generalisation of our results.

Regarding the accuracy of our measurements, we took precautions to ensure its validity, including following an established energy measurement methodology

[19] and installing in the computer an energy measurement device that has been validated against a gold standard, which guarantees its reliability [18]. This means that the exact measurements we obtained are specific to our DUT, and repeating this study in a different computer may yield different results. Another hazard is the fact that any measurement has the potential to be affected by a number of unknown variables inherent to the very conditions of the measurement environment, and this is where the measurement methodology comes into play: taking measurements in a systematic fashion and repeating each task several times helps to mitigate the effects of these uncontrollable factors. We then carried out a statistical analysis of the raw data we obtained from those measurements, which further guarantees that our results aren't tainted by outliers and other abnormalities.

On the topic of generalisation, the main obstacle is that being able to tell which specific characteristic of a dataset is the one affecting the behaviour of a model can be quite tricky. Therefore, we took great care in the choice and design of the datasets, driving us to use synthetic data generators in order to have full control of the resulting dataset and ensue that class balance was the only difference. These directives kept us in check and helped us to avoid errors and faulty measurements, so that we had a solid basis to draw conclusions from. Another important point in this regard is the relationship between synthetic data and model convergence (and model energy consumption in general), of which we have only a preliminary conclusion taken from our very first observations. We hope to study this question further in our future works.

7 Discussion and Conclusions

In this study, we have explored how the class balance in a dataset can impact the energy consumption of LRMs. By combining those results with the data from a previous study we can reach new conclusions about how these factors interact. After this analysis, we can say that class imbalance has a small influence on the model's energy efficiency. Our results show that, when class balance is the only variable, LRMs trained on unbalanced datasets require a little bit more energy in their training. This results also imply that the energy required to train a ML model does not depend just on its size and attributes, but that there are other dataset characteristics that can impact training efficiency.

In general, there appears to be a trade-off between the computational demands, the runtime and the energy consumption of the models: Newton-CG is fast but requires a lot of energy to finish a task, while LBFGS is the fastest and the least energy-consuming model, since have observed that in the worst-case scenario it consumes as much as SAG. Meanwhile, SAG is the slowest model and, at best, it consumes as much energy as LBFGS. This is consistent with our conclusions from the previous study, in which we also found LBFGS to be the most efficient solver and Newton-CG to be the least efficient. This new study has added nuance, finding out cases in which LBFGS can consume as much as SAG, and cases in which Newton-CG is fast but still energy-expensive.

We can also add a new conclusion from this study: neither class balance nor the nature of the dataset affect the relationships between LRM models in regards of runtime and energy consumption relative to each other. This means:

- If a LRM using a particular optimisation function is trained on an unbalanced dataset and outperforms all other models in energy efficiency, it will also outperform the other models at that task if its trained on a balanced version of that dataset.
- If a LRM using a particular optimisation function is trained on a synthetic version of a real dataset and outperforms all other models in energy efficiency, it will also outperform the other models at that task when it is trained on the real dataset.

Based on these findings, we can propose a guideline for reducing the total energy cost of fully training a LRM by taking advantage of the energetic behaviour of the synthetic datasets. This guideline is aimed at the hyperparameter tuning phase of the training cycle, since the size of the search space for model hyperparameters makes finding the optimal combination one of the most time-consuming parts of training a ML model. In order to at least cut off the energetic demands, we propose carrying out at least part of this phase with a synthetic "clone" of the real dataset and evaluate if this synthetic data has a faster convergence that could be taken advantage of, so as to quickly and efficiently find out which optimisation function is the optimal choice in terms of both energy consumption and model performance. As we have just seen in Sect. 5, synthetic data may be a contributing factor to improving the convergence and would not need to have the exact same class balance as the real dataset, and simply generating it to have the same instances and attributes would be enough to make an informed decision regarding the final LRM.

This way, we could decrease the energy required for the first training iterations, and once the optimization function has already been selected the developer can train the model on the real dataset to finish tuning the model. We think that this could help alleviate the stress that datacentres put on the environment by cutting down the amount of time and computational intensity of the most consuming phase of the ML life-cycle, which is training.

In future works we would like to continue exploring new ways of alleviating the environmental cost of the ML life-cycle. Since we seem to have found a viable possibility for reducing the energy consumption in some phases of the training cycle, new works could focus on testing other ML models (like decision trees, perceptrons and SVM) on synthetic datasets to see if they present the same consistency with regards to their energy consumption. We would also like to study the connection between energy consumption and model performance in more detail, especially from the angle of convergence during training and how it is affected by dataset characteristics such as synthetic data and the proportion of attributes to instances.

Acknowledgments. This work was supported by the following projects: OASSIS (PID2021-122554OB C31/ AEI/10.13039/ 501100011033/FEDER, UE); EMMA (Project SBPLY/ 21 /180501/ 000115, funded by CECD (JCCM) and FEDER funds); SEEAT (PDC2022-133249-C31 funded by MCIN /AEI/ 10.13039/501100011033 and European Union NextGenerationEU/PRTR); PLAGEMIS (TED2021-129245B-C22 funded by MCIN /AEI/ 10.13039/501100011033 and European Union NextGenerationEU /PRTR); UNION (2022-GRIN-34110).

References

1. Pimentel, L.D.A., et al.: Solving the train timetabling problem, a mathematical model and a genetic algorithm solution approach. In: 6th International Conference on Railway Operations Modelling and Analysis, RailTokyo2015, March 2015, Tokyo, Japan (2015). https://hal.science/hal-01338609. Accessed 10 Jan 2024
2. Brownlee, A.E.I., et al.: Exploring the accuracy - energy trade-off in machine learning. In: 2021 IEEE/ACM International Workshop on Genetic Improvement (GI), May 2021, pp. 11–18 (2021). https://doi.org/10.1109/GI52543.2021.00011. https://ieeexplore.ieee.org/document/9474356. Accessed 12 Jan 2024
3. Cai, E., et al.: NeuralPower: predict and deploy energy-efficient convolutional neural networks, 15 October 2017. arXiv arXiv:1710.05420 [cs,stat]. Accessed 12 Jan 2024
4. Castaño, J., et al.: Exploring the carbon footprint of Hugging Face's ML models: a repository mining study. In: 2023 ACM/IEEE International Symposium on Empirical Software Engineering and Measurement (ESEM), 26 October 2023, pp. 1–12 (2023). https://doi.org/10.1109/ESEM56168.2023.10304801. arXiv arXiv:2305.11164 [cs , stat]. Accessed 12 Jan 2024
5. Currie, C.S.M., et al.: How simulation modelling can help reduce the impact of COVID-19. J. Simul. **14**(2), 83–97 (2020). ISSN 1747-7778. https://doi.org/10.1080/17477778.2020.1751570. Accessed 10 Jan 2024
6. Devlin, J., et al.: BERT: pre-training of deep bidirectional transformers for language understanding, 24 May 2019. arXiv arXiv:1810.04805 [cs]. Accessed 08 Dec 2023
7. Dezen, N.: Microsoft creates new opportunities for partners through AI offerings and expansion of Microsoft Cloud Partner Program. The Official Microsoft Blog, 22 March 2023. https://blogs.microsoft.com/blog/2023/03/22/microsoft-creates-new-opportunities-for-partners-through-ai-offerings-and-expansion-of-microsoftcloud-partner-program/. Accessed 08 Dec 2023
8. Ferroni, P., et al.: Artificial intelligence for cancer-associated thrombosis risk assessment. Lancet Haematol. **5**(9), e391 (2018). ISSN 2352-3026. https://doi.org/10.1016/S2352-3026(18)30111-X. Accessed 10 Jan 2024
9. García-Martín, E., et al.: Estimation of energy consumption in machine learning. J. Parallel Distrib. Comput. **134**, 75–88 (2019). ISSN 0743-7315. https://doi.org/10.1016/j.jpdc.2019.07.007. Accessed 12 Jan 2024
10. Google: Sustainable Innovation & Technology - Google Sustainability. Sustainability (2023). https://sustainability.google/reports/google-2023-environmental-report/. Accessed 08 Dec 2023
11. Gutierrez, M., et al.: Dataset: the effects of class balance on the training energy consumption of logistic regression models, March 2024. https://doi.org/10.5281/zenodo.10823624

12. Gutiérrez, M., Moraga, M.A., García, F.: Analysing the energy impact of different optimisations for machine learning models. In: 2022 International Conference on ICT for Sustainability (ICT4S), June 2022, pp. 46–52 (2022). https://doi.org/10.1109/ICT4S55073.2022.00016.
13. Henderson, P., et al.: Towards the systematic reporting of the energy and carbon footprints of machine learning, 29 November 2022. arXiv arXiv:2002.05651 [cs]. Accessed 12 Jan 2024
14. Kucharski, A.J., et al.: Early dynamics of transmission and control of COVID-19: a mathematical modelling study. Lancet Infect. Dis. **20**(5), pp. 553–558 (2020). ISSN 1473-3099. https://doi.org/10.1016/S1473-3099(20)30144-4. https://www.sciencedirect.com/science/article/pii/S1473309920301444. Accessed 10 Jan 2024
15. Lacoste, A., et al.: Quantifying the carbon emissions of machine learning, 4 November 2019. https://doi.org/10.48550/arXiv.1910.09700. arXiv arXiv:1910.09700 [cs]. Accessed 12 Jan 2024
16. Li, P., et al.: Making AI less: uncovering and addressing the secret water footprint of AI models, 29 October 2023. arXiv arXiv:2304.03271 [cs]. Accessed 08 Dec 2023
17. Luccioni, A.S., Viguier, S., Ligozat, A.-L.: Estimating the carbon footprint of BLOOM, a 176B parameter language model, 3 November 2022. arXiv arXiv:2211.02001 [cs]. Accessed 08 Dec 2023
18. Mancebo, J., et al.: EET: a device to support the measurement of software consumption. In: Proceedings of the 6th International Workshop on Green and Sustainable Software, GREENS 2018, 27 May 2018, pp. 16–22. Association for Computing Machinery, New York (2018). ISBN 978-1-4503-5732-6. https://doi.org/10.1145/3194078.3194081. Accessed 19 Jan 2022
19. Mancebo, J., et al.: FEETINGS: framework for energy efficiency testing to improve environmental goal of the software. Sustain. Comput. Inf. Syst. **30**, 100558 (2021). ISSN 2210-5379. https://doi.org/10.1016/j.suscom.2021.100558. https://www.sciencedirect.com/science/article/pii/S2210537921000494. Accessed 04 Feb 2022
20. Mehdi, Y.: Announcing Microsoft Copilot, your everyday AI companion. The Official Microsoft Blog, 21 September 2023. https://blogs.microsoft.com/blog/2023/09/21/announcing-microsoft-copilotyour-everyday-ai-companion/. Accessed 08 Dec 2023
21. Microsoft: 2022 Environmental Sustainability Report. In: Global Sustainability (2022)
22. Joppa, L., Smith, B.: An update on Microsoft's sustainability commitments: building a foundation for 2030. The Official Microsoft Blog, 10 March 2022. https://blogs.microsoft.com/blog/2022/03/10/anupdate-on-microsofts-sustainability-commitments-building-afoundation-for-2030/. Accessed 08 Dec 2023
23. Srinivasa Rao, A.S.R., Vazquez, J.A.: Identification of COVID-19 can be quicker through artificial intelligence framework using a mobile phone-based survey when cities and towns are under quarantine. Infect. Control Hosp. Epidemiol. **41**(7), 826-830 (2020). ISSN 0899-823X, 1559-6834. https://doi.org/10.1017/ice.2020.61. Accessed 10 Jan 2024
24. Rodrigues, C.F., Riley, G., Luján, M.: SyNERGY: an energy measurement and prediction framework for Convolutional Neural Networks on Jetson TX1 (2018)

25. Rösler, D., et al.: Discerning primary and secondary delays in railway networks using explainable AI. Transp. Res. Procedia **52**, 171–178 (2021). 23rd EURO Working Group on Transportation Meeting, EWGT 2020, 16–18 September 2020, Paphos, Cyprus (Jan. 1, 2021). ISSN 2352-1465. https://doi.org/10.1016/j.trpro.2021.01.018. https://www.sciencedirect.com/science/article/pii/S2352146521000405. Accessed 10 Jan 2024
26. Shoieb, D., Youssef, S., Ahmed, W.: Computer-aided model for skin diagnosis using deep learning. J. Image Graph. **4**, 116–121 (2016). https://doi.org/10.18178/joig.4.2.122-129
27. Sisodia, D., Sisodia, D.S.: Prediction of diabetes using classification algorithms. Procedia Comput. Sci. **132**, 1578–1585 (2018). International Conference on Computational Intelligence and Data Science. ISSN 1877-0509. https://doi.org/10.1016/j.procs.2018.05.122. https://www.sciencedirect.com/science/article/pii/S1877050918308548. Accessed 10 Jan 2024
28. Strubell, E., Ganesh, A., McCallum, A.: Energy and policy considerations for deep learning in NLP. In: Proceedings of the 57th Annual Meeting of the Association for Computational Linguistics, ACL 2019, July 2019, Florence, Italy, pp. 3645–3650. Association for Computational Linguistics (2019). https://doi.org/10.18653/v1/P19-1355. https://aclanthology.org/P19-1355. Accessed 04 Feb 2022
29. Verdecchia, R., et al.: Data-centric green AI an exploratory empirical study. In: 2022 International Conference on ICT for Sustainability (ICT4S), Plovdiv, Bulgaria, June 2022, pp. 35–45. IEEE (2022). ISBN 978-1-66548-286-8. https://doi.org/10.1109/ICT4S55073.2022.00015. https://ieeexplore.ieee.org/document/9830097/. Accessed 12 Jan 2024
30. WEKA's RandomRBF. https://weka.sourceforge.io/doc.dev/weka/datagenerators/classifiers/classification/RandomRBF.html. Accessed 09 Jan 2024

Optimising Sustainability Accounting: Using Language Models to Match and Merge Survey Indicators

Vijanti Ramautar[1(✉)], Noah Ritfeld[1], Sjaak Brinkkemper[1], and Sergio España[1,2]

[1] Department of Information and Computing Sciences, Utrecht University, Princetonplein 5, 3584 CC Utrecht, The Netherlands
{v.d.ramautar,n.ritfeld,s.brinkkemper,s.espana}@uu.nl

[2] Valencian Research Institute for Artificial Intelligence, Universitat Politécnica de València, Valencia, Spain

Abstract. **[Context]** To assess the sustainability performance of companies, diverse environmental, social and governance accounting (ESGA) methods exist, each with their own set of topics and indicators. In earlier research, we have shown that several ESGA methods contain overlapping indicators. **[Aim]** We aim to develop a semi-automated approach for identifying the overlap between ESGA methods, and then merging the methods into a single combined method that has no redundant indicators. **[Method]** We have approached this goal as a model management challenge. We have surveyed companies to formulate the problem statement, conducted a literature study on model management operations, created ESGA method models according to our openESEA domain-specific language, and developed algorithms that leverage the power of language models to match and merge the methods. The matching threshold is determined by performing an experiment with 16 experts. Lastly, we validate our algorithms by merging 4 real-life ESGA methods. **[Result]** The algorithm has proven capable of successfully identifying overlap between ESGA methods. While we would prefer to further reduce the number of false positives, the results already provide valuable insights into the optimisation of sustainability accounting. Moreover, our findings demonstrate how language models can be used for model management.

Keywords: Model management · model merging · environmental · social and governance accounting · survey indicators · ICT for sustainability

1 Introduction

As part of the European Green Deal, the Corporate Sustainability Reporting Directive (CSRD) has entered into force [11]. More and more companies are

This work is supported by the SCENTISS project, funded by the Dutch Research Council (KICH1.MV01.20.018). Sergio España is supported by a María Zambrano grant of the Spanish Ministry of Universities, co-funded by the Next Generation EU European Recovery Plan.

J. Araújo et al. (Eds.): RCIS 2024, LNBIP 513, pp. 338–354, 2024.
https://doi.org/10.1007/978-3-031-59465-6_21

obliged to disclose information on environmental impacts, treatment of employees, protection of human rights, anti-corruption and bribery policies, and the diversity of their board members. Though this is highly necessary to foster more sustainable business, assessing and reporting this information is a tedious and time-consuming task. Through contemporary technological breakthroughs, we hope to simplify this task so companies can spend more resources on improvement planning, rather than on reporting. The power of language models has already been leveraged to create enduring, positive impacts [39]. Therefore, in this paper, we demonstrate how to use language models to optimise sustainability accounting.

There are numerous environmental, social and governance accounting (ESGA) methods [10]. These methods specify how the accounting should be performed, what data should be collected and how, and they often offer guidelines for reporting the results. Our survey results indicate that many companies use multiple ESGA methods to report on their sustainability performance. For example, Microsoft uses the CDP method to assess their carbon emissions, the LEED method to measure the environmental impact of their data centres, and the GRI Reporting Standards to report on their overall sustainability [20]. There are several reasons why some companies choose to apply several ESGA methods at the same time, such as to comply with laws and regulations, because the methods assess different complementary aspects of the organisation, or because each method offers them a different certification or label [10].

From a procedural point of view, there are two major issues in ESGA. The first issue is related to technical inadequacy. ESGA methods are typically supported by information systems (ISs). These ISs are often rigid and can only support one ESGA method, and cannot be extended with additional indicators or other methods [9]. To alleviate this problem, we have developed a model-driven technology called openESEA [26]. It can support any ESGA method that can be modelled with the openESEA textual domain-specific language (DSL). It also allows extending ESGA methods with additional topics and indicators. The second issue is related to work redundancy. ESGA methods usually define a set of indicators that companies should report on (such as greenhouse gas emissions, electricity consumption, and potable water usage). However, many ESGA methods contain some overlap of indicators that are equivalent. These two issues cause the ESGA process to be tedious and cumbersome. Given that we have already presented a treatment for the first issue [9,26], we now present a treatment to alleviate the second issue by reducing the redundancy and the extra efforts it entails. Therefore, our main research question is:

How to identify overlap in ESGA methods and create a merged method with no redundant indicators?

We have approached this challenge as a model management problem. Our contributions include an algorithm that (1) first identifies matches that identify overlap between two ESGA method models and (2) then automatically merges the models into one new model without redundant indicators. Moreover, we contribute a performance comparison of different language models and similarity metrics. We have validated the algorithm by matching and merging 4 ESGA methods in the market.

The outline of the paper is the following. The research method and subresearch questions are presented in Sect. 2. In Sect. 3 we present the problem investigation and highlight relevant aspects of the openESEA technology, provide background knowledge on model management and natural language processing (NLP), and present a set of similarity metrics. Our match and merge algorithm is explained and demonstrated in Sect. 4. Section 5 contains the results of our validation activity. We reflect on the results and present the key findings in Sect. 6.

2 Research Method

To answer the main research question, we answer the following subquestions.

RQ1: Which approaches exist to calculate similarities and differences between textual models, so as to match redundant indicators in two ESGA methods?
RQ2: How to develop a merging function that generates ESGA methods without redundant indicators, that comply with the openESEA DSL?
RQ3: What are the benefits and drawbacks of our proposed solution?

We treat the project as a design science project, and follow the corresponding phases [37]. Let us explain the activities that make up the research method.

Problem Investigation. We deploy a survey among 59 ESGA practitioners to investigate the problem. Find the survey design in [29] and the respondent demographics in [30]. After that, we implement model management operations that allow ESGA practitioners to match and merge ESGA methods, and operationalise the merged method in the openESEA interpreter. After establishing the problem, we perform a multivocal literature study on such model management operations, similarity metrics, and NLP techniques.

Treatment Design. As part of the treatment design, we define the requirements of the proposed solution. Based on the literature study insights and the requirements, we have opted to match the ESGA methods using similarity metrics. A priori, it is not possible to determine which similarity metric will yield the best performance in this domain, so we have decided to implement Cosine similarity, Euclidean distance, and Manhattan distance, and select one via a parameter. Likewise, we parameterise the language model that is used to create vectors of model fragments. Given that language models have different vector databases, the performance of the matching algorithm varies given the model. For now, we have implemented BERT [7] and Sentence-BERT (S-BERT) [31]. We run an experiment with 16 experts to determine the match threshold. Find the full experiment protocol in the accompanying technical report [30]. To merge the models, we have developed custom-made functions that ensure that the merged model complies with the rules of the openESEA DSL.

Treatment Validation. To validate our solution, we run an experiment where we let the algorithm compare and merge 4 ESGA methods in the market. The ESGA methods contain in total 1664 fragments (consisting of indicators, questions, and topics). We measure false positives, true positives, and execution time, for each combination of similarity metrics and language model. Moreover, we have asked an expert about their opinion of our proposed solution. The expert works as Operations & Innovations Lead for a company that creates ESGA methods focused on workplace culture. The company helps its clients apply ESGA and monitors clients' performance over time.

3 Problem Investigation

3.1 Redundancy Problems in ESG Accounting

Of the 59 survey respondents, 47% stated that the company that they work for uses more than one ESGA method, and that they experience redundant efforts because of overlapping activities or indicators in the ESGA methods they apply. We analysed commonly used ESGA information systems and found that there are no tools that can currently eliminate the redundancy on a large scale. Some ISs (such as the Impact Compass [19]) allow using a predefined ESGA method set at the same time. However, in these systems, the overlap between ESGA methods is determined manually. Therefore, only a few ESGA methods can be assessed with the same tool. Given that there is a multitude of ESGA methods, and thus a multitude of combinations of ESGA methods, creating mappings between methods manually is unfeasible. Therefore, we present an approach that allows matching and merging ESGA methods more efficiently. We demonstrate the approach using our model-driven, open-source ESGA IS, called openESEA.

We aim to identify similarities and dissimilarities between two ESGA methods (i.e. matching), and then create one combined model that only contains all overlapping questions and indicators once (i.e. merging). An elaborate example of two original methods, and the resulting merged method can be found in Fig. 1. The rationale for selecting the indicator that will be part of the merged method is explained in Sect. 4.1.

3.2 The OpenESEA Domain-Specific Language

The abstract syntax of the openESEA DSL is defined by a metamodel [28] and the concrete syntax is implemented as an Xtext grammar [3]. The DSL contains many constructs that are necessary for specifying the deliverable part of ESGA methods. The constructs that are relevant to this work are the following.

- **Topic:** A concept that groups indicators. *"Climate impact"* and *"Diversity, equity and inclusion"* are examples of topics.
- **Indicator:** The definition of a measure that is assessed and reported on during the accounting. Examples are *"Scope 1 greenhouse gas emissions"* and *"Gender pay gap"*.

Method A	Method B	Merged Method
A1. Total Greenhouse Gas Emissions (tonnes of CO2 equivalent) in: Scope 1 ✓	B1. Total gross Scope 1 GHG emissions, (weight in tons of CO2 equivalent) ✓	M1. Total gross Scope 1 GHG emissions, (weight in tons of CO2 equivalent) **B1 (A1)**
A2. Recycled waste disposed (in tonnes) during the last 12 months ✓	B2. Amount of waste recycled in kilograms last year o < 100 kg o between 100 kg and 500 kg o > 500 kg ✓	M2. Recycled waste disposed (in tonnes) during the last 12 months **A2 (B2)**
A3. How has your company integrated sustainability into supply chain management? X	B3. Specify your company's biodiversity and ecosystem conservation initiatives X	M3. How has your company integrated sustainability into supply chain management? **A3**
		M4. Specify your company's biodiversity and ecosystem conservation initiatives **A4**

Fig. 1. An example of how two methods are merged. The checkmark indicates when there is a match between two questions with the same ID number (e.g. A1 vs B1). The cross indicates that there is no match. For each question in the merged model, the ID of the original question is denoted. The ID in parentheses indicates that the selected question also represents the question in parentheses.

- **Question:** Asks for the value of an indicator. For instance, *"What was the total greenhouse gas emissions (metric tonnes of CO2 equivalent) in scope 1 during the reporting period?"*

For the indicators in the scope of this research, there is a one-to-one relationship between questions and indicators. Moreover, questions have a data type, which specifies the format of the data to be collected. When the data type is "integer" or "double", the answer to a question should be a numeric value with zero or more decimals, respectively. If the data type is "text" a string is required; for "Boolean" indicators, a true or false answer is expected; if the type is "date" a calendar date is required. A list of answer options is defined for "singleChoice" and "multipleChoice" question types. The ESG accountant should then select at most one of the answer options in case of a single-choice question, and zero or multiple options in case of a multiple-choice question.

As an example, other constructs required to model ESGA methods that are not used for the matching and merging operations are **Survey** (questionnaires used to elicit data from stakeholders, and **Certification level** (used to structure the requirements for certifying an organisation in cases where the method supports this). Find a complete specification of the openESEA DSL in [27].

3.3 Model Management as an Approach for Eliminating Redundancy

Since we are capable of specifying ESGA methods as models, this allows us to treat redundancy in ESGA methods as a model management problem. Two ESGA methods have redundancy when their models have equivalent elements. Such equivalent elements can be discovered by a matching process, and the redundancy can be eliminated by a merging operation. An early vision of model management is presented in [2], which describes the key elements of model management and defines model management operations. In this work, a model is defined as a textual specification that adheres to an underlying grammar (the openESEA DSL, in our case). We apply the match and merge operations on said models. The *match* operator establishes correspondences between elements in different models. It facilitates alignment, merging, and comparison of models. Merging is the activity of moving the content of an input model into a source model [2]. Therefore, the *merge* operator integrates several input models into one source model.

A systematic approach for comparing diverse merging approaches is demonstrated in [4]. It outlines the algebraic properties expected from an ideal *merge* operator while shedding light on related operators such as *match*. The importance of expressive mapping languages has been highlighted in [1]. The challenges and opportunities of matching and merging are presented in several systematic literature reviews [22,33] and a feature-based survey of model transformation approaches [6]. Advanced technologies such as machine learning and cluster analyses have been used to improve and automate the matching process [24,32]. Our work complements the body of knowledge by providing a new approach to identifying overlap on the instance level, using language models (or more specifically, contextual embeddings), in contrast to recent works that have used language models to perform ontology and entity matching [14,23].

3.4 Sentence Embedding as a Support for the Matching Operation

In NLP, often an initial modification of the text is required before any analysis can be performed. Our approach utilises tokenisation, which refers to the act of breaking down sentences into individual words (or tokens) [8]. Additionally, stopwords such as"and","the", or"it" could be removed. We have opted not to do this since removing stopwords resulted in worse matching performance. Presumably, the stopwords provide much-needed context in the case of ESGA methods. Next, the modified text should be transformed into a representation that is amenable to analysis. To do this we use contextual embeddings, instead of the more traditional word embeddings approach. In word embeddings, each word is assigned a unique vector, and the distance and direction between vectors capture semantic relationships between words [16]. The representation of a word is the same regardless of its context in a sentence. Contextual embeddings, on the other hand, capture different meanings of a word based on its usage in a specific sentence or context, which makes them more powerful in capturing nuances and word sense disambiguation [18]. The contextual embeddings used in our algorithms are BERT [7] and S-BERT [31].

After BERT or S-BERT has transformed the relevant ESGA model fragments into a numerical representation (also called embedding), the similarity between embeddings of different ESGA models can be calculated using similarity metrics. There are several similarity metrics, such as Cosine similarity [17], Manhattan distance [21], and Euclidean distance [15]. Cosine similarity is often used for comparing the direction of vectors, Manhattan similarity for comparing set overlap, and Euclidean distance for measuring the geometric distance between points in a multidimensional space. The applicability of the similarity metric depends on the data set.

4 Match and Merge Algorithms for ESGA Method Models

4.1 The Five-Step Approach for Eliminating Redundancy

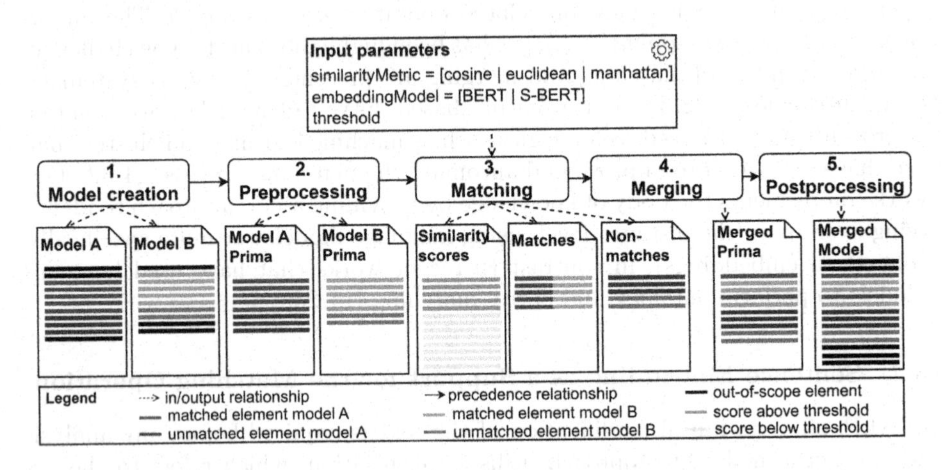

Fig. 2. The five steps necessary to match and merge ESGA methods

Merging ESGA methods consists of five steps. Figure 2 shows a concise and abstract representation of the steps and their outputs. Moreover, it depicts how model fragments flow between the different documents. This section discusses each of the steps briefly.

1) Model creation. Firstly, textual models of the ESGA methods are produced that are subjected to match and merge operations. These models should adhere to the rules specified in the openESEA DSL [26]. Our accompanying model editor can ensure the creation of valid models.

2) Preprocessing. The models are preprocessed. In our DSL, all topics are first defined, followed by the list of indicators, and finally, the list of questions. During the preprocessing the models are restructured, so that model fragments are formed that consist of one topic, one indicator and one question. This requires temporarily duplicating topics and breaking up the list of indicators and questions. We refer to each group of a topic, indicator, questions as "model fragment". Creating these fragments ensures that the match function can use all relevant information to identify overlap. Moreover, the model fragments that are not relevant for the matching process are temporarily removed, as can be seen in Fig. 2.

3) Matching. The preprocessed models (i.e. modelA Prima and modelB Prima in the figure) are uploaded to the matching function, which compares model fragments and identifies matches. The similarity metric is parameterised so users can select Cosine similarity, Manhattan distance, or the Euclidean distance. The contextual embedding is also parameterised, which currently allows users to select BERT or S-BERT. The set of contextual embeddings can be extended with additional language models. The matching function returns similarity scores for each combination of model fragments. All matched pairs that have a similarity score that passes a set threshold can be merged.

4) Merging. In the fourth step, the merging takes place. In Sect. 3.2 we explained that there are several question types in ESGA methods. Even if the match function has established a match between two fragments, a merge may not be possible. Find a list of all question types that cannot be merged in the accompanying technical report [30]. If a merge is possible, the merge function selects one of the model fragments and adds it to the merged model (as can be observed in Fig. 1). Let us briefly explain some of the cases that can be merged. When the question types are the same, the algorithm selects the question from the first model by default. If questionA is of type "integer", "double", "singleChoice", or "multipeChoice" and questionB is of type "text", questionA is selected because numbers can be converted to strings and selected answer options can be concatenated and presented as a string. When questionA is a "double" and questionB is an "integer", the merge function selects questionA because it can be converted to an integer. In all merge cases, we select the question type that allows converting the answer value to both original methods. Find all merge cases in the technical report [30].

5) Postprocessing. Finally, all non-matched model fragments are added to the merged model, to ensure that no model fragments are omitted (as is depicted in Fig. 1). To ensure that the openESEA interpreter can parse the merged model, it is restructured so it adheres to the openESEA DSL, and any contextual elements that were removed during the preprocessing phase are added back.

The Algorithm in Pseudocode. Listing L1 shows a simplified version of the match and merge algorithm as pseudocode[1]. For the sake of brevity, we have expressed that model fragments are matched and merged. In practice, all questions in model A are compared to all questions in model B, all indicators in model A are compared to all indicators in model B, and the same goes for answer options. Each comparison receives a similarity score and the total similarity score of two fragments is computed by averaging all subscores. Finally, we have omitted the intricacies of merging questions, given that this differs per question type.

[1] Code available at: https://github.com/sergioespana/openESEA/tree/master/openESEA%20model%20management/src.

L1. Pseudocode depicting the match and merge algorithm

Require: Inputs: *modelA* and *modelB*, specify the content of the ESGA method (e.g. questions, indicators, topics, etc.)
Require: *preprocessing(modelA, modelB)* ensures that all irrelevant constructs are removed from the models and ensures that method fragments consisting of an indicator, question, and topic are formed. See Figure 3 for an example.
Require: Inputs: *modelAPrima* and *modelBPrima* are the preprocessed models. The *threshold* is a predefined numeric value that determines when two fragments can be considered a match. The *embeddingModel* determines which contextual embedding (BERT or S-BERT) is used. The *similarityMetric* determines which similarity metric is used.
Require: Inputs: *vectorA* and *vectorB* are arrays of numerical values that capture the context of a token.
Require: *cosineSimilarity* calculates the similarity between two vectors using the Cosine similarity metrics. It returns a *similarityScore* that indicates the similarity between two fragments.
Require: *manhattanDistance* calculates the distance between two vectors using the Manhattan distance metric.
Require: *euclideanDistance* calculates the distance between two vectors using the Euclidean distance metric.

```
1  def preprocessing(modelA, modelB):
2    return modelAPrima, modelBPrima
3
4  def compareAndMerge(modelAPrima, modelBPrima, threshold,
       embeddingModel, similarityMetric):
5    mergedModel = []
6
7    for each fragmentA in modelAPrima:
8      for each fragmentB in modelBPrima:
9        similarityScore = 0.0
10
11       if embeddingModel == "BERT":
12         vectorA = BERTEmbedding(fragmentA)
13         vectorB = BERTEmbedding(fragmentB)
14       elif embeddingModel == "SBERT":
15         vectorA = SBERTEmbedding(fragmentA)
16         vectorB = SBERTEmbedding(fragmentB)
17
18       if similarityMetric == "cosine":
19         similarityScore = cosineSimilarity(vectorA, vectorB)
20       elif similarityMetric == "manhattan":
21         similarityScore = manhattanDistance(vectorA, vectorB)
22       elif similarityMetric == "euclidean":
23         similarityScore = euclideanDistance(vectorA, vectorB)
24
25       if (similarityMetric == "manhattan" or similarityMetric ==
26       "euclidean") and similarityScore < threshold:
27         mergedFragment = mergeFragments(fragmentA, fragmentB)
28         mergedModelPrima.append(mergedFragment)
29       elif similarityMeasure == "cosine" and
30       similarityScore > threshold:
31         mergedFragment = mergeFragments(fragmentA, fragmentB)
32         mergedModelPrima.append(mergedFragment)
33
34   return mergedModel
35
36 def cosineSimilarity(vectorA, vectorB):       return similarityScore
37 def manhattanDistance(vectorA, vectorB):      return similarityScore
38 def euclideanDistance(vectorA, vectorB):      return similarityScore
39 def mergeFragments(fragmentA, fragmentB):     return mergedQuestion
40 def postprocessing(mergedModelPrima):         return mergedModel
```

An Experiment to Determine the Matching Threshold. To determine the merge threshold, we have conducted an experiment with 16 experts in ESGA. During the experiment, the experts were presented with fragments of two ESGA methods. They had to identify matches and create a merged method. The same method fragments were fed through the algorithm. Here we found that the matches identified by the experts overlapped with the matches of the algorithm at a threshold of 87% for the combination of BERT and Cosine similarity, and 78% for the combination of S-BERT and Cosine. We have calculated the thresholds of the other combinations of contextual embeddings and similarity metrics using the same experiment data. These thresholds can be observed in Tables 1 and 2, and can be adapted based on the needs of the users.

4.2 Demonstration of the Algorithm

Figure 3 shows an example of how two equivalent questions of different types are matched and merged. In the example, the question in ModelA requires an integer as the answer, whereas the question in ModelB is a singleChoice question. First, in the preprocessing phase, all irrelevant elements are removed. The question and topic remain (we have omitted indicators for the sake of brevity), forming fragments. FragmentA and fragmentB are converted to embeddings using S-BERT. The Cosine similarity function returns a value of 0.85, which is higher than the threshold of 0.78. Hence, the merging function can merge the two fragments. The merge function retains fragmentA (i.e. the integer question) because the answer to questionA can be converted to one of the answer options of questionB. For instance, if the answer to the merged question (questionA) is "1.2 tonnes", the answer should be converted to the answer option "> 500 kg". Finally, the merged model is postprocessed, and all information that was temporarily removed during preprocessing is restored.

4.3 Backward Transformation to the Original Questions

When the user responds to a merged question, the two original questions have to be answered as well. This is particularly crucial as the answers to these questions serve as necessary input for obtaining certifications or labels. When two questions are equivalent, but they ask for different units, the answer of the merged question should be converted back to the original method(s). Therefore, we have implemented unit conversions, for instance from dollars to euros or from kilowatt-hours to megajoules.

When different question types are involved, the backwards transformations are more complex. We have conceptually designed these transformations, but not yet implemented. Find all backward transformation cases and their conceptual solution in the technical report [30]. Implementing the backward transformations is crucial for enhancing the value proposition of our solution.

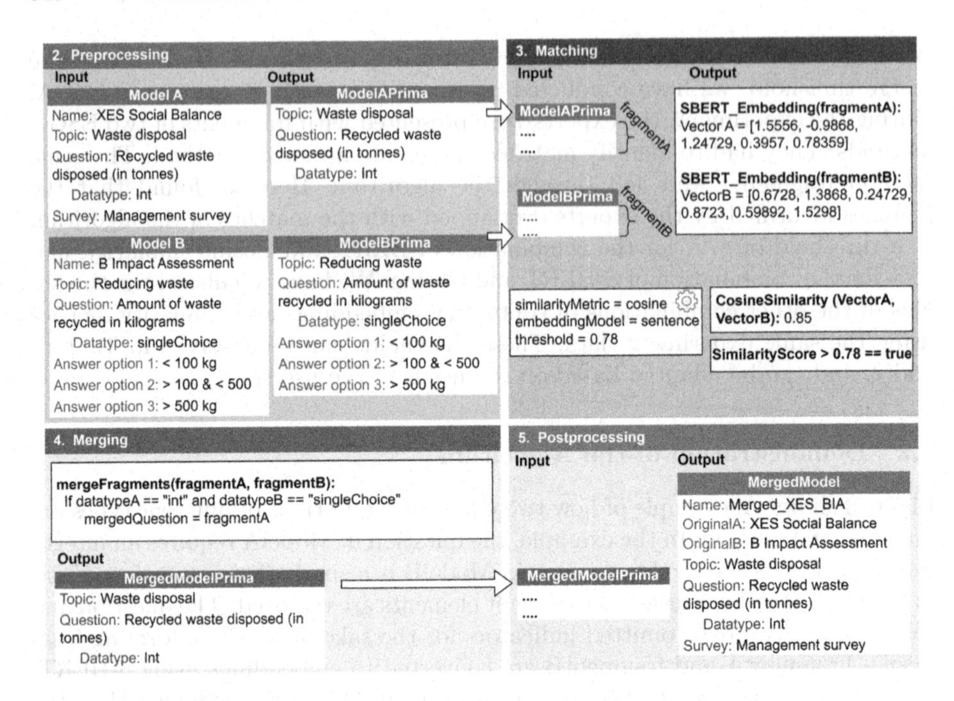

Fig. 3. A simplified example of how the match and merge algorithms work. In this case, S-BERT and Cosine similarity are used. Indicators are omitted for the sake of brevity. The activity numbers conform to the ones in Fig. 2.

5 Validation

To validate our proposed solution we run an experiment with 4 ESGA methods in the market. The methods are the B Impact Assessment [36], which consists of 360 fragments; the XES Social Balance [5], 246 fragments; UniSAF [13], 87 fragments; and STARS [35], 911 fragments. Find elaborate process models of each of the methods in a technical report [25].

Key Finding 1: *Our proposed solution can effectively merge several ESGA methods, which can then be operationalised in our model-driven tool called openESEA.*

Following the process in Fig. 2, with the 4 ESGA methods mentioned above as input, yields the results presented in Tables 1 and 2. The algorithm can successfully merge the models, based on the matches, resulting in one merged model that can be parsed by the openESEA interpreter.

Key Finding 2: *The S-BERT contextual embedding in combination with Cosine similarity yields the best results for matching ESGA methods.*

Following the results in Tables 1 and 2 we derive that the combination of the S-BERT model and Cosine similarity results in the best overall performance.

Table 1. The results of the matching function, using BERT as the contextual embedding. We present the true positives (TP) and false positives (FP)

Test nr.	Model A	Model B	Metric	Threshold	Matches	TP	FP
1	XES	BIA	Cosine	0.87	6	0	6
1	STARS	UniSAF	Cosine	0.87	22	6	16
2	XES	BIA	Euclidean	4.66	2	0	2
2	STARS	UniSAF	Euclidean	4.66	16	1	15
3	XES	BIA	Manhattan	103.26	4	0	4
3	STARS	UniSAF	Manhattan	103.26	16	2	14

Table 2. The results of the matching function, using S-BERT as the contextual embedding

Test nr.	Model A	Model B	Metric	Threshold	Matches	TP	FP
4	XES	BIA	Cosine	0.78	5	1	4
4	STARS	UniSAF	Cosine	0.78	18	5	13
5	XES	BIA	Euclidean	3.93	4	1	3
5	STARS	UniSAF	Euclidean	3.93	30	5	25
6	XES	BIA	Manhattan	60.20	4	1	3
6	STARS	UniSAF	Manhattan	60.20	26	3	23

We used this combination of techniques to investigate how to further increase the number of true positives. Table 3 shows that when the threshold is lowered, the number of true positives increases. For the comparison between STARS and UniSAF, the precision peaks on a threshold of 0.80 and for XES and the BIA at 0.77. These preliminary results indicate that the threshold determined using the experiment data is not ideal. Moreover, the ideal threshold is dependent on the ESGA models that the algorithm is used on.

Key Finding 3: *Using contextual embeddings and similarity metrics to identify overlap in ESGA methods results in considerable time savings.*

To put our outcomes into perspective we compare the time it would take a human to compare two models with the time that it took the algorithm to compare the same models. The algorithm took 2 hours, 44 minutes and 46 seconds to compare XES and the BIA (using S-BERT and Cosine similarity). Comparing STARS and UniSAF took 2 hours, 20 minutes and 7 seconds. Based on the results of the expert assessment, we derive that comparing two method fragments takes a human roughly 2 minutes. If a human had to compare the full XES and BIA methods, it would take `246 × 360 × 2 = 177.120 minutes` (or 2952 hours). After using the algorithm to compare XES and the BIA on a threshold of 0.77, a human will only have to compare 8 matches manually, which will take roughly 16 minutes. Based on this rough comparison we can conclude that using our proposed solution can save sustainability professionals a considerable amount of time to identify overlap in ESGA methods. Naturally, this estimation is an overstatement and the merges can be stored so that it is not necessary for each company to make its own comparison. In Sect. 6 we discuss how the precision of the algorithm can be further improved.

Table 3. Matches using the S-BERT model and Cosine similarity, at different thresholds

Test nr.	Model A	Model B	Threshold	Matches	TP	FP	Precision
7	XES	BIA	0.76	8	3	5	37.50%
7	STARS	UniSAF	0.76	20	6	14	30.00%
8	XES	BIA	0.77	8	3	5	37.50%
8	STARS	UniSAF	0.77	20	6	4	30.00%
9	XES	BIA	0.78	5	1	4	20.00%
9	STARS	UniSAF	0.78	18	5	13	27.78%
10	XES	BIA	0.79	5	1	4	20.00%
10	STARS	UniSAF	0.79	13	4	9	30.77%
11	XES	BIA	0.80	4	1	3	25.00%
11	STARS	UniSAF	0.80	10	4	6	40.00%

Key Finding 4: *An ESGA method engineer deems our solution useful, clear, and user-friendly.*

Additionally, to gain some insight regarding the usefulness of our algorithm, we interviewed an expert in ESGA. The expert found our solution intriguing due to its potential time-saving benefits and user-friendly interface. The expert highlighted the tool's ability to mitigate "survey fatigue" by eliminating redundancy and suggested that combining ESGA methods could enhance benchmarking opportunities. They also mentioned a drawback: companies that create ESGA methods often hesitate to share their ESGA methods because they consider the methods a core part of their value proposition. Lastly, the expert raised concerns about the complexity of merging ESGA methods and wondered if we will be able to address nuanced aspects such as defining conditions and contextual dependencies in questions.

6 Discussion and Conclusion

While validating the match and merge algorithm we found several approaches for improving the number of true positive matches, which can be considered as future work. Firstly, ESGA questions tend to be very elaborate and answering one question may require entering multiple data points. We refer to these questions as *compound questions*. An example of a compound question can be found in the example below.

```
Compound question
Q1: State the greenhouse gas emissions of scope 1, scope 2,
scope 3, and the total

Split up questions
Q1a: State the total greenhouse gas emission
Q1b: State the greenhouse gas emission of scope 1, scope 2,
and scope 3
```

There is an equivalent question in another method, which reads "Q48: State the total greenhouse gas emission". A match between Q1 and Q48 would be regarded as a false positive. However, the split-up questions Q1a and Q48 would have a higher similarity score and be considered a true positive. Upon further inspection of ESGA methods, we found that compound questions occur frequently. Hence, splitting up compound questions in the preprocessing phase will most probably result in more true positive matches.

Because ESGA methods assess specific and sensitive ethical, social and environmental topics, the language used in questions and indicators is highly influenced by the ethical values and political beliefs of the company that creates the method. At times this causes the language in the questions to be convoluted and complex. As an example, XES refers to "remunerations", rather than "salaries" or "wages". Apparently, "remunerations" and "salaries" are far apart in the vector space. Therefore, there is a low similarity between questions that ask for remunerations and questions that ask for salaries, while a human would probably consider these questions to be a match. To alleviate this issue, future work can consider lemmatisation, which maps similar words to a common root word (a.k.a. lemma).

Secondly, as a proof-of-concept, we used two different contextual embeddings, BERT and S-BERT. However, using contextual models with a larger vector space could potentially improve the results [12]. Therefore, future work could also entail comparing other models, such as GPT 4.0.

In terms of validity threats, we are convinced that our approach will work in a real-world scenario. We foresee one barrier in using our algorithm, which is that the ESGA methods have to be specified according to our DSL. Converting the methods into models that are compliant with the DSL is a time-consuming task. To overcome this problem we have created an Excel sheet that automatically converts indicators, questions and topics to the format specified in the openESEA DSL. Moreover, during the validation, we have not evaluated the 4 ESGA methods manually. As a result, we cannot identify true and false negatives and calculate the recall. The recall metric can give a better insight into the performance of the algorithm. We aim to perform a more thorough validation after we have further improved the preprocessing phase.

Our work demonstrates how the power of contextual embeddings can be leveraged to optimise sustainability accounting. So far, word embeddings and similarity metrics have been used for sentiment analysis [34], analysing social media posts, and calculating the similarity between documents [16]. However, using contextual embeddings to compare and merge documents remains a rather unexplored domain [38].

In conclusion, we have presented our five-step approach for matching and merging environmental, social and governance accounting methods, using language models. Our 4 key findings instil optimism regarding the potential improvement of sustainability accounting. This, in turn, can leave companies more time to evolve into socially responsible entities. Given the sustainability concerns surrounding language models, we cannot convincingly claim that the current developments in language models will result in a revolution. However, our work demonstrates how these new technologies can optimise sustainability accounting and open new routes towards more sustainable development.

References

1. Bernstein, P.A., Melnik, S.: Model management 2.0: manipulating richer mappings. In: SIGMOD, pp. 1–12 (2007)
2. Bernstein, P.A., Halevy, A.Y., Pottinger, R.A.: A vision for management of complex models. ACM SIGMOD Rec. **29**(4), 55–63 (2000)
3. Bettini, L.: Implementing domain-specific languages with Xtext and Xtend. Packt Publishing Ltd (2016)
4. Brunet, G., Chechik, M., Easterbrook, S., Nejati, S., Niu, N., Sabetzadeh, M.: A manifesto for model merging. In: GaMMa, pp. 5–12 (2006)
5. Crusellas, R.A., Padilla, R.S., Solidaria, X.D.: El Balance Social de la XES: 10 años midiendo el impacto de la ESS en Cataluña. In: UNTFSSE (2019)
6. Czarnecki, K., Helsen, S.: Feature-based survey of model transformation approaches. IBM Syst. J. **45**(3), 621–645 (2006)
7. Devlin, J., Chang, M.W., Lee, K., Toutanova, K.: Bert: pre-training of deep bidirectional transformers for language understanding. arXiv preprint arXiv:1810.04805 (2018)
8. Ding, Y., Ma, J., Luo, X.: Applications of natural language processing in construction. Autom. Constr. **136**, 104169 (2022)
9. España, S., Bik, N., Overbeek, S.: Model-driven engineering support for social and environmental accounting. In: RCIS, pp. 1–12. IEEE (2019)
10. España, S., Ramautar, V., Martín, S., Thorsteinsdottir, G., Anneria-Sinaga, Y., Pastor, Ó.: Why and how responsible organisations are assessing their performance: state of the practice in environmental, social and governance accounting. In: COPERMAN. Springer, Cham (2023)
11. EU: Directive (EU) 2022/2464 of the European Parliament and of the Council of 14 December 2022 amending Regulation. ELI (2022). http://data.europa.eu/eli/dir/2022/2464/oj. Accessed 11 Nov 2023
12. Garg, S.B., Subrahmanyam, V.: Sentiment analysis: choosing the right word embedding for deep learning model. In: Bianchini, M., Piuri, V., Das, S., Shaw, R.N. (eds.) Advanced Computing and Intelligent Technologies. LNNS, vol. 218, pp. 417–428. Springer, Singapore (2021). https://doi.org/10.1007/978-981-16-2164-2_33
13. Green Office Movement. https://www.greenofficemovement.org/sustainability-assessment/
14. Hertling, S., Paulheim, H.: Olala: ontology matching with large language models. In: K-CAP, pp. 131–139 (2023)

15. Krislock, N., Wolkowicz, H.: Euclidean Distance Matrices and Applications. Springer, Heidelberg (2012)
16. Kusner, M., Sun, Y., Kolkin, N., Weinberger, K.: From word embeddings to document distances. In: ICML, pp. 957–966. PMLR (2015)
17. Li, B., Han, L.: Distance weighted cosine similarity measure for text classification. In: Yin, H., et al. (eds.) IDEAL 2013. LNCS, vol. 8206, pp. 611–618. Springer, Heidelberg (2013). https://doi.org/10.1007/978-3-642-41278-3_74
18. Liu, Q., Kusner, M.J., Blunsom, P.: A survey on contextual embeddings. arXiv preprint arXiv:2003.07278 (2020)
19. mehrWerte. https://mehrwerte.at/impact-compass/
20. Microsoft: 2022 Environmental Sustainability Report (2023). https://query.prod.cms.rt.microsoft.com/cms/api/am/binary/RW15mgm. Accessed 25 Jan 2023
21. Mohibullah, M., Hossain, M.Z., Hasan, M.: Comparison of euclidean distance function and manhattan distance function using k-mediods. IJCSIS **13**(10), 61–71 (2015)
22. Nejati, S., Sabetzadeh, M., Chechik, M., Easterbrook, S., Zave, P.: Matching and merging of variant feature specifications. Trans. Softw. Eng. **38**(6), 1355–1375 (2011)
23. Peeters, R., Bizer, C.: Entity matching using large language models. arXiv preprint arXiv:2310.11244 (2023)
24. Rahm, E., Bernstein, P.A.: On matching schemas automatically. VLDB J. **10**(4), 334–350 (2001)
25. Ramautar, V., España, S.: Domain Analysis of Ethical, Social and Environmental Accounting Methods. arXiv preprint arXiv:2208.00721 (2022)
26. Ramautar, V., España, S.: The OpenESEA modeling language and tool for ethical, social, and environmental accounting. CSIMQ (34) (2023)
27. Ramautar, V., España, S.: The openESEA Modelling Language for Ethical, Social and Environmental Accounting: Technical Report. arXiv preprint (2023). https://doi.org/10.48550/arXiv.2205.15279
28. Ramautar, V., España, S., Brinkkemper, S.: Task completeness assessments in the evolution of domain-specific modelling languages. In: Indulska, M., Reinhartz-Berger, I., Cetina, C., Pastor, O. (eds.) CAiSE 2023. LNCS, vol. 13901, pp. 314–329. Springer, Cham (2023). https://doi.org/10.1007/978-3-031-34560-9_19
29. Ramautar, V., España, S., Pastor, Ó.: A survey on environmental, social and governance practices: technical report (2024). https://osf.io/y5ghq
30. Ramautar, V., Ritfeld, N., Brinkkemper, S., España, S.: Optimising sustainability accounting: technical report (2024). https://doi.org/10.17632/gg5cyr9scb.1
31. Reimers, N., Gurevych, I.: Sentence-bert: sentence embeddings using siamese bert-networks. arXiv preprint arXiv:1908.10084 (2019)
32. Rodrigues, D., Silva, A.D.: A study on machine learning techniques for the schema matching network problem. JBCS **27**, 1–29 (2021)
33. Sharbaf, M., Zamani, B., Sunyé, G.: Conflict management techniques for model merging: a systematic mapping review. Softw. Syst. Model. **22**(3), 1031–1079 (2023)
34. Thongtan, T., Phienthrakul, T.: Sentiment classification using document embeddings trained with cosine similarity. In: ACL, pp. 407–414 (2019)
35. Urbanski, M., Filho, W.L.: Measuring sustainability at universities by means of the sustainability tracking, assessment and rating system (STARS): early findings from STARS data. Environ. Dev. Sustain. **17**, 209–220 (2015)
36. Villela, M., Bulgacov, S., Morgan, G.: B Corp certification and its impact on organizations over time. J. Bus. Ethics **170**, 343–357 (2021)

37. Wieringa, R.J.: Design Science Methodology for Information Systems and Software Engineering. Springer, Heidelberg (2014). https://doi.org/10.1007/978-3-662-43839-8
38. Wu, J.M., Belinkov, Y., Sajjad, H., Durrani, N., Dalvi, F., Glass, J.: Similarity analysis of contextual word representation models. arXiv preprint arXiv:2005.01172 (2020)
39. Yu, P., Xu, H., Hu, X., Deng, C.: Leveraging generative AI and large language models: a comprehensive roadmap for healthcare integration. In: Healthcare, vol. 11, p. 2776. MDPI (2023)

Evaluation and Experience Studies

Adaptive Portfolio Management Based on Complexity Theory and Sociotechnical Design

Jose Antonio Ortega[1](✉), Oscar Pedreira[1], and Mario Piattini[2]

[1] Centro de Investigación CITIC, Universidade da Coruña, A Coruña, Spain
{jose.ortegab,oscar.pedreira}@udc.es
[2] ITSI, Universidad de Castilla-La Mancha, Ciudad Real, Spain
mario.piattini@uclm.es

Abstract. Current market contexts are pushing leaders to face the need for new business paradigms and to adapt their organisations to an ever-faster changing, dynamic and even unpredictable world. Organisations are requiring new ways of operating and new IT governance approaches that allow them to act quickly while remaining alert to the signals of the environment in which they operate. In this sense, we have focused our work on the design and implementation of an Adaptive Portfolio Management (APM) approach based on complexity and sociotechnical design principles in two Spanish organisations. After some Action-Research cycles, we conducted a situational analysis phase through several focus groups, the results of which are presented in this paper. Our findings show that the sociotechnical approach has brought to light the interrelationship between the governance model and other dimensions such as strategy, operating models, organisational design, people and culture, and technology. One of the key findings of the diagnosis is that, by implementing a governance model devoid of deterministic principles, we can readjust the other dimensions and introduce changes to facilitate the organisation's ability to adapt to dynamic environments.

Keywords: IT Governance · IT PPM · IT Project Portfolio · Business Agility · Complexity theory · Sociotechnical · Agile · Lean

1 Introduction

The events of recent years, ranging from the pressure for technological disruption to socio-economic uncertainties, are challenging companies to develop new organisational capacities enhancing their adaptability to ongoing changes. However, this need for organisational adaptability is not a new concept; in fact, it is one of the principles of organisational design studies. As such, Schumpeter indicated in [1], that for organisations to survive they have to adapt to their environment. More recently, as John Chambers of Cisco pointed out: "*if you don't reinvent yourself, change your organisation structure; if you don't talk about speed of innovation—you're going to get disrupted. And it'll be a brutal disruption, where the majority of companies will not exist in a meaningful way 10*

J. Araújo et al. (Eds.): RCIS 2024, LNBIP 513, pp. 357–375, 2024.
https://doi.org/10.1007/978-3-031-59465-6_22

to 15 years from now" [2]. This situation requires organisations to be proactive in detecting changes in the market, while at the same time having the ability to adapt and respond to their customers [3]. In this context, information technologies are called to be a vehicle to foster changes in business models or achieve greater efficiency in business processes [4]. The demand for flexibility and responsiveness in portfolio governance models has been occupying space both in the academic literature [3, 5, 6], and in the IT practitioner's literature. Some Gartner Reports [7] have been indicating the need for more adaptive governance models, moving away from the traditional Command-and-Control models.

This paper aims to collect the conclusions of a situational analysis phase, conducted at the end of the third cycle of a much larger research project, involving the design and implementation of an Adaptive Portfolio Management model (APM) in two Spanish companies, based on Action-Research (AR). This situational analysis phase, which was conducted through qualitative research, consisted of collecting data from several focus groups that were conducted with 15 professionals. To conduct these focus groups, we proposed the following research question: **What are the main challenges to consider when implementing the adaptive governance model in different organisations?**

The objective of these focus groups was to obtain contributions to enhance the design of the APM model and strengthen the adoption model, generating further insight and knowledge for both practice and academia. This paper is structured as follows. Section 2 provides an overview of the whole research project on the design and implementation of an Adaptive Portfolio Management (APM) model to facilitate understanding of the context of this paper. Section 3 contains a summary of the theories upon which the entire research project is built, and which also form the basis for the conclusions of this paper. Section 4 describes the research method. Section 5 describes the most important findings of the qualitative research. Section 6 outlines both the limitations of the study. Section 7 describes next steps for ongoing research and possible future research directions.

2 Overview of the Whole AR-Based Research Project

As mentioned above, it is important to highlight that in this paper, we present the findings that emerged during a reflection phase (using a focus group approach). This reflection is part of a much broader project, whose aim is to design an adaptive IT governance model (based con complexity theory principals) and its adoption model (based on sociotechnical design principals) that can be applied to any type of organisation. The entire research project is conducted using the Action Research (AR) method to enhance knowledge in real organisational contexts through collaboration between researchers and practitioners [8]. In these initial AR cycles, the designed APM model was implemented in two Spanish organisations (Table 1).

The first cycles of this research project were launched in the 4th quarter (Q4) of 2020, when the problem to be solved was established. During the first half of 2021, interviews were conducted in several large organisations (over 90 people) to surface the biggest challenges concerning the governance approach they had been using. In Q3 2021, a first draft Adaptive Portfolio Management model was designed, and implementation began in Q4 2021 in two organisations (Table 1). By the end of Q1 2022 (2nd AR Cycle), activities were initiated to collect findings in order to make adjustments to the model.

Finally, following the implemented adjustments during the third AR cycle, at the beginning of Q1 2023, several focus groups were conducted with up to 15 practitioners based on qualitative research methods. **The findings that emerged from these focus groups are presented in this paper**. The practitioners belong to a consultancy firm, and they were involved in the transformation projects in two Spanish organisations (Table 1). In these transformation projects the adaptive portfolio management model was designed and implemented. These projects were the basis of the entire research process.

For the design of the APM model itself, we have distanced ourselves from models such as COBIT [9] or ISO 38500 [10], due to their highly stable environment-oriented and command-and-control approaches. We have been inspired by the Lean Portfolio Management model from SAFe [11], from which we have drawn many ideas, but we observed a significant dependency on the SAFe framework itself. Fundamentally, we employed the Agile Portfolio Management approach by Krebs [12], upon which we constructed a series of layers that would serve us to have a generalizable model to any context. Moreover, the reflection on the APM model's design was always grounded on the principles of complex adaptive systems [8]: autonomy, purpose, diversity, self-organisation, adaptability, emergence, and resource exchange. The APM model was based on:

- Every portfolio component, including initiatives and programs, should focus on objectives, key results (OKRs), and value-driven outcomes to foster a culture of accountability, centring on initiatives with cross-functional teams over projects.
- A flexible framework for initiative implementation allows comparison without stifling adaptability.
- Emphasising roles over hierarchy and fostering accountability.
- Communication events improve portfolio visibility and encourage open dialogue, supporting a culture that measures and aligns initiative progress with Key Results.
- Portfolio monitoring occurs quarterly, aiding in strategic initiative evaluation for informed decision-making.
- Governance automation, minimizing bureaucratic decision-making barriers.

We developed the APM model's implementation strategy through Sociotechnical Systems Thinking [9], focusing on the integration of technical and social factors in organisational change. The aim is to ensure that changes, such as new technologies or news ways of working, consider their overall systemic effects to preserve effectiveness. It views organisations as interconnected systems, where changes in one part affect the entire structure. During the first two AR cycles of design and implementation of the APM model, through the research processes, we shaped the following sociotechnical framework for the adoption model (Fig. 1):

1. **Strategy and strategic planning**: considering that the strategy has to respond to a volatile and uncertain environment.
2. **Governance model**: an adaptive approach, balancing delivery, performance, autonomy and control with organisational capacity, and being a lever to remove blockages and risks.
3. **Operating models** that foster the evolution of business and IT ways of working to achieve greater adaptability and innovation.

4. **Organisational design**: enabling new evolutionary structures to gain liquidity and make centralisation/decentralisation part of the strategy to gain flexibility.
5. **People, talent and leadership**: implementing an adaptive governance model, away from classic command-and-control approaches, requires establishing reskilling strategies as well as team coaching approaches.
6. **Technology** is a key enabler to facilitate decision-making and to implement centralisation and decentralisation strategies.

This sociotechnical framework (Fig. 1) aims to serve as a guide for implementing the APM model in any organisation. **We wanted to use this framework as a structure for the focus groups, whose findings are collected in this paper**.

Table 1. Reference organisations for the implementation AR cycles of the governance model

Case / Sector	Size of the Company	Organisational model	Model of governance	Size of persons involved in Adaptative portfolio management
Spain-based international company in the utilities sector	Up to 15.000 employees WordWide	Hierarchical/ centralised. Implementing Agile practices and struggling with a change of mindset	Based on traditional PPM and PMI	Up to 250 people directly involved, from the People and Organization Department and from IT Areas and other transversal Areas
International Retail Company based in Spain	Up to 100.000 employees Worldwide	Flat hierarchical. Process-focused Agile initiatives. Team topologies based con Spotify model	Based on traditional PPM and PMI	Up to 500 people directly involved, from the Commercial Areas, IT Departments and other transversal areas like logistics, facilities etc..

3 Background

The entire research project has been based on a set of theories that are reflected in an important base of both academic and professional literature. Whilst the results presented in this paper are confined to the final stage of our research project, it is essential to review this background in order to provide a full context for the conclusions drawn in Sect. 5.

3.1 Project Portfolio Management

We consider Project Portfolio Management (PPM) as a fundamental practice of IT governance [10], whose objective is to maximize the business value of IT investments. Such an aim requires a continuous alignment between IT and Business [11]. Several kinds of research contribute to the implementation of portfolio management models. It is worth mentioning the work by Cooper [12], which identifies the main objectives of Portfolio Management. Other researchers proposed PPM approaches based on Agile principles [6, 13, 14]. Krebs [15] discusses challenges in three portfolio domains: (1) Project portfolio: too many active projects and wrong mix of projects; (2) Resource portfolio: lack of vision, too many projects while not enough (adequate) resources and lack of feedback; (3) Asset portfolio: legacy systems as obstacles and under-estimation of the total cost of ownership. According to him, the implementation of a project management office (PMO) and resource transparency is key to agile project management.

Traditional governance frameworks, such as COBIT [16] or ISO 38500 [17], serve as significant references but have fallen short in addressing the needs of complex or rapidly changing environments. In this context, for the design of the adaptive portfolio model, we drew inspiration from Lean Portfolio Management [13], which emphasizes the crucial identification of workflow streams, a task not always feasible from the outset in many organisations. Therefore, Krebs' Agile Portfolio Management [15] approach has been our foundation, upon which we built our initial model. We also considered other elements outlined by Stettina and Hörz [6], who propose four practice domains to structure a PPM process: (1) strategy and roadmap, (2) identify and funnel, (3) review prioritization and balance, (4) allocate and delegate. Ahmad [18] outlines different tools and methods for framing PPM processes in Lean and Agile organisational contexts. Puthenpurackal [11] identifies several aspects of PPM impacted by Agile delivery practices, such as alignment, continuous delivery, the adaptive nature, learning through feedback, the financial processes and performance indicators.

3.2 Complex Adaptive Systems

Traditional governance models are often based on mechanistic approaches, with a Command-and-Control orientation and based on prediction [3]. These approaches are failing to be helpful for organisations to adapt to current contexts [6]. An alternative is to consider organisations as Complex Adaptive Systems (CAS) [8, 19]. A CAS is composed of heterogeneous elements, called agents. The interactions of the agents allow the system to adapt to the context, displaying emergent properties and thus producing unintended consequences [20]. Therefore, the behaviour of a CAS is not explained by the individual behaviour of its agents, but by the interconnections between them. From the complexity point of view, there are a number of outstanding research works that have been an important source for this research. It is worth mentioning the work by Lichtenstein [21] focusing on the concept of emergence, which refers to the creation of order and new structures in complex systems, challenging the idea that complex systems always act on their own and arguing that self-organisation is not spontaneous and requires external constraints and structures. Uhl-Bien [22] opens the debate on the need for a new model of leadership in the knowledge age. The authors argue that traditional models of leadership, which are top-down and bureaucratic, are not suitable for a knowledge-based economy. Instead, they propose Complexity Leadership Theory, which sees leadership as a complex interactive dynamic that enables learning, innovation, and adaptability in complex adaptive systems. Sweetman and Conboy [20], consider the portfolio itself as a CAS, analysing the properties of projects as if they were agents in an adaptive complex portfolio moving in dynamic contexts. They introduce, as well, the concept of attractor states in which a PPM can exist, uncovering the factors that enable organisations to adapt in dynamic or adverse environments. The attractor state requires being on the edge of chaos so that a CAS can continue to adapt to the environment. Other authors, such as Kaufmann, Kock and Gemünden [23] introduce complexity into strategy formulation, combining intentional (top-down) strategy with emergent strategy from teams at the frontline (bottom-up). They suggest that the combination of strategy and complexity has a major impact on making agile portfolio management successful in business terms. Starsia [24] challenges traditional strategic planning, which is highly predictive. Its focus is on the need for a strategic planning model based on complexity theory that enables organisations to pivot strategy considering changing market conditions. Tsilionis et al. [25] proposed a framework for the agile adoption of strategic opportunities, assessing their value at multiple levels within the organisation.

3.3 Design of Sociotechnical Organisational Systems

Complexity theory led us to sociotechnical systems [9], as a perspective for the design of human-centred organisations. This perspective, which was born in the UK coal mines in the post-war period, conceives the organisation as an entity composed of three elements: (1) Social: consisting of people, culture, and their relationships; (2) Technical: capabilities, processes, tools and technology; and (3) The system itself concerning the ecosystem, which gives purpose to the whole system. This perspective allows us to understand that when we introduce changes in the system, the culture and people layer is the most complex one to change. This means that any adoption model must consider all layers if we want the system to change to another attractor state [26]. If new attitudes and behaviours in people and culture do not emerge, change will not thrive. According to this approach, organisational changes (which include, for example, the adoption of new management models or incorporating disruptive technologies etc.) are dense from a social perspective [27], which implies a greater emphasis on creating relationships within the system than on the specific methods of achieving change. For an organisation to be more adaptive, it needs to create a more human-centred organisational environment [28]. The 2023 Gallup report [29] indicates that, globally, 77% of employees in the organisations surveyed do not feel committed to their work. This data illustrates how difficult it is to achieve team accountability in an environment where organisations need to adapt quickly. 2022 Gartner Report on CIO Priorities [30] even calls for creating human-centred work environments, combining governance models, operating models and organisational design to create healthy work environments. There is a growing trend in business environments to create organisations that move away from mechanistic and hierarchical approaches. Examples include the organisations cited by Minnaar and De Morree [31], the case of Shell [32], or even such innovative organisational approaches as Haier [33, 34].

4 Research Design and Case Description

As previously mentioned, upon completing the 3rd Action Research (AR) cycle of our research project, we aimed to create a reflective space with 15 practitioners involved in the design and implementation of the Adaptive Portfolio Management (APM) model in the two case study organisations. Our goal was to reflect on the challenges and conclusions reached so far, with the purpose of identifying patterns that would allow us to pinpoint the challenges in generalising the implementation of this adaptive governance model across organisations from different contexts.

To achieve this, we employed the focus group technique to explore the knowledge and experiences of these practitioners in an interactive environment. Using a qualitative research methodology, the aim was to gather detailed information about the consultants' thoughts and behaviours, as well as to delve into the "why" and "how" of their opinions and actions [35].

To facilitate the process, we began with the sociotechnical framework (Fig. 1), where we transformed each framework factor into a "reflection section", as we had identified in previous AR cycles that these factors constituted the sociotechnical foundation of the evolutionary process of the case organisations. Therefore, the approach was to conduct two focus groups per factor, each consisting of 7/8 individuals. Dividing the participants into two groups allowed us to validate the conclusions of both groups for each factor. Additionally, we conducted a launch session and a final session for the convergence of results, identifying eight sections as follows:

- **A setup section**, to agree and reflect on the thematic sections with the participants.
- **Six thematic sections** (Fig. 1), based on the sociotechnical factors that we identified throughout the research project.
- **and a closure section**, to serve as a point of convergence for all the reflections made.

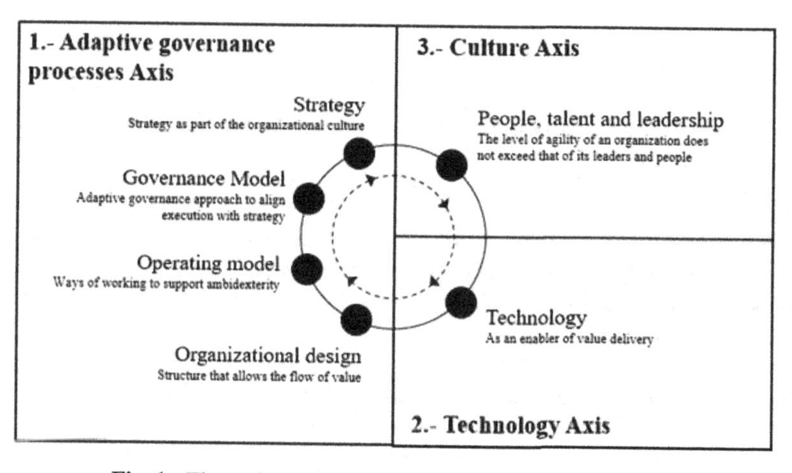

Fig. 1. Thematic section based on sociotechnical approach.

Therefore, with these eight sections, we conducted a total of up to 16 focus groups, each lasting a maximum of 1.5 h. All the professionals involved had more than 5 years of experience in transformation processes. They also all had previous awareness and training in complexity theory and sociotechnical systems design.

The **first setup section**, with its two corresponding focus groups, made it possible to validate, with the participants, the six sociotechnical elements (Fig. 1) that constitute each thematic section. Subsequently, **two focus group per thematic section** (12 focus group) were held to uncover insights. A **final or closure section** (with two focus group) was held to reflect on the general conclusions that form the basis of the contributions in this paper.

In all the thematic focus group, the reflections included the links between the dimension addressed and the others. It is worth noting that the governance model was a cohesive element in this systemic approach [11]. The design of the focus groups was crucial in order to bring out reflections of value to the research question. Therefore, each focus group was designed to have a first part of reflection on different challenges, related to attitudes, beliefs, and behaviours that the practitioner had observed in the two organisations. The challenges were intended to reflect how the sociotechnical perspective impacted the implementation of an APM. The second part of the focus group served to group and prioritise the findings, based on a criterion of greater impact on the generalisation of the adoption of the governance model. The last part of the focus consisted of discussing the most important findings and next steps. After each pair of sectional focus groups, work was conducted to verify and validate the patterns that tended to emerge from the reflections made by the practitioners, and which are presented in this paper.

5 Case Findings

The findings of the focus groups, which are described below, have been organized into the sections mentioned above:

- Setup section
- 6 Thematic sections: strategy, governance model, operating model, organisational design, people talent and leadership, and technology. These thematic sections are grouped according to the axes shown in Fig. 1.
- Closure section

5.1 Setup Section: Sociotechnical Dimensions of Change

In these firsts focus groups, a reflection took place on how the sociotechnical approach had made it possible to understand the existence of a connection between business strategy, governance models, operating models, organisational design, technological architecture, and organisational culture (Table 2).

Table 2. Setup section Insights

Practical Insights	Theoretical Insights
The consultants concluded that organisational elements form interdependent layers or dimensions, highlighting the importance of a systemic perspective. Alterations to one element can have unintended consequences for all other elements, underscoring the interconnected nature of organisational systems. The sociotechnical approach provided a solid construct for rethinking challenges, offering a different lens through which to view organisational issues	From a research perspective, it becomes evident that the application of a systemic view in diverse organisational contexts not only validates the enduring relevance of sociotechnical systems theory but also extends its utility beyond traditional domains. This approach underscores the theory's adaptability and its potential to inform and enhance understanding in varied settings, thereby enriching both theoretical and practical applications of socio-technical principles [9, 36]
Complexity theory gave the consulting team a different perspective, where the understanding of the organisational system could be interpreted from the connection and relationship between the different elements of Fig. 1, and not from the mere causal relationships	The focus group's insights underscore the CAS's emergence principle, referred to the unexpected properties and behaviours that arise from the interactions of individual agents, roughly described by the phrase "the action of the whole is more than the sum of the actions of the parts" [21, 22, 37]

5.2 Adaptive Governance Processes Axis

Strategy Section
See Table 3.

Table 3. Strategy section insights

Practical Insights	Theoretical Insights
Strategy and operation are perceived as divided, needing the integration of strategy into the organisation's culture to foster agility and adaptability	The sociotechnical approach highlights the interdependence of business strategy, governance models, operating models, organisational design, technological architecture, and organisational culture [36]
The combination of intentional (top-down) and emergent (bottom-up) strategies is essential for allowing greater agility and flexible adjustments to the organisation's strategy	The emerging strategy recognition has been related to portfolio project success [23]

(*continued*)

Table 3. (*continued*)

Practical Insights	Theoretical Insights
Autonomy given to teams enables them to act as radars for detecting market changes, but is hindered by a Command-and-Control mindset in uncertain environments	The Command-and-Control and predictive mindset that can work in stable environments does not give organisations space to deal with high levels of uncertainty [8, 22, 24]
The annual planning process is improved by adopting a lighter approach that allows for iterative review of strategy and initiatives in short cycle	Implementing an APM serves as a catalyst for managers to recognize the necessity for change. This approach encourages a shift from traditional, rigid planning processes to more flexible, responsive planning that can adapt to market dynamics more effectively [38]
To grant autonomy to teams effectively, it is necessary to ensure they have the necessary expertise and leadership to ensure results	This insight is shared by academics [23, 39], that autonomous teams, with the necessary mastery, and not afraid to experiment, tend to have better results
The use of OKRs as predictive and control tools produce poor result	Organisations were very attracted by stability and balance [26]. It is critical to create awareness among managers about a new management mindset to guarantee the adaptability of organisations [39]

Governance Model Section
See Table 4.

Table 4. Governance section insights

Practical Insights	Theoretical Insights
The APM model's non-prescriptive nature allows for adaptability to the organisation's current reality, emphasizing the balance between exploitation and exploration	The necessity of a balance between control and autonomy in adaptive governance models aligns with complexity theory, highlighting the dynamic interplay between structure and responsiveness within organisations [23, 26, 39]
Balancing control with autonomy requires a management layer that understands complexity and adaptive leadership, despite resistance from many leaders	This resistance underscores the deep-rooted nature of predictive management paradigms, validating the complexity theory's assertion about the challenges of shifting organisational mindsets towards adaptability [39]

(*continued*)

Table 4. (*continued*)

Practical Insights	Theoretical Insights
The challenge of synchronizing exploitation and exploration is exacerbated by high-risk initiatives led by individuals lacking the necessary mindset or expertise	This finding illustrates the practical challenges of implementing complexity theory in organisational settings, emphasizing the importance of leadership and expertise in navigating complex adaptive systems [22, 24, 40]
Implementing Krebs' approach of having multiple business portfolios necessitates synchronization across portfolios, including a portfolio of technological assets	This approach validates the sociotechnical perspective on organisational design, highlighting the interdependence between technology and business strategy in adaptive governance [36]
The implementation of an adaptive governance model is related to operating models, with a focus on transitioning towards value streams and identifying the value of initiatives	This shift towards value streams and the emphasis on value identification align with the principles of sociotechnical design, advocating for a holistic view of organisational effectiveness that integrates technical and social systems, with a focus on value and customer orientation [20, 23, 36, 41]
Management layers are attracted to prediction and Command-and-Control, seeking stability and balance, which may inhibit the adoption of adaptive governance models	This attraction to traditional management styles provides empirical evidence of the tension between stability and adaptability in organisations, a core theme in complexity theory and sociotechnical design [23, 39]

Operating Model's Section

See Table 5.

Table 5. Operating model section insights

Practical Insights	Theoretical Insights
Adoption of Lean and Agile practices was catalyzed by the implementation of an adaptive governance model, indicating a shift from traditional project management metrics (time, cost, scope) towards value-driven delivery	This observation underscores the theoretical premise that complex adaptive systems principles foster flexible and responsive operating models that prioritise value over predictive metrics, generating a positive relationship with Lean and Agile value systems and principles [23]

(*continued*)

Table 5. (*continued*)

Practical Insights	Theoretical Insights
The historical focus on project management created a predisposition towards measuring success through traditional metrics, leading to a potential misalignment with the value contribution. This necessitated a transition towards a Value Stream focus, particularly centered on customer needs	This aligns with the principles of Lean management and Agile methodologies, which emphasize customer value and responsiveness to change over adherence to pre-defined plans. It validates the theoretical adaptation of operational models to be more customer-centric and value-focused [3]
Consultants highlighted the importance of aligning initiatives with the organisation's strategic objectives and fostering the emergence of strategy from operational teams. This strategic alignment and emergent strategy facilitated organisational agility and adaptability	This reflects the theoretical framework of sociotechnical systems design, which advocates for the integration of social and technical elements to achieve strategic goals [36]. Moreover, it supports the complexity theory's view that adaptability emerges from the interaction of system components (e.g., teams, strategies) rather than from top-down control [39]

Organisational Design Section

See Table 6.

Table 6. Organisational design section insights

Practical Insights	Theoretical Insights
Implementing flexible or adaptive working models in hierarchical organisations presented challenges but also facilitated smoother adoption when staff allocation was based on roles and responsibilities rather than rigid organisational structures	This reflects sociotechnical systems theory, which emphasizes the importance of designing organisations that balance technical and social aspects. The move towards roles and responsibilities aligns with the principle of creating adaptive organisations capable of responding to environmental changes [23, 39]
The utilization of cross-functional teams and experimentation with various team topologies based on the nature of initiatives highlighted the need for different organisational designs to unlock initiative value while maintaining cost-effectiveness	This supports the concept within complexity theory that diversity and adaptability in team structure contribute to an organisation's resilience and innovation capability. The practical experimentation with team structures validates the theoretical premise that complex adaptive systems thrive on diversity and flexibility [20, 40]

(*continued*)

Table 6. (*continued*)

Practical Insights	Theoretical Insights
Some organisational units perceived the shift towards flexibility and self-organisation as a loss of power, indicating resistance to change from traditional hierarchical models	This observation underscores the challenges in applying complexity theory to organisational change, where existing power dynamics and resistance can impede the transition to more adaptive models. It highlights the need for managing change and addressing power dynamics within the context of sociotechnical system design [22, 39]
The necessity for raising awareness about complexity theory and organisational management among managers and staff to counteract the belief that removing hierarchical structures leads to spontaneous self-organisation	This aligns with theoretical insights that effective organisational design in complex systems requires a blend of top-down direction and bottom-up emergence, necessitating active leadership and a deep understanding of complexity principles among all organisational members [39, 42]
The goal of achieving organisational liquidity through staffing decisions based on value contribution rather than hierarchical positions	Reflects the theoretical understanding that in sociotechnical systems, optimizing the interaction between people and technology based on value contribution rather than traditional hierarchies enhances organisational agility and effectiveness [36]

5.3 Technology Axis

See Table 7.

Table 7. Technology insights

Practical Insights	Theoretical Insights
The integration of IT and non-IT participants in initiatives underscored the necessity for a mutual understanding of the IT value stream, highlighting the significance of features, defects, risks, and technology architecture debt on initiatives outcomes	This reflects the sociotechnical systems design principles, advocating for an integrated approach to organisational design that considers both the social and technical aspects of the work environment, emphasizing the importance of technology in enhancing organisational adaptability and efficiency [36]

(*continued*)

Table 7. *(continued)*

Practical Insights	Theoretical Insights
The consultants noted the essential role of technology in bridging the gap between business and IT professionals, fostering a deeper comprehension of how technology architecture impacts business operations and vice versa	Aligns with complexity theory's view that in complex adaptive systems, the interactions between different system components (in this case, IT and business domains) can lead to emergent, adaptive behaviours, enhancing the organisation's capacity to respond to environmental changes [8]
The push towards automating the governance model for promoting decentralization of control and enhancing team autonomy, with practical examples including the use of JIRA and low-code tools for implementing APM	This demonstrates the application of complexity theory in a practical setting, where the use of technology to decentralize decision-making and automate governance processes enables the system (organisation) to adapt more dynamically to internal and external changes, facilitating a more agile and responsive organisational model [18, 43]

5.4 Culture Axis (Including People, Talent and Leadership)

See Table 8.

Table 8. Culture insights

Practical Insights	Theoretical Insights
The transition from Command-and-Control to a complexity-based leadership approach is highly challenging, demanding significant energy and effort to facilitate a paradigm shift towards strategic agility and autonomy	Reflects the transition towards complexity leadership theory, which emphasizes adaptive, emergent leadership styles over traditional hierarchical models, aligning with the theoretical shift towards embracing complexity in organisational behavior and management [22, 39, 42]
Not all teams and leaders adapted equally to the demand for increased accountability and autonomy, with some expressing reluctance or lack of confidence in this new approach	These observations underlines the variability of the response of human systems to change, a basic tenet of sociotechnical systems theory, which posits that effective change management must consider the interaction between social dynamics and technological or structural change [36, 42, 44]
The sociotechnical perspective (Fig. 1) facilitated the analysis and adaptation of strategies, acknowledging the diversity in organisational systems and the importance of tailoring change processes to these variances	

(continued)

Table 8. (*continued*)

Practical Insights	Theoretical Insights
Leadership is viewed as an emergent quality that develops from a system's ability to adapt and innovate, moving away from the traditional singular leader model towards a more distributed and adaptive form of leadership	This aligns with the theoretical framework provided by complexity theory and sociotechnical systems design, suggesting that leadership in complex environments emerges from the interactions within the system rather than being imposed from above [36, 39, 42]
Implementing transparent communication mechanisms was crucial for fostering a culture of accountability, collaboration, and innovation across the organisation	It underlines the importance of transparency and open communication as key elements in building adaptive capacity in organisations, as emphasized in both complexity theory and socio-technical systems design principles [8, 36, 42]

5.5 Closure Section

See Table 9.

Table 9. Closure insights

Practitioners' Insights	Researchers' Insights
The consultants highlighted the multiple challenges across all sociotechnical dimensions in implementing the APM model. Their journey highlighted the importance of employing a sociotechnical perspective in change management, advocating a holistic view in which progress depends on all dimensions moving in unison. This perspective challenges traditional consulting paradigms, which often adopt a mechanistic approach, aiming to transform the state of the organisation towards an ideal, rather than evolving from the starting point of the organisational system. The consultants concluded that the AR process they went through in these two organisations did not only result in designing and implementing a more adaptive governance model, but that this was part of the change path. The actual outcome was that the organisations became more adaptable to market conditions, introducing major innovations in their organisational capabilities (practices, processes, strategies, structures and people)	For the researchers, these practical insights provide a rich empirical basis for theorizing the evolution of traditional governance models towards a new management paradigm that emphasizes organisational adaptability. The APM model, as evidenced by the consultants' experiences, pushes traditional organisations onto an evolutionary path, developing a management model that fundamentally prioritizes adaptability over the prescriptive nature of traditional governance frameworks such as COBIT or even Agile-oriented models that, despite their flexibility, impose considerable prescriptiveness in implementation. This shift not only highlights the limitations of conventional governance models in fostering organisational agility, but also underlines the potential of this APM to serve as a pivot for organisations aspiring to thrive in an era of unprecedented change

6 Threats to Validity: Limitations and Possible Generalisation

As in any study of this type, it is important to acknowledge the study's limitations. The conclusions drawn from the focus groups were based on an analysis of the state of the art after completing only three iterations of the AR cycle. Therefore, we can deduce that we would need more change cycles to mature the conclusions.

Overall, the AR was conducted in only two companies, therefore, the conclusions that emerged from the focus groups cannot be generalized, especially when we move in complexity. Nevertheless, the reference literature allows us to see patterns that are repeated in different studies [6, 43, 45, 46].

Nevertheless, we are aware of other factors that might pose challenges in generalising the results. Characteristics specific to types of organisations, for example, those with highly hierarchical and control-oriented organisational cultures, where power mechanisms can be inhibitors to adopting such governance models. Another factor could be the industry, for instance, in highly regulated sectors that significantly limit the capacity for adaptation. It is evident that this APM must be deployed in organisations from diverse contexts to enable us to validate whether it is generalisable or if its verticalisation is necessary.

Finally, the focus groups were conducted within a group of 15 consultants from the same company; they are carrying out the transformation projects. Therefore, the conclusions might be subject to influence due to the participants' long-standing collaboration and shared knowledge domains, potentially resulting in confirmation biases.

7 Future Work

Some of the future work we plan to address focuses on the following aspects:

- Future research could extend this study by applying the APM model across a broader range of organisations, including those of varying sizes, industries, and cultural contexts. This would not only enhance the ability to generalize the findings but also offer deeper insights into the model's adaptability in diverse settings.
- We had already identified the need to introduce the value stream concept in the APM model. At this point, we need to dig deeper into what set of interventions could be used as ways to trigger the change in this regard.
- Additionally, conducting further iterations of the action research cycle could validate and refine the model further, contributing to a more robust framework for adaptive governance.
- In this APM, we have shifted from a project portfolio to an initiatives portfolio paradigm to incorporate business considerations, marking a significant evolution given the project-oriented predisposition of the case organisations. However, we are finding that a further evolution is required towards product portfolios, which entails a significant change in the management approach.
- Future work must explore the evolution from managing technology assets to overseeing platform portfolios, considering the increasingly significant impact of digital transformation and AI on the organisation. This approach represents a shift in management focus towards an ecosystem platform portfolio approach.

- Incorporating the approach of the Complexity Leadership Theory [22] into the adoption model: This approach should help us to create a new management culture at the different management levels. It should also encourage the emergence of a culture of entrepreneurship, autonomy, and accountability throughout the organisation.

Acknowledgments. The present work has been funded by: PID2021-122554OB-C33 and PID2021-122554OB-C31 (OASSIS): partially funded by MCIN/AEI/10.13039/501100011033 and EU/ERDF A way of making Europe; GRC: ED431C 2021/53, partially funded by GAIN/Xunta de Galicia; and CITIC is funded by the Xunta de Galicia through the collaboration agreement between the Department of Culture, Education, Vocational Training and Universities and the Galician universities for the reinforcement of the research centers of the Galician University System (CIGUS).

We would also like to thank KELEA, and its consultant's team, for encouraging academic participation in these projects and bringing out the different lessons learned.

References

1. Schumpeter, J.A.: Economic Theory and Entrepreneurial History. Harvard University Research Center in Entrepreneurial History Change and the Entrepreneur, pp. 63–84. Harvard University Press (1949)
2. Chambers, J.: Cisco's John Chambers on the digital era (2016)
3. Horlach, B., Schirmer, I., Drews, P.: Agile portfolio management: design goals and principles. In: Twenty-Seventh European Conference on Information Systems (ECIS 2019), pp. 1–17 (2019)
4. Overby, E., Bharadwaj, A., Sambamurthy, V.: Enterprise agility and the enabling role of information technology. Eur. J. Inf. Syst. **15**, 120–131 (2006)
5. Luna, A., Costa, C., Moura, H., Novaes, M., Nascimento, C.: Agile Governance in information and communication technologies: shifting paradigms. JISTEM J. Inf. Syst. Technol. Manag. **7**, 311–334 (2010)
6. Stettina, C.J., Hörz, J.: Agile Portfolio Management: An Empirical perspective on the practice in use. Int. J. Project Manage. **33**, 140–152 (2015)
7. Gartner: Top Priorities for IT: Leadership Vision 2021 | Gartner (2021)
8. Cilliers, P.: What can we learn from a theory of complexity? Emergence **2**, 23–33 (2000)
9. Trist, E.: The evolution of socio-technical systems. a conceptual framework and an action research program | Toronto Ontario Ministery of Labour (1981)
10. De Haes, S., Van Grembergen, W.: An Exploratory study into IT governance implementations and its impact on business/IT alignment. Inf. Syst. Manag. **26**, 123–137 (2009)
11. Puthenpurackal Chakko, J., Huygh, T., De Haes, S.: Achieving agility in IT project portfolios – a systematic literature review. In: Przybylek, A., Miler, J., Poth, A., Riel, A. (eds.) LASD 2021. LNBIP, vol. 408, pp. 71–90. Springer, Cham (2021). https://doi.org/10.1007/978-3-030-67084-9_5
12. Cooper, R.G., Edgett, S.J., Kleinschmidt, E.J.: New product portfolio management: practices and performances. J. Prod. Innov. Manag. **16**, 333–351 (1999)
13. Leffingwell, D., Knaster, R.: SAFe 5.0 Distilled. Achieving Business Agility with the Scaled Agile Framework (2020)
14. Vähäniitty, J.: Towards Agile Product and Portfolio Management (2012)
15. Krebs, J.: Agile Portfolio Management. Microsoft Press, Washington (2008)

16. ISACA: COBIT 2019: Introduction and methodology (2018)
17. ISO: ISO / IEC DIS 38503 Information technology — Governance of IT — Assessment of governance of IT (2021)
18. Ahmad, M.O., Lwakatare, L.E., Kuvaja, P., Oivo, M., Markkula, J.: An empirical study of portfolio management and Kanban in agile and lean software companies. J. Softw. Evol. Process. **29**, 1–16 (2017)
19. Reeves, M., Levin, S., Harnoss, J.D., Ueda, D.: The five steps all leaders must take in the age of uncertainty. MIT Sloan Manag. Rev. **59**, 1–5 (2018)
20. Sweetman, R., Conboy, K.: Portfolios of agile projects: a complex adaptive systems' agent perspective. Proj. Manag. J. **49**, 18–38 (2018)
21. Lichtenstein, B.: Generative Emergence. A New Discipline of Organizational, Entrepreneurial, and Social Innovation (2014)
22. Uhl-Bien, M., Marion, R., McKelvey, B.: Complexity Leadership Theory: shifting leadership from the industrial age to the knowledge era. Leadersh. Q. **18**, 298–318 (2007)
23. Kaufmann, C., Kock, A., Gemünden, H.G.: Emerging strategy recognition in agile portfolios. Int. J. Project Manage. **38**, 429–440 (2020)
24. Starsia, G.: Plan to Pivot Agile Organizational Strategy in an Age of complexity. Morgan James Publishing, New York (2022)
25. Tsilionis, K., Wautelet, Y.: A model-driven framework to support strategic agility: value-added perspective. Inf. Softw. Technol. **141**, 106734 (2022)
26. Sweetman, R., Conboy, K.: Finding the edge of chaos: a Complex Adaptive System approach to information systems Project Portfolio Management. In: ECIS 2019 Proceedings (2019)
27. Pflaeging, N.: Essays on Beta, vol. 1 (2020)
28. Guest, D., Knox, A., Warhurst, C.: Humanizing work in the digital age: lessons from socio-technical systems and quality of working life initiatives. Hum. Relat. **75**, 1461–1482 (2022)
29. Gallup: State of the Global Workplace 2023 Report: The voice of the World's employees (2023)
30. Gartner: Top 3 Strategic Priorities for CIOs Leadership Vision for 2022 | Gartner (2022)
31. Minnaar, J., De Morree, P.: Corporate Rebels make work more fun (2020)
32. Pascale, Ri.T., Milleman, M., Gioja, L.: Surfing the Edge of Caos: The Laws of Nature and the New Laws of Business (2001)
33. Hamel, G., Zanini, M.: Humanocracy. Creating Organizations as amazing as the people inside them (2019)
34. Minnaar, J., De Morree, P.: Start-up Factory: Haier's RenDanHeYi model and the end of management as we know it. Corporate Rebels Nederland B.V. & HMI (2022)
35. Henriques, T.A., O'Neill, H.: Design science research with focus groups – a pragmatic meta-model. Int. J. Manag. Projects Bus. **16**, 119–140 (2023)
36. Davis, M.C., Challenger, R., Jayewardene, D.N.W., Clegg, C.W.: Advancing socio-technical systems thinking: a call for bravery. Appl. Ergon. **45**, 171–180 (2014)
37. Holland, J.H.: Complexity: A Very Short Introduction. Oxford University Press, Oxford (2014)
38. Sirkiä, R., Laanti, M.: Adaptive finance and control: combining lean, agile, and beyond budgeting for financial and organizational flexibility. In: Proceedings of 48th Hawaii International Conference on System Sciences (HICSS), pp. 5030–5037 (2015)
39. Uhl-Bien, M., Arena, M.: Leadership for organizational adaptability: a theoretical synthesis and integrative framework. Leadersh. Q. **29**, 89–104 (2018)
40. Goldstein, J., Hazy, J.K., Lichtenstein, B.B.: Complexity and the Nexus of Leadership. Palgrave Macmillan, New York (2010)
41. Müller, R., Martinsuo, M., Blomquist, T.: Project portfolio control and portfolio management performance in different contexts. Proj. Manag. J. **39**, 28–42 (2008)

42. Doz, Y.: Fostering strategic agility: how individual executives and human resource practices contribute. Hum. Resour. Manag. Rev. **30**, 100693 (2020)
43. Hoffmann, D., Ahlemann, F., Reining, S.: Reconciling alignment, efficiency, and agility in IT project portfolio management: recommendations based on a revelatory case study. Int. J. Project Manage. **38**, 124–136 (2020)
44. Hennel, P., Rosenkranz, C.: Investigating the "socio" in socio-technical development: the case for psychological safety in agile information systems development. Proj. Manag. J. **52**, 11–30 (2020)
45. Martinsuo, M.: Project portfolio management in practice and in context. Int. J. Project Manage. **31**, 794–803 (2013)
46. Vejseli, S., Proba, D., Rossmann, A., Jung, R.: The agile strategies in IT Governance: towards a framework of agile IT Governance in the banking industry. In: Twenty-Sixth European Conference on Information Systems (ECIS 2018), pp. 1–17. University of Portsmouth, Portsmouth (2018)

Empathy vs Reluctance to Challenge Misinformation: The Mediating Role of Relationship Costs, Perspective Taking, and Need for Cognition

Rabab Ali Abumalloh[1(✉)], Selin Gurgun[2], Muaadh Noman[3], Keith Phalp[2], Osama Halabi[1], Vasilis Katos[2], and Raian Ali[3(✉)]

[1] Department of Computer Science and Engineering, Qatar University, Doha, Qatar
rabab.abumalloh@qu.edu.qa

[2] Faculty of Science and Technology, Bournemouth University, Poole, UK

[3] College of Science and Engineering, Hamad Bin Khalifa University, Doha, Qatar
raali2@hbku.edu.qa

Abstract. Misinformation can harm individuals and societies, with social media and online communities amplifying its reach and impact. One effective strategy to counteract the spread of misinformation online is social corrections, in which people on social media actively challenge others who post or spread it. People hesitate to do so for reasons related to empathy, fear of affecting their relationships, futility, and subjective norms. This research aims to explore the impact of empathy on individuals' willingness to challenge misinformation. The research also investigates the mediation role of the personal factors of perspective-taking and the need for cognition, along with the perceived impacts on their relationships, on the relationship between empathy and the willingness to challenge. The data was collected from 250 UK-based social networking users and then analyzed using Partial Least Squares Structural Equation Modeling. The results of the analysis supported that perspective-taking ($\beta = 0.064$, $p = 0.011$), the need for cognition ($\beta = 0.022$, $p = 0.048$), and perceived relationship costs ($\beta = 0.035$, $p = 0.003$) all fully mediated the impact of empathy on the willingness to challenge misinformation. The results also show that empathy does not have a direct impact on willingness to challenge misinformation. Individuals with varying levels of empathy converge in their attitudes toward challenging misinformation influenced by a combination of cognitive processes and considerations of their relationships.

Keywords: Misinformation · Social Correction · Relationship Cost · Empathy · Perspective Taking · Need for Cognition

1 Introduction

The growth of online communities has facilitated widespread misinformation, negatively affecting individual well-being and societal progress [1]. Engagement with online misinformation has intensified following the COVID-19 crisis and the war in Ukraine.

J. Araújo et al. (Eds.): RCIS 2024, LNBIP 513, pp. 376–392, 2024.
https://doi.org/10.1007/978-3-031-59465-6_23

Several phrases, including misinformation, rumors, fake news [2], and disinformation have been used in tandem to describe inaccurate information [3]. Misinformation can be defined as a type of information that is incorrect, unsubstantiated, imprecise, perceived as unclear in a specific circumstance or setting [4], or not supported by proof, expert opinion, or evidence [5]. Rumors are often information whose accuracy is doubted [6]. Fake news differs in format from misinformation and usually refers to the news [6]. Disinformation refers to the process of spreading inaccurate data with the intention to mislead, whereas misinformation is inaccurate information that does not intend to harm [7].

To face the spread of misinformation within social media and online communities, corrections from users prove to be an effective approach [8]. Still, one notable obstacle to stopping the spread of misinformation is the lack of action taken by people who receive it to confront those who share it [6]. Chen, et al. [9] identified six main types of factors that impact the spread of misinformation within social media: emotions, cognition, personality traits, demographics, motivations, and worldviews. The authors also addressed a research gap concerning the relationship between these factors. Identifying the factors that hinder users from challenging misinformation [6] and the relationships between these variables is crucial for facing its spread [9], given the significant harm it causes and the proven effectiveness of social corrections in countering it [8].

The purpose of this study is to investigate the impact of several factors on people's willingness to challenge others who post misinformation on social media sites. Using a sample of social networking users, we investigate the factors that impact people's willingness to challenge misinformation; focusing on empathy as a main factor and the need for cognition, perspective-taking, and relationship cost as mediators.

Mediation in the context of our study refers to the extent to which the effect of empathy on willingness to challenge is explained by the set of intermediate variables (Relationship Costs, Perspective Taking, and Need for Cognition), referred to as the indirect effect [10].

The rest of this paper is structured as follows; Hypotheses Development is presented in Sect. 2, Study Design is presented in Sect. 3, Empirical Results are presented in Sect. 4, Discussion is elaborated in Sect. 5, Research Contribution, Limitations, and Future Work are elaborated in Sect. 6, and Sect. 7 concludes the study.

2 Hypotheses Development

Empathy indicates the ability to notice and respond to the feelings of other people, which is frequently combined with a desire to take care of the well-being of others [11]. Empathic individuals are more inclined to foster and sustain their connections, facilitating the adoption of social interaction strategies such as emotional expression, understanding others' emotional states, and recognizing the influence of one's actions on others [12]. Empathic individuals are also more willing to defend vulnerable from harm [13] and aid others [14]. As empathy has been linked to positive pro-social behaviors in the literature in several contexts [15], we hypothesize that:

H1: *Empathy has a positive impact on the Willingness to Challenge.*

Perspective-taking is the ability to accept the other's standpoint of view and attribute their ideas and emotions to them [16]. Perspective-taking is an intellectual procedure involving observers' willingness to see the world from others' perspectives as well as understanding their emotions or mental states [17]. Perspective-taking is an activity that requires information as well as cognitive picturing [18]. Despite their close association, perspective-taking and empathy are considered separate concepts. In contrast to the cognitive orientation of perspective-taking, empathy is focused on the emotional reactions to others, including feelings such as warmth, compassion, sympathy, and worry [19]. In several studies, perspective-taking is considered an intellectual and logical procedure that involves examining another person's experiences, while empathy is a more feelings-driven response [20]. Perspective-taking is essential to positive social interactions [21], prosocial behavior [22], and morality [23]. Hence, we argue that empathetic individuals with the ability to take others' perspectives are more willing to confront challenging situations and suggest that:

H2: *Perspective Taking mediates the relationship between Empathy and Willingness to Challenge.*

Need for Cognition (NFC) refers to a person's inclination to participate in cognitive functions, such as examining problems critically [24]. Individuals with high levels of NFC are more motivated to look out for and analyze information in a deliberate and organized manner [25]. NFC has been associated with reduced bias in decision-making across a variety of domains [26]. When challenged with difficult activities, individuals with high levels of NFC demonstrate more optimistic mindsets and emotions [27]. Thus, the next hypothesis is presented as follows:

H3: *The Need for Cognition mediates the relationship between Empathy and Willingness to challenge.*

Building on the research by Gurgun, et al. [28], we explore the perceived influence on individuals' relations as an important factor that determines their readiness to dispute misinformation. Individuals might hold off sharing disagreeing viewpoints or criticizing others owing to worries about hurting their connections [29]. They might perceive the interpersonal consequences of disputing misinformation, assuming that challenging misinformation will elicit negative reactions and damage their connections [30]. Based on that, we present the following hypothesis (Fig. 1).

H4: *Relationship Cost mediates the relationship between Empathy and Willingness to challenge.*

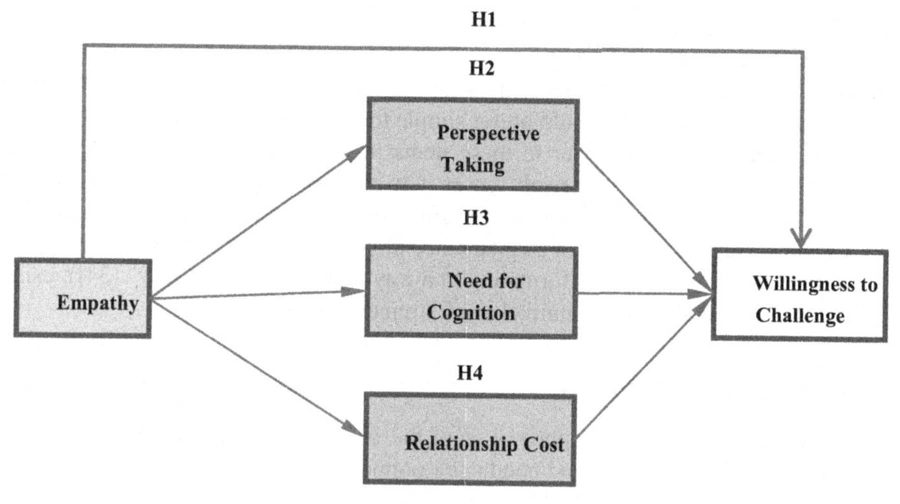

Fig. 1. The Initial Research Model

3 Study Design

3.1 Survey Development

To make sure that participants understood important concepts we gave descriptions for key terms like "misinformation" and "challenging". We deliberately provided the respondents with a question about challenging acquaintances. We clarified that acquaintances are contacts in the territory between strangers and intimates on Facebook, such as previous colleagues, neighbors, or anyone you know from a group on Facebook. This clarification was necessary because research reveals that individuals' behaviors differ when they deal with acquaintances compared to full foreigners. Given the possibility of influence on their social relationships, people may be more inclined to confront misinformation supplied by acquaintances.

The questionnaire measured participants' perceptions of how other people would react when being challenged, which we refer to as "Relationship Cost" throughout the paper. This part included measurements of the perceived relationship costs [31]. Relationship cost was rated on a 7-point Likert scale from Strongly Disagree to Strongly Agree. Sample items include "They would think that I am not empathetic", "They would think that I am aggressive", and "They would think that I am unfriendly". We are referring to the six items used for evaluating the relationship cost, denoted as RC1 through RC6.

Empathetic concerns and perspective-taking were measured using the scale by Davis [32]. Sample items of empathetic concerns include "I feel kind of protective toward them", "I sometimes don't feel very much pity for them", and " I often have tender, concerned feelings for people less fortunate than me". We are referring to the seven items used for evaluating the empathetic concerns, denoted as EMPC1 through EMPC 7. Sample items of perspective-taking include "Before criticizing somebody, I try to imagine how I would feel if I were in their place", and "When I'm upset at someone,

I usually try to "put myself in his shoes" for a while". We are referring to the seven items used for evaluating the perspective taking, denoted as PERST1 through PERST7. The need for cognition was measured using the scale by Thompson [33]. The scale contains six items such as "I would prefer simple to complex problems" and "Thinking is not my idea of fun". We refer to these items in this study by symbols that range from NFC1 to NFC6. Only one self-report item was used to assess the tendency of participants to challenge ("Think about a specific time when you saw misinformation on your Facebook feed, shared by an acquaintance in your Facebook network."How likely were you to challenge the misinformation in a way that others could see?" [34]), using a 7-point scale from Extremely unlikely to Extremely likely.

3.2 Data Collection Procedure

The research received ethical approval from the university's research ethics committee. The survey was designed and conducted online using Qualtrics. Participants were informed regarding the research objectives and required to provide consent before proceeding. Data collection took place between May 31st and July 7th, 2022, through ProlificTM (www.prolific.co), a well-established online platform for recruiting participants for research studies.

3.3 Participants

From 323 responses, a total of 250 participants were selected for this study. In the the size of the sample, we followed the 10 times rule of thumb by by Barclay, et al. [35], which indicates that the sample size should be equal to the larger of 10 times the largest number of structural paths directed at a particular construct in the structural model. This is equivalent to saying that the minimum sample size should be 10 times the maximum number of arrowheads pointing at a latent variable anywhere in the PLS path model. We excluded participants who failed to answer two attention-check questions correctly, provided nonsensical responses to open-ended questions, completed the survey too quickly, or provided uniform responses to questions where such uniformity was implausible due to the contradictory nature of the items. They were selected based on specific criteria: being 18 years or older, actively using Facebook with their identity (not anonymously), and having encountered misinformation on the platform. £4 compensation was provided for their participation in the survey.

3.4 Pilot Study

A pilot test was conducted before starting the actual data collection. An initial questionnaire was prepared and circulated to students and their networks for feedback. 19 participants completed the pilot questionnaire. Several changes in terms of language were made to make the items clearer. Face validity was assessed by prompting participants to share their thoughts and provide feedback for further refinement. The questionnaire's clarity and effectiveness were further verified by having multiple participants review it and explain their understanding of the questions, ensuring that the questions accurately reflected our intended purposes.

4 Results

4.1 Demographic Results

Participants were asked to specify their gender, age, and degree of education (Table 1). A total of 250 participants completed the online survey, with the majority of participants being females, in the age interval of 25–34 years old, and having a university degree.

4.2 PLS-SEM Analysis

Partial Least Square Structural Equation Modelling (PLS-SEM) is a method of statistical analysis employed to examine structural models and investigate complex interactions between many elements. PLS-SEM enables the evaluation of both measurement models (reflective and formative measurement) and structural models (path analysis and hypothesis testing). The variables and pathways of the research model were assessed using SmartPLS 4 to determine the validity and reliability of the model and the level of significance of the hypothesized pathways [36]. The results of the PLS algorithm and bootstrapping technique are reported in detail in the following subsections.

Table 1. Demographic Data

Factor	Item	Frequency	Percentage (%)
Gender	Female	143	57.2
	Male	104	41.6
	Others	3	1.2
Age	18–24 years old	44	17.8
	25–34 years old	92	37.2
	35–44 years old	66	26.7
	Over 45 years	45	18.2
Education	University	157	62.8
	College	57	22.8
	Secondary	36	14.4

4.2.1 Outer Model Assessment

Convergent validity (CV), Internal Consistency (IC), and Discriminant Validity (DV) are the three fundamental assessments that must be carried out in order to examine the study factors along with their linked indicators through SmartPLS 4 to make sure that the responses to the questionnaire have produced reliable and valid results.

Table 2 contains the findings of the CV analysis, in which we inspect the outer loading values and the AVE test. According to Hair et al. [37], the outer loading values refer to the level to which an indicator corresponds to other indicators in the same factor. The rule of thumb in this context is accepting outer loading values above 0.7 [38]. Outer loading values less than 0.4 need to be omitted, while the values that equal or above 0.4 but less than 0.7 should be inspected and considered for deletion if doing so will enhance the values of Composite Reliability or Average Variance Extracted (AVE) [38]. Still, obtaining values of outer loading that are less than 0.7 is common, particularly in social sciences research [37].

Based on the above, we deleted two indicators from Perspective Taking; PERST2 and PERST5. In terms of the Need for a Cognition factor, we deleted the NFC1 from the variable. Following that we examined the values of the AVE, and based on that we deleted EMPC3, EMPC4, and EMPC7 from the empathetic concern variable to improve the AVE value. Hence, we obtained values of AVE that met the threshold of 0.5 [38] for all research constructs. The second test is the IC, which is examined by values of Composite Reliability and Cronbach's alpha values, in which both tests met threshold values of 0.7 [36].

The degree of differentiation between the model's variables was assessed using a discriminant validity analysis. Heterotrait-Monotrait Ratio (HTMT) was used to assess the discriminant validity of the outer model (Table 3). Based on the HTMT values in the table being below 0.85 [39], there is evidence to support discriminant validity between the constructs, suggesting that they are distinct and measure different underlying constructs (Table 2).

4.2.2 Inner Model Assessment

The first stage in the assessment of the inner mode is to evaluate the model against collinearity issues using values of the Variance Inflation Factor (VIF). Hair Jr, et al. [38] suggest that a research model with values of VIF less than 5 is free of any multicollinearity issue, which was supported in this study. Common method bias [40] was also assessed using the VIF measures within the inner model. Values below 3.33 are typically considered indicative of the absence of common method bias, a conclusion that was confirmed by the analysis results. To examine hypotheses and determine the significance of correlations between study variables, the bootstrapping technique was used [37]. To evaluate the significance of the paths, we used p values, which refer to the probability of erroneously rejecting a true null hypothesis [38]. A p-value below 0.05 is considered significant in most settings [38]. As shown in Table 4, the Willingness to Challenge is strongly impacted by three variables included in our model (Perspective taking, Need for Cognition, and Relationship Cost). On the other hand, empathy influences the three factors (Perspective taking, Need for Cognition, and Relationship Cost). Table 5 presents the values of Q^2 and RMSE. The rule of thumb suggests that a Q2 value greater than zero for a specific endogenous variable confirms the predictive relevance of the path model for that variable [38].

4.2.3 Mediation Analysis

The mediation effect occurs when a mediator factor intervenes between two relevant factors. The analysis of the mediation effect follows a specific rule of thumb according

Table 2. Constructs' Reliability and Convergent Validity Test (N = 250)

Construct	Indicator	Outer Loadings	Cronbach's alpha	Composite Reliability	AVE
Empathy Concern (EMPC)	EMPC1	0.693	0.679	0.805	0.508
	EMPC2	0.651			
	EMPC5	0.748			
	EMPC6	0.755			
Need for Cognition (NFC)	NFC2	0.784	0.774	0.84	0.515
	NFC3	0.694			
	NFC4	0.652			
	NFC5	0.805			
	NFC6	0.635			
Perspective Taking (PERST)	PERST1	0.778	0.807	0.866	0.565
	PERST3	0.771			
	PERST4	0.733			
	PERST6	0.767			
	PERST7	0.707			
Relationship Cost (RC)	RC1	0.803	0.926	0.942	0.73
	RC2	0.885			
	RC3	0.917			
	RC4	0.866			
	RC5	0.861			
	RC6	0.787			

Table 3. Heterotrait-Monotrait Ratio

	EMPC	NFC	PERST	RC	WTC
Empathy					
Need for Cognition	0.235				
Perspective Taking	0.605	0.29			
Relationship Cost	0.232	0.135	0.163		
Willingness to Challenge	0.176	0.176	0.227	0.226	

to Hair, et al. [37]. The total indirect impact is a combination of the specific indirect impacts in the mediation study, whereas the total effect is the combination of the direct impact plus the total indirect impacts. The product of coefficients method is commonly

Table 4. Path Coefficient Analysis

Type of Effect	Hypothesis	Inner (VIF)	β	p-value
Direct Effect	Empathy - > Willingness to Challenge	1.301	0.025	0.639 ns
	Empathy - > Perspective Taking	1.000	0.461	0.000***
	Empathy - > Need for _Cognition	1.000	0.187	0.000***
	Empathy - > Relationship Cost	1.000	-0.182	0.000***
	Perspective Taking - > Willingness to Challenge	1.321	0.138	0.009**
	Need for _Cognition - > Willingness to Challenge	1.079	0.115	0.025**
	Relationship Cost - > Willingness to Challenge	1.040	-0.19	0.000***

Note: * $p < 0.05$, ** $p < 0.01$, *** $p < 0.001$, ns = not supported ($p > 0.05$)

Table 5. Prediction Summary

	Q^2 Predict	RMSE
Need for Cognition	0.027	0.992
Perspective Taking	0.201	0.898
Relationship Cost	0.028	0.989
Willingness to Challenge	0.016	0.995

used in PLS-SEM for estimating indirect impacts. To calculate the indirect impact, the route coefficients of the factors included in the mediating pathway are calculated.

To explain the concept of mediation, we consider Mi and Mj as mediators, the IV as the independent variable, and the DV as the dependent variable. For instance, the specific indirect effect reflects the potential of Mi to mediate the impact of IV on DV contingent on the presence of other mediators in the model [41], which is quantified as $p1.p2$, where $p1$ and $p2$ are the path coffecients from the IV to Mi and the Mi to the DV, respectively. Besides, $p3$ represents the direct effect between the IV and the DV. Similarly, if p4 and p5 are the path coefficients from the IV to Mj and the Mj to the DV, respectively, the total indirect effect is the sum of the specific indirect effects (i.e., $p1.p2 + p4.p5$). The total effect of IV on DV is the sum of the direct effects and the total indirect effects (i.e., $p3 + p1.p2 + p4.p5$) [38].

According to the results, the mediated impacts of perspective taking, need for cognition, and relationship cost are all substantial, while the direct influence of empathy on the willingness to challenge is not. This means that the mediator factors fully explain the connection between the exogenous factor (empathy) and the endogenous factor (willingness to challenge). Full mediation refers to the case in which the direct impact from the IV to the DV is not supported but the mediated impact is, hence the relationship between the IV and the DV is entirely explained by the mediator variable [42].

The results of the analysis of the specific indirect effects (see Table 6) supported the positive influence of empathy on willingness to challenge through perspective-taking ($\beta = 0.064$, $p = 0.011$). The outcomes also supported the positive impact of empathy on willingness to challenge through the need for cognition ($\beta = 0.022$, $p = 0.048$). Besides, the results of the analysis supported the positive influence of empathy on willingness to challenge through relationship costs ($\beta = 0.035$, $p = 0.003$). The total effect (Table 7) of empathy on the willingness to challenge was significant ($\beta = 0.145$, $p = 0.002$).

Table 6. Specific Indirect Effects

Specific Indirect Effects	β	p-value	95% Confidence Interval	
			Lower	Upper
H2: Empathy -> Perspective Taking - > Willingness to Challenge	0.064	0.011*	0.014	0.114
H3: Empathy -> Need for Cognition - > Willingness to_Challenge	0.022	0.048*	0.003	0.046
H4: Empathy -> Relationship Cost -> Willingness to_Challenge	0.035	0.003**	0.016	0.061

Note: * $p < 0.05$, ** $p < 0.01$, *** $p < 0.001$, ns = not supported ($p > 0.05$)

Table 7. Results of Total Effect and Direct Effect

Type of Effect	Hypothesis	β	p-value
Total Indirect Effect	Empathy - > Willingness to Challenge	0.12	0.000***
Direct Effect	Empathy - > Willingness to Challenge	0.025	0.639 ns
Total Effect	Empathy - > Willingness to Challenge	0.145	0.002**

Note: * $p < 0.05$, ** $p < 0.01$, *** $p < 0.001$, ns = not supported ($p > 0.05$)

5 Discussion

The study explored the mediation influences of three factors; need for cognition, perspective taking, and relationship costs on the relationship between empathy and willingness to challenge. In the following, we will discuss the results in detail.

First, the study rejected the direct impact of empathy on the willingness to challenge misinformation (H1). This result contradicts the outcomes presented by several studies that have demonstrated a strong link between empathy and pro-social behavior and attitudes [15, 43–45]. However, the result runs with the outcomes by Kim, et al. [46], in which affective empathy did not impact pro-social behavior. The authors justified this outcome by suggesting that the presence of both positive and negative emotions within affective empathy might have led to an insignificant impact on prosocial behavior.

Second, the study emphasizes the significance of perspective-taking as an essential factor through which empathy promotes people's willingness to reject and confront misinformation (H2). Embracing the viewpoint of others involves considering the impact of an event on them, which allows perspective-takers to establish a deeper connection with others [47]. This explains the robust connection between perspective-taking and pro-social behavior [48]. Referring to Ku, et al. [49], perspective-taking helps individuals deal with scenarios characterized by a mix of competing interests or motives. Referring to the study by Galinsky, et al. [50], individuals with high perspective-taking skills have an enhanced capacity to identify hidden agreements and secure valuable resources during negotiations. This effect can be linked to the idea that perspective-taking negotiators can strategically build arguments that allow them to claim more value for themselves by attempting to grasp their opponent's point of view [50]. By diving into their opponent's perspective, they get insight into what is important to the other party and can modify their negotiating strategy to grasp a larger portion of the resources. In the same study [50], it was evident that empathy had limited effectiveness in helping individuals identify hidden agreements and optimize their individual gains. In another study by Gilin, et al. [51], in the context of complex competitive situations, the authors indicated that individuals with high levels of perspective-taking excel through cognitive analysis of interpersonal interactions. This enables them to develop effective strategies for both cooperation and competition and to choose when to implement them [51]. In another study by Wang, et al. [52], the findings indicated that perspective-taking reduced stereotyping for both positively and negatively stereotyped groups.

Third, the results supported the full mediation influence of the individual's need for cognition on the relationship between empathy and willingness to challenge (H3). Challenging behavior isn't just defined by one's motives or desires; it is also influenced by the belief that one possesses the necessary skills to engage in such behavior [6]. This belief is particularly prevalent among individuals with a high need for cognition. According to Day, et al. [53] individuals with a strong desire for cognition tend to actively seek and analyze information and engage in complex learning practices. They focus on the process rather than on the outcome [54], including reasoning methods and steps involved in a task or decision-making process [53]. Nussbaum [55] further supports this argument and indicates that individuals with a high need for cognition place greater emphasis on the quality of reasoning when evaluating messages, as opposed to peripheral indicators such as message length or the opinions of others. As a result, these individuals spend more time and effort on intellectual activities, which helps them acquire advanced argumentation skills. Previous research has also shown that individuals with a high need for cognition analyze media content more attentively and critically [56]. They also employ rational and analytical thinking skills when consuming media content [57]. Our result also endorses the outcomes by Su, et al. [58], which indicates that the need for cognition serves as a significant factor in mitigating the spread of conspiracy beliefs during the COVID-19 pandemic.

Finally, the study endorsed the full mediation influence of individuals' perceptions of relationship costs on the path between empathy and willingness to challenge (H4). This result runs with the outcome by Gurgun, et al. [6], which identified social concern as one of the barriers to challenging misinformation on social media. Individuals use social media to sustain positive connections [59]. Hence, when they are worried about the negative effects of correcting misinformation, they would rather avoid challenging others [6]. We further refer to the spiral of silence theory to explain why and how people choose to speak up or keep silent [60]. The basic premise of the theory suggests that individuals tend to retain their opinions and become more silent, as they perceive that societal opinion differs from their opinions. This theory proposes that fear of isolation is the main driver of this tendency. The theory also specifies the one's reference group as a small circle of people that includes friends and relatives. The perceptions of this reference group have a significant impact on an individual's tendency to express their thoughts [61]. Oshagan [61] suggests that reference groups have a more significant impact on individuals' willingness to express themselves compared to the influence of societal opinion.

5.1 Research Implications

The results of this study can guide platform design decisions targeted at reducing the spread of misinformation on social media. Platforms can, for instance, include elements that promote perspective-taking before users share information. Policymakers can integrate effective communication strategies, tailored to various cognitive and emotional abilities, to accommodate diverse viewpoints in the design of these platforms. Consequently, there is an opportunity to create educational interventions and training programs focused on enhancing perspective-taking skills, promoting critical thinking, and fostering active participation in the fight against internet misinformation.

Challenging misinformation is an individual decision-making process that is impacted by several drivers and barriers. People's perceptions of others' opinions affect their decisions to speak up or stay silent within or outside these communities [62]. Effective decision-making necessitates individuals to evaluate various choices and anticipate their potential consequences, which demands structured thinking and resistance to persuasion [63]. Effective decision-making involves both cognitive and emotional elements [57], which requires a thorough exploration of the interplay among different possible outcomes [64]. Therefore, when designing social media platforms, it is crucial to integrate features that support these aspects of decision-making, creating an environment where challenging misinformation is not only possible but also a facilitated and encouraged activity.

5.2 Limitations

This research holds a few limitations that we will highlight briefly. First, the sample in this study consists of 250 respondents from the UK. This demographic might not precisely mirror the attitudes and behaviors found within diverse populations and cultures. To enhance the robustness of our findings, future investigations could delve into the same variables using more representative samples. This would help us explore the potential variations in perceptions and attitudes toward challenging misinformation across various cultures. Second, it is worth noting to emphasize that the variables examined within our study represent only a fraction of the exhaustive array of potentially influential variables. While our study sheds light on these important variables, there is a need for further comprehensive research to obtain a more nuanced understanding of the complex interplay between attitudes, behaviors, and cultural factors in challenging misinformation. The study employed a survey-based data collection approach, a method that comes with inherent limitations. Referring to Grimm [65], in survey-based studies people might tend to respond according to societal expectations instead of expressing their true feelings, as described by social desirability bias. The bias is more pronounced when it comes to socially sensitive issues, including politics, religion, environment, and personal factors.

6 Conclusion and Future Work

Preventing the spread of misinformation on social media is essential, yet numerous factors often deter people from actively challenging it. This study underscores the importance of expanding awareness and fostering a more cohesive social media environment. By doing so, it aims to create a setting where individuals feel empowered and unencumbered to tackle misinformation.

Future research can incorporate the level of media literacy and its impact on the individual's willingness to challenge misinformation. Understanding how media literacy affects user engagement with misinformation on social media could help in addressing it. The study focused on the willingness to challenge misinformation with a concentration on the individual's oriented variables, future studies can incorporate the impact of organizational-based factors. Such as the study by Bautista, et al. [66], which examined the impact of organizational support on the intention to correct misinformation in the medical field. The importance of diving into the function of organizational-based issues arises in this context, as it provides an additional level of depth to our understanding of how corrective measures against misinformation are implemented within a larger framework. One of the major threats associated with misinformation spreading is the risk of it becoming an established norm among communities. Individuals who embrace false news and receive backing from other individuals who match their views via social media will be more encouraged to spread misinformation. This is because they sense reinforcement and affirmation from others. More research should be conducted to determine why helpful interaction actually hinders or disempowers one's inclination to speak up in a social media situation.

Acknowledgment. This publication was made possible by the NPRP 14 Cluster Grant Number NPRP14C-37878-SP-470 from the Qatar National Research Fund (a member of Qatar Foundation). The results herein reflect the work and are the sole responsibility of the authors.

Data Availability. The raw data required to reproduce the above findings are available at: https://osf.io/uny7g/?view_only=365026b20d134902b6e23b4c98b16007.

References

1. Barua, Z., Barua, S., Aktar, S., Kabir, N., Li, M.: Effects of misinformation on COVID-19 individual responses and recommendations for resilience of disastrous consequences of misinformation. Progress Disaster Sci. **8**, 100119 (2020)
2. Belloir, N., Ouerdane, W., Pastor, O., Frugier, É., de Barmon, L.-A.: A conceptual characterization of fake news: a positioning paper. In: Guizzardi, R., Ralyté, J., Franch, X. (eds.) RCIS 2022, pp. 662–669. Springer, Cham (2022). https://doi.org/10.1007/978-3-031-05760-1_41
3. Wu, L., Morstatter, F., Carley, K.M., Liu, H.: Misinformation in social media: definition, manipulation, and detection. SIGKDD Explor. Newsl. **21**(2), 80–90 (2019). https://doi.org/10.1145/3373464.3373475
4. Karlova, N.A., Fisher, K.E.: A social diffusion model of misinformation and disinformation for understanding human information behaviour (2013)
5. Nyhan, B., Reifler, J.: When corrections fail: the persistence of political misperceptions. Pol. Beh. **32**(2), 303–330 (2010). https://doi.org/10.1007/s11109-010-9112-2
6. Gurgun, S., Cemiloglu, D., Close, E.A., Phalp, K., Nakov, P., Ali, R.: Why do we not stand up to misinformation? Factors influencing the likelihood of challenging misinformation on social media and the role of demographics. Technol. Soc. **76**, 102444 (2024). https://doi.org/10.1016/j.techsoc.2023.102444
7. Stahl, B.C.: On the difference or equality of information, misinformation, and disinformation: a critical research perspective. Inf. Sci. **9**, 83 (2006)
8. Bode, L., Vraga, E.K.: See something, say something: correction of global health misinformation on social media. Health Commun. **33**(9), 1131–1140 (2018)
9. Chen, S., Xiao, L., Kumar, A.: Spread of misinformation on social media: what contributes to it and how to combat it. Comput. Hum. Beh. **141**, 107643 (2023). https://doi.org/10.1016/j.chb.2022.107643
10. Grubic, N., et al.: Mediators of the association between socioeconomic status and survival after out-of-hospital cardiac arrest: a systematic review. Canadian J. Cardiol. (2024). https://doi.org/10.1016/j.cjca.2024.01.002
11. Decety, J., Michalska, K.J.: A developmental neuroscience perspective on empathy. In: Neural Circuit and Cognitive Development, pp. 485–503. Elsevier (2020)
12. Li, X., et al.: Indirect aggression and parental attachment in early adolescence: examining the role of perspective taking and empathetic concern. Personality Individ. Differ. **86**, 499–503 (2015)
13. Davis, M.H.: 23 Empathy, Compassion, and Social Relationships. The Oxford Handbook of Compassion Science, vol. 299 (2017)
14. Hafenbrack, A.C., Cameron, L.D., Spreitzer, G.M., Zhang, C., Noval, L.J., Shaffakat, S.: Helping people by being in the present: mindfulness increases prosocial behavior. Organ. Behav. Hum. Decis. Process. **159**, 21–38 (2020)
15. Fu, W., Wang, C., Chai, H., Xue, R.: Examining the relationship of empathy, social support, and prosocial behavior of adolescents in China: a structural equation modeling approach. Hum. Soc. Sci. Commun. **9**(1), 1–8 (2022)
16. Decety, J.: Dissecting the neural mechanisms mediating empathy. Emot. Rev. **3**(1), 92–108 (2011)
17. Galinsky, A.D., Ku, G., Wang, C.S.: Perspective-taking and self-other overlap: fostering social bonds and facilitating social coordination. Group Process. Intergroup Relat. **8**(2), 109–124 (2005)
18. Cole, G.G., Millett, A.C.: The closing of the theory of mind: a critique of perspective-taking. Psychon. Bull. Rev. **26**, 1787–1802 (2019)

19. Batson, C.D., Early, S., Salvarani, G.: Perspective taking: imagining how another feels versus imaging how you would feel. Pers. Soc. Psychol. Bull. **23**(7), 751–758 (1997). https://doi.org/10.1177/0146167297237008
20. Myyry, L., Juujärvi, S., Pesso, K.: Empathy, perspective taking and personal values as predictors of moral schemas. J Moral. Educ. **39**, 213–233 (2010). https://doi.org/10.1080/03057241003754955
21. Decety, J.: Perspective taking as the royal avenue to empathy. Other Minds How Hum. Brid. Div. Between Self Others **143**, 157 (2005)
22. Tamnes, C.K., et al.: Social perspective taking is associated with self-reported prosocial behavior and regional cortical thickness across adolescence. Dev. Psychol. **54**(9), 1745 (2018)
23. Decety, J., Cowell, J.M.: The complex relation between morality and empathy. Trends Cogn. Sci. **18**(7), 337–339 (2014)
24. Zhang, Y., Tian, Y., Yao, L., Duan, C., Sun, X., Niu, G.: Teaching presence promotes learner affective engagement: the roles of cognitive load and need for cognition. Teach. Teach. Educ. **129**, 104167 (2023)
25. Lavrijsen, J., Preckel, F., Verschueren, K.: Seeking, mastering, and enjoying cognitive effort: scrutinizing the role of need for cognition in academic achievement. Learn. Individ. Diff. **107**, 102363 (2023). https://doi.org/10.1016/j.lindif.2023.102363
26. Double, K.S., Cavanagh, M.: Need for cognition predicts the accuracy of affective forecasts. Person. Individ. Diff. **216**, 112399 (2024). https://doi.org/10.1016/j.paid.2023.112399
27. Cacioppo, J.T., Petty, R.E.: The need for cognition. J. Pers. Soc. Psychol. **42**(1), 116 (1982)
28. Gurgun, S., Arden-Close, E., Phalp, K., Ali, R.: Online silence: why do people not challenge others when posting misinformation? Internet Research, no. ahead-of-print (2022)
29. Cialdini, R.B., Trost, M.R.: Social influence: social norms, conformity and compliance (1998)
30. Gurgun, S., Cemiloglu, D., Arden-Close, E., Phalp, K., Nakov, P., Ali, R.: Challenging Misinformation on Social Media: Users' Perceptions and Misperceptions and their Impact on the Willingness to Challenge (2023). Available at SSRN 4440292
31. Zhang, Z.-X., Zhang, Y., Wang, M.: Harmony, illusory relationship costs, and conflict resolution in Chinese contexts. Cambridge University Press, Cambridge (2011)
32. Davis, M.H.: A multidimensional approach to individual differences in empathy (1980)
33. Thompson, M.E.: The impact of need for cognition on thinking about free speech issues. J. Mass Commun. Quart. **72**(4), 934–947 (1995)
34. Cohen, E.L., et al.: To correct or not to correct? Social identity threats increase willingness to denounce fake news through presumed media influence and hostile media perceptions. Commun. Res. Rep. **37**(5), 263–275 (2020)
35. Barclay, D., Higgins, C., Thompson, R.: The partial least squares (PLS) approach to casual modeling: personal computer adoption ans use as an Illustration (1995)
36. Hair, J.F., Ringle, C.M., Sarstedt, M.: PLS-SEM: indeed a silver bullet. J. Mark. Theory Pract. **19**(2), 139–152 (2011). https://doi.org/10.2753/MTP1069-6679190202
37. Hair, J., Hult, G.T.M., Ringle, C.M., Sarstedt, M.: A Primer on Partial Least Squares Structural Equation Modeling, pp. 184–185. SAGE Publications Inc, Thousand Oaks (2013)
38. Hair Jr., J., Hair Jr., J.F., Hult, G.T.M., Ringle, C.M., Sarstedt, M.: A primer on partial least squares structural equation modeling (PLS-SEM). Sage Publications (2021)
39. Clark, L.A., Watson, D.: Constructing validity: Basic issues in objective scale development (2016)
40. Kock, N.: Common method bias in PLS-SEM: a full collinearity assessment approach. Int. J. e-Collab. (IJEC) **11**(4), 1–10 (2015)
41. Xie, Y., Siponen, M., Laatikainen, G., Moody, G.D., Zheng, X.: Testing the dominant mediator in EPPM: an empirical study on household anti-malware software users. Comput. Secur. **140**, 103776 (2024). https://doi.org/10.1016/j.cose.2024.103776

42. Nitzl, C., Roldán, J., Cepeda-Carrion, G.: Mediation analysis in partial least squares path modeling: helping researchers discuss more sophisticated models. Ind. Manag. Data Syst. **116**, 1849–1864 (2016). https://doi.org/10.1108/IMDS-07-2015-0302
43. Van der Graaff, J., Carlo, G., Crocetti, E., Koot, H.M., Branje, S.: Prosocial behavior in adolescence: gender differences in development and links with empathy. J. Youth Adolesc. **47**(5), 1086–1099 (2017). https://doi.org/10.1007/s10964-017-0786-1
44. Bohns, V.K., Flynn, F.J.: Empathy and expectations of others' willingness to help. Person. Individ. Differ. **168**, 110368 (2021)
45. Persson, B.N., Kajonius, P.J.: Empathy and universal values explicated by the empathy-altruism hypothesis. J. Soc. Psychol. **156**(6), 610–619 (2016). https://doi.org/10.1080/00224545.2016.1152212
46. Kim, E.K., You, S., Knox, J.: The mediating effect of empathy on the relation between child self-expressiveness in family and prosocial behaviors. J. Child Family Stud. **29**(6), 1572–1581 (2020). https://doi.org/10.1007/s10826-019-01676-2
47. Sassenrath, C., Vorauer, J.D., Hodges, S.D.: The link between perspective-taking and prosociality — not as universal as you might think. Curr. Opin. Psychol. **44**, 94–99 (2022). https://doi.org/10.1016/j.copsyc.2021.08.036
48. Shih, M., Wang, E., Trahan Bucher, A., Stotzer, R.: Perspective taking: reducing prejudice towards general outgroups and specific individuals. Group Processes Intergroup Rel. **12**(5), 565–577 (2009)
49. Ku, G., Wang, C.S., Galinsky, A.D.: The promise and perversity of perspective-taking in organizations. Res. Organ. Beh. **35**, 79–102 (2015). https://doi.org/10.1016/j.riob.2015.07.003
50. Galinsky, A.D., Maddux, W.W., Gilin, D., White, J.B.: Why it pays to get inside the head of your opponent: the differential effects of perspective taking and empathy in negotiations. Psychol. Sci. **19**(4), 378–384 (2008)
51. Gilin, D., Maddux, W.W., Carpenter, J., Galinsky, A.D.: When to use your head and when to use your heart: the differential value of perspective-taking versus empathy in competitive interactions. Pers. Soc. Psychol. Bull. **39**(1), 3–16 (2013). https://doi.org/10.1177/0146167212465320
52. Wang, C.S., Ku, G., Tai, K., Galinsky, A.D.: Stupid doctors and smart construction workers: perspective-taking reduces stereotyping of both negative and positive targets. Soc. Psychol. Person. Sci. **5**(4), 430–436 (2014)
53. Day, E.A., Espejo, J., Kowollik, V., Boatman, P.R., McEntire, L.E.: Modeling the links between need for cognition and the acquisition of a complex skill. Personality Individ. Differ. **42**(2), 201–212 (2007)
54. Novak, T.P., Hoffman, D.L.: The fit of thinking style and situation: new measures of situation-specific experiential and rational cognition. J. Consum. Res. **36**(1), 56–72 (2009)
55. Nussbaum, E.M.: The effect of goal instructions and need for cognition on interactive argumentation. Contemp. Educ. Psychol. **30**(3), 286–313 (2005). https://doi.org/10.1016/j.cedpsych.2004.11.002
56. Xiao, X., Su, Y., Lee, D.K.L.: Who consumes new media content more wisely? Examining personality factors, SNS use, and new media literacy in the era of misinformation. Soc. Media+ Soc. **7**(1), 2056305121990635 (2021)
57. Austin, E.W., Muldrow, A., Austin, B.W.: Examining how media literacy and personality factors predict skepticism toward alcohol advertising. J. Health Commun. **21**(5), 600–609 (2016)
58. Su, Y., Lee, D.K.L., Xiao, X., Li, W., Shu, W.: Who endorses conspiracy theories? A moderated mediation model of Chinese and international social media use, media skepticism, need for cognition, and COVID-19 conspiracy theory endorsement in China. Comput. Hum. Beh. **120**, 106760 (2021). https://doi.org/10.1016/j.chb.2021.106760

59. Brandtzæg, P.B., Heim, J.: Why people use social networking sites. In: Ozok, A.A., Zaphiris, P. (eds.) OCSC 2009. LNCS, vol. 5621, pp. 143–152. Springer, Heidelberg (2009). https://doi.org/10.1007/978-3-642-02774-1_16
60. Noelle-Neumann, E.: The spiral of silence a theory of public opinion. J. Commun. **24**(2), 43–51 (2006). https://doi.org/10.1111/j.1460-2466.1974.tb00367.x
61. Oshagan, H.: Reference group influence on opinion expression. Int. J. Pub. Opinion Res. **8**(4), 335–354 (1996)
62. Chun, J.W., Lee, M.J.: "Understanding empowerment process of willingness to speak out on social media: amplifying effect of supportive communication. Telem. Inform. **66**, 101735 (2022). https://doi.org/10.1016/j.tele.2021.101735
63. Buijzen, M., Van Reijmersdal, E.A., Owen, L.H.: Introducing the PCMC model: an investigative framework for young people's processing of commercialized media content. Commun. Theory **20**(4), 427–450 (2010)
64. Yiend, J.: The effects of emotion on attention: a review of attentional processing of emotional information. Cogn. Emot. **24**(1), 3–47 (2010)
65. Grimm, P.: Social desirability bias. Wiley international encyclopedia of marketing (2010)
66. Bautista, J.R., Zhang, Y., Gwizdka, J.: Predicting healthcare professionals' intention to correct health misinformation on social media. Telematics Inform. **73**, 101864 (2022)

An Industrial Experience Leveraging the iv4XR Framework for BDD Testing of a 3D Sandbox Game

Fernando Pastor Ricós[1(✉)], Beatriz Marín[1], I. S. W. B. Prasetya[2], Tanja E. J. Vos[1,3], Joseph Davidson[4], and Karel Hovorka[4]

[1] Universitat Politècnica de València, València, Spain
fpastor@pros.upv.es
[2] Utrecht University, Utrecht, The Netherlands
[3] Open Universiteit, Heerlen, The Netherlands
[4] GoodAI, Prague, Czechia

Abstract. Industrial-grade games, like Space Engineers, must adopt swift development and testing processes to conform to rigorous quality standards. Nevertheless, the testing phase of these extensive and complex games heavily relies on manual effort from play-testers, leading to productivity constraints during development cycles. This experience paper reports a Behavior-Driven-Development (BDD) software development process for automated regression test scenarios that allows complement testers' work during development cycles. To enable BDD test scripts for the Space Engineers game, we have extended the iv4XR framework into a game plugin to connect and execute game actions. Additionally, we have integrated the Cucumber software to describe game test scenarios using natural language. This approach allows testers to create, maintain, and execute a subset of regression test scenarios by relying on a BDD *agent* that can autonomously verify Space Engineers game features, enabling seamless integration into the development cycle.

Keywords: Industrial game testing · Autonomous agents · BDD testing

1 Introduction

In the early 2020s, the video game industry captivated 2 billion global players and generated a revenue of 120 billion dollars [5]. This growth raised the expectations of its audience, demanding high-quality products with engagement to long-term releases [18]. As a result, the game industry has increasingly embraced agile development processes to facilitate rapid game feature enhancements. Nevertheless, there is a lack of test automation methods and tools for effective feature verification and efficient bug resolution [19]. Consequently, game testing predominantly relies on manual efforts, with testers dedicating countless hours to ensure user interactions produce the intended responses in virtual scenarios.

The game Space Engineers, developed by Keen Software House and GoodAI companies, faces a similar challenge. Its testers have devised a regression testing pipeline that comprises thousands of tests. However, the manual execution

J. Araújo et al. (Eds.): RCIS 2024, LNBIP 513, pp. 393–409, 2024.
https://doi.org/10.1007/978-3-031-59465-6_24

of these tests is time-consuming, which limits its seamless integration into the development cycle. Testers must collaborate with developers over several months to validate new features until the game is stable. Only then do testers have time to manually run the entire regression testing pipeline. This delay in regression testing can potentially delay bug detection, affecting the game release timelines.

To tackle these challenges and improve early bug detection in the game development cycle, the Space Engineers team joined the project Intelligent Verification/Validation for Extended Reality Based Systems (IV4XR 2019-2022) [21]. This project aimed to implement autonomous testing *agents* [20] to help game industry stakeholders complement and enhance manual testing efforts.

Behavior-Driven Development (BDD) is a software development process that fosters collaboration and communication between developers and testers [25]. Although there is a scarcity of BDD research linked to industrial practices [4], various studies have shown its increasing popularity and widespread use in diverse systems [3,15]. Recognizing the advantages of BDD, the Space Engineers team decided to embrace its adoption to streamline the regression test automation.

Contribution. This paper describes the testing practices employed by the Space Engineers development companies and presents an experience report of our efforts to introduce automated agent-based testing in combination with the BDD process in the Space Engineer's regression testing pipeline. Although we did not automate the execution of the complete regression test suite, an essential subset of game features was chosen for automation, yielding reliable results.

This contribution is valuable for game development researchers and practitioners as it offers insights into Space Engineers' game-testing methodology. The paper discusses the advantages and challenges of using automated approaches and highlights the potential integration of autonomous *agents* through the IV4XR project to improve automated test execution. Additionally, the paper provides guidance for practitioners who want to leverage the BDD process for effective and efficient automated execution of game-testing scenarios.

The paper is structured as follows. Section 2 presents the related work. Section 3 introduces the Space Engineers game and game testing practices. Section 4 describes the IV4XR framework and the Space Engineers-plugin. Section 5 presents the BDD process as an autonomous *agent* and Sect. 6 its industrial application. Section 7 describes the threats to the validity. Section 8 provides the lessons learned. Section 9 summarizes the conclusions and future work.

2 Related Work

In the last decades, there have been significant test automation advances for traditional desktop [16], web [8], and mobile [12] Graphical User Interface (GUI) systems [23]. However, the field of game systems still lacks well-established and widely adopted methodologies to automate the execution of complex 3D games.

In recent years, Machine Learning (ML) has been researched for game-testing automation. ICARUS framework [17] employs Reinforcement Learning (RL) to train agents that complete linear adventure games. Wuji's approach [30] uses Deep Reinforcement Learning (DRL) to train exploration agents that accomplish mission objectives. Wu et al. [29] evaluates multiple RL rewards for divergent behavior detection in two versions of a commercial game task. Gordillo et al. [9] apply curiosity-driven RL to identify areas that stuck players. Andrade et al. [1] combine metamorphic tests with RL techniques to train agents capable of detecting collisions and camera behavior failures. Ariyurek et al. [2] use RL to train agents that simulate different personas to discover alternative playstyles. Sestini et al. [24] use curiosity and imitation RL to train agents that explore goal-destination areas while uncovering collision bugs and glitches.

These studies showcase the capability of ML techniques to automate game play-testing through trained agents. However, most studies are more concerned with the performance of the ML models rather than developing methodologies that assist companies in automating the verification of test case goals and oracles used to determine if a test passes or fails [18]. This emphasis on ML often leads stakeholders to perceive game-testing automation approaches as complex solutions, raising concerns regarding usage and maintenance. Therefore, it is essential to research methodologies that empower non-technical testers to design, create, and maintain human-readable tests.

VRTest framework [27] provides an interface to integrate different testing techniques using rotation, movement, and click-trigger events in Virtual Reality (VR) scenes. However, this framework requires further work to support wide types of events and be evaluated with software projects not based on Unity.

RiverGame framework [14] uses a BDD approach for involving non-technical stakeholders in describing desired game features and evaluating their usability and correctness. They evaluate the accuracy of visual recognition techniques with demo applications and their voice testing detection rate approach using an industrial game. In contrast, our industrial experience focuses on automating regression test execution to enable testers to streamline the development cycle.

In summary, while ML techniques have demonstrated exploratory test effectiveness, these can be perceived as complex solutions for testers who require a scenario-based testing approach that guides specific actions to assess concrete functionalities. Thus, we emphasize the importance of creating solutions that seamlessly facilitate the versatile integration of various tester-friendly methods. To address this, we took two main steps: (i) we extended the IV4XR framework to establish a connection with the Space Engineers game environment, allowing to observe internal game objects and execute game actions; and (ii) we integrated a BDD testing process for designing test scenarios in natural language.

3 Context: Space Engineers Game

Space Engineers is an industrial 3D sandbox game that allows users to assume the role of an astronaut, a playable game character capable of interacting with realistic open-world scenarios. Since its initial alpha release in 2013 until late 2023, the game has continuously evolved through around 600 game updates and ongoing maintenance for feature updates and bug fixes.

3.1 Game Objects

In Space Engineers, users have the freedom to explore diverse planets and asteroids, where they can design and build space stations and spaceships for spatial exploration. Resource management is a vital game aspect, requiring players to organize and gather materials to thrive in space environments. Figure 1 shows the main game objects: the *astronaut*, the *items*, and the *blocks*.

Fig. 1. Space Engineers game objects

The *astronaut* character owns various attributes, including energy, hydrogen, oxygen, and health, which players must manage to ensure their survival. To explore space scenarios, the *astronaut* is equipped with movement, rotation, and jet-pack flying capabilities. Additionally, the *astronaut* can utilize *items* to build *blocks*, with these *items* falling into two main categories: tools and components.

The welder and the grinder are the primary tools. The welder empowers players to construct *blocks*, provided the *astronaut* has the necessary components. Conversely, the grinder enables the destruction of *blocks* to retrieve components.

Each game *block* resides in a specific position and orientation and has properties that represent their type, integrity, volumetric physics, mass, inertia, and velocity. Certain types of *blocks* possess functional capabilities, such as gravity and energy generation or restoring the astronaut's health. The construction of multiple interconnected *blocks* forms a *grid*, where different *blocks*, such as a medical room *block* and an oxygen generator *block*, can be connected via conveyor *blocks*. The wide variety of *blocks* exhibits dynamic functional interoperability with other *grid blocks* (e.g., a medical room connected to an oxygen generator can also restore the astronaut's oxygen). Consequently, Space Engineers is an extensive and complex game that poses significant testing challenges due to the vast range of blocks and their diverse and intricate interactions.

3.2 Current Testing Practices

During a Space Engineers *game release*, a development cycle begins with the team of developers working on changing the game to add, update, or fix features (Fig. 2). As developers finalize these feature changes, they open Jira tickets[1] to point testers to the development branch that contains features that require testing. Testers then process these Jira tickets and *manually* test the newly implemented, updated, or fixed features. This development cycle takes about 3 or 4 months, depending on the required changes of the new game version.

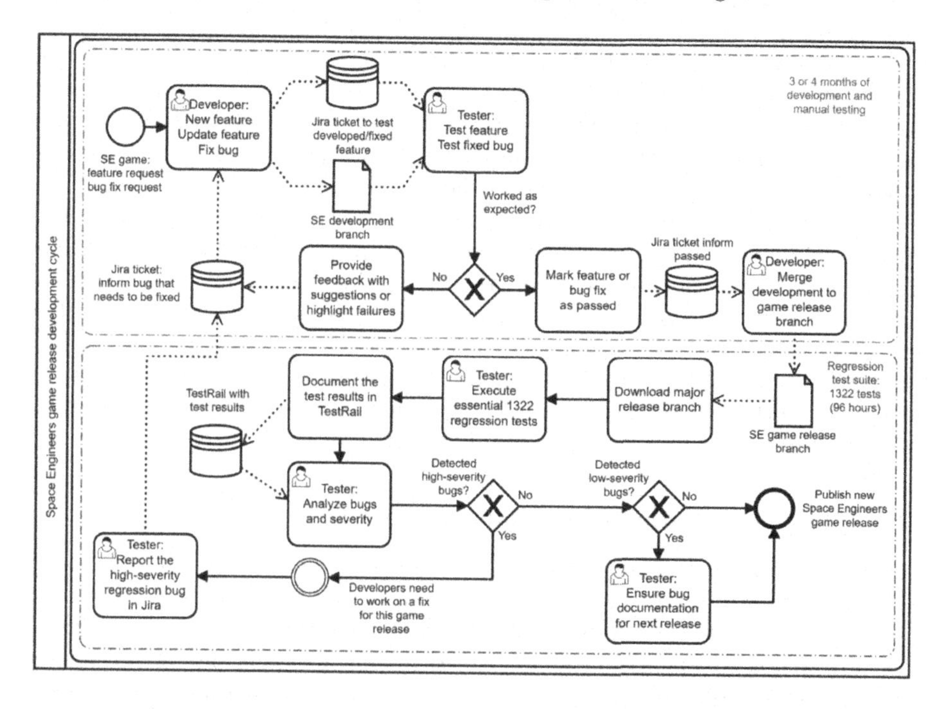

Fig. 2. Space Engineers development cycle diagram

Once developers have implemented all the required changes, testers assess whether the changed features have reached a stable state. If testers discover bugs in some of the changed features, they report the unstable features to the developers, who continue improving the development. After addressing any identified bug and confirming the stability of the development branch, testers proceed with executing a regression test suite.

This suite comprises 1322 tests, encompassing game features related to the astronaut's movements, attributes, items usage, interactions with blocks, as well as graphic or sound aspects. Testers manually execute the 1322 tests and document the results in TestRail[2], a management system that supports the organization of the testing process. Upon completing the regression test suite, testers

[1] https://www.atlassian.com/software/jira.
[2] https://www.gurock.com/testrail/.

analyze the results to determine if all tests passed without detecting any bugs. However, if any bugs are found during testing, testers assess each bug's severity.

If testers determine that a bug is of high severity, they open a Jira ticket, requiring developers to address it and fix the bug before proceeding with the game release. However, if the bug is classified as low-severity, the development branch can be published as a new *game release*, with the understanding that the low-severity bugs will be fixed in subsequent releases.

The regression testing phase described above is a time-consuming task that requires approximately 96 h of manual effort. Due to the substantial manual time investment, integrating the regression test suite into the development cycle of a *game release* is not feasible. Instead, the regression testing is executed at the end of the cycle, typically occurring every 3 or 4 months. This delay in regression testing may result in possible bugs that could have been detected and fixed in the iterative development process but are now postponed for several months, potentially impacting the final steps of the *game release* process.

The proficiency and perception of human testers are essential to ensuring the accurate verification of astronaut, item, and block functionalities, as well as graphic and sound aspects. However, the Space Engineers team recognized the possibility of automating a subset of functional regression tests that are repetitive and time-consuming. By automating this functional regression subset, nightly builds can seamlessly integrate into the development cycle. This integration will enable the early detection of possible regression bugs, significantly improving the overall effectiveness and efficiency of the *game release* process.

3.3 Challenges Automating Game-Testing Execution

The Space Engineers team evaluated the use of record-and-replay and visual recognition techniques to automate the execution of regression tests. However, they found a significant challenge as the recorded scripts proved to be unreliable in nearly all tests, even executed in the same scenario and game version. Reproducing the recorded astronaut movements and interactions did not yield the expected outcome, highlighting the lack of reliability in the recorded scripts.

Record-and-replay and other scripted GUI testing methods have shown fragility and unreliability, attributed to unpredictable GUI system behaviors and evolving GUI changes made between versions [6,10]. Despite this, the GUI testing field has seen considerable progress. For example, accessing the web DOM data to generate reliable test scripts [11]. However, this progress has limitations for sandbox 3D games like Space Engineers due to the inability of existing tools to obtain the required game system information.

Automated tools for game systems rely on recording keyboard and mouse inputs, lacking access to internal game data such as position, orientation, and object properties. Thus, successfully reproducing recorded test actions becomes challenging, as the traversed states may exhibit slight variations in each replay. Unlike traditional GUI testing, this is not due to a test script being vulnerable to GUI modifications but rather because the test lacks the necessary resilience to account for the inherent dynamics and non-determinism of the game. To achieve effective game test automation, it is crucial to have access to the game object's positions for precise navigation and to the object's properties for test oracles.

Recognizing the constraints of record-and-replay tools in game automation and the need for an automation solution that connects with the game's internal data, the Space Engineers team opted to incorporate two key technologies. They chose the IV4XR framework [21] for creating reliable tests integrated with the game's internal data and the principles of Behavior-Driven Development (BDD) [25]. The latter enables game testers to actively design, generate, and maintain evolving test scenarios. We will discuss each of them in subsequent sections.

4 Space Engineers-Plugin in IV4XR Framework

The IV4XR framework consists of a plugin architecture that includes a set of Java classes that streamline connection, information retrieval, and interaction with game objects. This framework supports the integration of an autonomous *agent* capable of connecting with and controlling a playable character, such as the astronaut in Space Engineers. To implement this functionality, we have developed a dedicated Space Engineers-plugin in Kotlin[3]. This plugin adheres to the architecture of the IV4XR framework, enabling seamless integration and automated test execution. Figure 3 shows an overview of the plugin's architecture (A) and a Kotlin test script example that employs the plugin's interfaces (B).

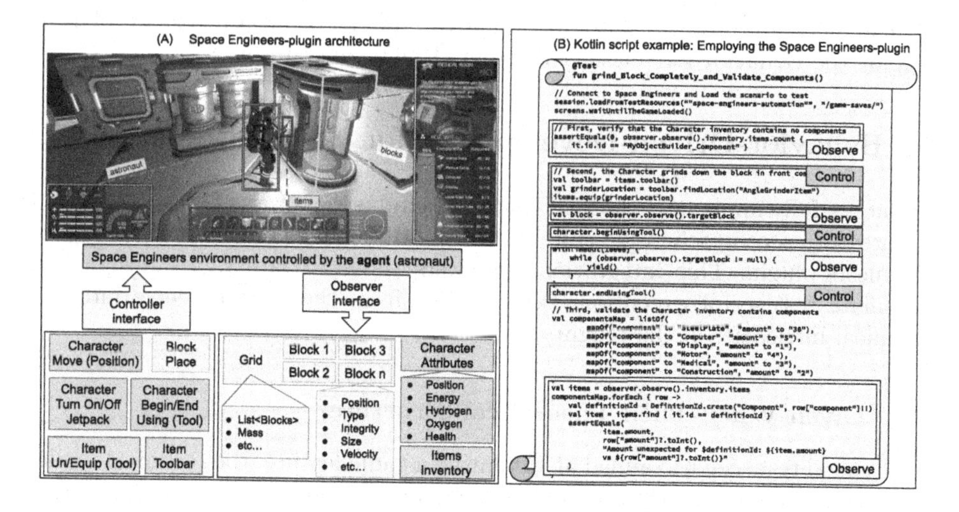

Fig. 3. Overview of the Space Engineers-plugin and Kotlin test script example

The *agent* gathers information about the surrounding game objects by *observing* the Space Engineers scenario. Since the astronaut character is always

[3] https://github.com/iv4xr-project/iv4xr-se-plugin.

present, the *agent* can continuously monitor its own character attributes and inventory. However, block observation depends on the proximity of the *agent* to different grids and block entities. For instance, if the *agent* is near a construction, it will be able to observe all of the blocks, whereas being far away would not allow it to observe them. In the Kotlin test script example (B), observing the game objects involves identifying the inventory and the target block aimed by the *agent.* The inventory is observed initially to ensure it contains no items. At the end of the test script, it is observed again to validate that a specific amount of items can be found. On the other hand, observing the target block allows for checking its existence or whether it has been destroyed.

To ensure a reliable astronaut *agent* control, the Space Engineers-plugin incorporates *controller* classes that interact with internal game functions. These classes translate actions into executable game commands empowering the *agent* to maneuver the astronaut to move and rotate to specific vector positions, equip or unequip various items such as tools, utilize equipped items, and place equipped blocks. In the Kotlin script example (B), first, the items toolbar is controlled to find and equip the desired tool for the character. Once the tool is equipped, the character is controlled to begin using the tool until the target block is destroyed. When the target block is destroyed, the character ends up using the tool.

The IV4XR Space Engineers-plugin addresses the technical side and enables the creation of reliable test scripts to observe and manipulate the game environment through the IV4XR *agent.* This feature enables the automation of the regression test suite. However, it is necessary to create reusable scripts that foster collaboration between developers and testers. To facilitate this, we plan to employ the Behavior-Driven Development (BDD) software process.

5 Behavior-Driven-Development for Test Automation

To successfully implement the BDD process in industrial practice, it was essential to integrate a comprehensive suite of tools and techniques on top of the Space Engineers-plugin. This integration empowers testers to abstractly design and automate test script execution, freeing them from the need to delve into the technical intricacies of the plugin.

5.1 Given-When-Then Theoretical Structure

In a typical test scenario aimed at validating a game feature, the technical implementation involves the creation of a test script class with the following steps:

1. **Load Game Scenario:** The test script should programmatically load a specific game scenario containing the desired state objects and properties that require testing. This step sets the initial conditions for the test.
2. **Execute Actions:** The next step involves running a specific set of actions that will alter the desired game object. These actions simulate user interactions or in-game events that trigger changes in the system.

3. **Validate Results:** After executing the actions, the test script should validate that the game object and its properties have responded adequately to the actions. This verification ensures the game features work as intended.

To implement these steps in a Space Engineers test scenario, testers can create a test script in Kotlin utilizing specific functions from the Space Engineers-plugin. However, crafting this Kotlin script requires technical knowledge of the existing plugin functions and valid game-plugin variables, which can pose challenges for non-technical testers in designing and maintaining test scripts.

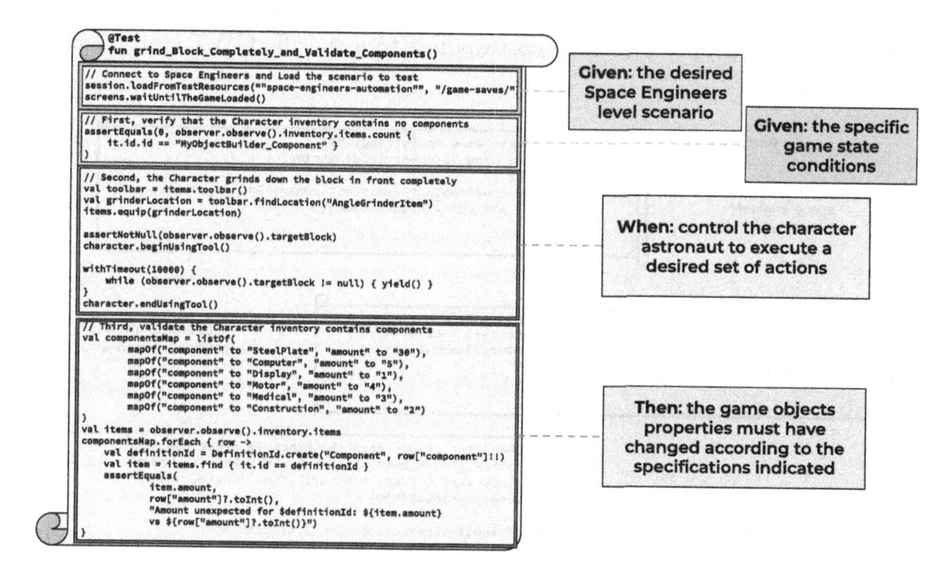

Fig. 4. Space Engineers Kotlin test script with GWT structure

The Given-When-Then (GWT) structure [13] is a standard format widely used in the BDD process. It enables the abstraction and description of system component functionality and expected behavior in a human-readable language:

- *Given* statement defines the Space Engineers level scenario and the game state that the test aims to validate. For instance, it can involve loading a specific level with the desired block to be tested.
- *When* statement enables to control the astronaut and execute a predefined set of actions that alter the Space Engineer's game objects. For example, the test instructs the astronaut to equip a grinder tool and destroy a block.
- *Then* statement verifies that the game object's properties have changed as expected, according to the specifications of the game feature being tested. For instance, it could check if the astronaut's inventory contains a specific amount of components after the block is destroyed.

Figure 4 illustrates how the Space Engineers test script written in Kotlin can be structured in a GWT structure. By adopting this structure in Space Engineers, testers can design and maintain regression test scenarios straightforwardly.

5.2 BDD in Practice with Cucumber

Cucumber [7] is an open-source tool that facilitates the practical implementation of the GWT structure in the context of a BDD process. This tool has been integrated as a bridge between human-readable GWT statements and the technical Space Engineers-plugin functions that interact with the game.

During the implementation with Cucumber, Kotlin functions are created to encapsulate the necessary invocations, aligning with the objectives specified in the *Given*, *When*, or *Then* statements. Figure 5 shows how the original Kotlin script can be disassembled into functions to load game scenarios, execute actions, or validate game properties. Each corresponds to a specific GWT statement.

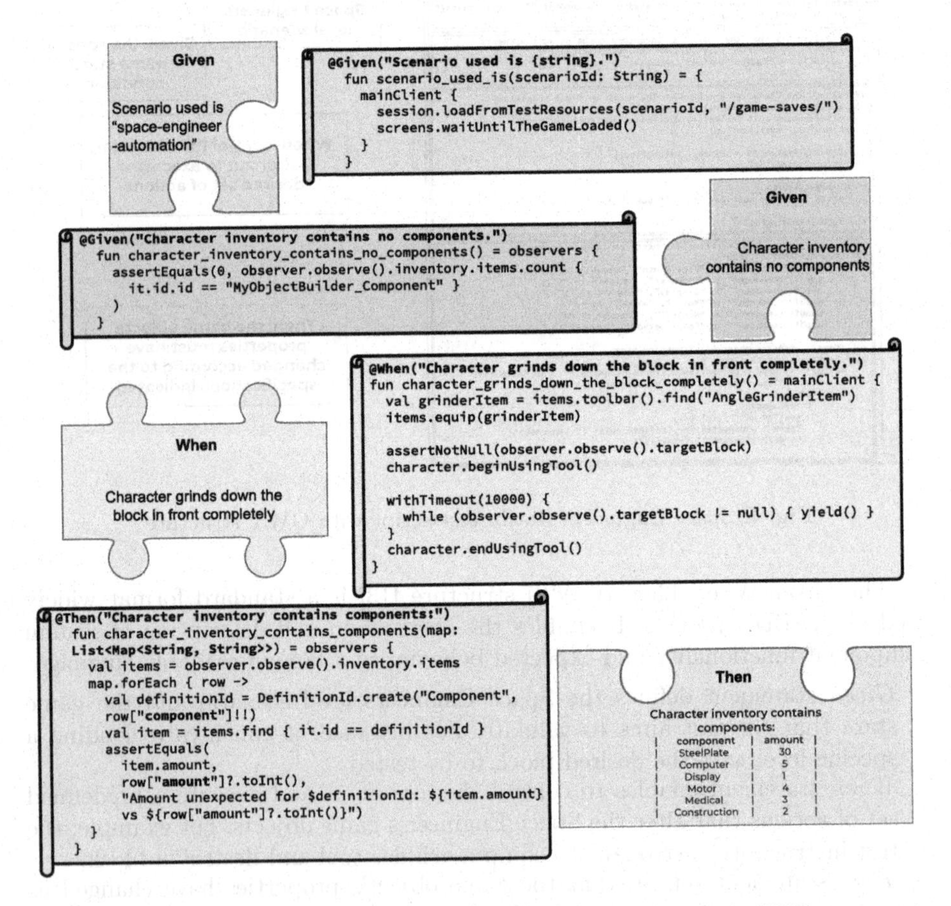

Fig. 5. Cucumber mapping of Space Engineers-plugin functions with GWT statements

Once these functions are implemented, they can be seamlessly aggregated and invoked using Cucumber's human-readable language. This approach offers

Space Engineer's testers an intuitive way to create, maintain, and execute test scenarios within the Space Engineers game environment.

In Fig. 6, we present a human-readable test script resulting from the combination of the BDD process with Cucumber. This script represents a concrete Space Engineers regression test scenario that non-technical stakeholders can craft to validate the grinding of a block to collect components by the astronaut. In the subsequent section, we delve into the association between these test scenarios and specific game objects, such as T254794, providing a comprehensive understanding of how these scenarios are applied in practical testing situations.

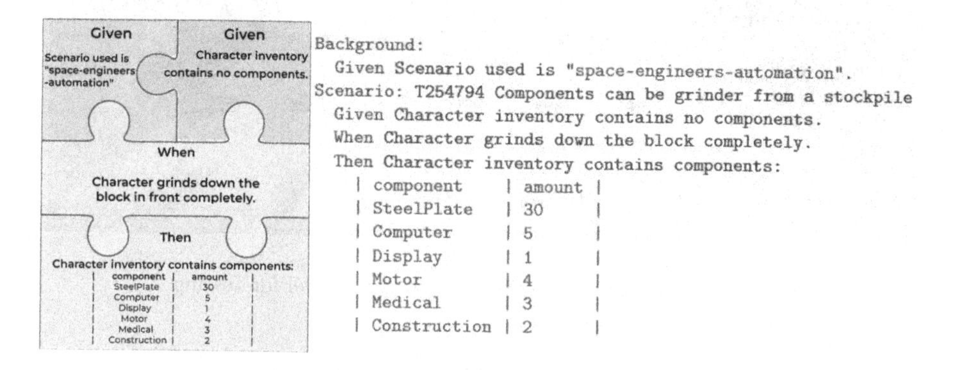

Fig. 6. Space Engineers BDD test scenario to validate astronaut inventory components

5.3 Space Engineers Pre-designed Game Level for Testing

Executing BDD regression tests in a sandbox open-world game like Space Engineers poses an additional challenge. Our goal is to keep astronaut actions (*When*) and validation oracles (*Then*) as minimalist as possible. To do this, testers have pre-designed a game level (see Fig. 7) that can be loaded using the *Given* statement. This level, divided into sections and stations, provides all the necessary blocks and items required by the astronaut in the diverse BDD test scripts.

Each section groups together generic features, such as astronaut movements, attributes, jet-pack, dampeners, and block grinding. In each section, multiple stations are set up with custom-arranged blocks designed to test specific functionalities. For example, a section is dedicated to validating the astronaut's health, oxygen, and energy attributes. In one medical station equipped with an oxygen tank, we validate the agent's oxygen replenishment. At another medical station without an oxygen tank, we validate that the agent cannot replenish his oxygen.

Each BDD test scenario is linked to the specific station that helps validate the intended functionality. To transit between test scenarios, the Space Engineers team utilizes a process that spawns the testing *agent* at the designated station.

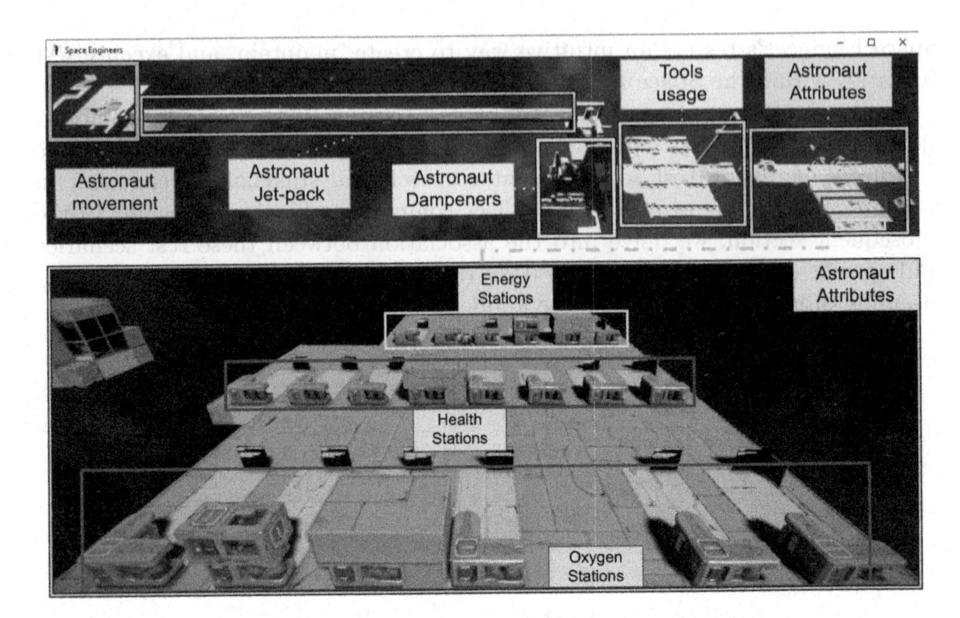

Fig. 7. Space Engineers pre-designed game level for testing

6 Industrial Application of BDD Test Scenarios

The Space Engineers team decided to initiate the automation process with a subset of 236 tests from the total of 1322 regression tests for the following reasons:

1. **Essential Features:** The selected subset comprises essential features for validating the astronaut's movements and attributes. These functionalities play a significant role in ensuring an immersive player experience.
2. **Plugin Classes:** The current state of the plugin covers all the necessary functions required for automating the control of the astronaut's movement and interactions with blocks using tools and items.
3. **Oracle Validation:** It is feasible to integrate a reliable Oracle validation, which involves observing the state of the game, including assessments of the astronaut's position and attributes. Test scenarios that involved validating game graphics or sounds currently present challenges for validation.
4. **Manual Effort Time:** While executing each individual test is not time-consuming, the entire subset requires a considerable amount of time.

The Space Engineers team has successfully automated a subset of 236 regression tests using the BDD *agent*. In contrast to previous experiences, the BDD approach has demonstrated *reliability* in controlling the astronaut and applying validation oracles. This automation effort offers an opportunity for seamless integration of these 236 regression tests into nightly builds, marking an enhancement from the prior practice of conducting them solely at the end of the 3 to 4-month development cycle. Instead, the BDD *agent* can complement the daily verification work that testers do manually. The Space Engineers team estimates that this automation initiative could be equivalent to saving approximately 17 h of manual effort each time the regression tests have to be executed.

Each BDD test scenario automatically generates a TestRail entry indicating whether the test passed or failed. If a test fails, the automated process generates a detailed report containing textual and visual information about the failed step. This seamless integration with TestRail has proven effective by successfully identifying a regression testing bug related to the engagement of the astronaut's magnetic boots. This success motivates further integration in nightly builds to streamline the regression test suite during the *game release* development cycle.

7 Threats to Validity

This section presents threats that could affect the validity of our results [22, 28].

Content Validity. While the number of successful automated regression tests can be quantitatively assessed, evaluating the satisfaction of the Space Engineers team is more subjective and difficult to measure from an analytical standpoint. In future work, we plan to continue automating regression tests using the BDD *agent* and evaluate team satisfaction through qualitative feedback.

Internal Validity. The selection of regression tests for automation relied on the expertise of the Space Engineers team. However, this experience could introduce selection bias and impact the representativeness of chosen regression tests. To mitigate this threat, the team fostered a collaborative environment where developers and testers with high expertise in the game led the selection process.

External Validity. This industrial study with Space Engineers serves as a testing reference for a complex sandbox game. However, it is essential to acknowledge that games significantly vary in implementation and design. Therefore, to assess the generalizability of the IV4XR and BDD approaches, it is recommended to evaluate their effectiveness and applicability across a wider range of games.

Conclusion Validity. The equivalence of saving 17 h of manual effort resulting from automating the initial 236 regression tests provides a compelling advantage. However, it is important to acknowledge the absence of a controlled experiment, which limits the generalizability and statistical robustness of the conclusions. Future research should prioritize conducting other empirical evaluations to validate and reinforce the observed benefits of automated regression tests.

8 Lessons Learned

The current collaboration to address the lack of automated regression test execution in an industrial environment provided us with valuable lessons learned:

Lesson one: Do not underestimate the complexity of automating oracles when testing 3D sandbox games. Multimedia aspects like visual textures, UI text displays, pixel particles, animations, lighting, and sounds are as crucial as correct block and item functionalities. However, the complexity of 3D perspectives, dynamism, and non-determinism (e.g., weather systems or NPC behaviors) make multimedia oracle verification a complex and unreliable endeavor that also introduces subjectivity in the form of diverse user experiences. For example, certain regression tests that have not been automated yet demand human testers to move the astronaut on the ground of planets or spatial structures or fly with the jetpack while continuously checking animations and textures. The complexity in verifying these animations and textures underscores the difficulties in automating human checks as technical oracles. Achieving this automation would require integrating an approach capable of recording a running video of the astronaut's 3D perspectives and accurately verifying the dynamic multimedia aspects.

The physics in 3D sandbox games, marked by non-deterministic, randomness, and unpredictability, complicates precise test oracle automation. For instance, Space Engineers testers are developing self-crafted environments that automate the execution of collision of rockets and vehicles while measuring range variables with in-game sensor entities. However, due to the non-deterministic physics, human validation is necessary to ensure the accuracy of the game's responses.

In multiplayer games, challenges intensify when verifying oracles from dedicated server and client player perspectives. The extensive game functionalities and multimedia aspects must be verified in dynamic multiplayer environments, where it is necessary to confirm the correct synchronization of game events (e.g., a bug provoked that astronaut faction changes were accurately synced from the client but were not triggering the desired changes when invoking from the server itself). Moreover, these complexities extend to validating network protocols and cross-platform compatibility (e.g., a recent issue resulted in console players being unexpectedly disconnected after minutes of playing on computer servers).

Lesson two: Automation efforts must concentrate on supporting human testers rather than aiming for their outright replacement. During the selection of regression test cases intended to integrate the BDD automation process, we learned a lesson that exceeds technical challenges: the value of human tacit knowledge [26].

Humans possess the capacity for comprehension, intuition, and accumulated understanding that is sometimes challenging, if not impossible, to express or formalize in technical scripts. Furthermore, this knowledge and experience is subjective, varying according to each individual perception and personal beliefs. Attempting to automate this human tacit knowledge as test scripts is an unrealistic goal that should be avoided when integrating a software automation process.

In the case of Space Engineers, which contains thousands of regression tests, automating astronaut controls and validating individual game properties through integrated test oracles is essential for reducing manual effort. However, the wide variety of game objects that can be used in dynamic and highly customizable environments makes it challenging to anticipate every possible scenario. For instance, although game blocks exhibited correct functionality when tested individually, a user discovered that removing a block connecting a self-constructed station to an asteroid caused the entire station to move inside the asteroid.

An expert human tester with a broad tacit knowledge background in testing Space Engineers can intuitively understand the combination of game blocks and physics causing the problem. Leveraging this tacit knowledge, they can notify developers of potential causes and apply their expertise to future developments affecting these specific buggy game features.

Therefore, the primary objective should not be focused on replacing human testers with artificially embedded tacit knowledge in a testing procedure. Instead, the focus should be on complementing and supporting human testers to streamline their work, reduce their workload, and provide them with information that enhances their productivity.

Lesson three: The dynamic nature of the industry requires flexibility and adaptability in academic-industry collaborations. In long-term projects, the industry is more likely to change team members by incorporating new personnel or reallocating existing members to projects deemed of higher priority. Consequently, in academic-industry collaborations, flexibility and adaptability are essential. Proactive anticipation of potential changes in team compositions, along with the recognition that industry members may face varied future workloads, enables smoother transitions and enhances the resilience of the collaboration. This entails strategic foresight, planning, and meticulous documentation for effective knowledge transfer. Within this collaborative environment, future industry members or external open-source contributors can quickly assimilate into the project, fostering a fluid and robust long-term environment.

9 Conclusions and Future Work

This paper reports the industrial experience of developing a game plugin and adopting a BDD process to automate a subset of regression tests for the game Space Engineers. A BDD *agent* has proven effectiveness and reliability in verifying the astronaut functionalities through natural language instructions, allowing the automation of 236 regression tests and showcasing its bug-detection capabilities. This automation is estimated to be equivalent to saving around 17 h of manual effort per executed regression subset. These outcomes have encouraged the Space Engineers team to develop a dedicated server for seamless integration of automated regression testing into nightly builds during the 3-4 month development cycle periods. We hope this experience motivates and guides other stakeholders in integrating a framework to map internal entity observation and game action execution with natural language statements to enable BDD testing.

In future work, the Space Engineers team will continue automating the remaining regression test scenarios and extend the regression test suite to validate the placement and properties of blocks. Additionally, we are working to overcome challenges such as (i) using intelligent IV4XR *agents* (e.g., with concepts such as belief, goal, and tactics) to deal with complex 3D paths and motion planning when navigating longer test scenarios, (ii) determine how internal game observations may differ from the user's GUI perspective (e.g., users must scroll the inventory to find an item), (iii) research approaches to integrate reliable graphics and sound oracles within the IV4XR framework. Furthermore, we plan to explore the benefits of BDD practices in other interactive system domains.

Acknowledgment. This work has been partially funded by: H2020 EU iv4XR grant nr. 856716 and ENACTEST ERASMUS+ grant nr. 101055874.

References

1. de Andrade, S.A., Nunes, F.L., Delamaro, M.E.: Exploiting deep reinforcement learning and metamorphic testing to automatically test virtual reality applications. STVR **33**(8), e1863 (2023)
2. Ariyurek, S., Surer, E., Betin-Can, A.: Playtesting: what is beyond personas. IEEE Trans. Games (2022)
3. Bahaweres, R.B., et al.: Behavior-driven development (BDD) cucumber katalon for automation GUI testing case CURA and swag labs. In: ICIMCIS, pp. 87–92 (2020)
4. Binamungu, L.P., Maro, S.: Behaviour driven development: a systematic mapping study. J. Syst. Softw. 111749 (2023)
5. Bucchiarone, A., Cooper, K.M., Lin, D., Melcer, E.F., Sung, K.: Games and software engineering: engineering fun, inspiration, and motivation. ACM SIGSOFT Softw. Eng. Notes **48**(1), 85–89 (2023)
6. Coppola, R., Morisio, M., Torchiano, M.: Mobile GUI testing fragility: a study on open-source android applications. Trans. Reliab. **68**(1), 67–90 (2018)
7. Dees, I., Wynne, M., Hellesoy, A.: Cucumber Recipes: Automate Anything with BDD Tools and Techniques. Pragmatic Bookshelf (2013)
8. García, B., Gallego, M., Gortázar, F., Munoz-Organero, M.: A survey of the selenium ecosystem. Electronics **9**(7), 1067 (2020)
9. Gordillo, C., Bergdahl, J., Tollmar, K., Gisslén, L.: Improving playtesting coverage via curiosity driven reinforcement learning agents. In: 2021 IEEE Conference on Games (CoG), pp. 1–8. IEEE (2021)
10. Hammoudi, M., Rothermel, G., Tonella, P.: Why do record/replay tests of web applications break? In: International Conference ICST, pp. 180–190. IEEE (2016)
11. Nguyen, V., To, T., Diep, G.H.: Generating and selecting resilient and maintainable locators for web automated testing. STVR **31**(3), e1760 (2021)
12. Nie, L., Said, K.S., Ma, L., Zheng, Y., Zhao, Y.: A systematic mapping study for graphical user interface testing on mobile apps. IET Software **17**(3), 249–267 (2023)
13. North, D., et al.: Introducing BDD. Better Softw. **12**, 7 (2006)
14. Paduraru, C., Paduraru, M., Stefanescu, A.: Rivergame-a game testing tool using artificial intelligence. In: International Conference ICST, pp. 422–432. IEEE (2022)
15. Pereira, L., Sharp, H., de Souza, C., Oliveira, G., Marczak, S., Bastos, R.: Behavior-driven development benefits and challenges: reports from an industrial study. In: ICASD Companion, pp. 1–4 (2018)
16. Pezze, M., Rondena, P., Zuddas, D.: Automatic GUI testing of desktop applications: an empirical assessment of the state of the art. In: ISSTA, pp. 54–62 (2018)

17. Pfau, J., Smeddinck, J.D., Malaka, R.: Automated game testing with ICARUS: intelligent completion of adventure riddles via unsupervised solving. In: CHI PLAY: Extended Abstracts, pp. 153–164. ACM (2017)
18. Politowski, C., Guéhéneuc, Y.G., Petrillo, F.: Towards automated video game testing: still a long way to go. In: 2022 IEEE/ACM 6th International Workshop on Games and Software Engineering (GAS), pp. 37–43. IEEE (2022)
19. Politowski, C., Petrillo, F., Guéhéneuc, Y.G.: A survey of video game testing. In: International Conference on Automation of Software Test (AST), pp. 90–99. IEEE (2021)
20. Prasetya, I., Dastani, M., Prada, R., Vos, T., Dignum, F., Kifetew, F.: Aplib: tactical agents for testing computer games. In: 8th EMAS Workshop (2020)
21. Prasetya, I., et al.: An agent-based approach to automated game testing: an experience report. In: A-TEST, pp. 1–8 (2022)
22. Ralph, P., Tempero, E.: Construct validity in software engineering research and software metrics. In: 22nd International Conference EASE, pp. 13–23 (2018)
23. Rodríguez-Valdés, O., Vos, T., Aho, P., Marín, B.: 30 years of automated GUI testing: a bibliometric analysis. In: Paiva, A.C.R., Cavalli, A.R., Ventura Martins, P., Pérez-Castillo, R. (eds.) QUATIC 2021, pp. 473–488. Springer, Cham (2021). https://doi.org/10.1007/978-3-030-85347-1_34
24. Sestini, A., Gisslén, L., Bergdahl, J., Tollmar, K., Bagdanov, A.D.: Automated gameplay testing and validation with curiosity-conditioned proximal trajectories. IEEE Trans. Games **16**(1), 113–126 (2022)
25. Smart, J., Molak, J.: BDD in Action. Simon and Schuster (2023)
26. Walker, A.M.: Tacit knowledge. Eur. J. Epidemiol. **32**(4), 261–267 (2017)
27. Wang, X.: Vrtest: an extensible framework for automatic testing of virtual reality scenes. In: 2022 IEEE/ACM 44th ICSE-Companion, pp. 232–236. IEEE (2022)
28. Wohlin, C., Runeson, P., Höst, M., Ohlsson, M.C., Regnell, B., Wesslén, A.: Experimentation in Software Engineering. Springer, Heidelberg (2012). https://doi.org/10.1007/978-3-642-29044-2
29. Wu, Y., Chen, Y., Xie, X., Yu, B., Fan, C., Ma, L.: Regression testing of massively multiplayer online role-playing games. In: 2020 IEEE ICSME, pp. 692–696. IEEE (2020)
30. Zheng, Y., et al.: Wuji: automatic online combat game testing using evolutionary deep reinforcement learning. In: 34th International Conference ASE, pp. 772–784. IEEE (2019)

Emotion Trajectory and Student Performance in Engineering Education: A Preliminary Study

Edouard Nadaud[1(✉)], Antoun Yaacoub[1], Siba Haidar[1], Bénédicte Le Grand[2], and Lionel Prevost[1]

[1] Learning, Data and Robotics (LDR) Lab, ESIEA, Paris, France
edouard.nadaud@esiea.fr

[2] Centre de Recherche en Informatique (CRI), Université Paris 1 Panthéon Sorbonne, Paris, France

Abstract. In this study, we aim to establish the connection between the emotional trajectory of students during a pedagogical sequence and their performances. The project aims to develop an affective and intelligent tutoring system for detecting students facing difficulties and helping them. We designed this experimentation during the 2022–2023 academic year with students in a French engineering school. We collected and analyzed two primary data sources: student results from the Learning Management System (LMS) and images captured by students' webcams during their learning activities.

It is known that basic (primary) emotions (like fear or disgust) do not reflect student affective states when facing pedagogical issues (like misunderstanding or proudness). Since such "academic emotions" are not easy to define and detect, we changed the paradigm and used a 2D dimensional model that describes better the wide spectrum of emotion encountered. Moreover, it allows to build a temporal emotion trajectory reflecting the student's emotional trajectory.

Firstly, we observed a correlation between these trajectories and academic results. Secondly, we found that high-performing student trajectories are significantly different from the others. These preliminary results, support the idea that emotions are pivotal in distinguishing highly performing students from their less successful counterparts. This is the first step to assess students' profiles and proactively identify those at risk of failure in a human learning context.

Keywords: Emotion Trajectories · Affective Computing · Early Difficulties Detection · Machine-Learning Applications · Ethical AI · Learning Analytics

1 Introduction

In recent years, higher education institutions have widely adopted online learning environments, including LMSs, to maintain educational continuity, particularly since the COVID-19 pandemic [1]. Web interaction platforms are now an essential tool for learning, offering students and teachers easy access to educational resources and learning activities, as well as a wide range of online features.

J. Araújo et al. (Eds.): RCIS 2024, LNBIP 513, pp. 410–424, 2024.
https://doi.org/10.1007/978-3-031-59465-6_25

Numerous studies show the importance of introducing LMSs [2] into an educational setting and highlight the importance of integrating AI to improve student learning [3–5]. The introduction of AI in education is deemed crucial due to its potential to revolutionize learning outcomes, reduce teachers' workload, and personalize educational experiences. It offers solutions to challenges in sustainable education, such as improving access and creating personalized learning experiences [5]. One of the main challenges facing these environments is student follow-up and the need for adequate support to prevent the risk of failure and encourage active participation.

Although LMSs offer possibilities for organizing and managing courses, distributing content, retrieving learning traces, and tracking learning outcomes, they sometimes lack effective methods for ensuring individual student supervision. Indeed, monitoring student progress in a virtual environment presents additional challenges compared to a traditional learning environment. Teachers may find it difficult to deal with signs of drop-out or difficulties due to the physical distance and asynchronous nature of online learning. Face-to-face interactions are replaced by online discussions and messages, making it more difficult to detect potential problems [6]. The question arises of how to identify students with learning difficulties, and how to do so in a way that limits disruption to the learning process.

In the following section, we will show that emotions exhibited by students during a pedagogical sequence are informational cues about their comprehension and final assessment. Our main - and original - proposal is to study the emotional trajectory of each student, depicted by an emotion trajectory in an affective space. In Sect. 3, we will describe in detail our experimentation, including data collection and cleansing, emotion prediction through the sequence finally leading to the visualization of the emotion trajectory. In Sect. 4, we will present a thorough analysis of these trajectories, their variability, and their potential link to student assessment. Finally, in Sect. 5 we will conclude our work and give some prospects.

2 Motivation and Related Work

We are going to study the link between artificial intelligence and emotion in education to introduce the concept of emotion trajectory.

2.1 Affective Computing and Education

Grades are indicators that may reflect a student's difficulties in the process of learning and assimilating knowledge. Unfortunately, most of the times, those indicators occur at the end of the learning process. Other elements, such as engagement or behaviour indicators, can also contribute to detecting difficulties before they have an impact on the learning process [7, 8]. Research demonstrated that emotions influence subjects learning process and performance [9–11]. Emotions can both facilitate or hinder learning. Research suggests that positive emotions contribute to students' enjoyment within learning environments. In contrast, negative emotions such as boredom, anger can delay the learning process [12].

Ekman described and modelized basic, primary emotions, such as anger, joy, disgust, sadness, fear, and surprise. They are innate and associated with specific facial expressions [13]. People from different cultures can recognize facial expressions of basics and primary emotions; however the amplitude of these expressions depends on the cultural context [14].

From the literature review, we retain that basic (primary) emotions do not cover the spectrum of emotion exhibited by students in classes. Since then, we decided to change the paradigm and use a 2D affective model from the seminal works of Russell.

In this sense, researchers are increasingly identifying the emotions that play a role in the educational context [15, 16]. Emotions are defined as: "Emotion is a response of the organism to a particular stimulus (person, situation, or event). Usually it is an intense, short duration experience and the person is typically well aware of it" [17]. These emotional responses are an unbidden occurrence, they happen without choosing them.

An alternative model uses a small number of latent dimensions to characterize emotions such as valence, arousal, control, power. In the dimensional approach, emotions are not independent of each other. One of the most preferred dimensional models is a two-dimensional model that uses arousal, activation, or excitement on one dimension, as opposed to valence, appreciation, or evaluation on the other. The valence dimension describes whether an emotion is positive or negative. The dimension of arousal defines the strength of the emotion felt. These models are more representative of the wide spectrum of human emotion [18].

Meanwhile, academic emotions are different, and they depend on the goal. In Pekrun's study [19], academic emotions are defined according to domains (achievement goal, avoidance goal and performance-approach goal). Achievement emotions refer to achievement activities or achievement outcomes (success and failure). These emotions are complex to evaluate. Pekrun and Perry defined these emotions using the control value theory of achievement emotions framework [20]. Scherer [21] based his work on this model to express academic emotions in an 2D space. He linked emotions into the valence, arousal space based on a study of 80 german terms that demonstrate the hypothesis that semantic space can be organized by evaluation criteria.

2.2 Pedagogical Emotions

Human are able, through experience, of determining a person's emotion through the combination of several senses [22]. Numerous studies have attempted to reproduce these combinations of senses, using numerous sensors to capture the signals of emotive students, to determine academic emotions with the help of computers [23, 24]. These studies are able to identify emotions more precisely, but they are costly and have an impact on the learning process [25].

A great deal of research has focused on identifying students in difficulty. Some of researchers have focused on measuring student engagement [26, 27] while other have looked at the correlation between emotions and physiological signs [28]. Other research in the field of affective computing shows the importance of positive emotions in learning and observes students' difficulties by detecting their negative emotions in order to warn the teacher [29]. These studies are interesting and demonstrate the value of detecting difficulties at an early stage to help subjects learn. However, they are hampered by the material required (electroencephalography) [30] as well as the numerous tracking algorithms to be calibrated. Many of existing studies use emotional models trained on Facial Emotion Recognition dataset [29, 31, 32]. These solutions use one-dimensional models and they have the advantage of being easy to train since large datasets are available. However, the use of a two-dimensional model makes it possible to increase the emotional palette of even the most introverted students.

Despite the clear relevance of emotions for education, most of these studies use Ekman emotion to profile students [29, 33] and to the best of our knowledge, no academic emotion dataset is freely available.

It would therefore be interesting to also look at the emotional profile and emotional variations to identify profiles with learning difficulties.

2.3 Emotion Trajectory

To our knowledge, no study has examined the profiling of students from the point of view of their emotional trajectory using a webcam. We define the emotion trajectory as the evolution over time of emotions in a 2D model. The point represents the emotion expressed by the student during one question labelled with the corresponding grade as shown in Fig. 1. Due to the sequential nature of the quiz, no backtracking is possible. The red line represents the evolution of emotion during the quiz. Studies in psychology have shown the importance of emotional narratives in the decision-making process of students in higher education [34]. Arousal can take on different values from one person to another, depending on how they express their emotions. (introvert/extrovert student). So, it's not just a question of looking at general emotion, which can vary from one individual to another for the same emotion. It's about the evolution of everyone's emotion.

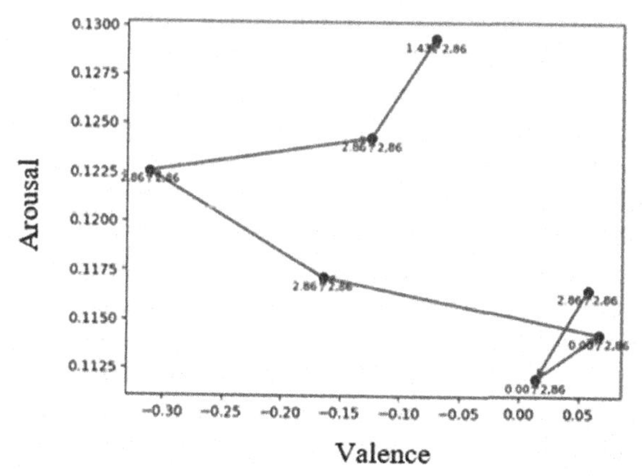

Fig. 1. Example of an emotional trajectory for a student during the quiz

3 Methodology

Our methodology unfolds in three distinct phases: data collection, pre-processing, and analysis. The initial phase involves gathering data, crucial for capturing a comprehensive view of the variables in play. In the pre-processing phase, we focus on refining the quality of the data, which is essential for the reliability of subsequent analyses. The final stage, analysis, concentrates on identifying specific types of emotion trajectories and their correlation with academic test results. This step is key in understanding how different emotional patterns relate to student performance, providing insights that could influence future pedagogical strategies.

3.1 Experimental Setting

The subjects are first year masters students at a French engineering school. They have IT expertise and are familiar with the use of a laptop and webcam. For the experiment, students bring their own laptop equipped with a webcam to a face-to-face course. In our preprocessing stage, we prioritize maintaining the naturalness of the test environment, thus refraining from giving students specific instructions regarding their posture or camera focus to ensure their attention remains focused on the test itself. The quiz takes place at the beginning of the course and is composed of 7 MCQ questions with no backtracking possible. The duration for the quiz is 15 min maximum and **students** should submit an answer to pass to the next question. At the end of the time the quiz is automatically corrected.

3.2 Data Collection

For our experimental setup, we deliberately minimize the hardware requirements to just a standard laptop with an integrated webcam, thereby eliminating the need for more sophisticated and costly equipment like ECG/EDM sensor bracelets. This choice not only reduces the complexity of the setup but also aligns with our subject's familiarity with digital tools. The experiment involves fourth-year higher education students who are subjected to a uniquely structured test, consisting of both multiple-choice questions and programming tasks. Designed to be sequential, the test permits only a single attempt per question, disallowing any revisitation of questions once moved past. Data collection is thus streamlined to include images captured at five-second intervals, which are then meticulously synchronized with specific test parameters such as question content, corresponding scores, and the exact timing of each captured image.

In compliance with the General Regulation on the Protection of Personal Data (GDPR) [35], our data collection process began with establishing a legally compliant procedure for obtaining user consent. We wrote a consent form, clearly stating the purpose of our experiments and the process for deleting personal data in case of consent withdrawal. This form was presented to students at the start of each test, and we secured 76 authorizations from a class of 359 students. To ensure full understanding of their rights and the experiment's purpose, we conducted meetings with all the groups involved.

Following consent acquisition, we design the real-time capture of student face via webcams during test activities, using the "Proctoring" plugin in Moodle [36]. This plugin, adapted from its original facial recognition purpose, allows us to capture images at chosen intervals, ensuring student identity verification throughout the test. Simultaneously, we gather comprehensive LMS data, including grades, time spent on each question, and question difficulty levels. This parallel process provides an integrated view of individual and class performance.

To ensure the security and privacy of this data, we subsequently transfer it from Moodle to a secure server exclusively used for our research and accessible only to authorized staff. This transfer includes only data from students who had given explicit consent. We also establish a dedicated email address to handle requests for data deletion, reinforcing our commitment to data security and subject privacy.

3.3 Data Pre-processing

Due to the naturalness of the test environment, a significant portion of the captured images were not immediately suitable for emotion recognition due to obscured or partially visible faces.

To address this, we initiated a cleaning process to filter out unusable images, focusing on retaining only those with clearly visible faces. We evaluated various face detection algorithms (*HAAR cascade classifier* from OpenCV, *Frontalface detector* from dlib, *MTCNN* and *YOLOv5*) ultimately selecting YOLOv5 [37] for its superior performance on our dataset. Images were first passed through this algorithm to identify clear facial images. For images where faces were not detected, we applied an enhancement process involving gamma correction and histogram equalization to improve image quality for subsequent face detection attempts. Images that still did not reveal a face post-enhancement were discarded. This rigorous preprocessing reduced our dataset from 76 to 56 subjects but significantly enhanced the quality of the data retained for analysis.

3.4 Data Analysis Methods

Emotion Prediction. To accurately identify affective states from approximately 12000 captured images and project them into Russell's circumplex model of affect, we bypassed traditional annotation methods due to their resource-heavy nature and inherent subjectivity, which often leads to low inter-annotator agreement and delays. The complexity and sheer volume of our image data rendered conventional machine learning methods insufficient, primarily because of their limitations in processing high-dimensional data without extensive manual feature engineering.

We therefore opted for a deep learning approach, despite facing difficulties with the availability and documentation of existing 2D emotion prediction models. We used the AffectNet dataset [38] to build our models as it is widely used in the affective computing community. We trained two models for predicting valence and arousal. Both models are compounded of five sequential convolution and subsampling layers (conv2D, conv2D, max pooling) and two dense layers for a total of 2M parameters. This approach enabled precise emotion predictions (validation loss lower than 0.1) from our experimental images, offering a deeper insight into the subjects' emotional states in line with

Russell's model as shown in Fig. 2. The representation of emotional expression on a 2D circle are representative of the active facial expression. The lower picture shows a more active facial expression (Fig. 2). His representation in the 2D circle is further to the right (his coordinates are represented by a square).

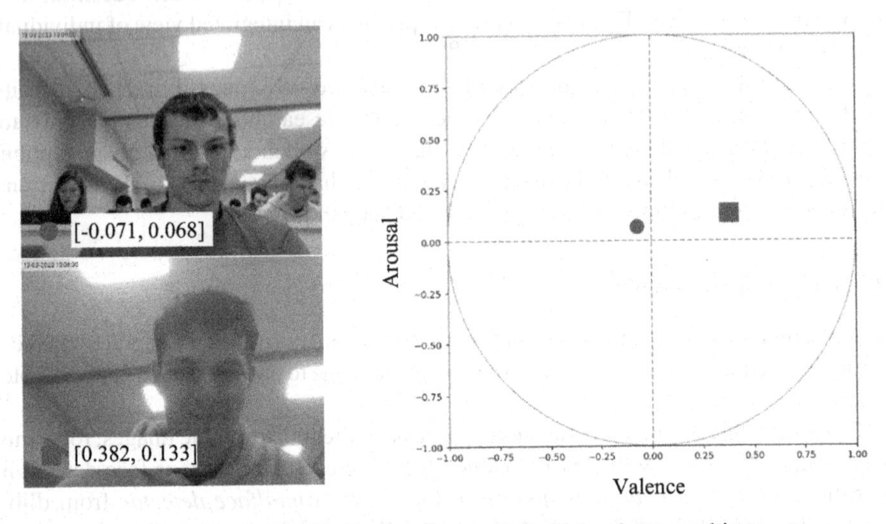

Fig. 2. Example of valence and arousal prediction for two subjects

Matching the Quiz Grade and the Average Emotion. In our study, we initially focused on the two-dimensional emotional states of students throughout the examination, aiming to correlate these average emotional states with their academic performance. We assumed that a poor grade is associated with an average negative emotional state. This approach involved computing the average valence and arousal levels for each student over the duration of the test, as illustrated in Fig. 2. However, this method did not yield a significant correlation between the averaged emotional states and the students' performance.

This finding prompted a reassessment of our approach, to consider the fluctuation of students' emotional states during the exam. We note that most of our valence and arousal values are positive, which confirms our idea of considering fluctuations relative to the subject and not generalizing to the whole population. We represent these fluctuations as the subject average emotional state. We only represent the valence values; arousal being more subjective and depends on the personality of the student, for which we do not have information in the context of achievement goals. This concept, while mathematically convenient, fails to capture the complexity and variability of the emotional experiences encountered. It became clear that a more detailed analysis was necessary, one that considers the dynamic emotional shifts students experience throughout the testing period to reflect their emotional trajectory more accurately (Fig. 3).

Emotional Centroids. In the context of our study, we observed significant variability in the valence and arousal data points for individual students throughout a single question. This variability can be attributed to various factors, including spontaneous

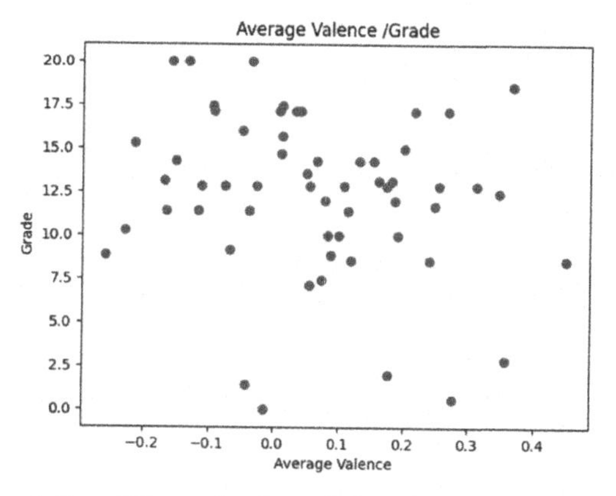

Fig. 3. 2D Representation of the student's grade (y-axis) and his average valence (x-axis)

movements or brief lapses in concentration. To date, there is a scarcity of literature addressing the aggregation of emotional states over a defined time interval within such a model. Consequently, we proposed a centroid-based approach to encapsulate the predominant emotional state experienced by a student in response to a particular question. This approach involves the aggregation of discrete emotional states detected during the question's duration.

To compute the average centroid (CM) for a student in relation to question q, which encompasses n images, we employed the following formula:

$$CM_q = \left(\frac{1}{n}\right)\sum_{i=1}^{n}(V_i, A_i)$$

where V_i and A_i represent the valence and arousal values for the ith image, respectively.

Given the susceptibility of centroid coordinates to outliers, we refined our approach by computing a final centroid (CF), which represents the mean position of the 50% of data points nearest to CM. This selection criterion was established after iterative experimentation to optimize the representation of a student's average emotional state for question q:

$$CF_q = \left(\frac{1}{m}\right)\sum_{j=1}^{m}(V_{j'}, A_{j'})$$

where $m = \frac{n}{2}$, and (V_j', A_j') are the coordinates of the points closest to CM_q.

Furthermore, we explored the concept of emotional amplitude (EA) for each question, defined as the Euclidean distance between CF and the most distant point considered in the centroid computation. This metric, EA_q, intended to quantify the breadth of emotional fluctuation within a single question:

$$EA_q = \max_{\mathrm{k}}\sqrt{(V_{CF} - V_k)^2 + (A_{CF} - A_k)^2}$$

The following Fig. 4 shows the centroids for each question for one subject.

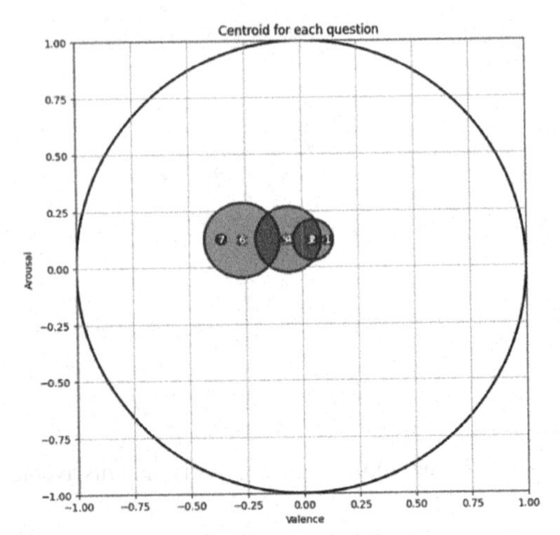

Fig. 4. Centroids for each question for one subject. The center represents CF_q and the radius is EA_q, where q is the number of questions.

However, preliminary analysis of the emotional amplitude (EA_q) and amplitude yielded inconclusive insights regarding students' academic challenges. As such, we elected to exclude this metric from subsequent phases of our research.

Naturally, the subsequent phase involves examining the variations in students' emotions throughout the test, which will be assessed using the calculated emotional centroids.

Emotion Trajectory. We conceptualize an emotion trajectory as the depiction of a student's emotions throughout the test. This trajectory is represented in the valence and arousal dimensions, with vectors connecting the centroids corresponding to each question. This method allows us to track the progression of emotions during the test. To facilitate this analysis, we compiled a dataset that integrates the learning traces with the calculated CF (Final Centroid) for each question. Therefore, for every student, we have a set of emotional coordinates, alongside the scores and question numbers.

Visualization. Upon formatting the data, our next step was to visualize these emotion trajectories to uncover analytical insights. We plotted these trajectories on valence and arousal axes, aiming to observe behaviours that could potentially impact performance. Each point on the plot represents the CF of a question, tagged with its corresponding score, and arrows connect these points in sequential order. This visual representation aids in analysing the temporal evolution of a student's emotions—indicated by the arrows—and correlating it with their performance. These visualizations (as Fig. 5) opened new avenues for exploration, particularly in terms of the trajectories' amplitude and shape.

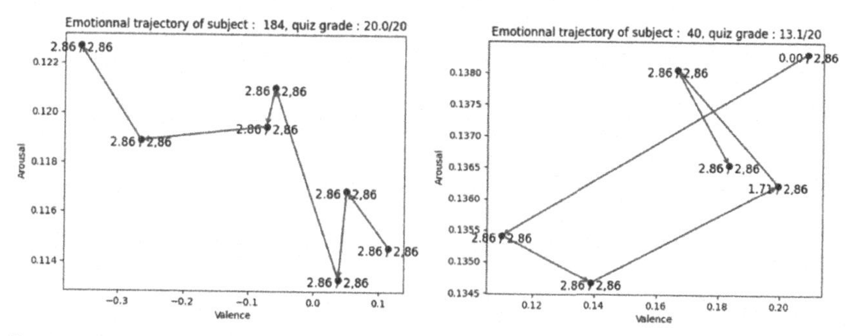

Fig. 5. Emotion Trajectories for subject 40 and 184. Points are the CF_q with the grade, where q is the question number and directed vectors are the emotion trajectory between questions.

4 Results and Discussions

4.1 Student's Emotional Range

Our initial focus was on quantifying the emotional amplitude for each subject. This involved calculating the difference between the maximum and minimum values of valence and arousal, taking the absolute value of this differential. The resultant amplitude value is then assigned a positive or negative sign, depending on the sequence of the min and max values. A negative sign is attributed if the maximum value precedes the minimum, indicating a decline in valence or arousal.

Let $p = \max_{i=1}^{n}(V_i)$ and $q = \min_{j=1}^{n}(V_j)$

$$Amplitude_V = \begin{cases} -|p - q| \text{ if } i > j \\ |p - q| \text{ if } i < j \end{cases}$$

We found a moderate correlation ($Corr_{pearson} = 0.40$) as show in Fig. 6 between the absolute amplitude values and student performance, with higher amplitudes observed in students who performed better. This could suggest that a deeper engagement with the material may elicit stronger emotional responses to errors, successes, or misunderstandings. However, the amplitude alone does not delineate a distinct emotional pattern or behaviour among students. Consequently, we shifted our focus towards exploring emotion trajectories and the potential circularity of emotions.

4.2 Emotion Trajectory Analysis and Circularity of Emotions

Emotion trajectories vary significantly among subjects. Throughout our experiments, we noted that some students exhibit subdued emotional reactions, while others are markedly more expressive. To standardize the interpretation of each student's unique emotion trajectory, we normalized the valence and arousal values across all subjects, scaling them according to their individual emotional range, defined by their minimum and maximum values. We assume that the range of emotions felt is comparable from one student to another, but that it is visible depending on the student's personality. Following normalization, we quantified the emotional trajectory during the test by calculating the total

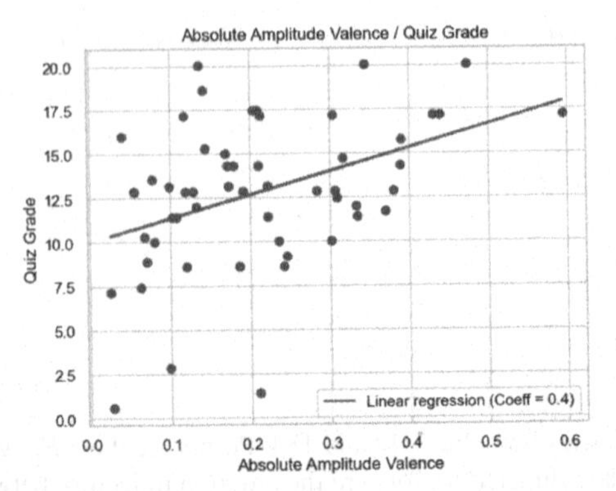

Fig. 6. Representation of absolute amplitude valence in relation to the quiz grade. The red line represents the linear regression.

emotional distance traversed, represented by the cumulative length of segments linking consecutive final centroids (*FC*). Additionally, we measured the distance between the initial and final centroids for each student's emotion trajectory. A shorter distance between the initial and final centroids implies a potential circularity in the emotion trajectory (as shown in Fig. 5), indicating a return to the initial emotional state.

In Fig. 7, we plot the quiz grade against the normalized distance covered. We coloured each point (subject) conditionally to to the distance between the initial and final emotional states. We compute the Pearson correlation coefficient between the extent of the emotional trajectory and student performance. We obtain ($Corr_{pearson} = 0.44$), that seems to underscore the significance of emotional engagement among high achievers. Notably, students who covered greater emotional distances tended to exhibit minimal deviation between their initial and final emotional states (upright, blue points), suggesting a dynamic emotional range but with a tendency to revert to the initial emotional state. In contrast, students with lower academic performance demonstrated limited emotional movement, evidenced by shorter emotional trajectorys (left points), yet significant shifts from their initial emotional state, indicative of either emotional deterioration or improvement (left red points) These findings align with previous insights, affirming the role of emotional variability as a contributory factor to academic performance.

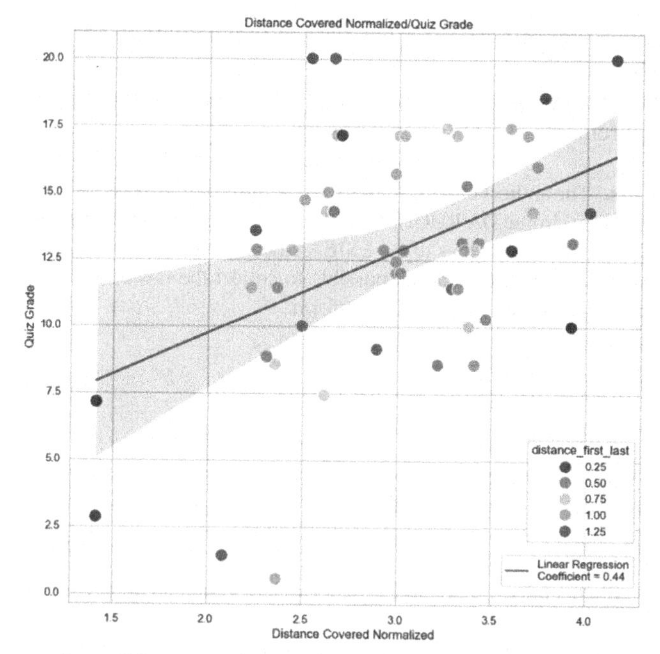

Fig. 7. Representations of distance covered normalized to student performance. Colors represent proximity between first and last centroid. (Color figure online)

5 Conclusion

We present in this paper an original idea, the "emotion trajectory" used to analyze student emotional trajectory during a pedagogical sequence. We built a powerful deep model to predict two affective variables, namely valence and arousal. This bidimensional affective model allows us to draw the temporal emotion trajectory of a student during the activity. After several processes (clustering, normalization), we analyze thoroughly several indicators like the trajectory extent and the emotional gap between initial and final state. We show that high-performing students are primarily those with intense – but not necessarily positive - emotional experiences. Students with longer emotional trajectories tend to cycle between emotional states, while low-performing students exhibit emotions that deteriorate or improve. It is necessary here to precise that we discuss correlation and not causality and that these preliminary results do not yet pretend to predict assessment from an emotional point of view. But it already confirms the role of emotional amplitude in explaining student outcomes.

Further works are planned to cluster trajectories and try to find more correlation between trajectories and grades to profile better student's needs in terms of help.

Thus, we envision that circularity of emotions could help in the detection of students in need of support.

A further direction is to investigate and to interpret the range of emotions triggered by the question characteristics such as the topic of the question, the difficulty, the average response time, etc. Moreover, when additional learning activities (beyond quizzes) are developed, another characteristic like the level of interactivity of the activity could be integrated to the analysis.

In the near future, we plan to share the first educational emotion in the wild dataset to meet the requirements of GDPR for diffusion.

Acknowledgments. The authors express their sincere gratitude to Mr. Loïc ROUSSEL, Managing Director of ESIEA, Mr. Jérôme DA RUGNA, Director of Pedagogy and Research, Mr. Jean-Pierre AUBIN, lecturer at ESIEA, as well as to the ESIEA students who generously participated in this study. Their valuable contribution made it possible to collect the raw educational data resulting from students' interactions with the Moodle platform.

References

1. Spitzer, M.W.H., Moeller, K.: Performance increases in mathematics during COVID-19 pandemic distance learning in Austria: evidence from an intelligent tutoring system for mathematics. Trends Neurosci. Educ. **31**, 100203 (2023). https://doi.org/10.1016/j.tine.2023.100203
2. Chaubey, A., Bhattacharya, B.: Learning management system in higher education. IJSTE Int. J. Sci. Technol. Eng. **2**, 158–162 (2015)
3. Lin, C.-C., Huang, A.Y.Q., Lu, O.H.T.: Artificial intelligence in intelligent tutoring systems toward sustainable education: a systematic review. Smart Learn. Environ. **10**, 41 (2023). https://doi.org/10.1186/s40561-023-00260-y
4. Chaudhry, M.A., Kazim, E.: Artificial Intelligence in Education (AIEd): a high-level academic and industry note 2021. AI Ethics **2**, 157–165 (2022). https://doi.org/10.1007/s43681-021-00074-z
5. Chichekian, T., Benteux, B.: The potential of learning with (and not from) artificial intelligence in education. Front. Artif. Intell. **5**, 903051 (2022). https://doi.org/10.3389/frai.2022.903051
6. Kraleva, R., Sabani, M., Kralev, V.: An analysis of some learning management systems. Int. J. Adv. Sci. Eng. Inf. Technol. **9**, 1190 (2019). https://doi.org/10.18517/ijaseit.9.4.9437
7. Büchele, S.: Evaluating the link between attendance and performance in higher education: the role of classroom engagement dimensions. Assess. Eval. High. Educ. **46**, 132–150 (2021). https://doi.org/10.1080/02602938.2020.1754330
8. Khelifi, T., Rabah, N.B., Grand, B.L., Daoudi, I.: EX-LAD: explainable learning analytics dashboard in higher education. In: Kambhampaty, K., Hu, G., Roy, I. (eds.) Proceedings of 36th International Conference on Computer Applications in Industry and Engineering, pp. 38–51. CAINE (2024). https://doi.org/10.29007/dsxd
9. Chishti, Z.N.S., Rahman, F., Jumani, N.B.: Impact of emotional intelligence on team performance in higher education institutes. Int. Online J. Educ. Sci. **3** (2020)
10. Asrar-ul-Haq, M., Anwar, S., Hassan, M.: Impact of emotional intelligence on teacher's performance in higher education institutions of Pakistan. Future Bus. J. **3**, 87–97 (2017). https://doi.org/10.1016/j.fbj.2017.05.003
11. Graesser, A., D'Mello, S.: Emotions during the learning of difficult material. Psychol. Learn. Motiv. Adv. Res. Theory **57**, 183–225 (2012). https://doi.org/10.1016/B978-0-12-394293-7.00005-4
12. Aspinwall, L.G.: Rethinking the role of positive affect in self-regulation. Motiv. Emot. **22**, 1–32 (1998). https://doi.org/10.1023/A:1023080224401
13. Ekman, P.: Basic emotions. In: Handbook of Cognition and Emotion, pp. 45–60. Wiley, Hoboken (1999). https://doi.org/10.1002/0470013494.ch3

14. Ekman, P.: An argument for basic emotions. Cogn. Emot. **6**, 169–200 (1992). https://doi.org/10.1080/02699939208411068
15. Andrés, M.L., et al.: Emotion regulation and academic performance: a systematic review of empirical relationships (2017). https://doi.org/10.4025/psicolestud.v22i3.34360
16. Activity Achievement Emotions and Academic Performance: A Meta-analysis | Educational Psychology Review. https://link.springer.com/article/10.1007/s10648-020-09585-3. Accessed 24 Jan 2024
17. Dzedzickis, A., Kaklauskas, A., Bucinskas, V.: Human emotion recognition: review of sensors and methods. Sensors **20**, 592 (2020). https://doi.org/10.3390/s20030592
18. Chiang, W.-W., Liu, C.-J.: Scale of academic emotion in science education: development and validation. Int. J. Sci. Educ. **36**, 908–928 (2014). https://doi.org/10.1080/09500693.2013.830233
19. Pekrun, R., Elliot, A.J., Maier, M.A.: Achievement goals and discrete achievement emotions: a theoretical model and prospective test. J. Educ. Psychol. **98**, 583–597 (2006). https://doi.org/10.1037/0022-0663.98.3.583
20. Perry, R.P., Raymond P.: Control-value theory of achievement emotions. In: International Handbook of Emotions in Education. Routledge (2014)
21. Scherer, K.R.: What are emotions? And how can they be measured? Soc. Sci. Inf. **44**, 695–729 (2005). https://doi.org/10.1177/0539018405058216
22. How Do We Feel the Emotions of Others? https://kids.frontiersin.org/articles/10.3389/frym.2017.00036. Accessed 25 Jan 2024
23. Ketonen, E.E., Salonen, V., Lonka, K., Salmela-Aro, K.: Can you feel the excitement? Physiological correlates of students' self-reported emotions. Br. J. Educ. Psychol. **93**, 113–129 (2023). https://doi.org/10.1111/bjep.12534
24. Jaques, N., Conati, C., Harley, J.M., Azevedo, R.: Predicting affect from gaze data during interaction with an intelligent tutoring system. In: Trausan-Matu, S., Boyer, K.E., Crosby, M., Panourgia, K. (eds.) ITS 2014. LNCS, vol. 8474, pp. 29–38. Springer, Cham (2014). https://doi.org/10.1007/978-3-319-07221-0_4
25. Cukurova, M., Luckin, R.: Measuring the impact of emerging technologies in education: a pragmatic approach. In: Voogt, J., Knezek, G., Christensen, R., Lai, K.-W. (eds.) Second Handbook of Information Technology in Primary and Secondary Education. SIHE, pp. 1181–1199. Springer, Cham (2018). https://doi.org/10.1007/978-3-319-71054-9_81
26. Gupta, S., Kumar, P., Tekchandani, R.: A multimodal facial cues based engagement detection system in e-learning context using deep learning approach. Multimed Tools Appl. **82**, 28589–28615 (2023). https://doi.org/10.1007/s11042-023-14392-3
27. (PDF) Measuring Student Emotions in an Online Learning Environment. https://www.researchgate.net/publication/339904943_Measuring_Student_Emotions_in_an_Online_Learning_Environment. Accessed 24 Jan 2024
28. Shi, G., Chen, S., Li, H., Tian, S., Wang, Q.: A study on the impact of COVID-19 class suspension on college students' emotions based on affective computing model. Appl. Math. Nonlinear Sci. **9** (2024)
29. Wang, C.-H., Lin, H.-C.: Emotional design tutoring system based on multimodal affective computing techniques. Int. J. Distance Educ. Technol. **16**, 103–117 (2018). https://doi.org/10.4018/IJDET.2018010106
30. Dorado, J., et al.: An affective-computing approach to provide enhanced learning analytics. Presented at the January 1 (2020). https://doi.org/10.5220/0009368401630170
31. Towards real-time speech emotion recognition for affective e-learning | Education and Information Technologies. https://link.springer.com/article/10.1007/s10639-015-9388-2. Accessed 24 Jan 2024

32. Fwa, H.L.: An architectural design and evaluation of an affective tutoring system for novice programmers. Int. J. Educ. Technol. Higher Educ. **15**, 38 (2018). https://doi.org/10.1186/s41239-018-0121-2
33. Pourmirzaei, M., Montazer, G.A., Mousavi, E.: ATTENDEE: an AffecTive Tutoring system based on facial EmotioN recognition and heaD posE Estimation to personalize e-learning environment. J. Comput. Educ. (2023). https://doi.org/10.1007/s40692-023-00303-w
34. Kellam, N., Gerow, K., Wilson, G., Walther, J., Cruz, J.: Exploring emotional trajectories of engineering students: a narrative research approach. Int. J. Eng. Educ. **34**, 1726–1740 (2018)
35. General Data Protection Regulation (GDPR) – Official Legal Text. https://gdpr-info.eu/. Accessed 19 Mar 2024
36. eLearning-BS23/moodle-quizaccess_proctoring (2023). https://github.com/eLearning-BS23/moodle-quizaccess_proctoring
37. Ieamsaard, J., Charoensook, S.N., Yammen, S.: Deep learning-based face mask detection using YoloV5. In: 2021 9th International Electrical Engineering Congress (iEECON), pp. 428–431 (2021). https://doi.org/10.1109/iEECON51072.2021.9440346
38. Mollahosseini, A., Hasani, B., Mahoor, M.H.: AffectNet: a database for facial expression, valence, and arousal computing in the wild. IEEE Trans. Affective Comput. **10**, 18–31 (2019). https://doi.org/10.1109/TAFFC.2017.2740923

Author Index

J. Araújo et al. (Eds.): RCIS 2024, LNBIP 513, pp. 425–427, 2024.
https://doi.org/10.1007/978-3-031-59465-6

Printed in the United States
by Baker & Taylor Publisher Services